Applied Bioelectricity

Springer

New York
Berlin
Heidelberg
Barcelona
Budapest
Hong Kong
London
Milan
Paris
Singapore
Tokyo

J. Patrick Reilly

Applied Bioelectricity

From Electrical Stimulation to Electropathology

With chapters by

Hermann Antoni
Michael A. Chilbert
James D. Sweeney

With 241 Figures

Springer

J. Patrick Reilly
Applied Physics Laboratory
Johns Hopkins University
and
Metatec Associates
12516 Davan Drive
Silver Spring, MD 20904, USA

Cover art by Anthony E. Randolph, The Johns Hopkins University Applied Physics Laboratory.

Library of Congress Cataloging-in-Publication Data
Reilly, J. Patrick.
  Applied bioelectricity : from electrical stimulation to
  electropathology / J. Patrick Reilly.
      p.   cm.
  Includes index.
  ISBN 0-387-98407-0 (hardcover : alk. paper)
  1. Electrophysiology.   2. Electric shock.   I. Title.
QP82.2.E43R437   1998
612'.01427—dc21   97-48860

Printed on acid-free paper.

Production coordinated by Chernow Editorial Services, Inc., and managed by Tim Taylor; manufacturing supervised by Thomas King.
Typeset by Best-set Typesetter Ltd., Hong Kong.
Printed and bound by Maple-Vail Book Manufacturing Group, York, PA.
Printed in the United States of America.

9 8 7 6 5 4 3 2 1

ISBN 0-387-98407-0 Springer-Verlag New York Berlin Heidelberg   SPIN 10657964

Preface

The use of electrical devices is pervasive in modern society. The same electrical forces that run our air conditioners, lighting, communications, computers, and myriad other devices are also capable of interacting with biological systems, including the human body. The biological effects of electrical forces can be beneficial, as with medical diagnostic devices or biomedical implants, or can be detrimental, as with chance exposures that we typically call electric shock. Whether our interest is in intended or accidental exposure, it is important to understand the range of potential biological reactions to electrical stimulation.

The subject of this book is applied bioelectricity, that is, biological reactions to electrical forces in situations of practical interest. For that purpose, I have attempted to treat biological reactions with emphasis on human reactions, from the just discernible to the clearly unacceptable. This book treats short-term biological reactions, with special attention to reactions of a detrimental nature. The term "electropathology" is used here to indicate any undesirable biological reaction to electrical stimulation. Whether a biological effect is judged beneficial or detrimental often depends on the context. With controlled medical applications of electrical stimulation, a reaction may be judged to be beneficial; with uncontrolled chance exposure, a similar reaction may be judged to be undesirable.

The present work is an adaptation of my previous book, *Electrical Stimulation and Electropathology*, which was published in 1992 by Cambridge University Press. Since that time, my continued research in the field of applied bioelectricity has led to the development of this edition. I have attempted to make this work useful to the biomedical scientist, as well as to the individual concerned with electrical safety. The material is directed to a reader who is familiar with basic physics and basic physiology, but who may not necessarily be an expert in either discipline. Chapters 2 to 5 provide a fundamental background by which later discussions on electrical stimulation and electropathology may be understood. Chapters 6 to 10 treat human reactions to electrical stimulation and are organized according to the nature of the response.

Electrical forces may be introduced into the biological medium through electrodes that directly contact the subject, or through electric and magnetic fields without direct electrode contact. In either case, the biological system responds to the same bioelectric relationships that are presented throughout the book. Chapter 9, however, applies these relationships specifically to electric and magnetic field exposure.

Chapter 11 of this edition treats rationale and standards for exposure to electromagnetic fields and electric currents in the frequency regime from zero to several GHz. The chapter emphasizes electromagnetic field exposure, although I have included a section on exposure to electric currents from consumer products. I have attempted to draw on the material of the preceding 10 chapters in discussing the rationale that underlies existing and proposed exposure standards.

As with *Electrical Stimulation and Electropathology*, this volume focuses on short-term reactions to electric currents, and electric and magnetic fields. This focus results in a book that primarily treats biophysical reactions to relatively high levels of electrical exposure, in contrast to the much lower chronic exposure issues that have been much discussed in other scientific and popular forums. The book emphasizes bioelectric mechanisms that are, for the most part, reasonably well understood, for which theoretical predictive models can explain a wide range of biophysical phenomena, and for which experimental evidence is available and largely unequivocal. I do not mean to imply that all the loose ends are tied up with respect to short-term bioelectric reactions—indeed there remain many issues that are poorly understood, and much research remain to be done. However, comparatively speaking, our grasp of biophysical mechanisms and safety issues associated with short-term, high-level exposure is much firmer than with chronic, low-level electrical exposure issues.

As will be evident in Chapter 11, exposure standards are largely developed with reference to relatively high-level exposures that approach thresholds of demonstrable short-term biological reactions in intact animals and humans. Nevertheless, public attention is often directed to chronic, low-level exposure issues. Accordingly, Chapter 11 includes a discussion of biophysical mechanisms that have been advanced to explain observed biological reactions to chronic, low-level electrical exposure. Although such mechanisms are not, for the most part, developed in the preceding chapters, I can recommend for the interested reader recent books on this subject (Blank, 1993; Frey, 1994; Gandhi, 1990; Klauenberg et al., 1995; Lin, 1989; Nordén and Ramel, 1992; Polk and Postow, 1996; Sagan, 1996; Wilson et al., 1990).

This book would not have been possible without the contributions of many people. I am particularly grateful for the help of those at The Johns Hopkins University Applied Physics Laboratory, and especially for the support of Stuart S. Janney Fellowships that supported a portion of my writing and research efforts. The Air Defense Systems Department (formally the Fleet Systems Department) of the Applied Physics Laboratory made available the resources of their art and editorial group. I appreciate the support of the Department Associate Director William Zinger for making these resources available. I thank the Department staff members Terry Joslin for editorial and word-processing work, and Anthony Randolph for new art work for this edition. Art work from the previous edition, also included here, was done principally by Jacob Elbaz. I am grateful to the library staff of the Applied Physics Laboratory, who went to the remote ends of the earth to track down many of my obscure citations.

Many colleagues provided valuable help. John Osepchuk of Full Spectrum Consulting reviewed Chapter 11 and made many valuable suggestions. I particularly wish to acknowledge Professor Willard Larkin of the University of Maryland for the many things I learned from him during the years we worked together researching electric shock. Professor Larkin is coauthor of many of the papers that have been adapted into Chapter 7. There were also many colleagues, too numerous to mention, who sent me papers and discussed research topics with me. I am grateful to them all.

Special thanks are due to those colleagues who contributed chapters to this book, namely Hermann Antoni of the University of Freiburg (Chapter 5), James D. Sweeney of the Arizona State University (Chapter 8), and Michael Chilbert of the General Electric Company (Chapter 10). I also acknowledge H.A.C. Eaton of The Johns Hopkins University Applied Physics Laboratory for his collaboration in magnetic brain stimulation, and for co-authoring Section 9.9.

As anyone who has written a technical book undoubtedly knows, the process requires a substantial commitment over a long period of time involving many personal sacrifices that are shared by one's family. I thank my wife Lynette for her patience and understanding during this project, and for the many ways that she assisted me.

Laurel, Maryland   J. PATRICK REILLY

Contents

Affiliations

HERMANN ANTONI
Physiological Institute
University of Freiburg
D 7800 Freiburg i. Br.
Germany

MICHAEL A. CHILBERT
General Electric Company
Medical Systems Division
Milwaukee WI 53201
USA

J. PATRICK REILLY
Applied Physics Laboratory
The Johns Hopkins University
Laurel, MD 20723
and
Metatec Associates
12516 Davan Drive
Silver Spring, MD 20904
USA

JAMES D. SWEENEY
College of Engineering and Applied Science
Department of Chemical, Biology, and Materials Engineering
Arizona State University
Tempe, Arizona 85287-6006
USA

1
Introduction

1.1 General Perspective

Electrical forces are vital for the functioning of living things, from the metabolism of individual cells to human consciousness derived from the activity of the brain. When electric currents are artificially introduced into a living organism, these natural forces can be activated or modified. The result is either detrimental or beneficial, depending on the circumstances. The term *electric shock* is generally used to describe the response of the body to inadvertent electrical exposure, and the consequences of electric shock are usually considered undesirable. However, the biological mechanisms responsible for electric shock may also be activated in a controlled fashion for beneficial medical purposes.

The fact that electricity can interact with biological processes has been known for more than 2,000 years (McNeal, 1977). The electrical discharge of the torpedofish was reported to have been used as early as 46 A.D. to treat pain. It is tempting to scoff at early claims of beneficial uses of electricity, but when one considers modern methods of pain management by electrical means, one is led to suspect that there might have been valid reasons for some of these early beliefs.

The beginnings of quantitative bioelectric science can arguably be ascribed to the investigations of Galvani (around 1790) and later of Volta. Galvani observed motion in severed frogs' legs when he touched them with metallic wires. He supposed that he was releasing stored "animal electricity," and that this was responsible for the observed muscle activity. It was later demonstrated that the dissimilar metals of the wires used in his procedures were in fact generating the electrical forces. His efforts did, however, encourage further investigations of electrical stimulation. Volta's invention of the "Voltaic pile" gave science a battery that could be used in systematic and controlled investigations. Later biological investigations by Faraday (around 1831) demonstrated that interrupted electric current was an effective means of electrical stimulation of nerves. The term "Voltaic" stimulation came to be used to indicate direct current stimulation, and "Faradaic" to indicate pulsed or interrupted stimulation.

1.2 Electrical Exposure

Electrical Fatalities

It has been long recognized that man-made electricity can be harmful and can even cause death. This understanding eventually led to the use of electrical executions of criminals to replace the "less civilized" means that had previously been devised. The first electrocution in the United States occurred in 1890 (Leyden, 1990). The number of electrocutions of criminals since that time is uncertain, but is estimated to number approximately 4,100 in the United States. Today, approximately 900 prisoners in 14 states await execution in the electric chair.

The National Center for Health Statistics publishes data on the mortality from electrical accidents in the United States. Figure 1.1 illustrates U.S. mortality statistics for the years 1975 to 1987 as summarized by Smith (1990). The figure shows data for all electrical fatalities, excluding lightning incidents, and for fatalities related to consumer products.

Because the mortality data do not specify the product involved in the accident, the number of fatalities related to consumer products was inferred by Smith on the basis of the reported locations of the accidents.

The downward trend of electrical mortalities is striking. And if one takes into account the increase in the U.S. population during the period reported in Fig. 1.1, the trend is even more impressive the per-capita rate of fatalities related to consumer products declined from 3.0 per million in 1975 to 1.3 per million in 1987. According to Smith, this trend could be the result of a number of factors, including (1) a steady increase in the number of residences having ground-fault current interrupters (GFCIs); (2) an increase in the number of locations having GFCIs within individual dwellings; (3) a decrease in citizen-band base-station antennas; and (4) an increase in the manufacture of double-insulated power tools. Details of some of these precautionary measures are given in Chapter 11. The downward trend illustrated in Fig. 1.1 contrasts with a slight increase in electrical mortality during the period 1960 to 1975 (Daiziel, 1978), a period when electrical safeguards were being developed but were not yet widely implemented.

Table 1.1 provides a further breakdown of electrical mortalities. The table also enumerates deaths by lightning, of which there are nearly 100 each year in the United States. Lightning accounts for roughly 10% of the total number of electrical fatalities. It has been estimated that one third of all lightning strikes to humans are fatal (Biegelmeier, 1986).

The worldwide experience of industrialized countries also shows a downturn in electrical fatalities (Kieback, 1988). The annual number of fatalities

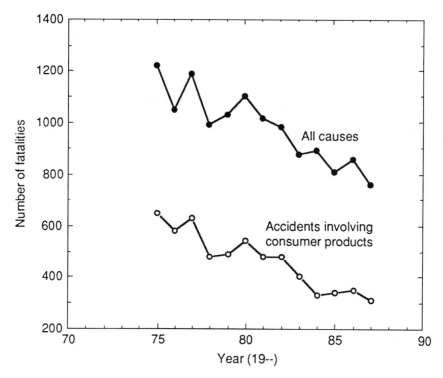

FIGURE 1.1. Mortality from electrical accidents in the United States for the years 1975–1987 (per data compiled by Smith, 1990).

per million inhabitants for 1982 (the latest year reported by Kieback) ranged from a low of 0.42 for the Netherlands to a high of 7.66 for Hungary. Kieback emphasizes that such comparisons should be treated with caution because of different reporting standards in various countries.

TABLE 1.1. Mortality because of electrical accidents.

	1987	1986	1985	1984
Electric current				
Total (all causes)	760	854	802	888
Domestic wiring and appliances	121	150	146	148
Power generating, distr., transmission	177	182	196	215
Industrial wiring, appliances, machinery	64	89	69	68
Other unspecified electric current	398	433	391	457
Lightning	99	78	85	—

Source: Vital Statistics of the united States, Volume II—Mortality, U.S. Department of Health and Human Services, National Center for Health Statistics, Washington, DC.

It is likely that the mortality data referred to above underreport the actual number of electrical fatalities. The data for the United States have been developed from death certificates. Not all electrical fatalities may have been identified in the certificates as such. Furthermore, deaths from injuries incidental to electric shock, such as falls, are probably not reported in the electrical category. Delayed fatalities, such as those caused by burns, also may have been omitted. Even if we allow for underreporting, the number of electrical fatalities remains small in comparison with other accidental causes in the United States. Death from choking on food or other ingested objects, for example, exceeds electrical fatalities by nearly five times.

Typical Electrical Exposures

Notwithstanding the relatively small number of electrical fatalities, the acceptability of electrical exposure by the public is a matter of growing concern. One area of concern is related to the increasing number of electrical consumer products. Although potentially fatal electrical exposures are seldom an issue in most consumer products, in some cases exposures might be unpleasant, or they might lead to injuries from startle reactions. Both consumers and manufacturers hope to avoid such exposures.

High-voltage transmission lines provide another source of public exposure. Increases in population and in individual electrical demand press utilities to increase electricity generation and transmission. The trend in transmission is to higher voltages because of economies in the cost of construction and operation of transmission facilities. One environmental consequence of high-voltage transmission is the potential for electric shock arising from the electric fields that the lines inevitably produce (see Chapter 9). The electric shock resulting from transmission line fields does not have the potential for injury, except possibly in certain unusual circumstances. Nevertheless, perceptible shock from transmission line fields can be unpleasant and can provoke fear in the exposed public.

Probably the most rapidly growing area of electrical stimulation is in biomedical technology. Electrical stimulation is being used increasingly as a tool for medical diagnosis, therapy, and prosthesis; examples of biomedical applications are listed in Table 1.2. Even this incomplete list is a testament to the numerous medical applications of electrical stimulation. Electrical stimulation for medical purposes may be introduced through electrodes in contact with the skin, through implanted electrodes, or through magnetic induction. The latter method precludes the necessity for any electrode contact whatsoever.

Electrical stimulation that might be considered detrimental in chance exposures can be beneficial when used in a controlled fashion. For example, cardiac arrhythmia's caused by chance electric shock are regarded as being potentially life threatening; these same responses can be life saving when used in implanted pacemakers. Furthermore, a response that is considered

TABLE 1.2. Examples of electrical stimulation in biomedical applications.

Restoration of muscle function after nerve injury
Preservation of muscle tone after nerve injury
Treatment of scoliosis
Diaphragm stimulation for respiration control
Electrical stimulation of sphincter for urinary control
Correction of foot-drop
Sensory aids for the blind
Cochlear prosthesis for the deaf
Management of intractable pain
Inhibition of intractable self-injurious behavior
Diagnosis of peripheral nerve function
Diagnosis of muscle function
Functional diagnosis and mapping of the brain cortex
Stimulation of the visual cortex
Electroconvulsive therapy
Automatic cardiac pacing
Automatic sensing and reversal of fibrillation (implants)
Defibrillation in emergency aid (external)
Bone healing
Electrical diathermy
Imaging of soft tissue

undesirable in one medical application may be considered beneficial in another. Electrically induced pain, for example, is considered a difficulty to be avoided in procedures involving the introduction of currents through electrodes on the skin. But painful electrocutaneous stimulation has also been used in a beneficial way to inhibit otherwise intractable self-injurious behavior (Newman, 1984).

In this work, the term "electropathology" is used to indicate any undesirable biological reaction to electrical stimulation. The study of electropathology can help to reduce the possibility of unacceptable exposures to electrical equipment and to understand better how to use electrical stimulation for beneficial reasons. Regardless of our particular orientation, it is useful to understand the range of potential human reactions and their underlying mechanisms.

1.3   Scales of Short-Term Reactions to Contact Current

Several scales of reactions are of interest in the study of electropathology and electrical stimulation. These scales can be best appreciated by way of example. Figure 1.2 illustrates possible reactions to 60-Hz alternating current flowing between the hand and feet. It is assumed that an adult is gripping a large electrode in one hand, while a return electrode is contacting a foot or both feet. The exposure duration is assumed to be approximately 5 s. Categories of reactions are identified in the five columns in the figure,

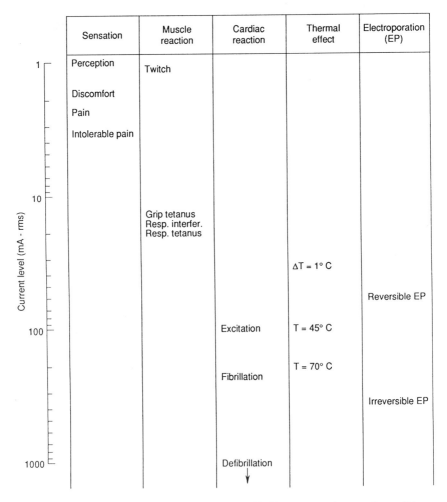

	Sensation	Muscle reaction	Cardiac reaction	Thermal effect	Electroporation (EP)
1	Perception	Twitch			
	Discomfort				
	Pain				
	Intolerable pain				
10		Grip tetanus Resp. interfer. Resp. tetanus			
				$\Delta T = 1° C$	
					Reversible EP
100			Excitation	$T = 45° C$	
				$T = 70° C$	
			Fibrillation		Irreversible EP
1000			Defibrillation		

Current level (mA · rms)

FIGURE 1.2. Potential short-term reactions to 60-Hz current. Assumed conditions: 5-s exposure; hand grip of large electrode; return electrode at feet; large adult subject; median response.

each involving a distinct continuum of reactions. Various thresholds of reaction are identified by descriptors whose vertical placement indicates an approximate median value of the electrical threshold, which is read on the left-hand scale. These thresholds are only approximate, and are subject to many variables. The intent of Fig. 1.2 is to provide a rough comparison of thresholds, rather than precise numerical values. As the indicated values are approximate medians of a statistical distribution, they would not normally be suitable as acceptability criteria unless some safety factor were applied.

The following is a commentary on each category. The commentaries call attention to particular chapters (6–10) where a detailed treatment may be found. Additional chapters (2–5) provide a foundation of bioelectric principles and models that explain the experimental data and related mechanisms described in later chapters.

Sensory Reactions (Chapter 7)

Electrical sensation involves the same pathways that are used to sense our external and internal environment. The dynamic range of electrical thresholds from perception to pain is, however, a small fraction of that for natural stimuli. The dynamic range for 60-Hz current is particularly small. Although pain thresholds can be measured with a reasonable degree of repeatability, the measurement of thresholds for "intolerable pain" may vary considerably with the experimental context. The tolerance threshold indicated in Fig. 1.2 indicates an experimental tolerance limit, described in Chapter 7, but does not necessarily represent the most extreme limit of human endurance. Indeed, subjects participating in hand-grip experiments (the "let-go" level in the next column) willingly endured considerably higher levels.

Muscle Reactions (Chapter 8)

The quantum of muscle reaction is called a "twitch." The threshold for the smallest measurable twitch for a hand-grip electrode is expected to be nearly equal to that for sensation. For the conditions represented in Fig. 1.2, the threshold twitch (around 1.0 mA) would occur in the small musculature of the hand. At much higher levels (around 15 mA), one encounters grip tetanus, where the hand "freezes" to the conductor (the so-called let-go threshold). The muscles responsible for grip tetanus lie much higher up in the forearm. It is probable that the small muscles of the hand would be tetanized at much lower current levels, but their effects on grip would be easily overcome by the finger extensors in the forearm. However, at current levels that cause grip tetanus, electrical stimulation of more powerful flexors located in the forearm would dominate.

Respiratory interference can also result from electrical exposure. Quantitative data supporting particular thresholds are rather sparse. Anecdotal reports gathered during grip tetanus tests indicate that respiratory interference may occur at levels somewhat above the grip tetanus threshold, and respiratory tetanus somewhat above that. The current levels indicated for respiratory effects in Fig. 1.2 are rough estimates based on these reports.

Cardiac Reactions (Chapters 5 and 6)

The cardiac excitation threshold indicated at about 100 mA in Fig. 1.2 applies to an extra beat elicited during the normally relaxed state of the heart. Such excitation is not necessarily life threatening, but is nevertheless treated as a potentially serious effect. At a higher current (around 240 mA), it is possible to induce ventricular fibrillation, in which uncoordinated contractions of heart muscle preclude pumping. Since the human heart rarely recovers spontaneously from electrically induced fibrillation, death will soon result unless defibrillation equipment can be applied to the victim.

Defibrillation occurs at still higher current levels (several amperes). Here, the fibrillating heart is forced into a uniform state of excitation, after which normal rhythmic activity becomes possible. Consequently, fibrillation is most probable within a band of current. Below the lower limit of the band, the current is too feeble to excite the heart; above the upper limit, the heart is defibrillated. This would explain why electrical fatalities due to cardiac failure occur most often at low voltages, whereas high-voltage injuries are principally from burns (see Chapter 10).

Thermal Reactions (Chapter 10)

The greatest temperature rise due to the current from a gripped conductor would probably occur in the high-current-density region of the wrist, or possibly in the hand at the edges of the conductor. With a 60-Hz stimulus, a 1 °C temperature rise would not be sensed, because the current necessary to produce it would also produce severe pain that would mask any thermal perception.

Tissue heating depends largely on the root-mean-square value of the current and relatively little on its waveform. Excitation of sensory nerves, on the other hand, is very sensitive to the stimulus waveform. If the waveform were inefficient for electrical stimulation, it would be possible for electrical perception thresholds to exceed thermal perception thresholds. Such a condition occurs with continuous sinusoidal current if the stimulus frequency exceeds 10^5 Hz. In that case, the current would be sensed as warmth, without electrical stimulation.

When skin or muscle tissue is heated to about 45 °C for prolonged periods, thermal damage can result. At that temperature, cutaneous nociceptors would probably be stimulated, resulting in pain (the body's natural defense against heat injury). Thermal damage, however, would not likely occur at 45 °C for a duration as short as 5 s. A temperature rise of somewhere around 70 °C would be needed to sustain permanent heat damage at that duration. Thermal perception for current frequency above 10^5 Hz is indicated at 35 mA for a touched contact (see Fig. 7.12). A large area grip contact will require an increased current to produce the same local temperature rise. Assuming that perception occurs with a temperature rise of 1 °C, the current for higher temperature rises can be estimated by using the relationship that the current is proportional to $\sqrt{\Delta T}$ [see Eq. (10.3)].

Electroporation (Chapter 10)

Current flowing within a biological medium will create potential differences across a cellular membrane. This effect is strongest in elongated cells (e.g., nerve and muscle cells) that are oriented in a direction parallel to the current flow. At the lowest levels on the scale in Fig. 1.2, the alteration of membrane potential will excite nerves and result in both sensory and muscular reactions. At much higher levels, the intense electric field that develops across the cellular membrane will promote the formation of pores. This process of electroporation (EP) is reversible at formative levels, but becomes irreversible at higher levels, leading to cellular death.

All cells normally maintain a natural "resting" potential, which for muscle cells is around 90 mV, with the inside negative relative to the outside (see Chapter 3). In response to an external current flowing parallel to the long axis of the cell, the membrane will be hyperpolarized at the cathode-facing end and depolarized at its anode-facing end. The thresholds indicated in Fig. 1.2 are predicated on a membrane voltage of 200 mV for reversible EP and 800 mV for irreversible EP (see Chapter 10). These thresholds correspond to hyperpolarization of 110 mV for reversible EP and 710 mV for irreversible EP in the hypothetical muscle cell used in this example. The corresponding current levels indicated in Fig. 1.2 are with reference to the maximum current density (and cellular polarization) occurring in the wrist.

1.4   Reactions to Electric and Magnetic Field Stimulation

The exposure conditions assumed in Fig. 1.2 apply to current conducted through direct contact with an energized electrode. It is also possible to produce short-term bioelectric reactions without direct electrode contact, as in the case of exposure to electromagnetic fields (EMF). With EMF exposure, an electric field and associated current density is induced within the biological medium by the external electric or magnetic field. Stimulation of excitable tissue is typically achieved most easily by magnetic fields at frequencies below 1 MHz, and thermal effects are most readily produced at higher frequencies. The principles governing EMF stimulation and thresholds of reaction are treated in Chapter 9.

Whereas the bioelectric mechanisms applying to contact current also apply to EMF stimulation, the measures of subject exposure may be quite different. To illustrate this point, Fig. 9.28 provides an example of scales of human reactions to short-term exposure of the head to a magnetic field at power frequencies (50, 60 Hz). The reactions listed include phosphenes, visual evoked potential effects, excitation of brain neurons, and seizures. Measures of exposure shown in Fig. 9.28 are the applied flux density in units of Tesla (T), the time rate of change of the flux density in Tesla per second (T/s), the induced electric field within the brain in volts per meter (V/m), and the associated current density in amperes per square meter (A/m^2). These four metrics of exposure can be tied together only for specified conditions which include the frequency of the incident field, the anatomical region of exposure, and the spatial distribution of the applied field.

The lowest threshold shown in Fig. 9.28 is associated with phosphenes, which are visual sensations produced by nonphotic stimuli. Experimental evidence on electrical phosphenes, discussed in Sect. 9.8, indicate that the site of electrical interaction is at synaptic processes within the retina. One significance of electrical phosphenes is that they are informative of synaptic interactions with in situ electric fields that may take place within the brain. Additional discussion of synaptic interactions is provided in Sect. 3.7.

At levels of exposure above the thresholds for phosphenes, Fig. 9.28 indicates excitation thresholds for brain neurons. The responsible bioelectric mechanisms are the same as those responsible for nerve and muscle excitation effects, which are represented in the first three reaction columns of Fig. 1.2. These mechanisms are discussed in Chapters 3 and 4.

Other short-term reactions to EMF exposure include auditory effects (Sect. 9.8) and heating effects (Sect. 11.5), both of which invoke thermal mechanisms related to the induced current density within the biological tissue. Thermal effects may become significant at much higher frequencies of EMF exposure than applicable to the example of Fig. 9.28. At the opposite end of the frequency spectrum, with static or extremely low frequency magnetic fields, short term reactions include magneto hydrodynamic effects, as discussed in Sect. 9.11.

1.5   Variables Affecting Thresholds

The thresholds discussed in Sects. 1.3 and 1.4 are illustrative values for a particular set of conditions. Many variables strongly affect these thresholds. These variables can be grouped into categories associated with the stimulus waveform, the spatial distribution of the stimulus, and the subject.

A significant part of the parametric sensitivity to electrical stimulation can be understood in the light of basic principles of bioelectricity. These basic laws are treated in the initial chapters of this book. Chapter 2 covers impedance and internal current distribution. In Chapter 2 the biological subject is considered for its passive electrical properties. Chapter 3 develops the principles that govern the electrical response of nerve and muscle. In Chapter 4 these principles are extended to computational models that allow one to study the excitatory effects of electrical stimulation. An "electrical cable" model is developed to help explain how electrical excitation is related to the temporal and spatial aspects of the stimulus. The electrical properties of the heart are developed in Chapter 5.

The excitation model of Chapter 4 has been extensively referred to in this book when discussing electrical excitation of the heart (Chapter 6), of sensory processes (Chapter 7), and of muscle (Chapter 8). The model is also used to derive excitation thresholds pertaining to electric and magnetic field exposures (Chapter 9).

Considering the extensive number of parameters that affect electrical sensitivity, one might wonder whether it is feasible to define "safe" or "acceptable" exposure levels. The subject of safety criteria is treated in Chapter 11, where performance criteria and electrical safeguards are discussed. The approach for setting electrical safety standards described in Chapter 11 is to make conservative assumptions regarding parameters, such as body size, impedance, stimulus waveform, and statistical threshold variations. While this approach does not necessarily protect against the most extreme combination of sensitivity factors, these standards are intended to provide an acceptable margin of safety.

In selecting criteria for the acceptability of electrical exposure, we are confronted with questions that cannot always be answered with scientific objectivity. Frequently there are policy or judgment issues that are settled on the basis of historical precedent, or on some other basis. One example of a judgment issue is the selection of the population percentile that should be assumed for a particular threshold reaction. Should a median sensitivity for a large segment of the population be used because it is "representative" in some sense? Should a lower percentile be selected to minimize the probability of an unacceptable exposure? If a safety factor is to be applied, on what basis should it be selected? The object of this book is not to define the "correct" criteria for the acceptability of electrical exposure. Rather, its object is to present the scientific data that will aid in the process of criteria selection.

The material of this book is the short-term human reactions to electrical stimulation, with particular attention to those of a detrimental nature. In this context, the term *electropathology* is meant to include those human reactions that might be considered undesirable in some context. It is necessary to understand the full range of reactions, from just-noticeable reactions to clearly undesirable or life-threatening ones. The intent of this book is to cover fundamental principles by which such responses may be understood.

2
Impedance and Current Distribution

This chapter examines tissue and body impedance as it affects the evaluation of electrical stimulation. It is concerned generally with current levels
exceeding $100\,\mu A$, and voltage levels exceeding $1\,V$. As a result, many of the
special problems associated with characterizing the body's response to
microampere currents or millivolt potentials will not be relevant. For
a thorough discussion of impedance in biological measurements at very
low voltages and currents, the reader is directed to the work of Geddes
(1972).

2.1   Dielectric Properties of Biological Materials

The bulk impedance properties of biological materials are important in
many applied problems of electrical stimulation. They dictate the current
densities and pathways that result from an applied stimulus. In order to
appreciate these bulk properties, consider dielectric properties in a more
general context.

Conductivity and Permittivity

Figure 2.1a illustrates a simple measurement of the electrical resistivity (ϱ)
or conductivity ($\sigma = 1/\varrho$) of a substance. Two electrodes of area A contact
a cylindrical block of biological material having length d. The resistance
between the electrodes, given by the ratio V/I, will be directly proportional
to the length of the material and inversely proportional to the area of the
electrodes and the material's resistivity in accordance with

$$R = \frac{\varrho d}{A} = \frac{d}{\sigma A} \tag{2.1}$$

Inversion of Eq. (2.1) yields a simple definition of the material's resistivity
or conductivity:

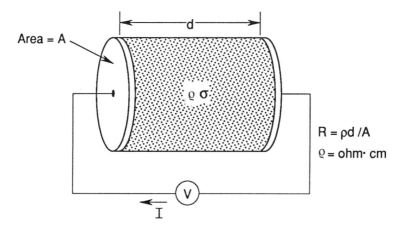

Area = A

d

ℓ σ

R = ρd /A
ℓ = ohm· cm

V

I

(a) Resistivity for volume material

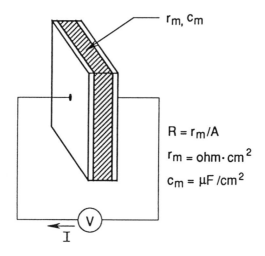

rm, cm

R = rm/A
rm = ohm· cm^2
cm = μF /cm^2

V

I

(b) Area proportional resistance of membrane

FIGURE 2.1. Electrical resistivity in volume and area materials: (a) resistivity for volume material; (b) area-proportional resistance of membrane.

$$\varrho = \frac{1}{\sigma} = \frac{RA}{d} \tag{2.2}$$

While the illustration in Fig. 2.1a indicates a simple conceptual method for measuring resistivity, a more accurate and practical method would use four electrodes—two to supply a current within the medium, and two intermediate electrodes to sample the voltage drop (Ruch et al., 1963). Resistivity is ordinarily cited in units of Ωm or Ωcm; the relationship between the two units of measurements is $\varrho(\Omega\text{cm}) = 100\varrho(\Omega\text{m})$. The inverse of resistivity in Ωm is conductivity, expressed in units of siemens per meter (S/m).[1]

Although the simple concept of conductivity expressed in Eq. (2.2) is adequate for many calculations, a more complete description of the dielectric properties of a material is often needed. The dielectric characteristics of a material are described in complex notation and include both conductive and capacitive properties. The concept of "polarizability" will help to explain the relationship among the variables and the remarkable electrical properties of biological materials.

Figure 2.2 illustrates a nonconductive material held between a pair of parallel-plate electrodes on which a potential of V volts is applied. The amount of charge accumulated on the electrodes is directly proportional to the product of applied voltage and capacitance in accordance with

$$Q = CV \tag{2.3}$$

where Q is charge in coulombs, C is capacitance in farads, and V is the potential in volts. If the plate separation, d, is much smaller than its linear dimensions, the capacitance is

$$C = \frac{\varepsilon_0 \varepsilon_r A}{d} \tag{2.4}$$

where ε_0 is the dielectric constant of free space ($8.85 \times 10^{-12}\,\text{F/m}$), and ε_r is the permittivity of the material relative to that of free space (expressed in dimensionless units, with $\varepsilon_r \geq 1$). The relative permittivity of air is very nearly equal to unity.

The relative permittivity constant, ε_r, is a measure of a material's ability to become polarized in response to an applied electric field. In the parallel-plate example, the electric field is simply given by $E = V/d$. The hypothetical material indicated in Fig. 2.2 is assumed to be nonconductive; that is, it lacks free electrons or ions. Consequently, there is no net current flow in response to the applied field. Nevertheless, the material is assumed to contain units of separated charge that are bound together into electrically neutral entities called *dipoles*. The displaced charge centers represent attractive forces within the dielectric medium, enhancing the internal electric field as indicated by the arrows in Fig. 2.2. Analogously, the dipoles act as if they were reducing the plate separation to an effective value of d/ε_r, and thereby increasing the capacitance.

The capacitive current is given by the change of charge versus time, dQ/dt. For a time-varying applied voltage, it follows from Eq. (2.3) that

$$I = C\frac{dV}{dt} \tag{2.5}$$

---

[1] Formerly called the "mho," the siemen is now the accepted international unit of conductance.

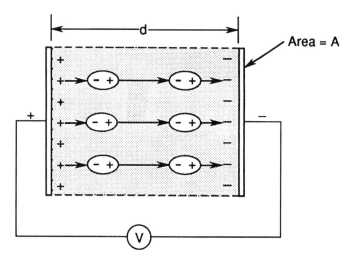

FIGURE 2.2. Dielectric property. Dipoles within material enhance internal field as indicated by flux lines (arrows).

For a sinusoidal voltage, $dV/dt = j\omega V$, where $\omega = 2\pi f$, f is the frequency of oscillation, and j is the phaser operator indicating $90°$ phase shift. Along with Eq. (2.4), the capacitive current can be expressed as

$$I = j\omega\varepsilon_0\varepsilon_r \frac{AV}{d} \tag{2.6}$$

When the material contains both dipoles and free charges, its dielectric description requires complex notation. Figure 2.3 illustrates an equivalent circuit of a partially conductive material. The total current is the sum of the resistive and capacitive components $I = I_R + I_C$, which is given by

$$I = \frac{V}{R} + C\frac{dV}{dt} \tag{2.7}$$

For a sinusoidal voltage, Eq. (2.7) can be expressed as

$$I = \frac{V\sigma A}{d} + V\left(\frac{\varepsilon_0\varepsilon_r A}{d}\right)j\omega$$

or, equivalently

$$I = \frac{VA}{d}\left(\sigma + j\omega\varepsilon_0\varepsilon_r\right) \tag{2.8}$$

The macroscopic electric field is $E = V/d$, and the current density is $J = I/A$. Consequently, Eq. (2.8) can be written as

$$J = \sigma^* E$$

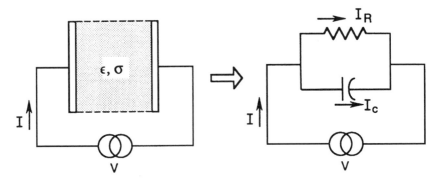

FIGURE 2.3. Complex permittivity. A partially conductive dielectric material is subjected to an alternating voltage V, resulting in resistive and capacitive current.

where σ^* is the complex conductivity given by

$$\sigma^* = \sigma + j\omega\varepsilon_0\varepsilon_r \tag{2.9}$$

Alternatively, Eq. (2.8) can be written as

$$I = \frac{j\omega\varepsilon_0 A}{d}\left(\varepsilon_r - \frac{j\sigma}{\omega\varepsilon_0}\right)V \tag{2.10}$$

By analogy with Eq. (2.6), the term in parentheses is the complex permittivity constant, given by

$$\varepsilon^* = \varepsilon_r - \frac{j\sigma}{\omega\varepsilon_0} \tag{2.11}$$

or, in conventional notation,

$$\varepsilon^* = \varepsilon' - j\varepsilon'' \tag{2.12}$$

in which $\varepsilon' = \varepsilon_r$, and $\varepsilon'' = \sigma/(\omega\varepsilon_0)$. Equations (2.9 and 2.11) are equivalent descriptions of the material. The form indicated by Eq. (2.11) is more often used; its symbolic representation is given as in Eq. (2.12) and is called the *complex permittivity*. Its imaginary part includes the conventional conductivity, and the real part is the relative dielectric constant.

In the above development, it is assumed that the dipoles orient themselves to the internal alternating field, requiring that they reverse their orientation every one half cycle of the applied voltage. The dipoles have a certain degree of inertia and cannot follow the field oscillation if it is too rapid. Therefore ε_r is at maximum at low frequencies. It drops when the frequency is raised above a critical value. With very high frequencies, the dipoles retain random orientation, and the relative permittivity of the material approaches unity.

The ability of a dielectric material to respond to an applied field can be expressed in terms of its *relaxation time constant* τ_r, or, equivalently, in terms of the relaxation frequency $\omega_r = 2\pi f_r = 1/\tau_r$. In a typical biological medium several mechanisms may exist for producing dipoles, each with a different relaxation time constant. In addition, the presence of boundaries between regions of differing permittivity will result in an equivalent relaxation time constant because of the buildup of charges at those boundaries (referred to as "interfacial" effects).

Figure 2.4 illustrates an example of the complex permittivity of a typical biological material (Pethig, 1979; Foster and Schwan, 1996). The dips in the curve arise from different mechanisms of polarization and are termed *dielectric dispersion*. At the lowest frequency, the so-called α dispersion has been attributed to electronic bilayers in organic molecules, ionic dispersion processes in micrometer-sized particles, active membrane conductance phenomena, and other membrane effects. The β dispersion is attributed to capacitive charging of cellular membranes (interfacial effects), and dipolar relaxation of proteins. The γ dispersion represents dielectric relaxation of water molecules.

Table 2.1 lists experimental values of conductivity (Part A) and relative permittivity (Part B) for various biological material. The tabulated values are geometric means of data published by Foster and Schwann (1996). Useful summaries have also been published by others (Geddes and Baker, 1967; Schwann, 1968; Stuchly and Stuchly, 1980; Stoy et al., 1982).

The relative permittivity of biomaterials at low frequencies can be on the order of 10^6. These values are remarkably large in comparison with

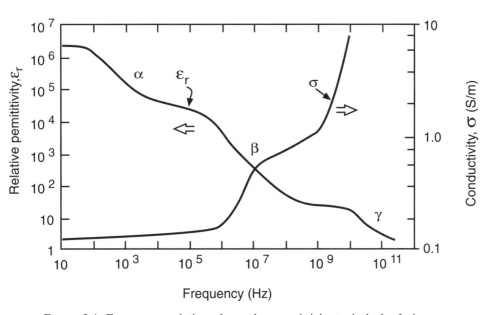

FIGURE 2.4. Frequency variation of complex permittivity typical of soft tissue.

TABLE 2.1. Dielectric properties of biological materials.

Frequency (Hz)	Skeletal muscle Parallel	Skeletal muscle Perpend.	Liver	Lung	Spleen	Kidney	Brain white matter	Brain gray matter	Bone	Whole Blood	Fat
Part A: Conductivity (values given in S/m)											
10	0.52	0.076	0.12	0.089	—	—	—	—	—	—	—
10^2	0.52	0.076	0.13	0.092	—	—	—	—	0.013	0.60	—
10^3	0.52	0.08	0.13	0.096	—	—	—	—	0.013	0.68	0.04
10^4	0.55	0.085	0.15	0.11	0.15	0.07	—	—	0.013	0.68	—
10^5	0.65	0.47	0.16	—	0.62	0.24	0.13	0.17	0.014	0.61	—
10^6	—	0.71	0.20	—	0.63	0.38	0.16	0.21	0.017	0.71	—
10^7	—	0.87	0.46	—	0.63	0.59	0.28	0.52	0.024	1.11	—
10^8	—	0.85	0.65	0.53	0.83	0.80	0.48	0.68	0.057	0.82	0.04
10^9	—	1.41	1.03	0.73	1.31	0.97	0.85	1.05	0.05	1.43	0.05
10^{10}	—	8.23	7.3	—	8.1	5.8	8.0	10.0	0.92	9.8	0.35
Part B: Relative Permittivity											
10	10^7	10^6	5×10^7	2.5×10^7	—	—	—	—	—	—	—
10^2	1.1×10^6	3.2×10^5	8.5×10^5	4.5×10^5	—	—	—	—	3,800	—	1.5×10^5
10^3	2.2×10^5	1.2×10^5	1.3×10^5	8.5×10^4	—	—	—	—	1,000	—	5×10^4
10^4	8×10^4	7×10^4	5.5×10^4	2.5×10^4	2.2×10^4	4.8×10^4	—	—	640	2,900	2×10^4
10^5	1.5×10^4	2.1×10^4	1.2×10^4	—	3,260	1.2×10^4	2,500	3,800	280	2,810	—
10^6	—	2,200	1,970	—	1,450	2,540	670	1,250	87	3,300	—
10^7	—	184	232	—	357	294	190	309	37	2,040	—
10^8	—	68	72	35	78	76	62	81	23	200	—
10^9	—	55	49	35	51	44	39	46	8	71	5
10^{10}	—	36	37	—	38	33	25	40	—	45	—

Source: Table adapted from Foster and Schwan (1996). Reprinted with permission from *CRC Handbook of Biological Effects of Electromagnetic Fields*, Copyright CRC Press, Inc., Boca Raton, FL.

synthetic dielectric materials, for which relative permittivities of 5 to 10 are typical. Despite the appearance of such large permittivity, the biological material remains overwhelmingly resistive, as can be seen from the following example. Assume the following values: $\sigma = 0.1\,\text{S/m}$, $\varepsilon_r = 10^6$, and $f = 100\,\text{Hz}$. In accordance with Eq. (2.9), the complex conductivity is $\sigma^* = 0.1 + j5.6 \times 10^{-3}$. In this example, the capacitive component of current is only 5.6% of the resistive component. In general, the bulk properties of biological tissue are dominantly resistive.

The conductivity of some biological preparations is markedly anisotropic. Skeletal muscle, for example, is shown in Table 2.1 to be approximately 6 to 7 times more conductive when low-frequency current is orientated parallel to the muscle fibers, as compared with a perpendicular orientation. The degree of anisotropicity of animal muscle tissue appears to vary greatly with the tested species (Chilbert et al., 1983). Anisotropic conductivity ratios ranging from about $5:1$ to $10:1$ apply to cardiac tissue, depending on the method of measurement (Plonsey and Barr, 1986). An anisotropic ratio of about $10:1$ has been attributed to nerve bundles (Nicholson, 1965; Ranck, 1963). For a critique of anisotropic impedance measurements in cardiac and other muscle tissue, the reviews of Plonsey and Barr (1986) and Roth (1989) are recommended.

The anisotropic behavior of muscle tissue can be understood by envisioning a collection of individual muscle fibers, as in Fig. 3.21. Current in a direction perpendicular to the fibers must travel in a circuitous path, whereas current parallel to the fibers travels in a shorter direct path—the path length difference accounting for the directional difference in resistivity. If the frequency of the current were high enough, the insulation afforded by the cellular membranes would become bypassed through capacitive coupling, and the anisotropic properties would disappear.

Cellular Membranes

The previous discussion has dealt with the bulk dielectric properties of composite biological materials. Characteristics of microscopic biological components, such as the cellular membrane, may also be important in applied studies of electrical stimulation. The cellular membrane consists of a bimolecular lipid structure whose impedance properties are usually expressed as area-proportional quantities—unit area resistance in Ωcm^2, and capacitance in $\mu\text{F/cm}^2$ as in Fig. 2.1b. These units stand in contrast to the bulk resistivity units (Ωcm) indicated in Fig. 2.1a for composite materials. The reason is that the thickness of the biological membrane cannot be subdivided without altering its basic structure. The capacity of a cellular membrane generally is in the region $0.5–1\,\mu\text{F/cm}^2$, resistivity in the range 10^2 to $10^4\,\Omega\text{cm}^2$, and relative permittivity around 2.5 (Pethig, 1979). The ionic permeability is specific to particular ionic species. For excitable membranes (nerve and muscle tissue), ionic permeability is highly dependent on the transmembrane voltage, as discussed in Chapter 3.

Skin Depth

The penetration depth of incident electromagnetic energy is often described in terms of the material's "skin depth." As described in Chapter 9, an incident magnetic field will set up eddy currents in a conducting material. The eddy currents, in turn, create their own magnetic field, which tends to oppose the incident field and resist its penetration into the material. Consequently, the current induced within the material will drop off in an exponential fashion from the surface. The distance at which the current density falls to e^{-1} of its surface value is known as the *skin depth*.

The skin depth of a material of arbitrary conductivity was described in 1888 by Oliver Heaviside (Nahin, 1987):

$$\delta = \cfrac{1}{2\pi f \left\{ \left(\mu\varepsilon/2 \right) \left[\sqrt{1 + \left(\sigma/2\pi f\varepsilon \right)^2} - 1 \right] \right\}^{1/2}} \qquad (2.13)$$

where $\mu = \mu_0 \mu_r$ is the magnetic permeability of the material, $\varepsilon = \varepsilon_0 \varepsilon_r$ is its dielectric permittivity, and f is the frequency of the induced current. The magnetic permeability of free space or air is $\mu_0 = 4\pi \times 10^{-7} \, \text{H/m}$; for all practical purposes, the permeability of biological materials is that of free space, that is, $\mu_r = 1$. For a good conductor, $\sigma/(2\pi f\varepsilon) \gg 1$, and Eq. (2.13) reduces to

$$\delta = \left(\pi f \mu \sigma \right)^{-1/2} \qquad (2.14)$$

which is the expression for skin depth found in most engineering texts.

The skin depth of biological materials for frequencies below about 10 MHz is generally much greater than any practically attainable material thickness. This can be seen by way of example. Assume that $\mu_r = 1$, $\varepsilon_r = 200$, $\sigma = 0.1 \, \text{S/m}$, $f = 10 \, \text{MHz}$. Then, from Eq. (2.13), $\delta = 0.8 \, \text{m}$. It is only when the frequency is well above 10 MHz that skin depth becomes a significant consideration in most cases. One interpretation of this result is that for frequencies below 10 MHz, a magnetic field will pass readily through biological material, and the internal magnetic field differs negligibly from the external field.

2.2   Skin Impedance

In most situations involving electrical stimulation, current is introduced through metallic electrode contact with the skin. The total circuit impedance will include contributions from the source, the electrode interface, the skin, and the internal tissues of the body. Of these, skin impedance is the most difficult to characterize. It is nonlinear, time-variable, and depends on environmental and physiological factors that are usually difficult to control in an experimental setting. Yet skin impedance is often the primary factor that limits current flow in the body, particularly where the applied voltage is moderately low (<200 V) and where the skin is undamaged.

Detailed Structure of Skin[2]

The skin consists of various layers, as indicated in Figs. 2.5 and 3.16. An outer layer (the *epidermis*) overlays the inner dermis. The epidermis consists principally of keratin, derived from dead cells of the lower layers, arranged in a flattened and irregular fashion. The bonding strength of these cells is low. They are constantly flaking away naturally and can be easily broken or removed. The germinating layer at the boundary of the epidermis contains a mixture of living and dead cells. The dermis contains living cells and a great density of blood vessels that are related to both nutrition of the skin and its thermoregulation. The dermis consists of bundles of *collagen* fibrils oriented in all directions, giving it strength and elasticity. The distribution of skin thickness varies greatly for different body areas. The epidermis itself ranges from about 10 to more than 100 μm. It is typically 10 times

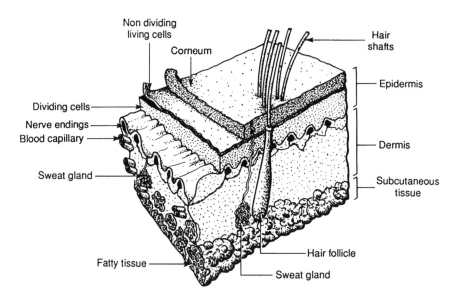

FIGURE 2.5. Structure of the skin.

---

[2] See also Edelberg (1971), Harkness (1971), and Tregear (1966).

as thick on the palms of the hands as compared with other body areas and is endowed with a much greater density of sweat glands.

The corneum (the outermost layer of dead skin cells) is a relatively poor conductor when dry. But when wet or sweaty, or when bypassed (as with an injury), the conductivity of the skin can rise dramatically. The contribution of the corneum to the total impedance can be studied by stripping the skin with cellophane tape (Harkness, 1971; Lykken, 1971; Tregear, 1966; Reilly et al., 1982; Clar et al., 1975). Figure 2.6 illustrates the drop of skin resistivity when the corneum is successively stripped away, showing an ultimate drop by a factor over 300. Impedance drop with corneal stripping has also been noted with high-voltage spark discharge stimuli (refer to Sect. 2.6). When a microelectrode penetrates the corneal layer, the resistance drops suddenly, as noted in Table 2.2 (Suchi, 1954).

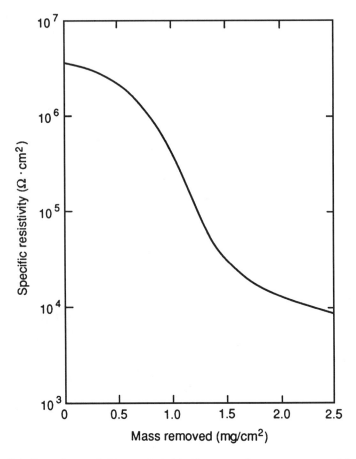

FIGURE 2.6. Impedance of skin as related to the mass of corneum removed from dry skin; sinusoidal current at 1.5 Hz. (Adapted from Tregear, 1966.)

TABLE 2.2. Impedance layers in the epidermis.

	Ball of thumb	Forearm
Dry, outer layer thickness (μm)	200	45
Penetration distance for		
impedance breakdown (μm)	350	50

Source: Suchi (1954).

Sweat is chiefly a 0.1 to 0.4% saline solution of sodium chloride, with a resistivity of 140 Ωcm at 37 °C (for a 0.3% solution). The density of sweat ducts (per cm^2) is approximately 370 on the palmar and plantar surfaces of the hands and feet, 160 on the forearm, 750 at the bend of the elbow, 150 to 250 on the breast, and 60 on the buttocks.

Current appears to be conducted in dry skin through discrete channels beneath a contact electrode. With a multipoint electrode, current is found to be preferentially conducted at one point (Mueller et al., 1953). In experiments with electroplating on the skin surface (Saunders, 1974), the pattern of silver deposition on the arm shows that current is conducted in discrete channels at a density of about 1 channel/mm^2. Other electroplating experiments reveal that the channel density on the palm is about three times that on the forearm (Panescu et al., 1993). These observations approximately correspond to the density and distribution of sweat ducts on the hand and forearm. The punctate nature of skin conductance was also explored with high-voltage discharges (Sect. 2.6).

The sweat ducts form electrical weak points in the epidermis, acting as conductive tubes into the well-conducting dermis and tissues below. Suchi (1954) explored the impedance role of sweat ducts using a silver microelectrode to scan various parts of the skin. When the electrode touched a duct filled with sweat, the impedance dropped by a factor of 10 as compared with adjacent areas of the skin. If the duct was dry, the drop was approximately a factor of two.

Equivalent Circuit Models

It would be desirable to represent the impedance of the skin and body as an equivalent circuit model that allows one to determine internal currents for a wide range of exposure conditions. Unfortunately, the skin is not readily expressible as a simple passive circuit—it is a distributed electrical system having markedly nonlinear and time-variant properties complicated by electrolytic interactions at the electrode interface. Nevertheless, impedance models are often valid under a sufficient range of circumstances to be of practical use.

A simple model sometimes used to represent skin impedance is a parallel network consisting of a capacitor and resistor, followed by a series resistor (cf. Yamamoto and Yamamoto, 1977; Burton et al., 1974). The parallel resistor and capacitor represent the resistivity and capacity of the skin, and the series resistance the well-conducting subepidermal medium. Evidence for this simple model can be seen when measuring the current response to a constant-voltage stimulus pulse applied to the skin as illustrated in Fig. 2.7 (from Lykken, 1971). The response, illustrated in Fig. 2.7b, shows an initial current spike that is limited by the series resistance, R_s. Afterward, the current decays to the value limited by $R_p + R_s$. After the corneum has been removed by abrading, the current, illustrated in Fig. 2.7c, is limited by R_s.

A more complete model considers the skin as composed of numerous layers of cells, each having capacitance and conductance, as illustrated in Fig. 2.8a (Edelberg, 1971; Lykken, 1971). The individual strings of elements are meant to represent the parallel paths beneath an electrode. In addition to the resistance and capacitance elements, the electrical model contains DC potential sources to account for the observed bulk potential of the skin of about 15 to 60 mV, with the surface negative relative to the underlying layers. These potentials are extremely small in comparison with the stimulus potentials typically needed for cutaneous electrical stimulation. The element R_B is treated as the body impedance, exclusive of the skin.

Figure 2.8b illustrates a model of intermediate complexity. The element Z_e represents the impedance at the electrode interface; it is seldom determined explicitly, but rather is lumped together into total impedance. The parallel RC circuit represents the epidermis, with the element R_{P2} added in the capacitive branch. It is shown in parentheses to indicate that it is

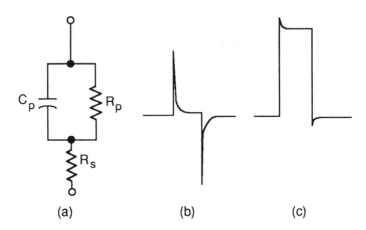

(a)   (b)   (c)

FIGURE 2.7. Current response of skin to square-wave constant-voltage pulses: (a) equivalent circuit; (b) response of intact skin; (c) response of skin with corneum removed. (Copyright 1971, The Society for Psychophysiological Research. Reprinted with permission of the publisher and the author from Lykken, 1971.)

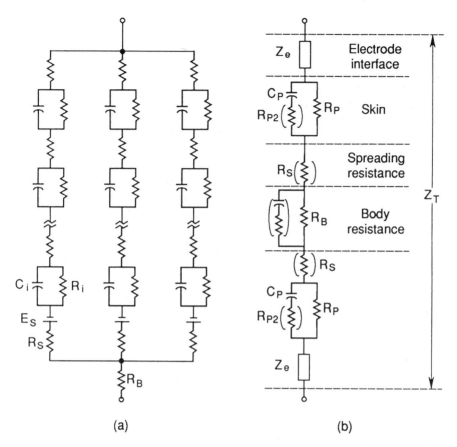

(a)   (b)

FIGURE 2.8. More complex impedance models: (a) multilayer model for skin impedance; (b) simplified body impedance model.

frequently ignored in impedance representations. The subepidermal layer
is shown as consisting of a spreading resistance element, R_s, which is inversely related to electrode size, and a body resistance element, R_B, which
increases with electrode separation. In most studies of body impedance, R_s
and R_B are lumped together into a single resistive term. The element R_B is
generally treated as a pure resistance, although in fact it contains a small
reactive component, indicated by the components in parentheses. The reactive component can be ignored for most practical applications. The circuit
representation includes additional terms to account for the second electrode. The total circuit impedance is designated Z_T. The circuit models
discussed here are merely approximations that are sometimes suitable; the
individual terms may behave in a more complex fashion than would a
laboratory component, as discussed below.

Area Proportionality of Skin Admittance

From geometric considerations, it might seem reasonable to suppose that skin admittance is directly proportional to contact area, as long as the electrode diameter is significantly larger than the skin thickness. Many investigators report equivalent circuit parameters as being area proportional. Lykken (1971), for example, notes an area-proportional admittance for DC currents with electrodes of $0.72\,\text{cm}^2$ and larger and for pulse stimuli with electrodes $2.4\,\text{cm}^2$ and larger. Tregear (1966) reports that the product of impedance and area is constant for wet skin only for electrodes having areas larger than $2\,\text{cm}^2$. For smaller areas, the impedance departs considerably from inverse area proportionality, with significant differences between wet and dry skin. Biegelmeier and Rotter (1971) evaluate equivalent circuit parameters for electrodes of 1.5 and $100\,\text{cm}^2$ (an area difference of $67:1$); they note a $4:1$ change in $R_{P2} + R_s$ as defined in Fig. 2.8b, and a $25:1$ change in R_P, showing that admittance rises more slowly than does contact area.

The area-dependent portion of the resistive path (R_s in Fig. 2.8b) comprises the region where the current still undergoes spherical spreading from the stimulating electrode. At the point where current no longer spreads out with distance, the resistance is no longer dependent on electrode area, but is more a function of electrode separation and body geometry. Thus, R_B in Fig. 2.8b is taken to be independent of electrode contact area.

Area-proportional parameters are usually determined by dividing admittance by electrode contact area. This simple calculation is confounded by the fact that the distribution of current beneath a contact electrode may be very nonuniform, as demonstrated in theoretical (Caruso et al., 1979; Rattay, 1988) and experimental (Lane and Zebo, 1967) studies. Figure 2.9 illustrates the theoretical current density for a three-layer conductivity model representing a layer of skin, fat, and muscle tissue; a two-layer model consisting of fat and muscle; and a single-layer model consisting of muscle tissue (Caruso et al., 1979). For the three-layer model, the calculated current density near the electrode edge is nearly a factor of 10 greater than that at the center. For the single-layer model, the current density from center to edge differs by a factor of about 2.2. Area conductivity is further complicated by the fact that current travels laterally beyond the confines of the electrode, such that the effective contact area may be larger than the physical contact area. Rather than rely on area-proportional values, contact impedance at high-currents can be expressed more accurately in terms of both the area and perimeter of the contact, as given by Equation 10.5.

With these caveats in mind, Table 2.3 lists some of the area-proportional values reported for low-frequency (0- to 65-Hz) skin impedance (Tregear, 1966). Because of the low frequency, the list applies primarily to the resistive component $R_P + R_S$ defined in Figs. 2.7a and 2.8b. The values are a

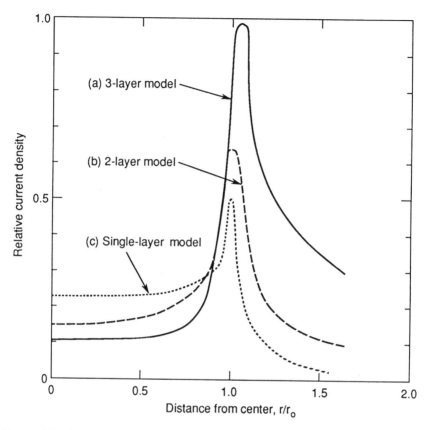

FIGURE 2.9. Current density beneath contact electrode; r_0 = electrode radius; vertical axis on dimensionless scale. (a) three layer model with skin, fat, and muscle; (b) two layer with fat and muscle; (c) single layer with muscle. (From Caruso et al., 1979.)

TABLE 2.3. Low-frequency resistivity of human and animal skin.

Species	Hydration	Body Location	Frequency (Hz)	Impedance ($k\Omega\,cm^2$)
Human	Dry	Arm	0	100–1,000
Human	Dry	Arm	1.5–10	600–1,200
Human	Dry	Fingertip	0–1	120–130
Human	Dry	Palm	30–65	60–80
Human	Wet	Forearm	1.5	880
Pig	Wet	Flank	1.5	12
Rabbit	Wet	Flank	1.5	18

Source: Adapted from Tregear (1966); data compiled from a variety of sources.

compilation from a variety of sources and do not necessarily represent data measured in a consistent manner.

Skin Capacitance

A variety of testing methods indicate that the skin's capacity lies in the range 0.02 to $0.06 \mu F/cm^2$ (Edelberg, 1971); by conventional calculations, this is considered very high. Consider, for example, a corneum thickness of $10 \mu m$ and a dielectric constant of 2.5 for biological membranes. With these values, the capacity is calculated to be about $2 \times 10^{-4} \mu F/cm^2$—a small fraction of the experimental values.

Skin capacitance will be affected by *polarization capacitance*—the phenomenon of stored charges that appear around an electrode in an electrolytic medium, forming an effective ionic capacitor that is dependent on excitation frequency (Schwan, 1966). Lykken (1971) argued that the apparent frequency-dependent property of skin capacitance is simply a consequence of the choice of equivalent circuit and that a representation with a resistively coupled capacitor (C_P and R_{P2} in Figure 2.8b) will demonstrate a fixed value versus frequency.

Much of the skin's capacity lies in the corneum. If the corneum is stripped away, the skin capacitance is reduced with each successive stripping operation (Edelberg, 1971). When the corneum is removed entirely, the capacity drops to a small fraction of its intact value (Lykken, 1971; vanBoxtel, 1977). These observations are counter to a model in which the corneum is simply a dielectric separating the electrode and the underlying conductive dermis. If that model were correct, we would expect to see an increase of capacitance as the thickness of the corneum is reduced. Biegelmeier and Miksch (1980) postulate that the skin's capacity is derived from membrane capacitance of the sweat gland duct. However, this explanation does not account for the leading-edge current spikes that are observed when constant-voltage pulses are applied to the sweat-gland-free skin of Rhesus monkeys (Bridges, 1985).

An alternative explanation for the skin's high capacity was provided by Tregear (1966), who treated the skin capacitance as being due to individual cell membranes as in Fig. 2.8a. If we assume that each cell's membrane may have a capacity as large as $5 \mu F/cm^2$, and acknowledging that each cell accounts for two membrane layers, then 200 cell layers could account for a capacitance of $0.05 \mu F/cm^2$. A related mechanism that might account for skin capacitance has been described as an ionic bilayer surrounding individual corneal cells (Clar et al., 1975).

Time-Variant and Nonlinear Aspects of Skin Impedance

When an electrode is placed on dry skin, impedance gradually falls with time, as noted in Fig. 2.10 (Mason and Mackay, 1976). During the first 2 min,

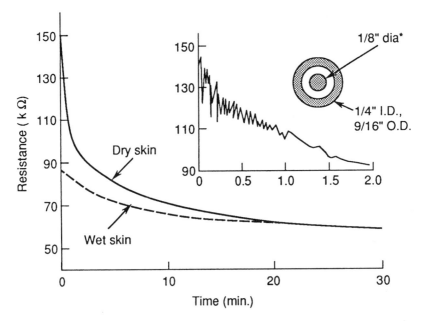

FIGURE 2.10. Variation of skin resistance with time. Concentric electrode as illustrated (inner diameter not clearly described in source). Wet skin treated with tap water. Inset shows expanded scale for first 2 min. Stimulus current: 1 mA. (Adapted from Mason and Mackay, 1976, ©1976 IEEE.)

the impedance undergoes rapid fluctuations superimposed on a relatively sharp overall drop (see insert). Thereafter, the impedance continues to drop, reaching an apparent asymptotic value after some 20 or 30 min. This effect may be explained by gradual hydration of the corneum from sweat buildup beneath the electrode. In the same experiments, skin pretreated with conductive electrode paste started with an initially lower impedance, rose somewhat, and attained a final value only slightly below its initial value after about 1 min.

Past studies have shown that the electrical properties of the skin are nonlinear. Edelberg (1971) suggests that the linear region for current density lies in the range below $2 \, mA/cm^2$. Lykken (1971) concludes that it is applied voltage, and not current density, that determines an upper limit of linearity around 2V. Biegelmeier and Rotter (1971) note variations in equivalent circuit values of over 2:1 for voltage changes from 4 to 45V. Stevens (1963) defines the DC current conducted in dry skin in terms of $I = aV + bV^2$, in which the V^2 term dominates above 3V.

Nonlinear phenomena in living skin was studied by Grimnes (1983a). He hypothesized that the time scale of impedance changes could be explained by either a nondielectric mechanism or a dielectric breakdown mechanism, depending on the applied voltage in the range of 600 to 1,000 V. In this work, the current was limited to a few microamperes, and the results may not necessarily apply directly to higher current densities.

A dramatic display of skin nonlinearity is seen in the sudden break-down of impedance above a critical voltage. Dielectric breakdown of dry, excised corneum has been observed by Mason and Mackay (1976) at 600V for a 15-μm-thick sample and at 450 V by Yamamoto et al. (1986). Dielectric breakdown of intact skin from high-voltage spark discharges appears at similar voltages, as described in Sect. 2.6.

With intact, living skin, a sudden impedance breakdown has been observed at significantly lower voltages than that needed to break down excised corneum. A discrete thermal model has been developed to account for nonlinear skin impedance, including breakdown (Panescu et al., 1994a,b). Pricking pain has been associated with the impedance breakdown (Nute, 1985; Gibson, 1968; Mason and Mackay, 1976; Mueller et al., 1953). During breakdown, current becomes concentrated in discrete channels. According to Mason and Mackay, microscopic examination of the electrode site reveals small blackened punctures under a dry-skin electrode, but no such evidence when the skin has been pretreated with water or electrode paste.

The major source of skin nonlinearity lies in the corneum. Equivalent circuit values R_P and R_S were evaluated by van Boxtel (1977) from the transient current response to constant-voltage pulses. With intact skin, R_P varied significantly with the applied voltage, whereas R_S remained independent of the stimulus level (up to a voltage of 40 V, and 40 mA final current). With the corneum removed, both R_P and R_S remained independent of the stimulus level.

Figure 2.11 illustrates nonlinear impedance response for pulses of 5-μs duration (Saunders, 1974). At these short durations, the corneal layer is probably bypassed through capacitive coupling to the dermis. However, the skin exhibits marked nonlinear properties, as noted in the knee of the curves at the higher current and voltage values—nonlinear breakdown occurs with voltages from about 150 to 250 V.

Freiberger (1934) describes breakdown as occurring at small contact areas within a fraction of a second for voltages above 100 V. Biegelmeier and Miksch (1980) noted breakdown above 200V for large hand-held electrodes (refer to Sect. 2.3). With small electrodes and voltages above 100 V, breakdown may occur within fractions of a second with dry-skin contact.

Nonlinear response to sinusoidal voltage shows up as distortions from a sinusoidal current waveform. Short-term distortions appear as an instantaneous nonlinear response within individual cycles; longer term nonlinearities appear as a gradual reduction of impedance. Figure 2.12 illustrates impedance breakdown with 50-Hz AC stimulation on a relatively long time scale (Freiberger, 1934). An initial precipitous drop occurs within

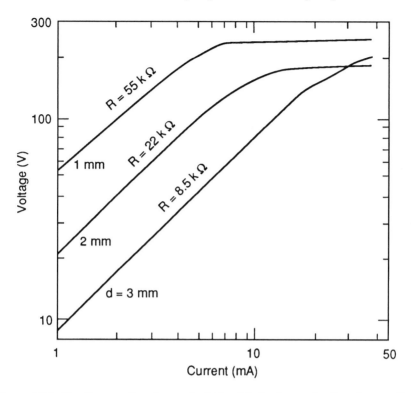

FIGURE 2.11. Nonlinear voltage/current relationship for 5-μs cathodic pulses applied to abdomen; resistance (R) applies to linear regions of curves; d = electrode diameter. (Adapted from F.A. Saunders, 1974 in *Conference on Cutaneous Communication Systems and Devices*, pp. 20–26, reprinted by permission of Psychonomic Society, Inc.)

a small fraction of a minute, a second drop during the first minute, and a final gradual drop during the next 7 or 8 min.

2.3   Total Body Impedance: Low-Frequency and DC

The previous section showed that body impedance depends on a variety of factors, including size and location of contact electrodes, skin hydration, and applied voltage. These factors were studied by Biegelmeier (1985a). Measurements consisted of the peak and root-mean-square (rms) voltage and current, dissipated power, and the phase angle between the voltage and current applied to the subject. The subject grasped cylindrical electrodes of 8 cm diameter, providing a hand-to-hand current path. The contact area on each hand was estimated to be 82 cm². After covering the cylindrical electrodes with insulating material, smaller hand contact areas consisted of 12.5,

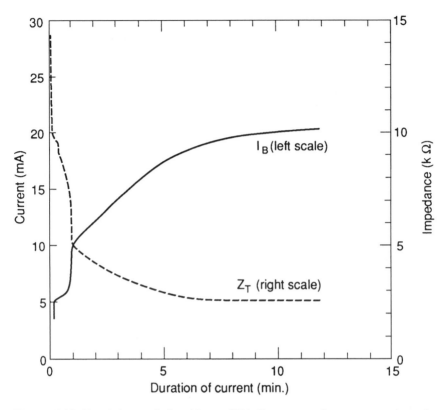

FIGURE 2.12. Breakdown of the skin at 50 V. Forearm-to-forearm current path; circular electrodes, 12 cm^2; dry skin. (From Freiberger, 1934.)

1.0, 0.1, and 0.01 cm^2 on each hand. Foot electrodes consisted of copper foil inserted into the shoes.

Figure 2.13 illustrates example current and voltage waveforms at the onset of a stimulus having a steady-state voltage and current[3] of 200 V and 118 mA, respectively (Biegelmeier, 1985a). Corresponding peak values were 280 V and 370 mA. Total body impedance (Z_T) and internal body resistance (R_B) were determined by

$$Z_T = \frac{V_T}{I_B} \tag{2.15}$$

$$R_B = \frac{V_{TP}}{I_{BP}} \tag{2.16}$$

---

[3] When referring to sinusoidal voltage and current, cited magnitudes indicate rms values, unless specifically stated otherwise. This convention applies throughout this book.

where V_T and I_B are steady-state (rms) values of applied voltage and body current; V_{TP} and V_{BP} are the peak values of voltage and current. In the example of Fig. 2.13, $Z_T = 1.7\,k\Omega$, and $R_B = 0.75\,k\Omega$.

Figures 2.14 to 2.16 illustrate the relationship between impedance and the applied voltage for large-area contacts and different electrode placements (from Biegelmeier, 1985b). Figure 2.14 applies to hand-to-hand contacts and indicates Z_T and R_B separately; Z_T is further subdivided into wet and dry contacts. Each measurement point is the average of six procedures with a single subject. The dry-contact data are indicated by filled circles; the error bars show the mean and extreme measurements. In Fig. 2.14, the wet-contact conditions were obtained by soaking the bands in either tap water or a saline solution prior to the measurement. The tap water treatment lowers the impedance somewhat relative to the dry condition;

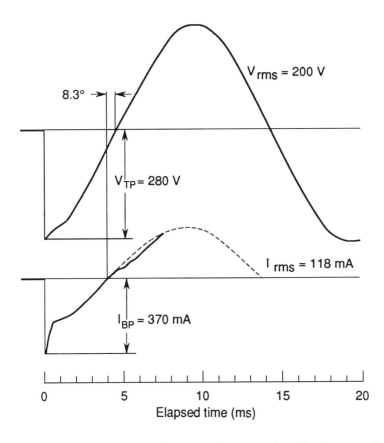

FIGURE 2.13. Oscillograms of applied voltage (upper trace) and body current (lower trace), with contact at the approximate peak of the voltage waveform; applied voltage = 200 V rms. (From Biegelmeir, 1985a.)

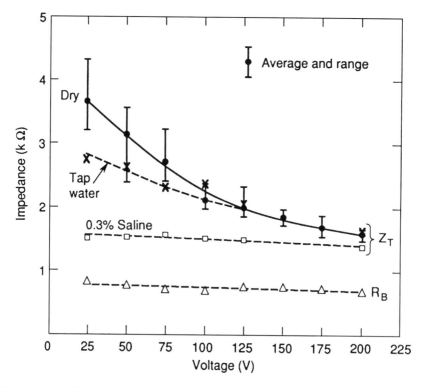

FIGURE 2.14. Impedance versus voltage for large-area hand-to-hand contacts. (From Biegelmeler, 1985b.)

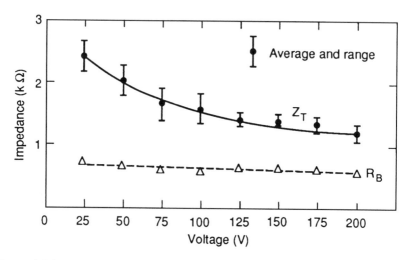

FIGURE 2.15. Impedance versus voltage for large-area hand-to-feet contacts. (From Biegelmeier, 1985b.)

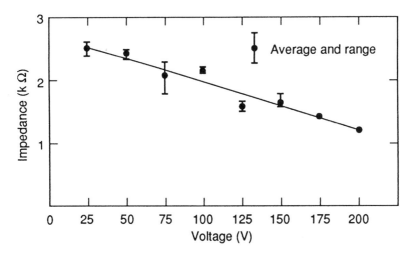

FIGURE 2.16. Impedance versus voltage for large-area foot-to-foot contacts. (From Biegelmeier, 1985b.)

treatment with saline lowers the impedance much more. At 200 V, there is little difference between wet and dry electrodes.

Figures 2.14 and 2.15 also show internal body resistance, R_B, for the indicated paths. The value of R_B drops only slightly as the applied voltage is raised from 25 to 200 V, dropping to 650 Ω for the hand-to-hand path and to 550 Ω for the hand-to-foot path. These values are reasonably close to the value of approximately 500 Ω determined for R_B from high-frequency measurements (refer to Sect. 2.4), and also with high-voltage capacitive discharges (refer to Sect. 2.5).

The cited values of R_B are similar to the resistance of a solution of NaCl of physiological concentration and of geometric dimensions similar to that of the body (Biegelmeier, 1986). If the resistivity of the NaCl solution is taken to be 80 Ω cm, Eq. (2.1) indicates that the resistance of a cylindrical cross-sectional area of 25 cm^2 and of length 150 cm (for two arms in series) is approximately 480 Ω. This simple calculation correlates quite well with measured values of R_B.

Figure 2.17 illustrates further results of Biegelmeier's (1985a) experiments showing the relationship between total impedance and applied voltage for various contact areas. At 25 V, impedance is nearly inversely proportional to contact area. As the voltage is increased, impedance drops rapidly, and area dependence is markedly reduced. At 225 V, total impedance varies by only about 4:1 for a change in electrode contact area of 8200:1.

Impedance-versus-voltage characteristics depend very much on the body locus of skin contact. With tests on cadavers, Freiberger found impedance

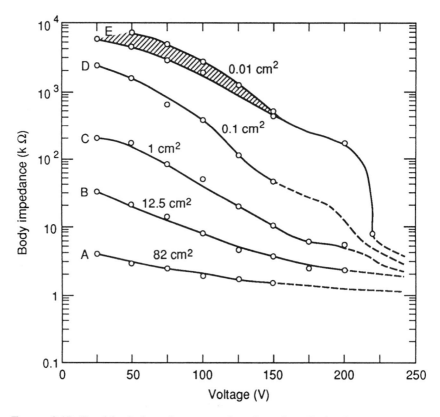

FIGURE 2.17. Total body impedance as a function of applied voltage for various contact areas; dry hand-to-hand electrodes. The various curves are identified by the contact area on each hand. (From Biegelmeier, 1985b.)

of dry contacts to drop precipitously for voltages above 200 V on the palmar and plantar surfaces of the hands and feet. On the forearm, the point of impedance breakdown was about 50 V.

Figure 2.18 illustrates measurements of Freiberger (1934) showing impedance values for large-area hand-to-hand or hand-to-foot contacts at 50 Hz. The illustrated values are reasonably consistent with those of Biegelmeier (Fig. 2.17) within the voltage range 25 to 200 V. In Fig. 2.18, impedance declines with voltage, reaching a minimum plateau somewhat beyond 500 V. Freiberger attributed the data in Fig. 2.18 to living persons, although the measurements above 50 V were actually conducted on corpses, and those below 50 V on living persons. He adjusted the measurements on corpses to living persons using a correction procedure that will be described presently.

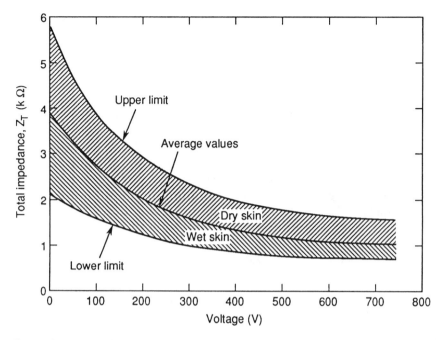

FIGURE 2.18. Total body impedance attributed to living persons; large-area hand-to-hand or hand-to-foot contacts. Measurements above 50 V conducted on cadavers, and corrected for living persons. (From Freiberger, 1934.)

Distribution of Current and Impedance Within the Body

The pathway and density of current within the body, as well as the total impedance presented to the stimulating device, depend critically on the size and placement of the electrodes. Figure 2.19 (from Roy et al., 1986) illustrates the current distribution measured for electrodes of various sizes within a saline-filled tank. A large (400-cm²) "indifferent" electrode and a smaller "active" electrode were separated by 16 cm. The current density was measured as a function of distance from the active electrode; the applied current was 3 mA. The current density near the active electrode depends substantially on electrode size. But when measured farther into the medium (beyond 6 cm in this example) the current density depends but little on electrode size. This is an illustration of "spreading" resistance, mentioned in connection with Fig. 2.8. When the electrodes are widely spaced on the body, the total impedance will depend on the separation of body contact points, plus the skin and spreading resistance. For high voltages, large contact areas, and hydrated skin, the total impedance will be dominated by internal body impedance.

A substantial body of knowledge concerning body impedance was provided in early experiments by Freiberger (1934). Even today, after the passage of more than 60 years, we can still learn much from his remarkable work. Much of Freiberger's research was conducted on cadavers. He determined the contribution of skin impedance by removing the cornium from the electrode site using a process of inducing heat blisters on the skin. The impedance after this treatment was called "body" impedance. Figure 2.20 illustrates the distribution of body impedance for various current paths. The numbers indicate internal impedance for various electrode placements as a percentage of the total hand-to-foot impedance. These data apply specifically to the internal component, analogous to R_B in Fig. 2.8. Roughly 50% of the internal impedance for hand-to-hand or hand-to-foot contacts resides in the wrists or ankle; these high-impedance regions are dominated by relatively poorly conducting bone and ligament. The impedance distribution of Fig. 2.20 is consistent with the investigations of Taylor (1985), who used high-voltage capacitive discharges on living people (see Sect. 2.6). The internal body impedance for hand-to-hand or hand-to-foot paths as measured by Freiberger averaged $1,000\,\Omega$ for 60 corpses—male and female, of

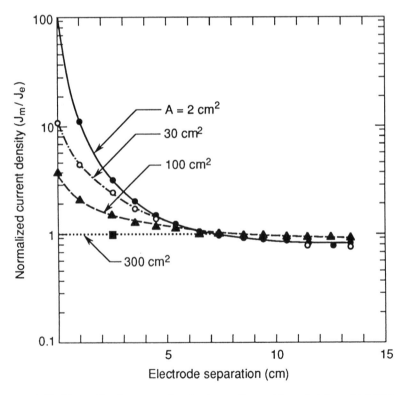

FIGURE 2.19. Normalized current density in medium with resistivity of $1,860\,\Omega\,cm$; J_m/J_e is current density in medium divided by average current density of electrode. Indifferent electrode area is $400\,cm^2$. (From Roy et al., 1986.)

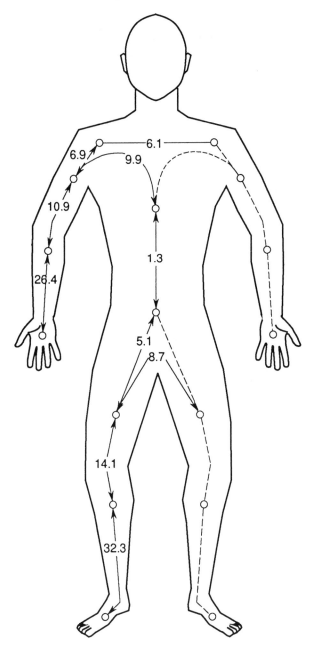

FIGURE 2.20. Distribution of internal impedance of the human body. Numbers indicate percentages of total internal impedance for hand-to-foot contacts. (Adapted from Freiberger, 1934.)

various ages and body structure. This value is somewhat larger than the previously discussed measurements (500–750 Ω) in living persons.

Figure 2.21 shows a simplified representation of body impedance for evaluation of electrical accidents in which the current path is from one extremity to another. If the contact area is large and if the skin is sweat-or saline-soaked, the impedance to low-frequency currents is largely resistive—a typical example value would use $Z_T = 1,000\,\Omega$. As a further approximation, we can consider the impedance in each extremity (Z_{TE}) as being equal, and the impedance of the body trunk (R_{BT}) to be negligible in comparison. The internal body impedance across both extremities lies in the range 500 to 750 Ω. If $Z_T = 1,000\,\Omega$, then the skin impedance at each electrode must be 125 to 250 Ω. Using the model of Fig. 2.21, the example indicates that approximately 1,000 Ω would be measured across any combination of two extremities. If a person were to grasp equipment with both hands while sitting on a conductive floor, the impedance would be reduced to 250 Ω. For other current paths, calculation of total impedance can be

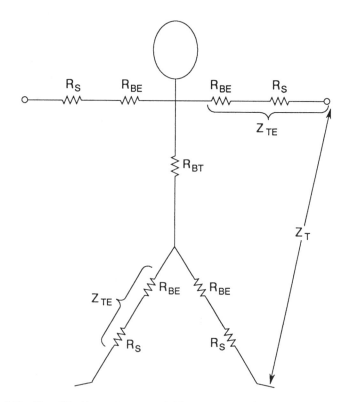

FIGURE 2.21. Simplified impedance model for current paths across extremities—wet skin, large-area contacts. Example values: $Z_T = 1,000\,\Omega$; $Z_{TE} = 500\,\Omega$; $R_{BE} = 250\text{–}375\,\Omega$; $R_S = 125\text{–}250\,\Omega$; $R_{BT} = 50\,\Omega$.

carried out by reducing R_B in accordance with Fig. 2.20. If the skin is damaged, the value of R_S would also be reduced.

Current Through the Heart

Freiberger applied 50-Hz currents to corpses to determine the percent of current conducted through the heart. After opening the chest of a corpse, he encircled the heart with a ring-shaped current transformer, and measured the current flowing through its cross-sectional area ($51\,cm^2$) as a function of electrode placement on the body. Measurements made in this manner are sensitive only to current in a direction along the long axis of the heart. The percentage of current flowing through the ring transformer, listed in Table 2.4, is quite small for current paths through the limbs. The greatest percentage was 8.5% for a right-hand-to-feet current path. For a foot-to-foot path, the current through the heart does not surpass 0.4% of the total body current. It is likely that a more favorable orientation of the ring transformer was responsible for the larger currents measured from the right-hand in comparison with the left-hand contact. Tests similar to those of Freiberger have been carried out on dogs (Kouwenhoven et al., 1932), with similar results.

Applicability of Measurements on Corpses

Tests on biological tissue reveal impedance changes after death. The dielectric permittivity of frog muscle, for example, is reported to decrease by a factor of 2 at 10 Hz within 2.5 h after death, but changes above 100 Hz are much less pronounced (Schwan, 1954). Permittivity changes would have little effect on tissue conductivity, since most tissue is dominantly resistive, even at low frequencies.

Freiberger measured systematic increases in body resistance with time after death. The internal body impedance increase after 5 h, for example, was about 20%. More gradual increases continued up to 30 h, at which time the increase from time of death averaged about 45%. These increases were

TABLE 2.4. Percentage of longitudinal heart current for various current paths.

Current Path	Heart current (%)		
	Min	Av.	Max
Hand to hand	1.9	3.3	4.4
Left hand to feet	1.2	3.3	5.1
Right hand to feet	4.8	6.7	8.5
Foot to foot	<0.1	—	0.4
Head to feet	4.8	5.5	5.9

Source: Data from Freiberger (1934).

attributed almost entirely to the cooling of body temperature to room temperature after death. Freiberger determined that the resistivity of physiological saline varies with temperatures in a manner that would account for the observed change in body impedance. Similar changes after death in the resistivity of muscle tissue of animals have been observed; resistivity changes in fatty tissue, however, are found to be minimal (Chilbert et al., 1983).

Changes in the processes of blood circulation and sweat activity after death should have the greatest influence on skin impedance. Considering that skin impedance is most significant at low voltages, we expect the impedance difference between living persons and corpses to be small when the voltage is raised above 200 V. Skin hydration should play a prominent role in the impedance changes after death. It is very difficult to measure skin hydration in a meaningful way. Freiberger instead studied the correlation of impedance with body temperature after death, and used a temperature correction factor to relate the impedance of corpses to living persons. These correction factors were applied in Fig. 2.18 to the measurements on corpses for voltages above 50 V; below 50 V, the data were derived from measurements on living persons.

Statistical Distribution of Impedance

Table 2.5 provides statistical data from Swiss and Austrian measurements at 50 Hz (Biegelmeier, 1985c). Data are shown for total impedance (Z_T) with large-area contacts on dry skin, and also for the body impedance component (R_B). Freiberger also determined the statistical distribution of 50-Hz impedance on corpses above 50 V and on living persons below 50 V. The statistical data of Freiberger have been adapted by the International Electrotechnical Commission (IEC) to produce an impedance model applicable to living persons. (Biegelmeier, 1986). Table 2.6 presents the IEC model at the 5, 50, and 95 percentile ranks, and for voltages ranging from 25 to 1,000 V. The asymptotic values are the presumed values of internal body impedance (R_B).

Freiberger tested 25 subjects using direct or 50-Hz alternating voltages ranging from 5 to 50 V. The current path was hand to hand, with large

TABLE 2.5. Statistical impedance measurements.

Voltage (V)	Number	Total (Z_T)		Body (R_B)	
		Avg. (kΩ)	S.D. (kΩ)	Avg. (kΩ)	S.D. (kΩ)
25	100	3.52	1.40	0.78	0.11
15	50	3.72	1.12	0.64	0.10

Large area contacts, 50 Hz, dry skin, hand-to-hand.
Data from Swiss and Austrian measurements, as summarized by Biegelmeier (1985c).

TABLE 2.6. Statistical data for total body impedance (Z_T) adopted by the IEC for 50-Hz currents.

Touch voltage (V)	Total body impedance (Ω) at the indicated percentile rank		
	5%	50%	95%
25	1,750	3,250	6,100
50	1,450	2,625	4,375
75	1,250	2,200	3,500
100	1,200	1,875	3,200
125	1,125	1,625	2,875
220	1,000	1,350	2,125
700	750	1,100	1,550
1,000	700	1,050	1,500
Asymptotic value	650	750	850

Large area contacts, hand-to-hand, dry skin.
Source: Data from IEC (1984).

cylindrical electrodes on dry skin. At 5 V, the average impedance was about 4.8 kΩ for DC voltages and 3.8 kΩ for AC voltages. At 50 V, the impedance dropped to about 70% of the 5-V values. Maximum and minimum impedances at DC were typically a factor of 2 and 0.5 times the averages. At AC, the extremes were 1.5 and 0.53 times the averages. These extremes probably represent approximately the 5 and 95 percentile ranks of a cumulative distribution. The mean DC impedance exceeded the AC value by a typical factor of 1.25, although there was considerable overlap in the distributions. Lower AC values are attributed to 50-Hz capacitive coupling through the high-impedance corneal layer.

Statistical impedance data were developed by Underwriters Laboratories (UL) for DC currents and low voltages (approximately 12 V) (Whitaker, 1939). Figure 2.22 illustrates statistical distributions for 40 adults; Fig. 2.23 applies to 46 children. Table 2.7 summarizes the data of adults and children. Adult weight ranged from 45.4 to 94.4 kg (median = 62.2 kg), and age ranged from 18 to 58 years (median = 30 years). For children, weight ranged from 14.1 to 58.1 kg (median = 31.8 kg), and age ranged from 3 to 15 years (median = 10 years). Hand electrodes consisted of two No. 10 AWG bare copper wires twisted together; the foot electrodes were large copper plates. Under "wet" conditions, the hands and feet of subjects were initially soaked in a 20% NaCl solution. Preliminary tests showed that resistance was independent of measuring current in the range of 1 to 15 mA, provided a constant contact area and pressure were maintained and the subject's hands were wetted in the solution before each measurement. The test voltage was adjusted up to a maximum of 12 V to maintain a current of

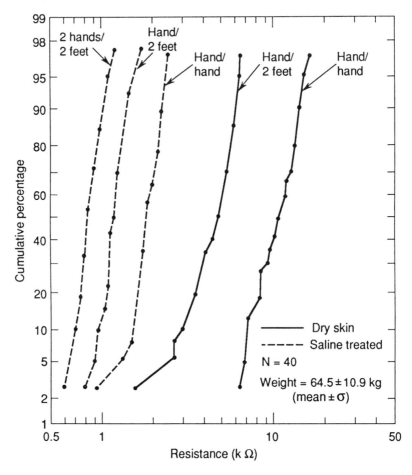

FIGURE 2.22. Cumulative distribution of DC resistance of adults, with various electrode paths. Hand electrodes = 2 No. 10 Awg twisted wires, feet on plate, I = 5 mA. (Data from Whitaker, 1939.)

1 mA with children and 5 mA with adults, except when body resistance was too great to allow these values.

The curves plotted in Figs. 2.22 and 2.23 are approximately straight lines, which indicate the log-normal distribution on this plotting format. The slopes of the distribution curves are substantially less for wet than for dry conditions, and dry-skin slopes are greater for children than for adults. In general, children's resistance is greater than that of adults. Apparently, the shorter current paths of children is more than offset by their reduced volume. This can be appreciated by a simple calculation treating the limbs as conducting cylinders of length L and cross-sectional radius r. According

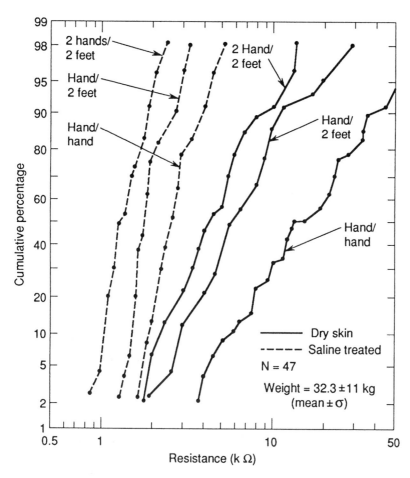

FIGURE 2.23. Cumulative distribution of DC resistance of children, age 3–15 yrs. Electrodes as in Fig. 2.22, $I = 1$ mA. (Data from Whitaker, 1939.)

to Eq. (2.1), the limb resistance would be calculated by $R = L\varrho/(\pi r^2)$. If we assume that the ratio L/r is independent of body size, then it is concluded that resistance would vary inversely with r, or equivalently, inversely with body weight to the 1/3 power.

2.4   Impedance at Higher Frequencies

Skin impedance decreases as the frequency of current is increased, as noted in Table 2.8 (Schwan, 1968). The capacitive component of skin impedance is primarily responsible for its frequency dependence. Figure 2.24 illustrates

TABLE 2.7. Statistical summary of DC Body impedance measurements. (Data listed in kΩ).

	Adults			Children		
	5%	50%	95%	5%	50%	95%
A. Dry conditions						
Hand/hand	6.96	11.45	15.69	4.04	14.35	51.10
Hand/two feet	2.62	4.00	6.51	2.62	5.70	18.72
Hand/two feet	—	—	—	2.00	4.25	12.68
B. Wet conditions						
Hand/hand	1.28	1.86	2.45	1.70	2.55	4.47
Hand/two feet	0.93	1.20	1.67	1.43	1.80	3.02
Two Hands/two feet	0.63	0.84	1.16	0.90	1.30	2.04

(a) Hand electrodes: two No. 10 Awg twisted copper wires.
(b) Voltage: 12 V DC; current ~1 mA (children), ~5 mA (adults).
(c) Wet conditions apply to treatment with 20% NaCl solution.
(d) Children's ages: 3–15 years; adults ages: 18–58 years.
Source: Data from H. B. Whitaker (1939).

the frequency dependence of impedance over the range 0 to 2,000 Hz, using large hand-to-hand electrodes on dry skin (from Biegelmeier, 1986). At the highest frequency, impedance is reduced to about 750 Ω, a value approximately that for the internal body impedance with hand-to-hand contacts.

Figure 2.25 illustrates the measurements of Osypka (1963) in the frequency range 0.3 to 100 kHz, using low voltages (approximately 10 V) and large copper-cylinder hand-to-hand contacts. The test results with dry and wet hands show that above 5 kHz, the effects of skin hydration become negligible. Table 2.9 shows impedance measurements at 0.375 and 1 MHz (Schwan, 1968). The hand-to-hand data are similar to the values for internal body impedance as discussed in Sect. 2.3.

TABLE 2.8. Dry-skin impedance versus frequency.

Frequency (kHz)	Magnitude (Ωcm^2)	Phase angle (deg)
1	14,000	—
5	3,000	−70
10	1,800	−65
20	1,000	−55
50	500	−30
100	300	−20
200	250	−10

Phase angle given by $\tan^{-1} X/R$, where X is capacitive reactance, and R is resistance.
Source: Schwan (1968).

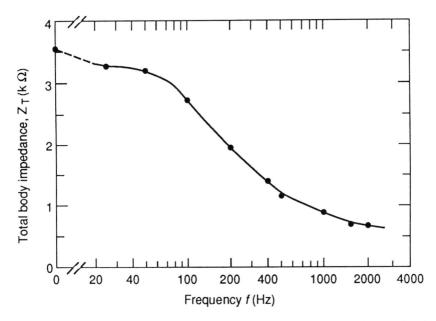

FIGURE 2.24. Frequency dependence of impedance with large electrodes, dry hand-to-hand current path. Applied voltage = 25 V rms. (From Beigelmier, 1986.)

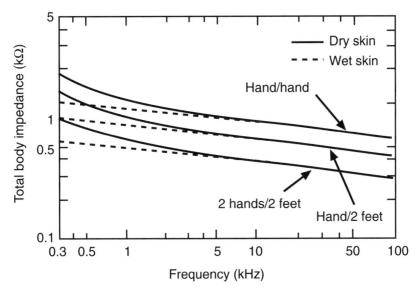

FIGURE 2.25. Total body impedance for large-area contacts in the frequency range 0.3–100 kHz. (From Osypka, 1963.)

TABLE 2.9. Body impedance at 0.375 and 1 MHz.

Electrode location	Area (cm^2)	Impedance	
		$f = 0.375$ MHz (Ω)	$f = 1$ MHz (Ω)
Hand to hand	90	475	460
Finger to other arm	10/65	500	470
Across left arm	32	34	21
Across elbow joint	32	37	21
Across shoulder joint	32	47	31
Across neck	32	36	18
Forehead to neck	32	82	57
Chest to back	150	31	20
Across thorax	150	29	19
Right wrist to left leg	75	248	234
Left wrist to right leg	75	274	266

Area applies to each electrode, except for finger/arm, where
two different electrode sizes were used.
Source: From Schwan (1968).

Chatterjee and colleagues (1986) measured impedance in the range
of 10 kHz to 10 MHz. Adult subjects numbered 367 (170 M, 197 F). The
hand electrode was a brass rod of diameter 1.5 cm; subjects stood barefoot
on a copper plate. Prior to measurements, the subject's hand was moistened
with 0.9% physiological saline. The subject's feet were not treated with
saline. Figure 2.26 shows results from these experiments. The authors assumed that impedance was related to body size, rather than sex per se, and
that impedance was inversely proportional to subject's height. The phase-
angle data show that the impedance is largely resistive in the indicated
frequency range. The magnitude of impedance drops steadily with increasing frequency, to a value nearly equal to the internal body impedance
reported in Sect. 2.3 for hand-to-feet contacts.

2.5   Impedance Through Foot Contact

The current in an electrical accident may travel to ground through a person's
feet, and the path resistance will include the contact resistance of the feet
against the ground, and the resistance of the footwear. A conservative
assumption treats the feet as bare or wearing conductive shoes—
accordingly, the feet are treated as metallic discs (IEEE, 1986), for which the
contact resistance can be calculated from the equations of Sunde (1968) as

$$R_f = \frac{\varrho}{4r} \tag{2.17}$$

$$R_M = \frac{\varrho}{4\pi d_f} \tag{2.18}$$

where R_f is the contact resistance of a single foot, ϱ is the resistivity of the surface material, r is the equivalent radius of the foot print, R_M is the mutual resistance of the two feet, and d_f is the separation distance of the two feet. We are often interested in the resistance of the two feet in human contact accidents. A pathway from a hand to two feet would involve the two feet in parallel. Alternatively, one may be interested in a current path where the

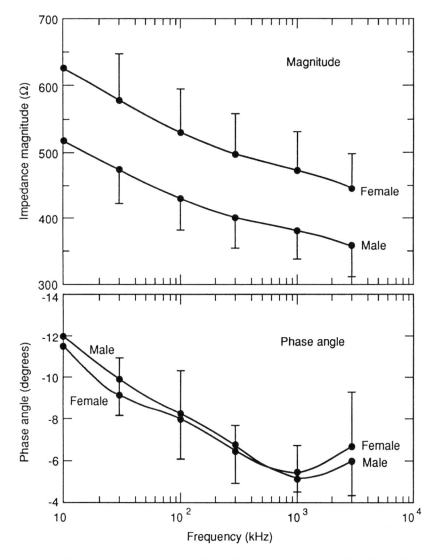

FIGURE 2.26. Magnitude and phase of impedance measured on 367 subjects, hand-to-feet contact. Hand electrode: 3.5-cm rod; wet hand contacts; bare feet on plate. Vertical bars show standard deviations: (From Chatterjee et al., 1986 © 1986 IEEE.)

TABLE 2.10. Resistivity of several materials—uniform composition.

Surface		Resistivity (Ωm)
Generic soil types		
Wet, loamy soil		10
Moist soil		100
Dry soil		1,000
Bed rock		10,000
Specific surface types		
Concrete,	moist	30–60
	air dry	910
Gravel,	dry	10^6
	wetted w/ground water	8,500
	wetted w/salt water	24

Source: IEEE (1986); Hammond & Robson (1955).

two feet are in series with a potential difference on the ground, as with so-called "step potentials" that arise from current flow within the earth. The contact resistance of two feet can be calculated by

$$R_{2fs} = 2\left(R_f - R_M\right) \tag{2.19}$$

$$R_{2fp} = \frac{1}{2}\left(R_f + R_M\right) \tag{2.20}$$

where R_{2fs} and R_{2fp} are the resistance of the two feet in series and parallel respectively.

Table 2.10 lists resistivity of several surface materials (IEEE, 1986). Contrary to common expectations, concrete can actually be a relatively good conductor—being a hygroscopic material, it absorbs moisture if it is in exposed to water, or is in contact with moist soil. Tables 2.11 and 2.12

TABLE 2.11. Resistivity of soil by water content.

Water content (% by weight)	Resistivity (Ωm)	
	Top soil	Sandy loam
0	$>10^7$	$>10^7$
2.5	2,500	1,500
5	1,650	430
10	530	185
15	190	105
20	120	63
30	64	42

Source: IEEE (1982).

TABLE 2.12. Variation of soil resistivity with temperature (Sandy loam, 15.2% moisture content).

Temperature (°C)	Resistivity (Ωm)
20	72
10	99
0 (liquid water)	138
0 (ice)	300
−5	790
−15	3,300

Source: IEEE (1982).

show that temperature and especially water content can greatly affect resistivity.

As an example, consider a foot print with area $200\,cm^2$ (equivalent radius $= 0.08\,m$), and a separation distance between the two feet of $0.5\,m$. Table 2.13 lists the series and parallel contact resistance of two feet for soil resistivity $\varrho = 10$, 100, and $1,000\,\Omega m$—attributed to wet-loamy soil, moist soil, and dry soil respectively (IEEE, 1986).

Footwear can add significantly to the total path resistance, as illustrated in Fig. 2.27 (Reilly, 1979b). This figure plots the distribution of DC resistance of individuals standing on various surfaces, with a current pathway from a large electrode held in the hand, to a nearby driven ground rod in the soil. The measurement voltage was $500\,V$. In all cases, footwear was dry, except for surface moisture on which the person stood. In general, leather soles are much more conductive than rubber soles. If leather soles become wet, their resistance can fall greatly. Grass blades that touch the sides of the shoes can also significantly lower their resistance. The curves labeled "damp grass" apply to individuals standing on short grass; in the "wet grass" condition, subjects first stood briefly in 1 cm of water before stepping on the grass. In tall grass (e.g., 8 cm), we would expect to see much larger percentages of low resistances.

TABLE 2.13. Examples of contact resistance of two feet.

	Contact resistance (Ω)		
	Wet soil	Moist soil	Dry soil
R_{2fs}	56.2	562	5,620
R_{2fp}	17.2	172	1,720

$\varrho = 10$, 100, $1,000\,\Omega m$ for wet, moist, and dry soil, respectively.

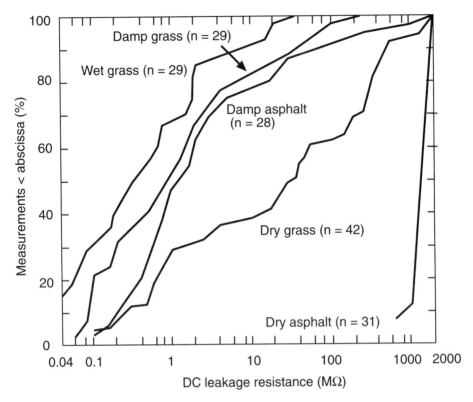

FIGURE 2.27. DC resistance of people through shoes, standing on various surfaces. (From Reilly, 1979b.)

2.6 High-Voltage and Transient Properties

We studied current and voltage relationships under high-voltage conditions with living subjects (Reilly et al., 1983, 1984, 1985). These experiments could be safely conducted on living subjects because the duration and energy of the stimuli were limited by the use of capacitive discharges. These experiments provided the first detailed account of high-voltage current response of humans in the microsecond time scale.

The experimental apparatus is illustrated in Fig. 2.28. A high-voltage transformer T was followed by a resistor R_C (variable from 1 to 128 MΩ) and a capacitor C_0) (variable from 100 to 6,400 pF in steps of 100 pF). An internal transformer resistance of 1 MΩ, and permanent series resistance of 2 MΩ, limited the current for human safety. The voltage at the energized electrode could be varied from 0 to ± 15 kV in the DC mode, or 15 kV peak in the AC mode.

The charge stored on C_0 could be discharged to a subject by one of several methods: by actively touching the energized electrode; by bringing to some location on the body a probe energized by C_0; or by closing a switch between C_0 and a passive electrode already in contact with the body. Stimulus voltage and current could be simultaneously sampled at a maximum rate of 50 ns, and stored in a dual-channel digital oscilloscope which was connected to a digital computer for manipulation and plotting of the data. The return electrode was silverplated, with a contact area of 15 cm^2. It was covered with conductive electrode paste and generally worn near the site of stimulation, such as the forearm for stimulation on the fingertip. Because of the size and treatment of the return electrode, its impedance was negligible compared with that at the stimulus site.

What may be thought to be a simple stimulus—a single capacitive discharge—actually has complexities that distinguish it from other electrical stimuli discussed previously. These complexities include a separation of the discharge into spark and contact components, a plateau voltage, marked impedance nonlinearities, differences in response to positive and negative stimuli, and significant waveform differences at different body locations.

Charge from a capacitor can be transferred to the skin by two modes: the stimulus can be presented through a metallic contact on the body

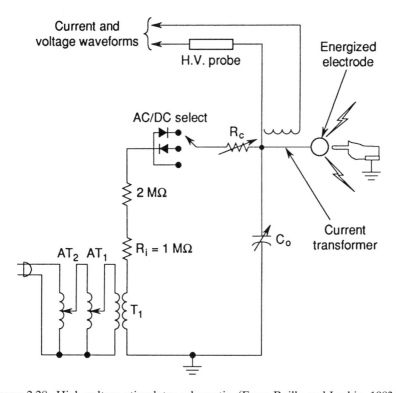

FIGURE 2.28. High-voltage stimulator schematic. (From Reilly and Larkin, 1983.)

or through a spark contact. When a subject touches an energized electrode, there can result both a spark (if the initial voltage is sufficient) and a subsequent discharge when mechanical contact is made. To understand the characteristics of the spark and contact components, we studied discharge waveforms under a variety of stimulating conditions.

Spark and Contact Components

When a voltage is applied across the air gap between two electrodes, the free electrons and negative ions that normally exist in air will move toward the anode (positive electrode). The positive ions move toward the cathode (negative electrode). If the electric field produced by the voltage is sufficient, the free electrons will be accelerated to velocities such that their collisions with neutral atoms or molecules create more free electrons. This process is known as *electron avalanche.*

When one of the electrodes is the human body, the nature of the electrical breakdown will be greatly affected by the nonlinear impedance properties of the skin. The nature of the breakdown phenomenon is evident in Fig. 2.29, where a subject lightly tapped with his finger an electrode having a 1-mm-diameter tip that protruded by 0.5 mm above an insulated housing. For each set of conditions depicted in the figure, four successive stimuli were applied; the four corresponding voltage and current waveforms were then averaged.

In this sequence of four recordings, the discharge capacitance is 400 pF; the initial voltage is varied. In Fig. 2.29a, the initial voltage of 405 V is too small to produce a spark because of the dielectric protection of the skin. As a result, current is conducted entirely by direct contact when the finger touches the electrode. In Fig. 2.29b, the initial voltage of 515 V is just barely at the point of a spark discharge at the initiation of the current trace as revealed by the small initial current transient at about $t = 75 \mu s$. In Fig. 2.29c, the initial voltage has been raised to 586 V (a 14% increase over Fig. 2.29b), and the magnitude of the spark discharge grows by a factor of 10. In Fig. 2.29d, with the initial voltage at 989 V, the spark discharge component of the stimulus has increased by another factor of 10, and we can see the convergence of voltage to a plateau around 450 V, followed by the contact at about $t = 130 \mu s$.

In this sequence of waveforms, an initial voltage of about 500 V delineates a level above which spark discharges occur and below which they do not. This is also approximately the plateau voltage level that is discussed in the succeeding paragraphs. The experimental data suggest that the breakdown voltage for spark discharges to the skin is at about the level of 500 V. This is consistent with the dielectric strength of the excised human corneum reported in Sect. 2.2.

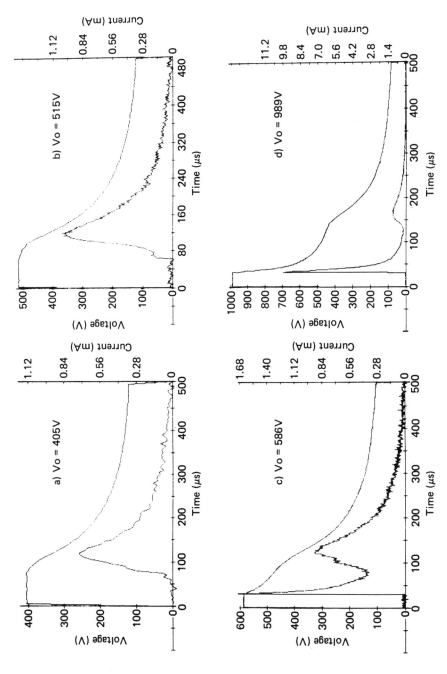

FIGURE 2.29. Stimulus waveforms for light tap force, capacitive discharges with $C = 400\,\text{pF}$. (From Reilly et al., 1982.)

Plateau Voltage

In the example illustrated in Fig. 2.29, separate voltage plateaus can be seen for the spark and the contact phases of the discharge. The contact plateau is approximately 100 V, and the spark plateau is approximately 450 V. These plateaus suggest that as the voltage declines, the impedance becomes very large in both phases of the discharge. The presence of a voltage plateau reveals that a charged capacitor cannot be totally discharged through a spark to the skin. It also delineates a voltage level above which spark discharges can occur.

The plateau voltage for a spark stimulus can be studied by bringing an energized electrode slowly to the body to produce a spark discharge that is not followed by a contact. Figure 2.30 illustrates four successive discharge waveforms for a pencil-shaped probe (0.8-mm-diameter tip) that is brought slowly to the same point on the finger pad. Discharge to the same point was ensured by the use of a dielectric mask with a 0.5-mm hole. The initial voltage has been varied from 750 to 1,900 V. In each case, a plateau voltage in the range of 450 to 500 V is evident. The presence of this plateau indicates that the spark component does not completely discharge the capacitor and that the discharge impedance converges to a very large value in the region where the discharge current approaches zero. When tested over various subjects and body locations, the plateau voltage was found to range from about 450 to 650 V, with an average of 525 V. Stripping away the corneal

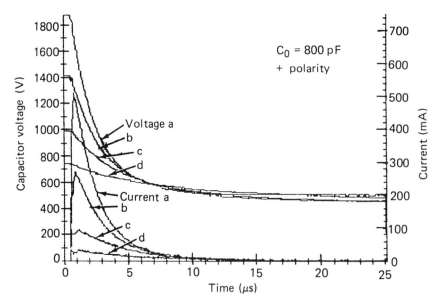

FIGURE 2.30. Four successive discharges to the same point (second finger, point A), with variable initial voltage. (From Reilly and Larkin, 1983.)

layer of skin on the forearm lowers the plateau voltage to about 330 V as described below.

Spark Discharges and Corneal Degradation

The contribution of the corneum can be studied by stripping the skin with cellulose tape. With each stripping operation, a portion of the corneum adheres to the tape. This procedures was applied to a portion of the volar forearm to study the effects of spark discharges (Reilly et al., 1982). A charged capacitor (400 pF @ 1,000 V) was discharged through a small probe to a precise point on the skin defined by a small hole in a dielectric mask. Altogether 11 stimuli were applied to each of four target points on both stripped and control skin, and a total of 60 strippings were performed. At the end, the skin appeared glossy, and was sensitive to the stripping procedure.

Table 2.14 shows experimental data; each entry is the average of measurements at four distinct points on both stripped and control skin. Listed are the plateau voltage V_p, and the initial impedance R_i, defined by the ratio of the initial voltage and peak current. V_p in the stripped area dropped from 480 V to a low of 330 V. At the same time, R_i dropped by a factor of over 5:1. At the control skin, we see a slight increase in V_p and a 16% decrease in R_i. The slight drop in R_i in the control area probably results from burning away a small portion of the corneum; the slight increase in V_p remains a mystery. The data also indicate that subcorneal layers also exhibit a significant plateau voltage. V_p mirrors the minimum voltage from a static or low-frequency source needed to support a spark discharge to the skin.

Tests were conducted with repeated spark discharges on the dry skin of the leg (Reilly et al., 1982) from a 200 pF capacitor charged to 2,000 V (stored energy = 0.4 mJ). With 100 repeated discharges to the same point on the skin, a small but noticeable change in skin impedance was produced, along with a small reddened spot about 0.5 mm in diameter. Similar discharges applied to the fingertip did not result in observable skin changes due to the greater thickness and conductivity of the palmar corneum.

The energy of a spark discharge may be dissipated in a very small volume in the skin. The effective area of a spark contact is less than 1 mm^2 as evidenced by the fact that a spark to the finger varies substantially over distances of about 1 mm—a distance that is similar to the density of sweat ducts in the palm of the hand. The previously presented data on initial impedance suggest that the effective diameter of contact is roughly a few tenths of a millimeter, and possibly as small as 0.1 mm.

We can compare these discharges with ordinary carpet sparks. Taking the capacitance of the human body as 100 pF and the voltage in the range of 2 to 3 kV (a common range for carpet sparks), the energy in a typical carpet spark would be in the range of 0.2 to 0.45 mJ. We see that the minimum

TABLE 2.14. Plateau voltage and minimum resistance for epidermal stripping.

Stimulus number	Epidermal stripping number	Stripped skin		Control skin	
		V_p (V)	R_i (kΩ)	V_p (V)	R_i (kΩ)
1	0	*	*	*	*
2	1	480	22.5	470	30.5
3	2	435	15.5	450	35.9
4	3	450	24.1	440	32.5
5	5	410	10.9	500	36.4
6	8	415	8.8	475	26.3
7	12	*	*	510	23.9
8	20	385	8.1	500	24.4
9	30	400	4.9	525	25.0
10	46	390	4.8	*	*
11	60	370	4.3	*	*
1**	60	330	3.7	*	*

* Data unavailable.
** Last data set applies to simulus in stripped area, but at four points not previously stimulated. Each measurement represents an average over four spatial points. (From Reilly et al., 1982).

energy per spark cited above for skin erosion is similar to the energy commonly encountered in carpet sparks.

In an alternating field, each half cycle of oscillation presents a new opportunity for a spark. The maximum rate of sparking will increase as we raise the frequency of the energizing field. At power frequencies, the spark discharge phenomenon is very much like that for static discharges (see Chapter 9). For frequencies substantially above 60 Hz, the breakdown voltage of air becomes lower as a result of the transit time of positive ions and electrons with respect to the gap length and oscillation frequency (Craggs, 1978). At frequencies of several MHz, and small metallic electrode gaps (<1 mm), the reduction in breakdown voltage at atmospheric pressure is approximately 15 to 20% compared with that at 60 Hz. At much higher frequencies in the GHz regime, breakdown voltages can be more significantly reduced. Section 11.4 discusses standards for human exposure to spark discharges induced by radio-frequency electromagnetic fields.

Discharge Impedance

The dynamic properties of discharge impedance are pertinent to transient high-voltage shock exposure. With capacitive discharges, for example, the impedance affects the time constant of the discharge. The time constant, in turn, significantly affects the neural excitation potency of the discharge, as discussed in Chapters 3 and 4. In other applications, knowledge about body impedance is needed to define the limiting value of current in a high-voltage exposure—such as with fault conditions in high-voltage installations.

Figure 2.31 expresses impedance data for Fig. 2.30 as the ratio voltage/current. These dynamic impedance functions reveal that the skin impedance to a spark discharge exhibits a nonlinear relationship with voltage in which peak current is not directly proportional to the initial voltage. The variation in initial impedance with initial voltage is evident in Fig. 2.31. The impedance attains a minimum value in a fraction of a microsecond after the stimulus onset and increases during the time course of the stimulus, during which time the stimulus voltage is decreasing. The large excursions beyond $16\mu s$ are the result of digital quantization effects.

The impedance to a capacitor discharge consists of two contributions: the voltage drop across the arc itself and the voltage drop across the subject. We measured the arc voltage/current relationship with the human body in direct contact with one of two metallic electrodes forming the arc gap, and in series with the electrical path. The voltage drop across the arc for this case varied from nearly zero at the point of arc initiation to approximately 50 to 75 V at the point of arc extinction. As a result, when two metallic electrodes form the discharge path, the arc does not contribute significantly to the overall measured impedance. We cannot be certain that the same arc voltage drop will apply when the spark terminates at the human skin rather

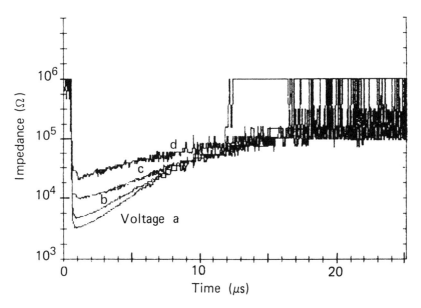

FIGURE 2.31. Impedance versus time for four successive discharges to the same point (second finger, point A) with variable initial voltage. Voltage and current waveforms as shown in Fig. 2.30. (From Reilly et al., 1982.)

than a metallic electrode. It does not seem possible to make a more direct measurement of the arc impedance when the discharge terminates at the human skin without significantly altering the skin's electrical response. If, however, the arc voltage drop is similar to that when two metallic electrodes are used, then the arc voltage drop must be small compared with that across a human subject.

Variation with Stimulus Location

Waveforms for discharges to different discrete body locations can differ significantly even though the stimulated points may be separated by only 2 mm. Figure 2.32 illustrates voltage and current waveforms for anodic (positive polarity) spark discharges to four points on the fingertip separated by 2 mm. To localize the spark stimulation, a dielectric mask with a 0.5-mm-diameter hole was placed on the finger. But repeated positive polarity stimuli at the same point produce nearly identical waveforms. This closely spaced pattern of skin impedance may reflect the mosaic of sweat ducts.

The fact that identical nonlinear current response is seen with repeated discharges to the same point on the finger suggests that, with the tested discharge parameters, the impedance breakdown is not permanent. Perhaps ionic disassociation rather than membrane rupture is involved.

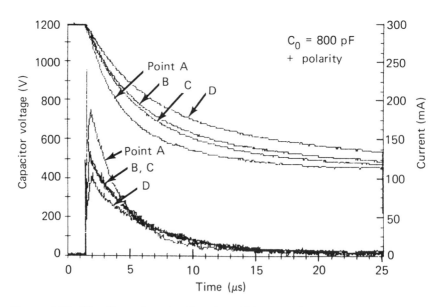

FIGURE 2.32. Waveforms for stimuli to four different points on the same fingertip (second finger, points A-D); each point separated by 2 mm. (From Reilly et al., 1982.)

Repeated discharges at similar energy levels on the dry skin of the calf on the leg, however, eventually resulted in visible erosion of the corneum.

Polarity Effects

Positive-polarity current transients appear relatively smooth and repeatable, but negative-polarity waveforms often are bistable and have lower initial impedance than the positive waveform, as illustrated in Fig. 2.33. In this example, four cathodic spark discharges are applied to point A (also used for the anodic discharges shown in Figs. 2.31–3.33).

The ionized discharge process may be responsible for these polarity differences. Electrons move so rapidly that by the time the avalanche has reached the anode, the newly created positive ions are virtually still in their original positions, forming a positive space charge conically concentrated at the anode, and tapering or decreasing toward the cathode as shown in Fig. 2.34a (Howatson, 1965). This distribution of positive ions modifies the field near the anode as shown in Fig. 2.34b. The conical spreading of the plasma discharge from the cathode to the anode results in a larger area of a spark contact at the anode than at the cathode. The lower initial minimum impedance for negative polarity may result from the greater ion contact area with the body as an anode than as a cathode. The bistable character of negative polarity discharges may result from an unstable space charge distribution and its effective area of anodic contact.

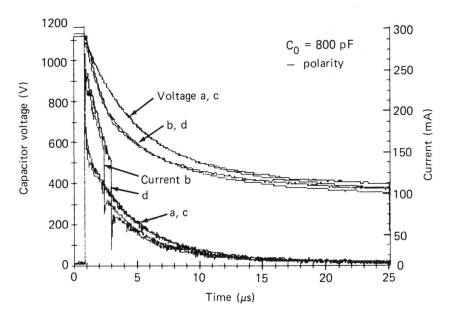

FIGURE 2.33. Waveforms for successive discharges to the same point (second finger, point A); negative polarity. (From Reilly et al., 1982.)

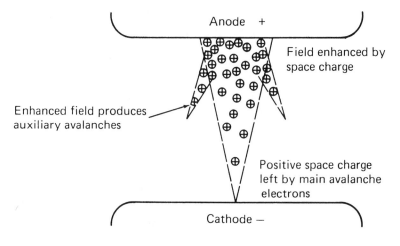

Anode   +

Field enhanced by
space charge

Enhanced field produces
auxiliary avalanches

Positive space charge
left by main avalanche
electrons

Cathode −

a) Formation of a steamer

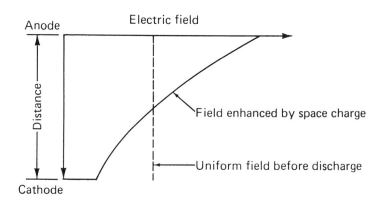

Anode

Electric field

Distance

Field enhanced by space charge

Uniform field before discharge

Cathode

b) Electric field

FIGURE 2.34. Spark discharge formation: (a) formation of a steamer; (b) electric field. (Reprinted with permission from A. M. Howatson, *An Introduction to Gas Discharges*, © 1965, Pergamon Press PLC.)

Stimulus Time Constants

A capacitance C discharged to an ideal linear resistance R will have an exponentially decaying current and voltage with a time constant $\tau = RC$, which defines the time at which the voltage or current decays to e^{-1} of its maximum value. As evident in the previous examples, many of the discharge patterns can be approximated by exponential functions with current decaying to zero and voltage decaying to the plateau value. Although an ideal RC circuit results in an exponential discharge pattern, the converse statement should not be assumed: an exponential current decay does not necessarily imply constant resistance. This can be appreciated with reference to the discharge patterns of Fig. 2.30: both current and voltage waveforms are approximately exponential functions, but, because of the nonzero voltage plateau, the impedance varies from a minimum value at the onset of the discharge to a very large value at its extinction.

The instantaneous impedance to a capacitor discharge departs significantly from what would be expected in a linear model. Linear circuit models for skin impedance, commonly invoked in studies of low-voltage cutaneous currents, do not adequately represent the measured impedance response. However, the measured transient response might be accounted for if the model parameters were voltage- or time-dependent.

We define time constants for capacitor discharge stimuli as the time required for the voltage to decay to $e^{-1} \Delta V$, where ΔV is the difference between the initial and plateau voltages. The decay time is measured from the onset of the current waveform. The measured plateau for the touch component was generally in the vicinity of $100\,V$ (as in Fig. 2.29), in contrast to the plateau of about $500\,V$ for the spark component. The value of ΔV defines the amount of charge passed, in accordance with $Q = C\Delta V$.

Figure 2.35 depicts time constants for spark and contact components as a function of voltage for a single subject who discharged a capacitor by actively touching an energized electrode. The discontinuity around 500 to $600\,V$ occurs because spark discharges could be produced only above this range. The time constants decrease as the initial voltage on the capacitor is increased and as the capacitance is decreased. In the test procedure a subject discharged a capacitor by actively touching an electrode. The resulting effective discharge contact area is very small for both spark and contact phases. We hypothesize that at the instant of a tapped electrode contact, an initially small point of conduction establishes a preferred current channel. As a result, both spark and contact phases may utilize a small contact area, regardless of the electrode size. In contrast, discharges to an electrode held in contact with the skin appear to utilize multiple current channels. The result is a much lower impedance at the electrode/skin interface, and significantly smaller discharge time constants. For example, discharges from a

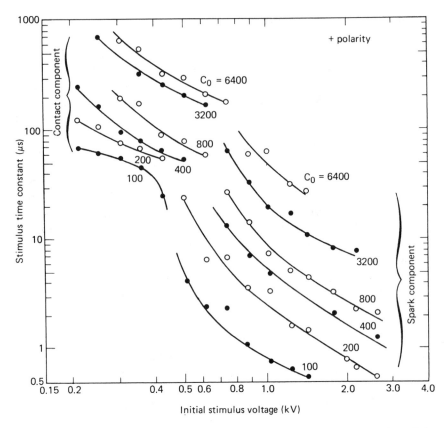

FIGURE 2.35. Time constants for monophasic capacitor discharge stimulation of the fingertip. Parameter shown is the capacitance in picofarads. (From W. D. Larkin and J. P. Reilly, *Perception and Psychophysics* 36 (1): 68–78, 1984, reprinted by permission of Psychonomic Society, Inc.)

6,400-pF capacitor to a 1.3-cm-diameter contact electrode result in discharge time constants under $3 \mu s$, even when the initial voltage is as low as 50 V.

Minimum Impedance with Spark Discharge[4]

The minimum impedance to a spark discharge occurs at the onset of the stimulus, when the applied voltage is greatest. Figure 2.36 shows the minimum impedance when the spark discharge and return electrode are on the calf of the leg. The curves are second-order least-squares fits to the experimental data. Two striking features are the voltage dependence and the

---

[4] Adapted from the work of R.J. Taylor, as reported in Reilly et al. (1982, 1983) and Taylor (1985).

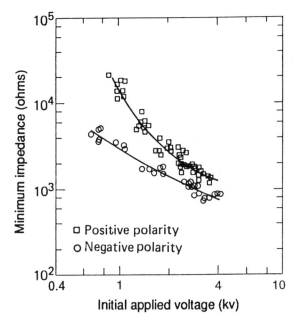

FIGURE 2.36. Minimum impedance of leg (volar calf) versus voltage, for spark discharges to skin; leg grounds. (From Reilly et al., 1982)

polarity sensitivity. The voltage dependence reflects the nonlinear skin impedance as discussed in Sect. 2.2. Additionally, the diameter of spark contact area may increase with voltage—at higher voltages, the distance between the skin and the discharge electrode is greater, allowing the base area of the space charge cone (Fig. 2.34) to increase. When the spark discharge terminates at an electrode already in contact with the skin, the impedance is much lower than when the spark terminates directly on the skin.

Internal Body Impedance[5]

The internal body impedance can be estimated from transient measurements of capacitor discharges to large contact electrodes, where the large contact area and high voltage minimize skin impedance. Figure 2.37 shows an example of the ratio of voltage and current during a capacitor discharge to a 3-cm-diameter cylinder electrode grasped by a subject with the left hand while standing with the bare left foot on a copper plate; the discharge capacitance was 800 pF, and initial voltage was 500 V. The discharge process

---

[5] Adapted from the work of R.J. Taylor, as reported in Reilly et al. (1982, 1983) and Taylor (1985).

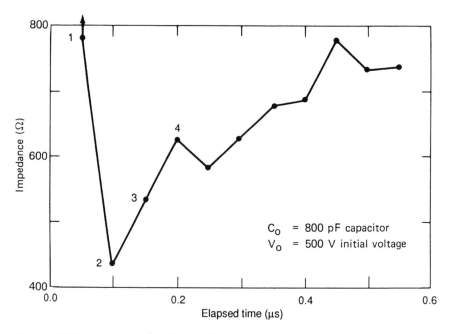

FIGURE 2.37. An example of body impedance measurements versus time; 3-cm-diameter electrode in left hand, left bare foot on plate. Spark discharge was initiated between 0.0 and 0.05 μs. (From Taylor, 1985.)

was essentially complete during the interval illustrated in the figure. Data point 1 is not considered meaningful because of the limited rise time of the instrumentation (0.1 μs). At data point 2, the impedance is 434 Ω (434 V and 1 A current). Data point 2 is strongly influenced by the distributed capacitance of the body to ground as shown in Figure 2.38; an initial current surge is required to supply the body's surface charge in addition to the current that passes internally. The impedance measurement at data point 2 changed as much as 50% as the proximity between the body and grounded objects was varied. The measurements at data point 3 changed only a little (about 10%) with variations in distributed capacitance. The measurements at data point 3, termed "initial body impedance," were considered by Taylor to represent the core body impedance.

Table 2.15 shows initial body impedance from various electrode configurations (from Taylor, 1985). The treated skin had conductive electrode paste applied to it. The impedance for dry and treated skin differs but a little, thus supporting the hypothesis that skin impedance is negligible with this measurement technique. The data in Table 2.15 agree well with internal body impedance data determined by other methods (refer to Sect. 2.3).

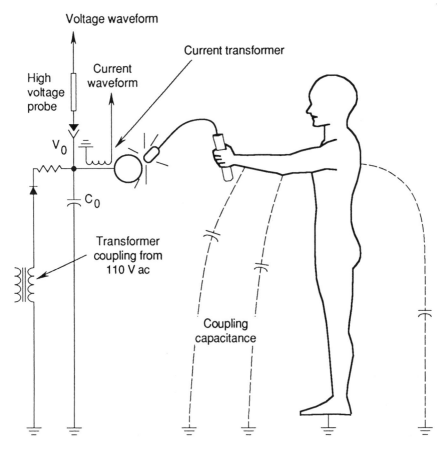

FIGURE 2.38. Body impedance measurement with capacitive coupling of body to ground. (From Taylor, 1985.)

TABLE 2.15. Initial total body impedance (Ω).

Path	Mean	Standard deviation
Dry skin		
Left hand/left foot	533	52
Right hand/left foot	521	37
Right hand/left hand	508	29
Treated skin		
Left hand/left foot	516	55
Right hand/left foot	507	36
Right hand/left hand	490	36

Source: Data based on measurements of seven individuals (from Taylor, 1985).

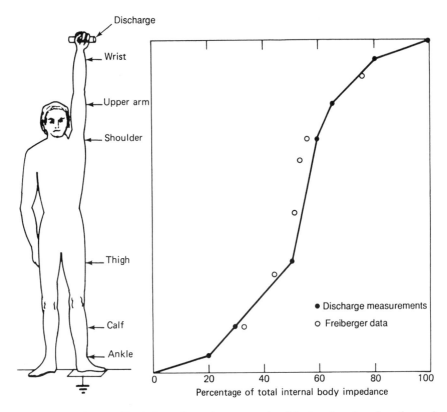

FIGURE 2.39. Internal body impedance between sole of foot and various locations of the active electrode along the body. Open points from Fig. 2.20. (From Taylor, 1985).

Figure 2.39 and 2.40 illustrate the internal body impedance for different locations of the active electrode, measured as a percentage of the total hand-to-hand and hand-to-foot internal impedance. The filled circles represent the transient discharge measurements, averaged over seven adult individuals. The open circles are derived from the model of Freiberger (Fig. 2.20). The agreement between the two experiments is apparent.

2.7 Impedance of Domestic Animals

Farmers are sometimes concerned about electrical exposure of farm animals. The term *stray voltage* has been coined to indicate unwanted electrical potentials to which farm animals may be exposed. Stray voltage is usually related to the grounding of equipment on the farmstead, on- and off-site transformers, and associated ground currents. Potential differences applied

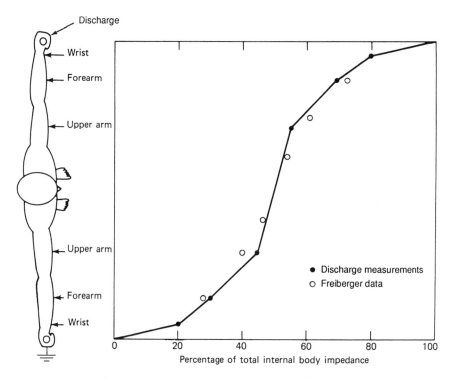

Discharge

Wrist

Forearm

Upper arm

Upper arm

Forearm

Wrist

● Discharge measurements
○ Freiberger data

Percentage of total internal body impedance

FIGURE 2.40. Internal body impedance from one hand to various locations of the active electrode along the arms. Open points from Fig. 2.20. (Adapted from Taylor, 1985.)

to a farm animal can result in unpleasant shock, which can lead to animal handling problems and aversion to feed or water (Lefcourt, 1991). Concern about stray voltage is particularly strong in the dairy industry, where many researchers, consultants, and legal practitioners have focused on this problem.

Potential differences of only a few volts can be disturbing to cows (see Sect. 7.12) while being unnoticed by the farmer. The animal's sensitivity to such low voltages does not mean that the cow is acutely sensitive to electrical stimulation, but rather that her electrical impedance is typically much lower than that of the farmer. This low impedance is a consequence of the cow's size and weight, and the fact that she contacts surfaces that are wet and contaminated with urine and manure. And like the human skin, cows' hooves are good conductors when impregnated with such effluents.

The animal may access potential differences across various points on its body—for instance, from a metal water or feed bowl to the feet, from foot to foot, or from a shoulder contacting a stanchion to the feet. One can determine impedance for various points of contact using the impedance

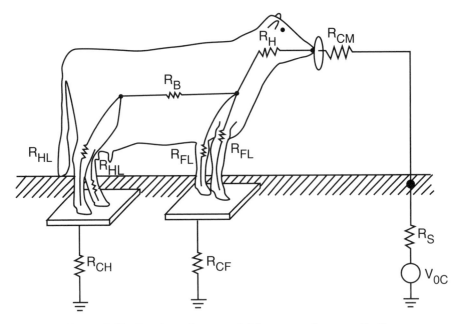

FIGURE 2.41. Cow impedance model for stray voltage application.

model of Fig. 2.41 (Reilly, 1994). The figure shows impedance components, front and hind foot contact resistance (R_{CF} and R_{CH}) and muzzle contact resistance (R_{CM}). Table 2.16 lists median values of cow impedance components from the data of Norell et al. (1983). Foot contact impedance is listed in Table 2.17 using Eqs. (2.19 and 2.20). The contact area of each hoof is 120 cm^2 (equiv. radius = 6.2 cm); distance between right and left feet is 0.4 m; distance between front and rear feet is 1.0 m.

To illustrate cow impedance from the model of Fig. 2.41, consider moist soil conditions ($\varrho = 100\,\Omega$m). With a muzzle to all feet pathway, the model

TABLE 2.16. Cow impedance model values.

Symbol	Impedance Component	Resistance (Ω)
R_H	Head & neck	183
R_B	Body	30
R_{FL}	Front leg	882
R_{HL}	Hind leg	525
R_{CH}, R_{CF}	Hind, front foot contact	see text
R_S	Source impedance	variable
R_{CM}	Muzzle contact	variable

Cow impedance componets derived form median data of Norell et al. (1983).

TABLE 2.17. Cow foot contact impedance.

| Soil condition | Soil resistivity (Ωm) | R contact | |
		One foot (Ω)	Two feet parallel (Ω)
Wet	10	40	22
Moist	100	400	220
Dry	1,000	4,000	2,200

Assumptions: Contact area each foot $= 120\,cm^2$; distance between right and left feet $= 0.4\,m$; distance between from and hind feet $= 1.0\,m$.

indicates an impedance of 472 Ω, excluding the muzzle and source impedance (R_{CM} and R_S). If the cow lifts one front foot, the impedance increases to 549 Ω. The impedance between the two front and two rear feet would be 1,173 Ω; with one front foot raised, the impedance would be 1,794 Ω. If the animal stands on a perfectly conductive surface, the impedance from muzzle to four feet would be 358 Ω. These data are attributed to an adult cow, for which the average weight at 2 years is approximately 500 kg (ASAE, 1993).

Much smaller cow impedances were reported by Lelcourt and Akers (1982). However, in Lefcourt's procedures, the leg electrodes were placed on shaved areas above the hock or knee, and these areas were treated with electrode paste. By placing the electrode above the knee rather than on the hoof, one eliminates more than 50% of the leg impedance. This occurs because a cross-section of the leg above the knee contains a large bulk of high-conductivity muscle, but the cross-section below the knee is much smaller, and consists largely of low-conductivity bone and ligament, as suggested by Fig. 2.42.

Impedance of growing or finishing pigs of average body weight (89.2 kg) were measured when standing on a metal grid (Gustafson et al., 1986). The mean impedance from mouth to four feet was 789 Ω, with a standard deviation of 262 Ω. For constant body proportions, impedance varies inversely with body weight to the one third power (see Sect. 2.4). When compared with the corresponding impedance of cows (358 Ω), one finds that the cow and pig resistances compare roughly as the inverse cube-root of their respective body weights, even though their body configurations are different.

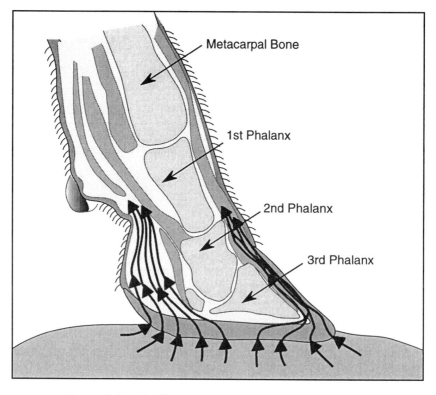

FIGURE 2.42. Distribution of current with cow foot contact.

3
Electrical Principles of Nerve and Muscle Function[1]

3.1 Introduction

This chapter begins with some basic electrical properties of biological cells and then examines the function of a special class of cells that have electrically excitable membranes. It shows how these properties are organized for sensory and muscle function. This information will help explain how externally applied currents can modify or interfere with normal function.

Consider a familiar example that points out several pertinent electrophysiological functions. You touch a hot object with your finger—after a brief delay, you feel pain, and jerk your hand away. This sequence of events involves a number of electrical functions of nerve and muscle. First, the skin's temperature rise is converted to an electrical potential by a specialized transducer or *receptor* in the skin—an example of a pressure-sensitive receptor is shown in Fig. 3.1. The receptor's voltage response, called a *generator potential*, initiates a nerve impulse, called an *action potential* (AP). The AP travels along an electrical cable known as a *nerve axon*, which runs from the receptor to the spinal column, where connections, called *synapses*, are made with additional nerve cells; some of these ultimately carry information to the brain. The receptor, axon, and synaptic terminus comprise a single nerve cell (also called a *neuron*). Bundles of neurons are commonly called *nerves*. In our example, the length of the nerve cell, from the fingertip's receptor to the spinal synapse, will be approximately 1 m.

Thus far, we have discussed an *afferent* nerve cell, that is, one that carries information from the body's sensory system to the spinal cord and then to the brain. Our example sensory nerve cell is also a slowly conducting type in which the conduction velocity may be only a few meters per second. Because of the delay in the conduction and synaptic processes, a substantial fraction of a second may pass before the brain is notified of the finger's

---

[1] For background material, see Kandel et al. (1991); Ruch et al. (1968); Stein (1980); and Plonsey (1969).

73

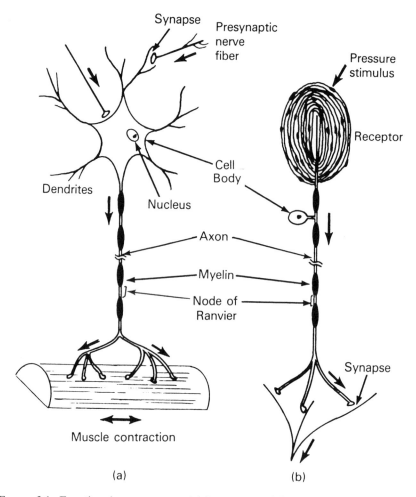

FIGURE 3.1. Functional components of (a) motor and (b) sensory neurons. Arrows indicate the direction of information flow. Signals are propagated across synapses through chemical neurotransmitters and elsewhere by membrane depolarization. Synapses are inside the spinal column. The sizes of the components are drawn on a distorted scale to emphasize various features.

temperature rise and pain is consciously registered. Even before that, however, a reflex action may be set in motion within the spinal column. In the reflex action, sensory inputs, and possibly other inputs from the brain, are processed within the spinal column, and, if the inputs satisfy threshold criteria, AP signals are sent to the muscles of the hand and arm—traveling along *efferent* (motor) neurons, that is, ones that carry signals from the central nervous system (CNS) to the muscles (Fig. 3.1a). When the efferent APs reach the neuron terminus at the muscles, they initiate a series of

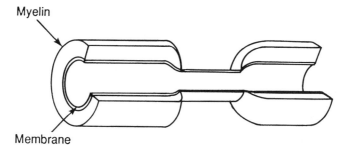

FIGURE 3.2. Detail of myelinated fiber at a node.

events leading to electrical excitation and contraction of muscle fibers. The
result is that the hand pulls away in a reflex action.

The mechanisms responsible for normal function also respond to externally applied currents. With appropriate control, electrical stimulation can
provide medical assistance and diagnostic benefits, such as muscle control,
pain relief, sensory prosthesis, and diagnosis of nerve and muscle pathol-
ogy. However, in chance encounters with electric currents (electric shock),
the result can be pain, hazardous muscle reactions, heart disturbances, and
ultimately death.

Figure 3.1 illustrates several functional components of a sensory and a
motor (muscle) neuron. In this example, the neuron is *myelinated*, that is,
covered with a fatty layer of insulation called *myelin*, and has exposed *nodes
of Ranvier*. Other neurons are unmyelinated. The conducting portion of the
neuron is a long, hollow structure known as the axon (illustrated in Fig. 3.2).
The axon plus myelin wrapping is frequently referred to as a nerve fiber.
The arrows in Fig. 3.1 indicate the direction of information flow. For the
motor neuron in Fig. 3.1a, APs propagate from synapses in the spinal
column to the terminus at the muscle. For the sensory neuron in Fig. 3.1b,
APs originate from one of a variety of specialized receptors (a *pacinian
corpuscle* is illustrated) and proceed to a synapse in the spinal column.
Communication across the synapses is accomplished through chemical
substances known as *neurotransmitters*.

3.2   Cellular Membranes

Cells are the basic building blocks of both plant and animal life. The
functional boundary of the cell is a thin (about 10-nm) bimolecular lipid and
protein structure. One role of the membrane is to regulate chemical exchange from within the cell (the *plasm*) to its surroundings (the *interstitial
fluid*). The electrochemical forces at the membrane are intimately involved
in the regulation.

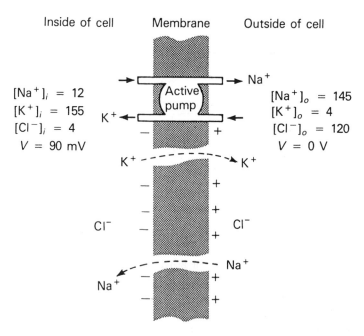

Inside of cell   Membrane   Outside of cell

$[Na^+]_i = 12$
$[K^+]_i = 155$
$[Cl^-]_i = 4$
$V = 90\ mV$

$[Na^+]_o = 145$
$[K^+]_o = 4$
$[Cl^-]_o = 120$
$V = 0\ V$

FIGURE 3.3. Schematic of a typical cell membrane. The channels allow the passage of ions. Numerical values indicate approximate steady-state concentrations ($\mu mol/cm^3$) for typical mammalian muscle cells. An active metabolic pump drives Na^+ out of the cell and K^+ into the cell. The interior potential is about $-90\,mV$ relative to the outside.

The plasm and interstitial fluids are composed largely of water containing ions of different species. The concentration of ions inside and outside the cell differs leading to the electrochemical forces across the cell membrane. The membrane is said to be *semipermeable*; that is, it is basically a dielectric insulator that allows some ionic interchange. Figure 3.3 represents a membrane as a barrier, with channels that permit the passage of ions; the individual channels may be very selective with respect to the ionic species that are allowed to pass. Typical concentrations inside and outside a cell are shown in Table 3.1 for Na^+, K^+, and Cl^- ions. The concentrations indicated in Table 3.1 are markedly different inside and outside the cell. The differences lead to two forces that tend to drive ions across the membrane: a concentration gradient and a voltage gradient. In order to understand these forces, first consider an environment where only one ionic species, substance S, is present.

The Nernst Equation

In a solution where the concentration varies from one region to another, there will be a net flux from the region of higher concentration to the lower.

TABLE 3.1. Example cellular ionic concentrations.

	Concentration ($\mu M/cm^3$)		Nernst potential
Species	Inside	Outside	(mV)
A. *Mammalian muscle cells*			
Na^+	12	145	66
K^+	155	4	−97
Cl^-	4	120	−90
Resting potential			−90
B. *Squid axon*			
Na^+	50	460	59
K^+	400	10	−98
Cl^-	40–100	540	−45 to −69
Resting potential			−60

Source: Data from Ruch et al. (1968) and Katz (1966).

The concentration potential energy difference, W_C, is work required to move a mole of S against the gradient. This quantity is proportional to the logarithm of the concentration difference in accordance with

$$W_c = RT\left(\ln[S]_i - \ln[S]_o\right) \tag{3.1a}$$

$$= RT \ln\frac{[S]_i}{[S]_o} \tag{3.1b}$$

where $[S]_i$ and $[S]_o$ represent the concentrations of S inside and outside the cell, R is the gas constant, and T is absolute temperature. The product RT has units of energy per mole.

If S is ionized, an electrical potential difference will occur between the two regions of differing concentration. The electrical potential energy, W_e, is

$$W_e = ZFV_m \tag{3.2}$$

where Z is the valence of S, F is the Faraday constant (the number of coulumbs per mole of charge), and V_m is the potential difference across the membrane.

The total electrochemical potential difference is the sum of the concentration and electrical potentials:

$$\Delta W = W_c + W_e \tag{3.3}$$

Substituting the quantities from Eqs. (3.1b) and (3.2) results in

$$\Delta W = RT \ln\frac{[S]_i}{[S]_o} + ZFV_m \tag{3.4}$$

When $\Delta W = 0$, S is at equilibrium across the membrane, that is, there is no net force in either direction, and the net flux across the membrane is zero. Under conditions of equilibrium, the membrane will attain the potential:

$$V_m = \frac{RT}{FZ} \ln \frac{[S]_o}{[S]_i} \qquad (3.5)$$

Equation (3.5) is known as the *Nernst equation*. It is a statement of the membrane potential for an ionic substance in electrochemical equilibrium. Using the values $R = 8.31 \, \text{J/mol} \, \text{K}$, $T = 310 \, \text{K}$ (37°C), $F = 96,500 \, \text{C/mol}$, and $Z = +1$ (for a monovalent cation), converting to the base 10 logarithm, and expressing V_m in millivolts, we obtain:

$$V_m = 61 \log \frac{[S]_o}{[S]_i} \quad (mV) \qquad (3.6)$$

For a system with more than one permeable ionic species, the equilibrium voltage will depend on the concentration and relative permeability of the individual ions. For a system consisting of K^+ and Na^+, for example, the expression is

$$V_m = 61 \log \frac{P_K[K^+]_o + P_{Na}[Na^+]_o}{P_K[K^+]_i + P_{Na}[Na^+]_i} \qquad (3.7)$$

where P_K and P_{Na} are the permeabilities (expressed in units of centimeters per second) for K^+ and Na^+, respectively. An alternate expression for Eq. (3.7) uses the ratio $q = P_{Na}/P_K$ to obtain

$$V_m = 61 \log \frac{[K^+]_o + q[Na^+]_o}{[K^+]_i + q[Na^+]_i} \qquad (3.8)$$

One can get a feel for the changes in V_m during excitation by considering the simplified circuit diagram in Fig. 3.4. The membrane permeability is represented by conductances g_{Na} and g_K, and the electrochemical gradients as potential sources E_{Na} and E_K. For an excitable membrane in the resting state, $g_{Na} \ll g_K$ and the membrane potential approaches the Nernst potential for K^+, as indicated by Eq. (3.8). In the excited state, $g_{Na} > g_K$, and the switches in Fig. 3.4 would be connected in the alternate positions, forcing the membrane to move toward the Nernst potential for Na^+.

Consider the individual Nernst potentials for the ionic species listed in the right-hand column of Table 3.1. The equilibrium potential for Na^+ (66 mV) is far removed from the membrane potential (−90 mV), K^+ is slightly out of equilibrium, and Cl^- is essentially in equilibrium. The magni-

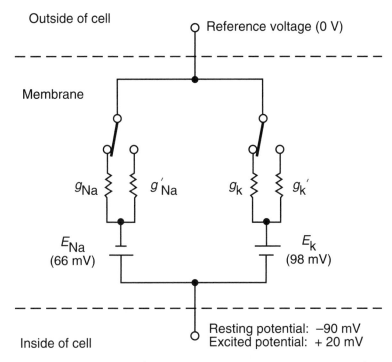

FIGURE 3.4. Circuit diagram representing membrane conductance for Na^+ and K^+ ions. In the resting condition, the inside of the cell is at a potential of -90 mV. In the excited state, the inside of the cell is at a potential of 20 mV.

tudes and signs of the potentials show that a strong electrochemical force tends to drive Na^+ into the cell and a relatively weaker force tends to drive K^+ out of the cell. Given that the membrane is at least somewhat permeable to the ions discussed here, these forces ought eventually to bring the species into equilibrium. Clearly, another force is working to maintain disequilibrium. The responsible force, is the so-called sodium pump, an active system that pumps Na^+ out of the cell and K^+ into the cell. The energy for the pump is derived from the cell's metabolism. A dead cell would eventually reach equilibrium potential.

The electrical forces on the membrane are quite large. Considering the membrane potential ($\approx 10^{-1}$ V), and thickness ($\approx 10^{-8}$ m), the electric field developed across the membrane is about 10^7 V/m. Conductivity properties of the excitable membrane are intimately tied to the membrane electric field; disturbances from the resting condition can lead to profound changes in the membrane's electrical properties. These changes ultimately initiate and sustain the functional responses of nerve and muscle.

3.3 The Excitable Nerve Membrane

Nerve and muscle cells possess membranes that are excitable, such that an adequate disturbance of the cell's resting potential can trigger a sudden change in the membrane conductance. The resulting membrane voltage change will affect adjacent portions of the membrane, and in a nerve, will propagate as a nerve impulse. The response of the excited membrane is an *action potential* (AP).

To illustrate the properties of excitability and propagation, consider the experiment illustrated in Fig. 3.5, in which a small stimulating electrode (SE) is near a nerve fiber, and two small recording microelectrodes (RE) pierce the membrane; the return electrodes are assumed to be immersed in the conducting medium some distance away. The stimulating electrode is connected to a current source. In Fig. 3.5, the small arrows represent the distribution of conventional current flow. The voltage disturbance due to the stimulating electrode reduces the membrane potential (depolarization) near the cathode, and increases the potential (hyperpolarization) elsewhere along the axon.

Figure 3.6 illustrates the response of the membrane to the rectangular current pulses shown in the upper part of the figure. Six possible current

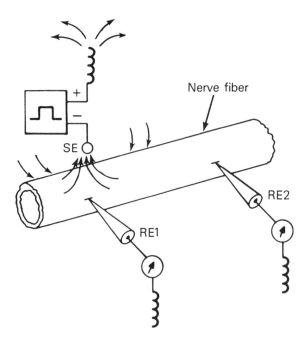

FIGURE 3.5. Nerve excitation and measurement arrangement. Excitation is initiated near the cathode of the stimulating electrode. Arrows indicate direction of conventional current flow.

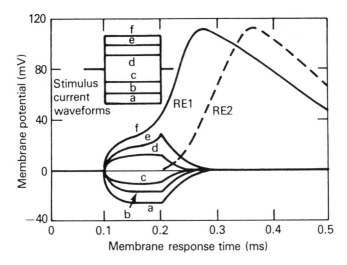

FIGURE 3.6. Excitation of a nerve fiber by an applied current, as in Fig. 3.5.

magnitudes labeled *a* to *f* are shown. Pulses *a* to *c* apply if SE is an anode; pulses *d* to *f* apply if SE is a cathode. The membrane response is shown as measured by RE1 and RE2, where 0 V represents the resting potential. Responses *a* to *c* are in the direction of hyperpolarization, and responses *d* to *f* are in the direction of depolarization. Responses *a* to *d* exhibit the characteristic of a linear *RC* network (see Chapter 4). Response *e* is slightly below the excitation threshold, and response *f* illustrates a fully developed action potential. The signal at RE2 is a delayed version of the AP, demonstrating conduction along the axon. The membrane response is often referred to as "all-or-nothing," because the peak AP response is not normally graded—the membrane is either excited or it is not.

To develop a quantitative model of excitation of the neuron by externally applied currents, nonlinear membrane models and electrical cable theory must be used. The membrane itself acts as the dielectric, and the ions on either side of the membrane act as the conductive plates of the capacitor. The nature of the leakage channels in the dielectric distinguishes the excitable membrane from the ordinary cellular membrane.

The Hodgkin–Huxley Membrane

With a series of ingenious experiments that eventually lead to a Nobel Prize in Physiology, Hodgkin and Huxley (1952) provided the first detailed description of the electrical properties of the excitable membrane of unmyelinated nerve cells. This work was later extended by Frankenhaeuser and Huxley (1964) to describe the myelinated nerve membrane. For brevity, we shall refer to the Hodgkin–Huxley and Frankenhaeuser–Huxley work as

HH and FH equations, respectively. The system of HH and FH equations is largely empirical.

Figure 3.7 illustrates the HH membrane schematically. The electrical model consists of membrane capacitance, nonlinear conductances for Na^+ and K^+, and a linear leakage element. For a parallel combination of capacitance and conductance, the current through the membrane is related to the capacitive and leakage currents by

$$J_m = c_m \frac{dV}{dt} + \left(J_{Na} + J_K + J_L\right) \tag{3.9}$$

where J_m is the membrane current density, c_m is the membrane capacity, V is the membrane voltage, and J_{Na}, J_K, and J_L are the ionic current densities. The ionic terms are expressed by

$$J_{Na} = g_{Na}\left(V - V_{Na}\right) \tag{3.10}$$

$$J_K = g_K\left(V - V_K\right) \tag{3.11}$$

$$J_L = g_L\left(V - V_L\right) \tag{3.12}$$

where g_{Na}, g_K, and g_L are the ionic conductances, and V_{Na}, V_K, and V_L are the ionic Nernst potentials. The g_L conductance is a linear conductance; the other two conductances are more complex nonlinear functions of the form

$$g_{Na} = \bar{g}_{Na}m^3 h \tag{3.13}$$

$$g_K = \bar{g}_K n^4 \tag{3.14}$$

where $\bar{g}_{Na}$ and $\bar{g}_K$ represent the maximum conductance values; and m, n, and h are so-called activation and deactivation variables that modulate the maximum conductances. The m, n, and h variables are governed by the first-order differential equations:

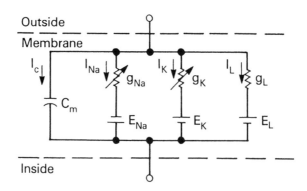

FIGURE 3.7. Hodgkin–Huxley membrane model.

$$\frac{dn}{dt} = a_n(1 - n) - \beta_n n \tag{3.15}$$

$$\frac{dm}{dt} = a_m(1 - m) - \beta_m m \tag{3.16}$$

$$\frac{dh}{dt} = a_h(1 - h) - \beta_h h \tag{3.17}$$

The a and β terms in Eqs. (3.15) to (3.17) are functions of the membrane voltage. At the experimental HH temperature of 6 °C, the a and β constants are

$$a_n = \frac{0.01(10 - \Delta V)}{\exp\left[(10\Delta V)/10\right] - 1} \tag{3.18}$$

$$\beta_n = 0.125\exp\left(-\frac{\Delta V}{80}\right) \tag{3.19}$$

$$a_m = \frac{0.1(25 - \Delta V)}{\exp\left[(25 - \Delta V)/10\right] - 1} \tag{3.20}$$

$$\beta_m = 4\exp\left(-\frac{\Delta V}{20}\right) \tag{3.21}$$

$$a_h = 0.07\exp\left(\frac{-\Delta V}{20}\right) \tag{3.22}$$

$$\beta_h = \left[\exp\left(\frac{30 - \Delta V}{10}\right) + 1\right]^{-1} \tag{3.23}$$

In Eqs. (3.18) to (3.23), ΔV is the change in membrane potential, expressed in millivolts, relative to the resting potential. Solutions to Eqs. (3.15) to (3.17) for a step change in ΔV can be obtained in the form:

$$n(t) = n_\infty - (n_\infty - n_0)e^{-t/\tau_n} \tag{3.24}$$

$$n_\infty = \frac{a_n}{a_n + \beta_n} \tag{3.25}$$

$$\tau_n = \frac{1}{a_n + \beta_n} \tag{3.26}$$

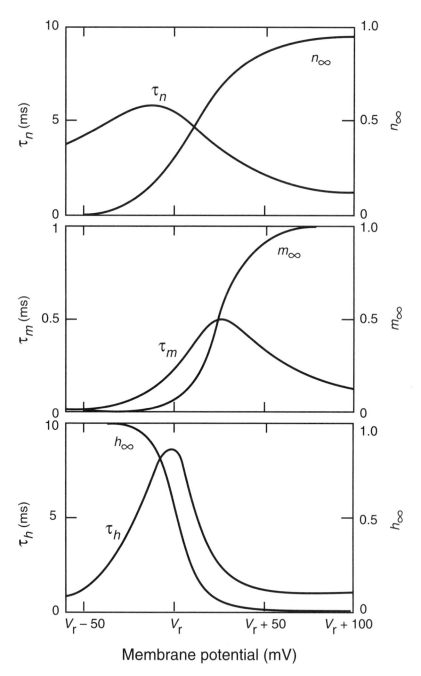

FIGURE 3.8. Relationship of m, n, and h constants to the membrane voltage. V_r represents the resting potential. (Adapted from Stein, 1980.)

where $n(0) = n_0$, and $n(t \to \infty) = n_\infty$. Expressions similar to Eqs. (3.24) to (3.26) are obtained for $m(t)$ and $h(t)$. The m, n, and h variables are constrained between 0 and 1, and can be regarded as the fraction of ion gates open at any one time. As indicated by Eqs. (3.13) and (3.14), these gates modulate the maximum conductance of Na and K. The σ_m, σ_n, and σ_h, variables determine the rate at which the gates can open and close. The asymptotic values of the m, n, and h variables, and the associated time constants, are all functions of membrane depolarization voltage as illustrated in Fig. 3.8 (after Stein, 1980).

Depolarization of the membrane is necessary for excitation. Figure 3.9 illustrates the membrane events accompanying excitation of the HH

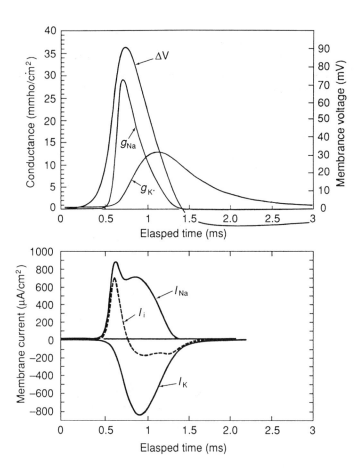

FIGURE 3.9. Membrane events during propagating action potential. The top figure shows the membrane voltage change (ΔV) and the conductances g_{Na} and g_k. The bottom figure shows the sodium current (I_{Na}) potassium current (I_K) and total ionic current (I_i). The positive axis indicates the ionic current influx. (Adapted from Hodgkin and Huxley, 1952.)

membrane. The upper part shows the AP voltage waveform along with the membrane conductance. The lower part shows membrane current—the positive axis refers to ionic current influx. An initial surge of Na^+ influx serves to further depolarize the membrane; this influx is followed by K^+ efflux, which repolarizes the membrane.

The Frankenhaeuser–Huxley Membrane

The FH equations for the myelinated membrane use four ionic terms, in contrast with the three for the HH unmyelinated membrane. The FH ionic current densities are

$$J = J_{Na} + J_K + J_p + J_L \qquad (3.27)$$

The terms J_{Na}, J_K, and J_L have the same interpretation as they do with the HH membrane. The term J_p was described as a "nonspecific" ionic component that responds to the concentration gradient of Na^+. Whether J_p represents a second type of Na^+ channel, or another ionic species, was left unresolved by the FH researchers. In the FH membranes, the individual ionic current densities have the expressions:

$$J_{Na} = \overline{P}_{Na} h m^2 \left(\frac{EF^2}{RT} \right) \frac{[Na]_o - [Na]_i \exp(EF/RT)}{1 - \exp(EF/RT)} \qquad (3.28)$$

$$J_K = \overline{P}_K n^2 \left(\frac{EF^2}{RT} \right) \frac{[K]_o - [K]_i \exp(EF/RT)}{1 - \exp(EF/RT)} \qquad (3.29)$$

$$J_p = \overline{P}_p p^2 \left(\frac{EF^2}{RT} \right) \frac{[Na]_o - [Na]_i \exp(EF/RT)}{1 - \exp(EF/RT)} \qquad (3.30)$$

$$J_L = g_L (V - V_L) \qquad (3.31)$$

where $E = V - V_r$, V is the membrane potential, and V_r is the resting potential. The variables m, n, h, and p are defined by differential equations of a form identical to Eqs. (3.15) to (3.17). The α and β variables for the m, n, h, and p rate terms of the FH equations have the form

$$\alpha_m, \beta_m, etc. = a(V - b)\left[1 - \exp\left(\frac{V - c}{d} \right) \right]^{-1} \qquad (3.32)$$

where the a, b, c, and d parameters are provided by Frankenhaeuser and Huxley (1963). Specific constants for the FH equations are given in Table 3.2.

TABLE 3.2. Constants for FH equations.

Constant	Value	Description
$\bar{P}_{Na}$	8×10^{-3} cm/s	Sodium permeability constant
$\bar{P}_K$	1.2×10^{-3} cm/s	Potassium permeability constant
$\bar{P}_P$	0.54×10^{-3} cm/s	Nonspecific permeability constant
g_L	30.3 mS/cm^2	Leakage conductance
V_L	0.026 mV	Leakage equilibrium potential
$[Na]_0$	114.5 mM	External sodium concentration
$[Na]_i$	13.7 mM	Internal sodium concentration
$[K]_0$	2.5 mM	External potassium concentration
$[K]_i$	12 mM	Internal potassium concentration
F	96,514 C/g mol	Faraday constant
R	8.3144 J/K mol	Gas constant
T	295.18 K	Absolute temperature
V_r	−70 mV	Resting potential
c_m	2 μF/cm^2	Membrane capacitance per unit area

Initial conditions: $m(0) = 0.0005$, $h(0) = 0.8249$, $n(0) = 0.0268$, $p(0) = 0.0049$.
Source: Frankenhaeuser and Huxley, 1963.

Differences between the FH and HH equations reflect the specific prop-erties of the myelinated and unmyelinated membranes. The activation vari-ables have different powers, and the FH equations have an additional ionic term. The ionic terms appear more complex in the FH membrane as com-pared to the HH membrane. Despite the seemingly greater complexity of the FH membrane, the electrical properties associated with the myelinated membrane's AP development are very similar to the HH membrane. One observed difference includes a somewhat longer AP duration for the HH membrane. In addition, the HH equations respond to a prolonged current stimulus by producing multiple APs, whereas the FH membrane produces a single AP.

Other differences are seen between myelinated (A fiber) and unmyeli-nated (C fiber) response when the overall excitatory behavior of the neuron is considered. Some of these differences include faster conduction rates and lower thresholds to external currents for A fibers (Ruch et al., 1968). The A fiber is an ideal one to model for electrical stimulation studies. Because of its lower excitation threshold for external current stimulation, the A fiber class will generally determine the limiting value for threshold currents. Nevertheless, it is important to consider the potential role of both fiber classes in electrical stimulation, and their functional properties in sensation and muscle response.

Species Dependence of Action Potential Dynamics

The HH and FH equations were developed from experiments on non-mammalian species—the HH experimenters used the giant axon of a squid, and the FH experimenters used the myelinated fiber of a toad. These species were chosen primarily for the large size of their axons, and relative ease of experimentation. The principles of membrane dynamics developed in that work have been found to be applicable to other species, including mammals. A quantitative description of membrane currents in rabbit myelinated nerve was developed by Chiu et al. (1979) using the HH experimental approach and mathematical framework. They showed that, as in squid and frog nerve, a transient inward sodium current was responsible for the initially rapid membrane depolarization. However, in contrast to squid and frog nerve, repolarization of the rabbit nerve was the result of outward flow of only a single passive ionic component, and potassium current outflow was virtually absent. Despite the simpler description of the rabbit nerve, its calculated AP was very close to that of the frog nerve in both amplitude and time course.

Propagation of Nerve Impulses

Action potential propagation can be understood by referring to Fig. 3.10. Consider that point A on the axon is depolarized. The local point of depolarization causes ionic movement between adjacent points on the axon, thus propagating the region of depolarization. If depolarization were initiated from an external source on a resting membrane at point A, an AP would propagate in both directions away from the site of stimulation. Normally, however, an AP is initiated at the terminus of the axon and propagates in only one direction.

After the membrane has been excited, it cannot be reexcited until a recovery period has passed. This period is termed the refractory state of the membrane. Before full recovery, the membrane becomes partially refractory; that is, it requires a stronger depolarizing force to become excited. The refractory property is principally the result of the prolonged decrease of the sodium deactivation variable, h (Fig. 3.8), which effectively turns off the membrane's sodium conductance until the passage of a recovery period that extends beyond the repolarization process.

Refractory behavior in frog nerve is illustrated in Fig. 3.11 (adapted from Katz, 1966), which shows the membrane response to an external current stimulus. An initial stimulus, applied at $t = 0$, results in response a. Successive stimuli lead to responses b to g. The absolute refractory period exists from the AP spike to somewhat within the negative after potential. During this period, a second stimulus, no matter how strong, fails to produce a response. In a mammalian nerve at body temperature, the neuron is absolutely refractory for typically about 0.5 ms. Afterwards, for a period of several milliseconds, the nerve is relatively refractory. During this period, an increased stimulus is needed to produce a response, and that response will initially be feeble, as noted by responses b to e. After several milliseconds, the neuron fully recovers from its less excitable state.

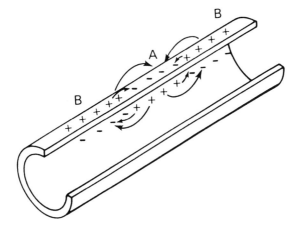

FIGURE 3.10. Spread of the depolarization wavefront. Depolarization occurring in region *A* results in charge transfer from the adjacent regions.

The refractory recovery period sets an upper limit on the number of APs per second that the membrane can support. With externally applied electrical stimuli, an AP rate cannot be produced beyond about 2,000 per second. In natural conditions in the body, the repetition rate rarely exceeds 500 per second and is more typically in the range from 10 to 100 per second (Brazier, 1977, Hoffmeister et al., 1991).

The AP velocity depends on the rate at which electrical charge is transferred from the locus of excitation to the region of membrane ahead of the AP. The charge-transfer rate, in turn, depends on the membrane capacity and the longitudinal resistance of the axon. Depending on the assumptions

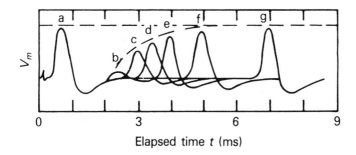

FIGURE 3.11. Illustration of the refractory period in frog nerve. The initial stimulus was applied at $t = 0$, resulting in response *a*. Subsequent stimuli were applied at various time delays, resulting in responses *b* through *g*. (Adapted from B. Katz, *Nerve, Muscle, and Synapse*, 1966, reproduced with permission of The McGraw-Hill Companies.)

TABLE 3.3. Characteristics of A and C fibers.

	Fiber class	
	A	C
Fiber diameter (μm)	1–22	0.3–1.3
Conduction velocity (m/s)	5–120	0.6–2.3
AP duration (ms)	0.4–0.5	2.0
Abs. refractory period (ms)	0.4–1.0	2.0
Velocity/diameter ratio (m/sμm)	6	~1.7
Myelinated	Yes	No

Source: Adapted from Ruch et al. (1968).

made, theoretical arguments suggest that conduction velocity ought to vary either as the square root of, or linearly with, fiber diameter (Paintal, 1967). Experimental evidence demonstrates that conduction velocity indeed increases with fiber diameter, and experimental data are frequently represented in terms of the ratio of conduction velocity to fiber diameter (Paintal, 1973).

The myelinated fiber is nature's means of obtaining fast conduction velocity without requiring unduly large fibers. The effective area requiring capacitive charging is limited mainly to the myelin-free internodes. Furthermore, the depolarization process is *saltatory*; that is, it jumps from node to node. As a result, propagation can proceed at a much faster rate than would be the case with an unmyelinated fiber of the same diameter.

Table 3.3 lists some general characteristics of A and C fibers. The myelinated A fibers are sometimes subdivided into diameter classes designated A_δ, A_β, and A_α. The A_δ fibers are typically related to cutaneous pain and temperature sensation, the A_β fibers are related to mechanoreception, and the A_α fibers are related to proprioception and contraction of striated muscle (Li and Bak, 1976). Figure 3.12 illustrates the distribution of afferent A fibers (Kandel et al., 1991). Unmyelinated fibers have a range more typically from 0.3 to 1.3μm. The distribution of efferent neurons shows prominent clusters from 12 to 20μm, and from 2 to 8μm, with a pronounced nadir in the range from 8 to 12μm (Ruch et al., 1968).

The distribution of myelinated fibers will differ substantially in the Central Nervous System (CNS) as compared with peripheral nerves. For instance, in the human pyramidal tract, 89.6% nerves were found in the diameter range from 1 to 4μm, 8.7% from 5 to 10μm, and 1.7% from 11 to 20μm (Lassek, 1942).

3.4   Action Potential Models for Cardiac Tissue

The electrical properties of excitable cardiac tissue have been defined using the experimental techniques and mathematical formalism that lead to the HH model. The membrane electrodynamics are specialized for different cardiac tissue. One common feature is a prolonged excited state, as compared with a much shorter period in nerve tissue.

The HH experimental techniques and mathematical formalism were applied to cardiac Purkinje fibers by Noble (1962). The Noble model used three ionic components, including a nonlinear sodium term similar to that in the HH equations, a nonlinear potassium term that exhibited a rectification process, and a nonspecific linear "anion" term. This model was later modified and extended by McAllister, Noble, and Tsien (1975) (referred to as the MNT model). The MNT Purkinje model is much more complex than the Noble Purkinje model—it includes nine separately described ionic components, along with five activation and deactivation variables. Some of the ionic terms in the MNT model are analogous to the HH nonlinear components; others are analogous to the HH linear leakage terms. The MNT model was modified and adapted for more efficient computer simulation by Drouhard and Roberge (1982b).

Further modifications and elaboration's of the MNT model were made by DiFrancesco and Noble (1985) (DN model). This model included no less

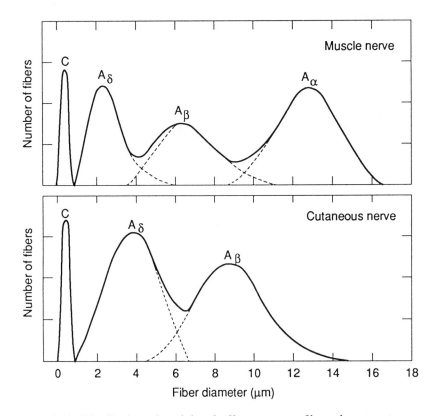

FIGURE 3.12. Distribution of peripheral afferent, nerve fibers: lower, cutaneous; upper, muscle. (Adapted from Boyd and Davey, 1968, with additional data by Martin, 1991.)

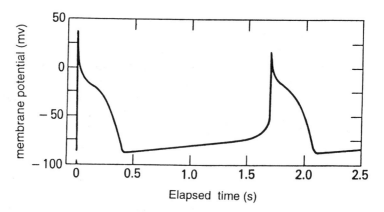

FIGURE 3.13. Computed action potential of cardiac Purkinje cell based on DiFrancesco-Noble model (From DiFrancesco and Noble, 1985.)

than 11 separate ionic components. A further feature of the DN model consisted of nonlinear differential equations governing the intracellular and extracellular ionic concentrations. None of the other models, including the various nerve membrane models, includes this feature. Figure 3.13 illustrates the simulated AP of the DN model, showing pacemaker activity.

A review of the ionic processes in the Purkinje fiber and their role in the development of computational models was presented in a tutorial paper by Noble (1984). He draws attention to the growth over time of the complexity of Purkinje computational models, and remarks: "It is sincerely to be hoped that this is not a case of indefinite exponential growth, or, by the year 2000, we shall all have great difficulty in explaining cardiac excitation to ourselves, let alone to students with formidable memories" (p. 42).

The various Purkinje models may be regarded as representing differing degrees of precision in the treatment of ionic processes. From the point of view of electrical stimulation studies, it is not clear how much of this precision is required. At a minimum, we would like to reproduce the pacemaker AP process and simulate evoked extrasystolic excitation. Even the simplest model (Noble, 1962) gives a good reproduction of the Purkinje AP and its pacemaker action. Evoked extrasystolic excitation has been demonstrated in both the Noble (1962) and MNT (McAllister et al., 1975) papers. Figure 3.14 illustrates the response of the Noble model to repeated stimulation at a rate of 3/s. The alteration in the duration of the evoked APs seen in Fig. 3.14 is similar to experimental observations (Trautwein and Dudel, 1954; Geddes et al., 1972).

The DN equations described above were modified by Noble and Noble (1984) to describe electrical activity of the sinoatrial node. Beeler and Reuter (1977) used the HH formalism to develop a mathematical model for the action potential of ventricular myocardial fibers (referred to as the BR

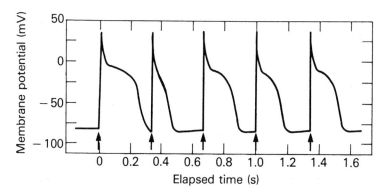

FIGURE 3.14. Effect of repetitive stimulation on the computed AP in the Noble Purkinje model. Arrows indicate application of stimulus at the rate 3/s. (From Noble, 1962.)

model), and later improved by Mogul and colleagues (1984). An interesting property of the BR model is its ability to produce oscillatory potentials when depolarized by steady outward current, as seen in Fig. 3.15. The authors point out that although the electrical properties of the myocardium seldom produce spontaneous activity, numerous experimental studies with steady depolarizing current have demonstrated oscillatory AP behavior, such as seen in Fig. 3.15.

To adequately model cardiac excitation, one needs to spatially couple the temporal electrodynamics of individual cells to account for the spatial and

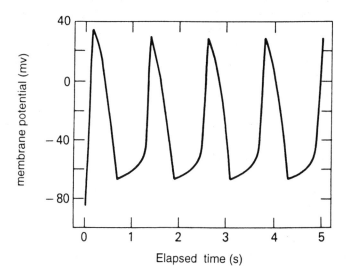

FIGURE 3.15. Oscillatory AP response of Beeler–Reuter model for ventricular myocardial fiber in response to steady outward current. (From Beeler and Reuter, 1977.)

temporal distribution of the stimulus. In the case of nerve cells, one can use cable models, such as the SENN model described in Chapter 4. For cardiac tissue, the bidomain model treats cardiac tissue as two coupled, continuous domains, one for the intracellular space, and one for the interstitial space (Henriquez, 1993).

3.5   Sensory Transduction[2]

The body is equipped with a vast array of sensors (receptors) for monitoring its internal and external environment. The receptor converts a stimulus to an electrical potential that initiates a propagating AP. Receptors are specialized to respond most efficiently to a specific type of stimulus, although they may also respond to a variety of stimuli if the intensity is great enough. We shall be concerned here with the *somatosensory* system, i.e., the system of receptors found in the skin and internal organs. There are other specialized receptors in the visual and auditory systems, chemical receptors for taste and smell, and other special chemical receptors by which neurons communicate with one another.

The somatosensory receptors can be classified as: *mechanoreceptors, thermoreceptors, chemoreceptors*, and *nociceptors*. Numerous specializations of mechanoreceptors respond to specific attributes of mechanical stimulation. Thermoreceptors are specialized to respond to either heat or cold stimuli. Nociceptors are unresponsive until the stimulus reaches the point where tissue damage is imminent, and are usually associated with pain. Many nociceptors are responsive to a broad spectrum of noxious levels of mechanical, heat, and chemical stimuli (Ruch and Patton, 1979). Both myelinated and unmyelinated nociceptors have been identified.

Figure 3.16 illustrates several cutaneous mechanoreceptors. The *pacinian corpuscle* is a so-called rapidly adapting receptor because it responds to the onset or termination of a pressure stimulus. As such, it transduces acceleration of skin displacement (Schmidt, 1978). The *Meisner corpuscle* is a less rapidly adapting receptor, and is most responsive to the velocity of skin displacement. The *Rufini ending* and *tactile disks* are slowly adapting, and respond most efficiently to steady pressure. Figure 3.16 also shows hair receptors that respond to displacement of hair follicles. Other mechanoreceptors, such as stretch receptors and muscle spindles, respond to movement and position of the muscles, and serve to monitor and regulate body posture and movement. An additional class of receptors, known as free nerve endings, appear as branching structures on the end of the neuron. Free nerve endings are involved in a variety of specialized transduction processes in both nociceptive and non-nociceptive classes. It is not always

---

[2]The following are useful references for this section: Kandel and Schwartz (1981); Lamb et al. (1984); Ruth et al. (1968), Ruch and Patton (1979).

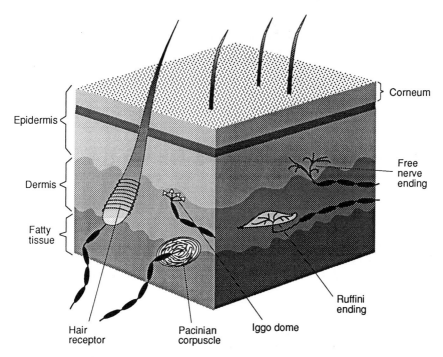

FIGURE 3.16. Morphology of several mechanoreceptors in the hairy skin.

clear from morphological evidence what distinguishes the transduction specializations of free nerve endings.

Intensity Coding

When a receptor is stimulated, it produces a voltage change called a *generator potential* at the terminal ending of its axonal connection. Unlike the all-or-nothing response of the axon, the generator potential is graded: if you squeeze a pacinian corpuscle, it produces a voltage at the axon terminus; if you squeeze it harder, it produces a greater voltage.

The generator potential initiates one or more APs that propagate along the axon. The AP repetition rate will depend, in part, on the intensity of the stimulus. Figure 3.17 illustrates the response rate of a slowly adapting (Fig. 3.17a) and an intermediately adapting (Fig. 3.17b) receptor (from Schmidt, 1978). The slowly adapting receptor responds to a constant pressure stimulus, although the AP rate is greatest at the onset of the stimulus. The intermediately adapting receptor AP rate is more or less constant during a constant-velocity pressure stimulus. For a rapidly adapting pacinian corpuscle, a single AP is produced at the onset or termination of a pressure stimulus. However, for a vibratory stimulus, a train of APs is produced at the rate of the vibration frequency. The intensity of the threshold amplitude

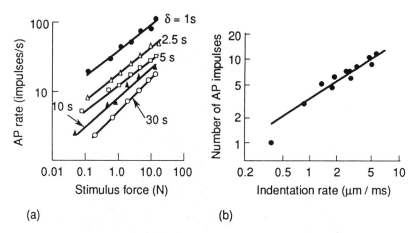

FIGURE 3.17. Response of (a) slowly adapting receptor and (b) intermediate adapting receptor to constant force stimulus. Part (a) shows AP rate of various time delays (δ) after onset of pressure stimulus. Part (b) indicates total number of APs during 0.5-s constant velocity stimulus. (Adapted from Schmidt, 1978.)

of the sinusoidal mechanical stimulus on the skin is shown in Fig. 3.18. Sensitivity to vibratory mechanical stimuli on the skin is greatest in the 200 to 300-Hz range. Anesthetization of the skin elevates thresholds only for frequencies below about 50 Hz. This can be explained by two modes of transduction: low-frequency sensations are the result of more superficial

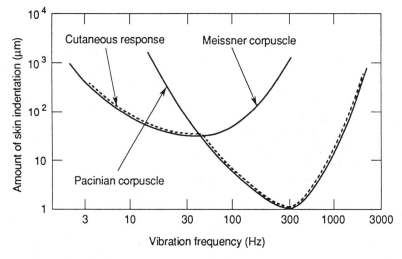

FIGURE 3.18. Threshold response for vibratory mechanical stimulus applied to the skin. (Reprinted by permission of the publisher from Martin, 1991. © 1991 by Elsevier Publishing Co. Inc.)

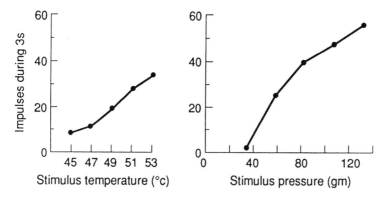

FIGURE 3.19. Response of myelinated nociceptive afferents innervating monkey hand. Stimulus duration is 3 s for both types of stimuli. (From Campbell et al., 1979.)

hair follicle receptors or Meissner corpuscles; high-frequency vibration is detected by pacinian corpuscles located in deeper strata (Martin and Jessell, 1991).

Figure 3.19 illustrates AP rate versus stimulus intensity for unmyelinated nociceptors (Campbell et al., 1979), showing a monotonic relationship between the average AP rate and intensity of both heat and pressure stimuli. Other studies (LaMotte et al., 1984) demonstrate a monotonic relationship between heat intensity and both AP rate and pain rating in nociceptive C fibers. Although AP rate is important in sensory encoding of stimulus intensity and painfulness, there are a number of other important factors as well. These include the spatial and temporal patterns of stimulation, and the number of neurons brought to excitation (recruitment). Chapter 7 provides additional discussion of the role of these factors in electrical stimulation.

3.6   Muscle Function

The previous section discussed the role of afferent neurons in conveying information from the body's sensory system to the central nervous system (CNS). Efferent neurons carry information from the CNS to the muscles to effect contraction. Figure 3.20 illustrates structural features of the neuromuscular junction at the *end plate* of the muscle (after Birks et al., 1960; Woodbury et al., 1966). When the AP reaches its terminus at the end plate, a chemical neurotransmitter is released across the nerve/muscle gap, which causes depolarization of the muscle cells. The result is that the muscle membrane is excited, and a depolarization wave is propagated in the muscle away from the end-plate region.

Figure 3.21 illustrates the anatomical structure of muscle, showing progressively smaller systems of structural elements. When the muscle is

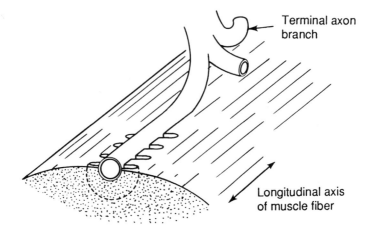

Terminal axon
branch

Longitudinal axis
of muscle fiber

FIGURE 3.20. End-plate region of frog muscle fiber. (After Birks et al., 1960.)

excited, the individual fibrils (shown in the lower part of the figure) slide together resulting in muscle contraction. The muscle illustrated in Fig. 3.21 is called *striated* because of the microscopically visible striations that arise from the arrangement of contractile elements. Striated muscle is found in the skeletal system. Skeletal muscle is under voluntary control, and is innervated by the somatic nervous system. *Smooth muscle* differs from striated muscle in that the characteristic cross-striation pattern is absent. Smooth muscle is involuntary, and is found in vessel artery walls, air passages of the lungs, and in various tissues of the intestines and reproductive system.

An action potential launched along a motor neuron results in a single contractile quantum called a *twitch*. When a succession of APs is produced on the motor neuron, the individual twitch quanta fuse as illustrated in Fig. 3.22. In this example, maximum muscle tension is achieved at an AP rate of about 80/s, leading to a condition of maximum fusion termed *tetanus*. Gradation of muscle tension results from fusion of individual twitch quanta and from recruitment (the excitation of additional neurons). Normally, the APs in individual motor neurons are asynchronous, and muscle tension can be finely graded. However, with stimulation by externally applied currents of a repetitive or oscillatory nature, the twitch quanta of various muscle fibers may be synchronized. Experimental data show that tetanus of the muscle group occurs at current levels that are only moderately above the twitch threshold value (Oester and Licht, 1971).

Externally applied electric currents can stimulate muscle by exciting motor neurons, or the muscle fibers themselves (Fig. 8.3). Direct stimulation of muscle fiber requires much higher currents than does stimulation of enervated muscle (Harris, 1971; Walthard and Tchicaloff, 1971). As a result, stimulation of muscle by external electric currents will usually take place most efficiently through neural excitation. See Chapter 8 for details of electrical stimulation of muscle and motor neurons.

The propagating excitation wave along the muscle, when recorded through cutaneous electrodes, is called an *electromyogram* (EMG). The EMG signal has been used in various experimental and clinical applications

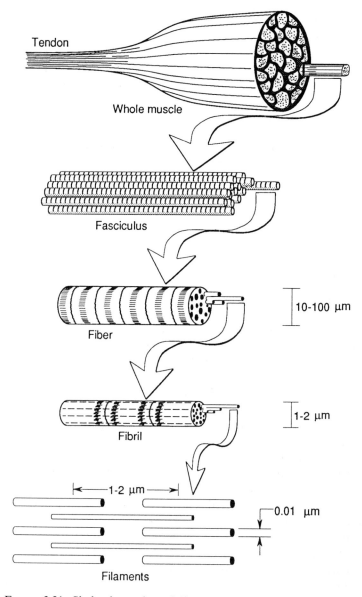

FIGURE 3.21. Skeletal muscle and filament structure of striated muscle.

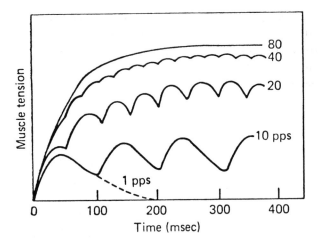

FIGURE 3.22. Effects of AP rate on muscle tension. [Adapted from McNeal and Bowman (1985a). Reprinted by permission from *Neural Stimulation*, vol. II CRC Press Inc., Boca Raton, FL.]

as an indicator of muscle activity, and can often be more easily detected than direct observation of muscle movement or tension.

3.7   Synapses

General Properties

Excitable cells communicate with one another across junctions called *synapses*. An action potential that has traveled along the presynaptic cell affects the postsynaptic cell through the release of specialized chemicals called *neuro transmitters* in the case of chemical synapses, or electrically (*ephaptic* transmission) across gap junctions in the case of electrical synapses. Chemical synapses are the more sensitive and common of these two modes of intracellular communication. Figures 3.1 and 3.20 illustrate a muscle synapse at the motor neuron end plate. Figure 3.1 also illustrates central synapses at the dendrites within the CNS. The neurotransmitter that is released from the chemical synapse flows across the gap, where it binds to receptor sites in the postsynaptic cell and opens ionic channels that alter the postsynaptic membrane potential. Through these means, a small presynaptic process which generates only a weak ionic current can depolarize a large postsynaptic cell. If the postsynaptic potential is sufficiently depolarized, an action potential will be launched.

   Both temporal and spatial integration can contribute to the postsynaptic potential (Dudel, 1989). With temporal integration, repeated action potentials from the presynaptic neuron can have an additive effect on the

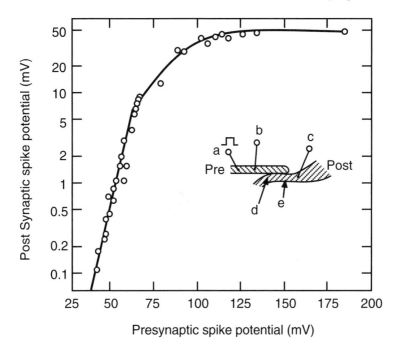

FIGURE 3.23. Example of relationship between pre- and postsynaptic potentials with injected current (t = 1 ms). Length of synapse (d–e): 0.8 mm; (a) current electrode; (b) pre-recording electrode; (c) post recording electrode. [Adapted from Katz & Miledi (1967).]

postsynaptic membrane potential. This occurs because the postsynaptic membrane has a much longer membrane time constant than does the presynaptic cell. With spatial integration, many individual neurons can synapse on a single postsynaptic cell. The additive effects of the spatially integrated action potentials can be either excitatory or inhibitory in the postsynaptic cell.

Experiments using the giant squid axon demonstrate the relationship between pre-and postsynaptic potentials in chemical synapses as in Fig. 3.23 (Katz and Miledi, 1967). The insert in the figure shows the experimental arrangement. In this example, a change in the presynaptic spike potential from 55 to 65 mV (an 18% change) results in a ten-fold change in the postsynaptic potential from 1 to 10 mV.

Synaptic Interactions with In-situ E-fields

Polarization of presynaptic processes due to an *in-situ* electric field can result in enhancement or inhibition of postsynaptic action potentials. An

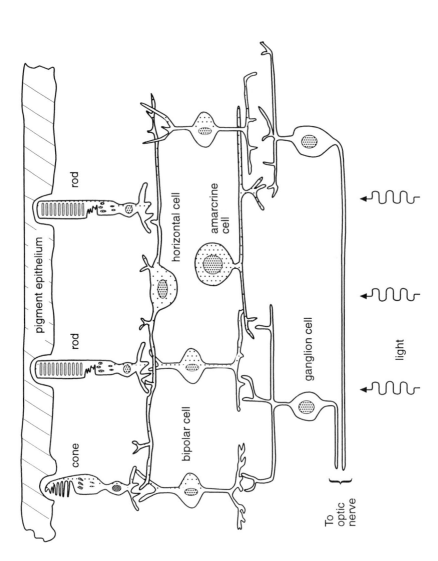

FIGURE 3.24. Structure of the retina showing various layers of neurons consisting of photoreceptors (rods and cones), horizontal cells, bipolar cells, amarcrine cells, and ganglion cells. (Adapted from Tessier-Lavigne, 1991; Dowling and Boycott, 1966).

example of this effect is attributed to the phenomenon of electro- and magnetophosphenes, which are the visual effects resulting from electric currents or magnetic fields applied to the head. Experimental evidence discussed in Sect. 9.8 suggests that phosphenes are generated through modification of synaptic potentials in the receptors or neurons of the retina. The retina is rich in the synaptic processes of photoreceptors and neurons that comprise a visual processing system. Figure 3.24 (adapted from Tessier-Lavigne, 1991; Dowling and Boycott, 1966) illustrates the organization of neurons within the retina, showing neuronal structures that have both tangential and radial orientations with respect to the retina. Phosphenes are most sensitive to current or an in-situ E-field that is oriented in a radial direction. This finding suggests that the E-field interacts with radially oriented neurons in accord with cable theory (Sect. 4.3).

The electrical thresholds and the temporal response of synaptic effects within the retina differ significantly from corresponding properties of nerve and muscle excitation. Considering the observed magneto-phosphene thresholds at the most sensitive frequency (20 Hz), the minimum in-situ electric field that produces phosphenes is approximately a factor of 100 below the rheobase thresholds for nerve excitation (refer to Chapter 4). Also, the strength-duration time constant for phosphene stimulation is approximately 100 times greater than that for nerve excitation and 10 times greater that for excitation of muscle cells. It does not necessarily follow that the low thresholds associated with phosphenes necessarily apply to other neural synapses because of the highly specialized configuration of neurons in the retina. However, experimental evidence discussed in Sect. 9.8 shows that short-term CNS responses can be seen with *in-situ* E-fields that are below the threshold of neural excitation. Although synaptic polarization could have important consequences in electrical stimulation of the CNS, the implication of these findings on neural reactions to externally applied currents or fields has been little studied.

3.8   The Spinal Reflex

Muscle movement can be electrically stimulated through excitation of motor neurons or direct excitation of muscle fibers. Reflex activity represents a third mechanism whereby muscle movement may result from electrical stimulation. Here, stimulation may be initiated at cutaneous or muscle sensory afferents, with muscle movement ensuing through the reflex arc. When we touch a hot object, for example, our hand jerks away before we have consciously appraised the situation. Afterward, deliberate action takes place. The initial response, in this case, takes place outside of conscious control through the *spinal reflex*.

Figure 3.25 illustrates organizational features of the spinal reflex arc. Sensory inputs, originating from cutaneous or muscle receptors, communi-

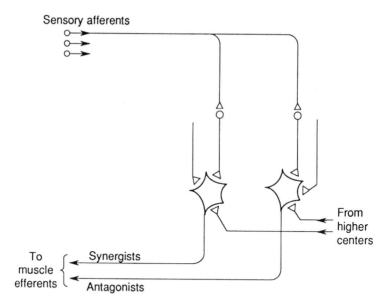

FIGURE 3.25. Organizational features of reflex arc. Small circles represent receptors; triangles represent synaptic terminals. Synaptic summation occurs at dendritic processes within spinal column.

cate with muscle efferents through synaptic processes in the spinal column. Intermediate neurons (*interneurons*) may also exist within the reflex arc. Inputs from higher centers may also be included. These modify the reflex action based on central processes—for example, we can deliberately inhibit the reflex action to maintain contact with a hot object. The synaptic summation may combine inputs as either inhibitory or excitatory, such that the presence of a synaptic output will depend on a weighted summation of inputs, thus forming a computing system within the spinal column. The output of this system consists of excitatory and inhibitory signals to muscle groups to cause coordinated, patterned movement.

The knee-jerk reflex is a simple example of a *monosynaptic* type in that it omits the interneurons such that the synapses of sensory inputs directly activate motor neurons. The motor neurons, in turn, produce AP signals that cause contraction in synergistic muscle groups, and inhibit AP signals to cause relaxation in antagonistic muscle groups.

More complex reflex activity is conditioned by experience. Some reflex actions are shared by individuals, regardless of prior experience, such as deep tendon reflexes, the eye blink response, and startle reflexes. Section 7.11 discusses experimental data on startle reactions to electrical stimuli.

4
Excitation Models

4.1 Introduction

Variations in the stimulus waveform or in the electrode arrangement can result in vastly different stimulation thresholds. To fully understand the factors responsible for electrical stimulation, computational models must connect features of the stimulus current with properties of excitable tissue described in Chapter 3. However, a simpler approach based on a linear, spatially limited membrane can provide insight into underlying electrical processes, and also provide closed-form mathematical expressions for some excitation relationships. This chapter begins with simple linear analysis models, and proceeds to a more complex, spatially extended nonlinear model.

The models treated here are essentially electrical cable representations of single fibers. Clearly, electrical stimulation normally involves a much more complex macroscopic system. Both sensory and motor responses depend on the spatial and temporal patterns of stimulation; cardiac responses involve a three-dimensional dynamic system. Nevertheless, the response due to large-scale excitation is a result of individual excitable elements. As will become apparent in subsequent chapters, electrical responses attributable to individual fibers can explain many observed macroscopic responses. Thus, single-fiber excitation models provide powerful analytic tools for evaluating a wide range of electrical stimulation properties.

4.2 Linear Strength-Duration Model

A simple analysis model treats an isolated segment of an excitable fiber as a linear electrical circuit as shown in Fig. 4.1. The membrane is assumed to consist of capacitance c_m and resistance r_m. The simplified analysis circuit is presumed to apply to a small patch of excitable membrane. The stimulus current flowing across the membrane is depicted as originating from a

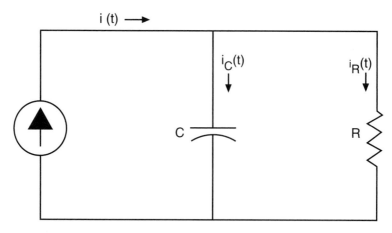

FIGURE 4.1. Simplified linear circuit model for isolated patch of excitable membrane.

current source. The resting potential is treated as $0\,V$, and depolarization is analyzed as the voltage rise across the electrical network. In a simplified analysis, r_m is treated as a constant up to the threshold of the action potential. In actuality, because of the nonlinear membrane properties described in Chapter 3, r_m is relatively constant only up to about 80% of the action potential (AP) threshold (McNeal, 1976).

The equations governing the response of the RC model are

$$i(t) = i_c(t) + i_R(t) \tag{4.1}$$

$$\frac{1}{c_m}\int i_c(t)dt = i_R(t)R \tag{4.2}$$

where $i(t)$ is the total current, $i_c(t)$ is the capacitive displacement current, and $i_R(t)$ is the current flowing in the resistance R. Consider that this simple circuit is excited by a step current pulse having the form

$$i(t) = \begin{cases} I & t \geq 0 \\ 0 & t < 0 \end{cases} \tag{4.3}$$

where I is the peak current. Equations (4.1), (4.2), and (4.3), may be readily solved (such as with Laplace transform methods) as

$$v_m(t) = i_R(t)R = IR\left(1 - e^{-t/\tau_m}\right) \tag{4.4}$$

where $\tau_m = r_m c_m$; τ_m is called the *membrane time constant*.

Assume that there is a single depolarization voltage, V_T, needed for excitation. This is a simplification, since the threshold depolarization voltage rises for very short pulses (e.g., $<10\,\mu s$) (Dean and Lawrence, 1983), or for biphasic stimuli (Reilly et al., 1985). For a pulse of duration t, the maximum depolarization voltage is evaluated from Eq. (4.4). The threshold current (I_T) required to drive the transmembrane potential to V_T can be derived from Eq. (4.4) as

$$I_T = \frac{V_T/R}{1 - e^{-t/\tau_m}} \tag{4.5}$$

The threshold current attains a minimum value (I_0) for an infinitely long pulse. As $t \to \infty$, $I \to I_0 = V_T/R$, and Eq. (4.5) may be expressed as

$$\frac{I_T}{I_0} = \frac{1}{1 - e^{-t/\tau_m}} \tag{4.6}$$

The threshold charge, Q_T, necessary for achieving the depolarization voltage, V_T, is

$$Q_T = I_T t = \frac{I_0 t}{1 - e^{-t/\tau_m}} \tag{4.7}$$

The threshold charge attains a minimum value Q_0, for short-duration pulses. As $t \to 0$,

$$Q_T \to Q_0 = I_0 \tau_m \tag{4.8}$$

The expression for normalized depolarization charge is

$$\frac{Q_T}{Q_0} = \frac{t/\tau_m}{1 - e^{-t/\tau_m}} \tag{4.9}$$

The energy dissipated by the rectangular stimulus is given by $I^2 R t$ (or, equivalently, IQR), where R is the resistance in the current path. It follows that the threshold energy, E_T, is given in normalized form as

$$\frac{E_T}{I_0^2 R} = \frac{t}{\left(1 - e^{-t/\tau_m}\right)^2} \tag{4.10}$$

As demonstrated by Pearce and colleagues (1982), the stimulus energy attains a minimum value, E_0, when $t = 1.25\tau_e$

$$E_0 = 2.46 I_0^2 R \tau_e \tag{4.11}$$

Figure 4.2 illustrates the normalized threshold current, charge, and energy versus the normalized duration (t/τ_m).

A parallel treatment has been derived (Blair, 1932a, 1932b; Reilly et al., 1983) for exponentially decaying current pulses of the form

$$i(t) = \begin{cases} Ie^{-t/\tau} & \text{for } t \geq 0 \\ 0 & \text{for } t < 0 \end{cases} \tag{4.12}$$

In Eq. (4.12), τ represents the time constant of the exponential current pulse. Such exponential currents are encountered with capacitive discharge

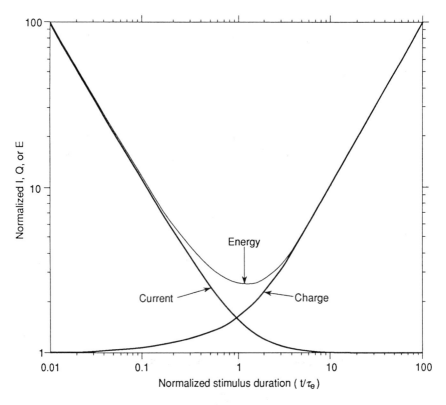

FIGURE 4.2. Calculated strength-duration relationships for square-wave mono-phasic current.

stimuli (see Chapter 2). The solutions for normalized threshold current and charge are

$$\frac{I_T}{I_0} = p^{1/(p-1)} \tag{4.13}$$

$$\frac{Q_T}{Q_0} = p^{p/(p-1)} \tag{4.14}$$

where $p = \tau/\tau_m$. When rectangular and exponential stimulus criteria are compared, Eqs. (4.9) and (4.14) both converge to the same minimum charge thresholds for short durations, and Eqs. (4.6) and (4.13) converge to the same minimum peak current threshold for long durations (Reilly and Larkin, 1983). This observation is equally valid when a more complete nonlinear model is used to represent the excitable fiber (see Fig. 4.14).

The simple linear model points out some important features concerning electrical thresholds for monophasic currents. For short-duration stimuli, the amount of charge transferred is the fundamental quantity defining stimulus potency. For brief monophasic transients, the details of the wave-shape are relatively unimportant both rectangular and exponential stimuli converge to the same minimum threshold charge. For long-duration monophasic currents, the peak current is the main determinant of electrical thresholds. Stimulus energy alone is not a meaningful index of excitability, as has been supposed by some in the past.

Empirical Strength-Duration Relationships

An empirical strength-duration (S-D) relationship for excitation of nerves was first derived by Weiss (1901). We can marvel today at the ingenuity of this early researcher, who, in those days of only crude electronics and measurement devices, was able to perform delicate and precise experiments that retain validity today. Expressions equivalent to the Weiss formulation can be stated as

$$I_T = I_0\left(1 + \frac{\tau_e}{t}\right) \tag{4.15}$$

and

$$Q_T = Q_0\left(1 + \frac{t}{\tau_e}\right) \tag{4.16}$$

where I_T and Q_T are threshold current and charge, respectively, I_0 is the minimum threshold current for long pulses $(t \to \infty)$, Q_0 is the minimum threshold charge for short pulses $(t \to 0)$, t is the duration of the pulse, and τ_e is an experimental parameter related to the time response of the tissue being excited. Equation (4.16) is the expression derived by Weiss; Eq. (4.15) follows from the relationship $Q_T = I_T t$. Equation (4.15) is sometimes re-ferred to as a "hyperbolic" relationship.

Lapicque (1907) studied the data of Weiss and proposed what he called a "logarithmic" relationship. Lapicque's formulation can be expressed as

$$I_T = I_0\left[1 - \exp\left(\frac{-t}{\tau_e}\right)\right]^{-1} \tag{4.17}$$

Equation (4.17) is now referred to as an "exponential" relationship. It is the same as that derived for the simple linear model [Eq. (4.6)], with the exception that Eq. (4.17) uses the empirical time constant, τ_e, rather than the membrane time constant, τ_m. The value of the time constant derived from the simple product of membrane resistivity and capacity, $r_m c_m$, is very different from the value determined by fitting experimental data to Eq. (4.17). More will be said about experimental time constants.

Lapicque introduced the term *rheobase* to describe the minimum thresh-old current achieved for very long pulses. He also defined *chronaxie* as the duration of a threshold current having a magnitude twice rheobase. These terms have become standard lexicon today. In both the parabolic and exponential formulations,

$$\text{Rheobase} = I_0 \qquad (4.18)$$

Chronaxie may be related to τ_e in the two formulations as

$$\text{Chronaxie} = \tau_e \quad \left(\text{hyperbolic formula}\right) \qquad (4.19a)$$

and

$$\text{Chronaxie} = \tau_e \ln 2 = 0.693\tau_e \quad \left(\text{exponential formula}\right) \qquad (4.19b)$$

In both formulations, the parameter τ_e may be determined by the ratio of minimum charge to minimum current:

$$\tau_e = \frac{Q_0}{I_0} \qquad (4.20)$$

Equation (4.20) suggests a simple and practical means of determining the parameter τ_e in both the hyperbolic and exponential relationships. It suggests that one may determine τ_e without having to trace out the entire S-D curve—rather, it is sufficient to evaluate the stimulus threshold at only two durations (one very long and one very short).

Figure 4.3 compares the hyperbolic and exponential expressions for current thresholds. A third curve applies to a myelinated nerve model, which will be described presently. Both hyperbolic and exponential forms of the S-D relationship have been used to represent experimental data. Experimental curve fits for the two formulas have been compared for nerve (Bostock, 1983) and cardiac (Mouchawar et al., 1989) excitation. Over the range of stimulus durations studied, the two formulas provide reasonably good curve fits, although one or the other may be marginally preferred in specific instances. One attraction of the exponential formulation is that it can be related to the linear RC network model (Fig. 4.1), which provides a heuristic, albeit crude, physiological model of neural excitation. A more accurate analysis model would include the nonlinear and spatially extensive properties of the excitable membrane as described in the next section.

τ_e can depend strongly on the spatial distribution of stimulus current. The apparent time constant generally becomes longer as the membrane current is distributed in a more gradual fashion. This is clearly demonstrated in Sect. 4.7, and also by Jack et al., (1983, p. 33) using a linear cable model. These studies show that the apparent time constant with membrane current injected at a discrete point may be less than half the value observed when the current is distributed more gradually along the fiber.

Large variations in τ_e are seen for different types of excitable tissue. An average for nerve excitation is around 0.27 ms, with a wide experimental range (see Table 7.1). The average for cardiac excitation is around 3 ms, also with a wide experimental range (see Table 6.2).

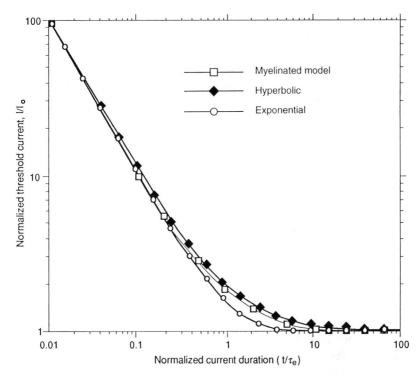

FIGURE 4.3. Strength-duration curves for neural excitation by rectangular mono-phasic current pulses. Hyperbolic and exponential curves are from Weiss and Lapicque formulations; thin curve applies to myelinated nerve model.

Experimental data using cutaneous stimulation show that chronaxy val-ues or S-D time constants obtained with a regulated current stimulator can be significantly greater than those obtained with a regulated voltage device (Harris, 1971). It is likely that this difference can be explained on the basis of the electrical response of the skin as explained in Chapter 2. When using a constant voltage device, one observes the current waveform to exhibit a leading spike due to the capacitive coupling of the skin and a nonlinear response in which impedance drops as voltage is increased. Both of these phenomena would tend to distort a voltage-regulated S-D curve relative to a current-regulated curve.

4.3   Electrical Cable Representations

While the simple linear model of the previous section provides useful insight into the neural excitation process, it is of limited value in attempting to describe the full range of interrelations between the excitable membrane and stimulus current. The linear model does not account for the effects of the spatial distribution of stimulus current, differential sensitivity to anodal versus cathodal currents, response to biphasic currents, or response to sequences of pulsed stimuli. To study these factors adequately, we need a more complex model involving the spatial and temporal interrelationships of the excitable membrane.

One-Dimensional Cable Models

Excitation of nerve and muscle cells can be analyzed using electrical cable theory. The cable equations to be applied to this problem were originated by O. Heaviside in 1876 in connection with the analysis of the first trans-Atlantic telegraphy cable (Nahin, 1987). This section briefly introduces cable theory as applied to the excitable membrane. More extensive treatments of this subject can be found in other publications (Rall, 1977; Jack et al., 1983; Plonsey and Barr, 1988).

An assumption of cable theory is that at any longitudinal position x along the cable, the inside and outside potentials do not vary with the position along the cable's circumference. Considering the small diameters of the cells to be studied ($<20\,\mu$m), this assumption is a reasonable one for biological applications, and it permits the cable to be modeled in a one-dimensional fashion, as in Fig. 4.4. The illustration shows the inside as having a resistance per unit length r_i (Ω/cm), a distributed membrane resistance from inside to outside having unit length value r_m (Ω cm), and a distributed length-proportional capacity c_m (μF/cm). I_e and I_i are the currents flowing externally and internally, V_e and V_i are the external and internal potentials, r_e is the external resistance per unit length of the cable because of the surrounding medium, and i_m is the membrane current.[1]

The currents and voltages are assumed to vary with the longitudinal distance along the cable. The cable equation is expressed in differential form as

$$\frac{\partial^2 V_m}{\partial x^2} = \left(r_i + r_e\right)i_m \tag{4.21}$$

where V_m in the membrane voltage defined by $V_i - V_e$. An additional relationship is that membrane current consists of both ionic and capacitive components; that is,

---

[1] Unit length resistances are defined such that the internal and external resistance's increase with length, L, in accordance with $R_i = r_i L$ and $R_e = R_e L$, and leakage resistance decreases with length as $R_m = r_m/L$ (Both R_i and R_m have units of ohms). The capacitance between inside and outside over the length L is given by $C_m = c_m L$.

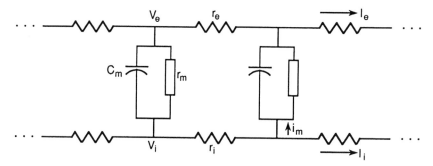

FIGURE 4.4. Electrical model for linear cable.

$$i_m = I_i + c_m \frac{\partial V_m}{\partial t} \tag{4.22}$$

For a linear cable, $I_i = V_m/r_m$, where r_m is treated as a constant, independent of V_m. When the cable equations are applied to an excitable membrane at or near the excited state, the ionic term is instead described by the nonlinear equations introduced in Chapter 3.

An alternative form of cable equation is obtained by combining Eqs. (4.21) and (4.22), and expressing in normalized coordinates as

$$\frac{\partial^2 V_m}{\partial X^2} - V_m - \frac{\partial V_m}{\partial T} = 0 \tag{4.23}$$

where $X = x/\lambda$ and $T = t/\tau_m$ are dimensionless space and time variables scaled to the space constant λ and time constant τ_m of the cell membrane. The scaling constants are given by $\lambda = [r_m/(r_e + r_i)]^{1/2}$, and $\tau_m = r_m c_m$. Since r_m has units of (Ω cm), and r_e and r_i have units of (Ω/cm), it follows that λ has units of centimeters; τ_e has units of seconds. λ is also called the *electrotonic distance* of the membrane; it defines the distance along the membrane that a steady-state voltage disturbance due to point current injection will decay to e^{-1} of the value at the disturbed location. τ_m similarly defines the time response of an isolated piece of membrane when a step voltage is applied. In general, $r_i \gg r_e$ and it follows that $\lambda \approx \sqrt{r_m/r_i}$. Taking an example from Jack et al., (1983, pp. 23–24), for a 20-μm-diameter fiber with $r_i = 30\,\mathrm{M\Omega/cm}$ and $r_m = 1.6\,\mathrm{M\Omega\,cm}$, the space constant is calculated to be 0.23 cm. Space constants for invertebrate nerve are in the range 0.23 to 0.65 cm (Rall, 1977).

Consider a cable of finite length $2L$ placed in a longitudinal static field of strength E. The steady-state solution for membrane voltage in response to a nonfluctuating field is given by (Sten-Knudsen, 1960)

$$V_m(X) = -E\lambda \left(\frac{\sinh X}{\cosh L/\lambda} \right) \tag{4.24}$$

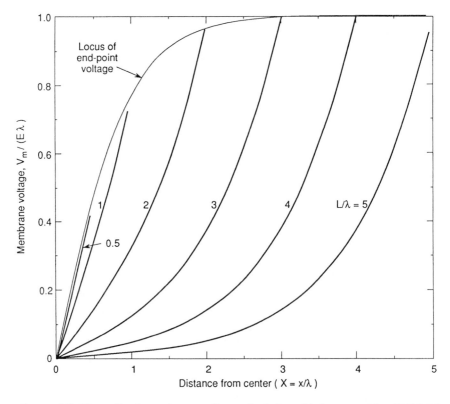

FIGURE 4.5. Normalized membrane voltage of a finite cable immersed in a DC field of strength E. Cable length $= 2L$. Voltage has odd-function symmetry about $X = 0$.

where $X = 0$ is taken as the center of the fiber, and the ends are at $\pm L$. Transient solutions of the membrane voltage are presented by Rall (1977). Figure 4.5 illustrates Eq. (4.24) as a function of distance for several cable lengths. V_m has odd-valued symmetry about $X = 0$, and only one-half of the function is displayed in Fig. 4.5. The symmetrical property means that the fiber is hyperpolarized along its anode-facing half, and is depolarized along its cathode-facing half. The maximum membrane voltage is attained at the ends of the fiber, and has the value

$$V_m = -E\lambda \tanh \frac{L}{\lambda} \tag{4.25}$$

For very long cells ($L \to \infty$), the terminus membrane voltage is $E\lambda$. But even for fibers of only modest length, that value is closely approached. For example, with $L/\lambda = 2$ (total length $= 4\lambda$), the membrane voltage at the two ends is $\pm 0.964E\lambda$. For an ideal nonconducting membrane, $\lambda \to \infty$, and Eq. (4.25) reduces to $V_m = EL$, which is the maximum possible voltage that can be developed across the membrane of an elongated cell of length

$2L$ by a static or low frequency extra cellular electric field. For an ideal spherical cell of radius r, the membrane voltage is $V_m = 1.5\ Er$ (Carstensen, 1987).

Figure 4.6 gives a physical interpretation of the distribution of current flow around an elongated cell that is placed in a medium having a uniform electric field (i.e., uniform current density). The fiber is presumed to be oriented parallel to the undisturbed field. The flux lines indicate that the current through the membrane is greatest near the ends of the fiber. At a sufficient distance from the ends, internal and external fields become equal, and there is no further current crossing the membrane.

Results from cable theory show that a longitudinal electric field (or current flow) is required to excite a nerve fiber. Other theoretical studies (McNeal, 1976) demonstrate that current flow oriented perpendicular to the long axis of a nerve fiber is relatively ineffective compared with a longitudinal orientation. These expectations are verified by a variety of experimental studies (Ranck, 1975; Bawin et al., 1986) showing that both peripheral and CNS neurons are very much less sensitive to a transverse field, as in Fig. 4.7a, as compared with a longitudinal field, as in Fig. 4.7b.

Excitatory effects with CNS neurons are also found to occur with a parallel alignment of the field and the dendrosomatic axis, but to be insensitive to a perpendicular orientation (Bawin et al., 1986). However, the same experimenters found that synaptic interactions do not depend on the orientation of the dendrosomatic axis, presumably as a result of the branching structure of the dendritic tree. This finding stands in contrast to electrical stimulation of synapses within the retina, in which a parallel orientation of the field and cell axis is required to produce phosphenes (see Sect. 9.8).

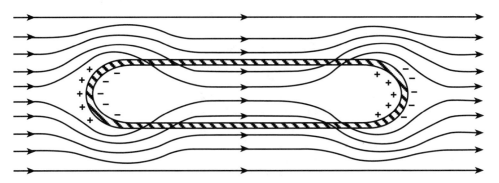

FIGURE 4.6. Representation of current flow around elongated cell placed in a medium having a uniform electric field (i.e., uniform current density). The membrane is assumed to be semipermeable to current flow.

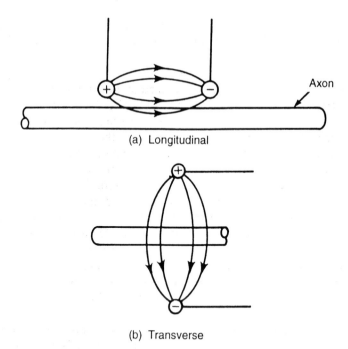

(a) Longitudinal

(b) Transverse

FIGURE 4.7. Longitudinal and transverse current excitation. An excitable cell is much less sensitive to a transverse field (b) as compared with a longitudinal field (a).

Nonlinear Models

Various computational models have been used to study the excitation properties of nerve fibers. Excitation properties have been studied for a spatially isolated segment of membrane, in which the current density crossing the membrane is the driving force. In these models, the membrane conductance is modeled by the Hodgkin–Huxley or Frankenhauser–Huxley equations that were introduced in Chapter 3. Despite simplifications, isolated membrane models have provided substantial insights into neuronal excitation by externally applied currents (Bütikoffer and Lawrence, 1978; Dean and Lawrence, 1983, 1985; Motz and Rattay, 1986).

More complete representations include assumptions about mutual interactions among adjacent segments of the excitable membrane. One such model, developed by Cooley and Dodge (1966), uses a discrete cable model to represent an unmyelinated fiber, with stimulation by an intracellular electrode. In this model, membrane conductance is governed by the Hodgkin–Huxley equations. Myelinated fiber models studied by Fitzhugh (1962) and Bostock (1983) include in a cable model the passive circuit role of the myelin internode, in addition to the active FH conductance's at the nodes. These models have also been configured to study excitation by current injection at a single point on the axon.

A difficulty with the aforementioned nonlinear models is that they presume knowledge of the current waveform and density crossing the membrane. In electrical stimulation problems we may be able to calculate the current within the medium containing the neuron, but not necessarily the current crossing the membrane. The waveform of current crossing the membrane can differ substantially from that in the surrounding medium (McNeal, 1976). Furthermore, the force driving current into the membrane is the external field distribution along the axon, which cannot be described by the current density at a single point. These difficulties have been removed in the myelinated fiber model of McNeal, described in the following section.

4.4 Myelinated Nerve Model

Myelinated A fibers are distinguished from unmyelinated C fibers by faster conduction rates, shorter action-potential (AP) durations, and lower electrical thresholds (Ruch et al., 1968). Because of the lower thresholds, the myelinated fiber is a good choice for electrical stimulation studies.

Figure 4.8 illustrates electrical stimulation of a myelinated nerve fiber. The current emanating from the stimulus electrode through the conducting medium causes external voltage disturbances ($V_{e,n}$) at the nearby nodes. These disturbances force current across the membrane.

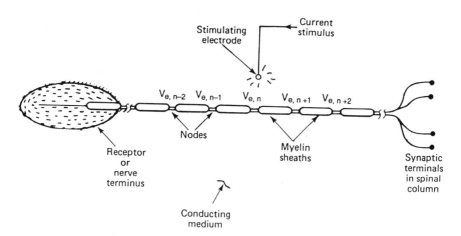

FIGURE 4.8. Representation of electrical stimulation of myelinated nerve. The current stimulus results in voltage disturbances (V_e) at the individual nodes. These in turn cause a local depolarization of the nerve membrane. (Adapted from Reilly and Larkin, 1984.)

Figure 4.9 illustrates the representation of the myelinated nerve as formulated by McNeal (1976). The individual nodes are shown as circuit elements consisting of capacitance (C_m), resistance (R_m), and a potential source (E_r), which maintains the transmembrane resting potential. The voltages $V_{e,n}$ are the external nodal voltages as in Fig. 4.8. In the McNeal representation, the myelin internodes are treated as perfect insulators. This framework could be expanded to include passive myelin properties; however, such expansion would add significantly to the complexity of the model. The model is therefore a compromise between relatively simple single-node models and a more complete node-plus-myelin representation. Despite the compromise, the model is able to account for a variety of sensory and electrophysiological effects.

In Fig. 4.9, the current emanating from the nth node is the sum of capacitive and ionic currents, and is related to internal axonal currents by

$$C_m \frac{dV_n}{dt} + I_{i,n} = G_a\left(V_{i,n-1} - 2V_{i,n} + V_{i,n+1}\right) \tag{4.26}$$

where C_m is the membrane capacitance of the node, V_n is the trans-membrane potential at the nth node, $I_{i,n}$ is the internal ionic current flowing in the nth node, and $V_{i,n}$ is the internal voltage at the nth node. In this expression, V_n is taken relative to the resting potential and positive V_n applies to membrane depolarization. Further relationships are given by

$$G_a = \frac{\pi d^2}{4\varrho_i L} \tag{4.27}$$

$$G_m = g_m \pi d W \tag{4.28}$$

$$C_m = c_m \pi d W \tag{4.29}$$

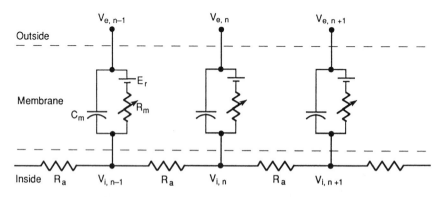

FIGURE 4.9. Equivalent circuit models for excitable membranes. The response near the excitation threshold requires that the membrane conductance be described by a set of nonlinear differential equations. (After McNeal, 1976.)

where d is the axon diameter at the node, ϱ_i is the resistivity of the axoplasm, L is the internodal gap, g_m is the subthreshold membrane conductance per unit area, c_m is the membrane capacitance per unit area, and W is the nodal gap width.

In Equation (4.26), V_n is the voltage difference across the membrane:

$$V_n = V_{i,n} - V_{e,n} \tag{4.30}$$

where $V_{i,n}$ and $V_{e,n}$ are the internal and external nodal voltages, respectively, with reference to a distant point within the conducting medium outside the axon. Substituting Eq. (4.30) into Eq. (4.26) results in

$$\frac{dV_n}{dt} = \frac{1}{C_m}\left[G_a\left(V_{n-1} - 2V_n + V_{n+1} + V_{e,n-1} - 2V_{e,n} + V_{e,n+1}\right) - I_{i,n}\right] \tag{4.31a}$$

Equation (4.31a) may be analogously expressed in continuous form as

$$\tau_m \frac{\partial V}{\partial T} - \lambda^2 \frac{\partial^2 V}{\partial x^2} + V = \lambda^2 \frac{\partial^2 V_e}{\partial x^2} \tag{4.31b}$$

where V and V_e are membrane voltage relative to the resting potential and external voltage respectively at longitudinal position x. A form of Eq. (4.31b), with the constants τ and λ explicitly expressed for a myelinated fiber, has been presented by Basser and Roth (1991). The main way that Eq. (4.31b) differs from the cable equation (4.23) is the inclusion of the right-hand term, in which the external field within the biological medium acts as a driving force on the membrane voltage.

Equation (4.3lb) can be derived from first principles, or can be obtained from (4.31a) by substituting $C_m = c_m \pi d\Delta x$, $G_a = \pi d^2/(4\varrho_i \Delta x)$, $G_m = g_m \pi d\Delta x$, where d is the fiber diameter, Δx is the longitudinal increment, ϱ_i is the axoplasm resistivity (in Ωcm) internal to the fiber, c_m is capacitance per unit area, and g_m is conductance times unit area. Continuous and discrete spatial derivatives are connected by $\partial^2 V/\partial x^2 \approx (V_{n-1} - 2V_n + V_{n+1})/\Delta x^2$; $\partial^2 V_e/\partial x^2 \approx (V_{e,-1} - 2V_{e,n} + V_{e,n+1})/\Delta x^2$; τ_m is the membrane time constant given by c_m/g_m; λ is the membrane space constant given by $\lambda = (r_m/r_i)^{1/2} = (d\varrho_m/(4\varrho_i))^{1/2}$, and ϱ_m is the membrane specific resistance (in Ωcm^2). An additional relationship is $I_{i,n} = V/G_m$.

If one treats λ as a constant, then (4.31b) describes the membrane response only during its sub threshold (linear) phase. For membrane depolarization approaching the threshold of excitation, membrane conductance of ionic constituents becomes highly nonlinear, leading to nerve excitation.

One conclusion that can be drawn from Eqs. (4.31a) and (4.31b) is that a second spatial derivative of voltage (or equivalently a first derivative of the electric voltage) must exist along the long axis of an excitable fiber in order to support excitation. The second spatial derivative of the external voltage has been included in an "activation function" in order to emphasize this essential aspect of stimulation (Rattay, 1986, 1989; Plonsey and Barr, 1995). In a typical analysis, the fiber is considered long and straight. However, excitation is nevertheless possible where the fiber is terminated or where it bends within a locally constant electric field (see Sect. 4.5). The orientation change or termination creates the equivalent of a spatial derivative of the applied field. Stimulation at "ends and bends" can, in fact, be the dominant mode of excitation in many cases (see Sect. 4.5).

The ionic current term in Eq. (4.31a) can be expressed for either a linear or a nonlinear membrane:

$$I_{i,n} = \begin{cases} G_m V_n & \text{(linear)} & (4.32a) \\ \pi d W \left(J_{Na} + J_K + J_L + J_P \right) & \text{(nonlinear)} & (4.32b) \end{cases}$$

Equation (4.32a) is a simple statement of Ohm's law for a linear conductor. Equation (4.32b) applies to the nonlinear ionic current expressions for the FH membrane given by Eqs. (3.28) to (3.31).

In Eq. (4.31a), the $V_{e,n}$ values are specified from knowledge about the stimulus current distribution, and the V_n values are unknowns for which solutions must be found. Procedures for evaluating Eq. (4.31a) typically consider $V_{e,n}$ as being independent of the membrane currents, that is, the current crossing the membrane is assumed not to perturb the voltages on the exterior of the fiber. For subthresholid conditions this is a reasonable assumption, but in an excited state, the membrane currents can be large enough to perturb the external voltages. As a consequence, the model will be less accurate in the excited state. Despite this limitation, the model accurately reproduces a wide range of excitation phenomena.

The most accurate representation would treat all the nodes in the model as nonlinear. However, such treatment can result in excessive computing time, which can be reduced by limiting the number of nonlinear nodes. In McNeal's original work, he studied an 11-node array with one central node as nonlinear and all the others as linear. In his study, excitation current was introduced via an electrode near the central nonlinear node. He defined excitation as occurring when the nonlinear node reached a peak depolarization value of 80 mV. For the range of stimuli studied by McNeal, this arrangement was entirely satisfactory. However, for a more general range of stimulus parameters, some modifications are required.

The model described in this chapter extends the McNeal model by including FH nonlinearities at each of several adjacent nodes. Additional modifications include a test for excitation based on AP propagation,[2] the

---

[2] The use of a single depolarization voltage is not always an adequate indicator of excitation when brief oscillatory stimuli are used. In that case, a threshold test based on propagation is needed (Reilly et al., 1985).

TABLE 4.1. Base-case parameters for SENN model.

Fiber diameter (D)	Variable
Axon diameter (d)	$0.7\,D$
Nodal gap (G)	$2.5\,\text{mm}$
Axoplasmic resistivity (ϱ_i)	$100\,\Omega\text{cm}$
External medium resistivity (ϱ_e)	$300\,\Omega\text{cm}$
Membrane capacity (c_m)	$2\,\mu\text{F/cm}^2$
Membrane conductivity (g_m)	$30.4\,\text{mS/cm}^2$
Internodal distance (L)	$100\,D$

Source: McNeal, 1976.

ability to model arbitrary stimulus waveforms, representation of stimulation at the neuron terminus, and representation of stimulation by uniform electric fields. This modified representation of McNeal's model has been referred to as a *spatially extended nonlinear nodal* (SENN) model (Reilly et al., 1985). Unless specifically noted, myelinated nerve model parameters used to obtain the data in this chapter are as noted in Table 4.1.

To exercise the model, the spatial distribution of voltage along the axon as a result of the stimulating current must be specified. For example, consider the voltage distribution for isotropic current propagation from a point electrode placed in a uniform medium. The "indifferent" electrode is taken to be in the conducting medium, far from the axon. The voltage at a radial distance, r, from the electrode is given by

$$V(r) = \frac{\varrho I}{4\pi r} \tag{4.33}$$

where ϱ is the resistivity of the medium. In the general case, I and V are functions of time that follow the stimulus waveshape.

The examples presented in this section are for a point electrode stimulating a 20-μm-diameter fiber having an internodal spacing of $2\,\text{mm}$; the electrode is assumed to be located one internodal distance ($2\,\mu$m) from the fiber, and centered above the central node. For fibers smaller than $20\,\mu$m, excitation thresholds (at a given electrode distance) would be greater as noted in Sect. 4.7. Section 4.5 presents additional results for excitation by a uniform electric field within the conducting medium.

Threshold Criterion

The membrane response of the myelinated nerve model to a rectangular current stimulus is illustrated in Fig. 4.10. The example is for a small electrode that is $2\,\text{mm}$ radially distant from a 20-μm fiber. The transmembrane voltage, ΔV, is scaled relative to the resting potential. The solid curves show the response at the node nearest the stimulating electrode.

Responses to three different cathodic pulse magnitudes, all of the same duration (100 µs), are depicted. Response a is for a pulse at 80% of the threshold current, stimulus pulse b is at threshold, and pulse c is 20% above threshold. The threshold stimulus pulse in this example has an amplitude (I_T) of 0.68 mA. Figure 4.10 also shows the membrane response to a threshold pulse at the node nearest the electrode and at the next three adjacent nodes (broken lines). The time delay from node to node implies a propagation velocity of 43 m/s.

It is possible for a developing AP to be reversed and entirely abolished by the phase reversal in a biphasic waveform (Bütikoffer and Lawrence, 1978). For this reason, a simple membrane voltage test for AP excitation can be misleading, particularly for stimulus durations less than 100 µs. The difficulty is illustrated in Fig. 4.11, which shows the response to a single cycle of a sinusoidal stimulus having a period of 20 µs and an initial cathodic phase.

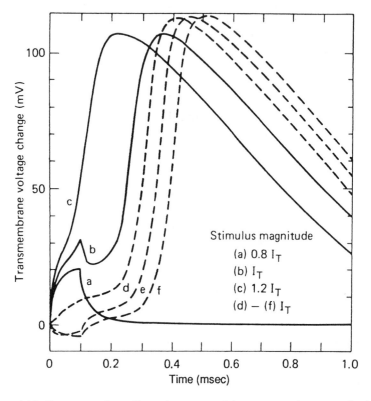

FIGURE 4.10. Response of myelinated nerve model to rectangular monophasic current of 100 µs duration, 20-µm-diameter fiber, point electrode 2 mm from central node. Solid lines show response at node nearest electrode for three levels of current. I_T denotes threshold current. Broken lines show propagated response at next three adjacent nodes for a stimulus at threshold. (From Reilly et al., 1985.)

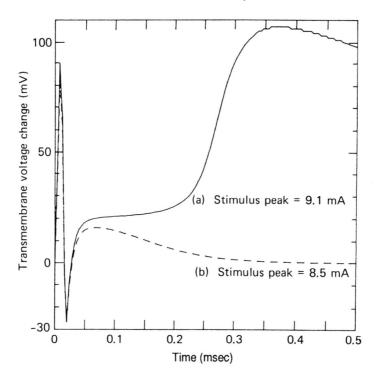

FIGURE 4.11. Nerve model response to one cycle of a sinusoidal pulse with initial cathodic phase and a period of $20\,\mu s$. Stimulus (a) is suprathreshold, stimulus (b) is subthreshold. (From Reilly et al., 1985.)

Stimulus pulse *a*, having a peak amplitude of 9.1 mA, is above threshold: a propagated AP develops after about 0.3 ms. Stimulus pulse *b*, having a peak amplitude of 8.5 mA, is below threshold: the membrane response decays rapidly to the cell's resting potential (0 mV). Both responses, however, exceed a depolarization voltage of 80 mV, at a time corresponding to the peak of the cathodic phase of the stimulus.

In the SENN nerve model, excitation is tested on the basis of unambiguous AP propagation. The computer algorithm recognizes this condition by testing for adequate depolarization (80 mV) that propagates to the third node beyond the point of initial excitation. The threshold current value is determined by iterating between a level causing excitation with a level not causing excitation. The iteration is continued until the threshold and no-threshold levels differ by no more than 1%.

Position and Number of Nonlinear Nodes

The most general form of the nerve model would invoke the FH equations at each of a large number of nodes. However, to reduce computation time,

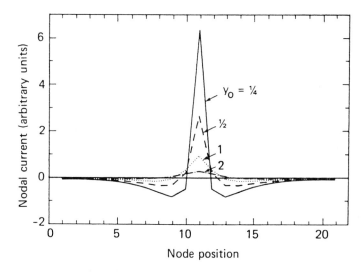

FIGURE 4.12. Steady-state nodal current distribution for myelinated nerve model. Stimulating electrode centered over node 11. For cathodic stimulation, positive nodal current indicates anionic current efflux (cationic flux is in opposite direction). Y_0 denotes distance between electrode and axon as a multiple of internodal distance. (From Reilly et al., 1985.)

one can minimize the number that are nonlinear, that is, those governed by the FH equations. The necessary number of linear and nonlinear nodes depends on the stimulus waveform, the electrode/neuron geometry, and the extent to which AP propagation is modeled.

In previous work, McNeal (1976) evaluated a cathodic point electrode placed one-half an internodal distance radially from the axon. Using one nonlinear node closest to the electrode and five linear nodes on either side, he found that the excitation threshold could be determined within a percent or so relative to a much longer axon. However, this arrangement is not acceptable for anodic or biphasic stimuli, or for more distant electrodes.

To understand the number of nodes required in the model, consider the distribution of membrane current resulting from a stimulus. Figure 4.12 shows a spatial pattern of nodal current in response to a cathodic stimulus. The current distribution plotted here represents a steady-state condition, that is, the convergent response to a long stimulus pulse. Current efflux[3] is represented by the positive current axis, and current influx by the negative

---

[3] In this chapter, unless otherwise specified, "current" flow refers to anionic current according to conventional engineering usage. Cationic current flow is, however, in the opposite direction.

axis. For an anodic stimulus, the figure would be inverted. An action potential results only from the depolarizing effect of nodal current efflux, represented by the positive direction of the vertical axis.

In Fig. 4.12, maximum current efflux with cathodic stimulation occurs at the node nearest the electrode. This point of maximum response is called the excitation node, that is, the node where an AP would be initiated with a threshold-level stimulus. By inverting Fig. 4.12, it can be seen that there are two potential excitation nodes for an anodic stimulus; these move to more distant locations along the axon as the electrode is positioned farther away. In addition, the nodal current distribution becomes more gradual. In modeling anodic stimulation, it is necessary to ensure that nonlinear nodes are included at the excitation location.

Care must also be exercised in the selection of the total number of nodes. The number needed for modeling anodic stimulation is greater than with cathodic stimulation. If there are not enough nodes, the nodal current distribution can be altered relative to a longer axon, and significant errors can be induced. These distortions can be illustrated by calculating the polarity sensitivity ratio (P), defined as the ratio of threshold current for anodic to that for cathodic stimulation (absolute values). Figure 4.13 illustrates P as a function of the electrode placement x_0 measured longitudinally from the terminus of the node array, for various radial distances y_0 from the axon. The array is assumed to be terminated at a node. The parameters x_0 and y_0 are expressed as multiples of the internodal spacing. With this normalization, P is only weakly dependent on the ratio of nodal to axonal resistivity (R_m and R_a in Fig. 4.9), and is otherwise independent of parameters related to fiber size. The pulsewidth is assumed to be long; this incurs no loss of generality.

The value of x_0 at which P oscillates about fixed limits can be interpreted as the array length at which the axon model appears to be infinitely long. This point occurs when x_0 is 9, 13, and 37 nodes from the axon terminus for values of y_0 equal to 2, 4, and 8, respectively. Because these distances represent one half the necessary extent of the axon model, we form the rule of thumb that for general applications, the model should be at least nine times longer than the radial distance to a longitudinally centered electrode. For the computations reported in this chapter, the array length is at least 10 times y_0 but not less than 21 nodes. To study AP propagation or related phenomena (such as AP collision effects), the model requires nonlinear nodes at least as far as propagation is to be simulated.

Suprathreshold Response

An assumption in the myelinated nerve model is that the voltages external to the nodes are defined solely by the stimulus current. However, once the nerve is excited, these external voltages will be perturbed by the nodal current from the propagating AP, and these perturbations are not included

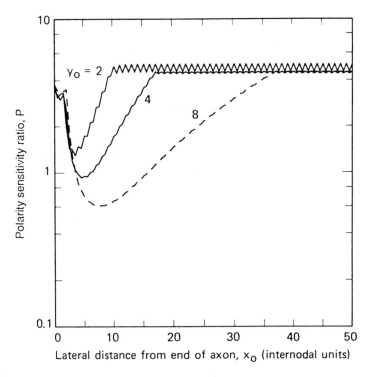

FIGURE 4.13. Ratio of anodic to cathodic threshold current when stimulating electrode is near one terminus of a 91-node axon model. Y_0 denotes distance between electrode and axon as a multiple of internodal distance. Computations were done at multiples of half the internodal distance, resulting in a sawtooth pattern. (From Reilly et al., 1985.)

in the model. Nevertheless, these perturbation potentials will fall to unexited levels at those nodes where the propagating AP has not yet arrived, or has already passed. Consequently, many aspects of suprathreshold behavior can be adequately addressed with the model, such as repetitive or sinusoidal stimulus effects, which will be discussed below.

4.5   Response to Monophasic Stimulation

Strength-Duration Relations

Figure 4.14 illustrates thresholds evaluated with the myelinated nerve model for anodic and cathodic rectangular stimuli with pulse durations of from 1 μs to 10 ms. Figure 4.14 also shows the S-D curve for a cathodic exponential stimulus having the form $Ie^{-t/\tau}$. The curves in Fig. 4.14 are similar to those for the simplified linear model, described by Eqs. (4.6) and

(4.9) for rectangular stimuli, and by Eqs. (4.13) and (4.14) for exponential stimuli. The threshold charge reaches a minimum at small stimulus durations for both exponential and rectangular stimuli; for long-duration stimuli, the peak current (equivalent to rheobasic current) is minimized. The exponential and rectangular stimuli have the same minimum charge threshold, indicating that, at brief durations, sensitivity is not affected by the fine structure of the monophasic pulse waveform.

The shape of the S-D curve for a linear RC network model of a single node can be characterized by the membrane's exponential time constant, given by the product $R_m C_m$ as noted in Sect. 4.2. The myelinated nerve model also has a response time that depends on both linear and nonlinear membrane properties and on the spatial distribution of the electric field external to the axon. Consequently, no single linear circuit parameter specifies the curves in Fig. 4.14. It is therefore useful to define an equivalent strength-duration time constant (τ_e) as the RC time constant of the linear single-node model having an S-D curve shape that best matches the shape of a given empirical curve. The value $\tau_e = 92.3\,\mu s$ is obtained using a linear least-squares fit of Eq. (4.6) to the rectangular current threshold curve in

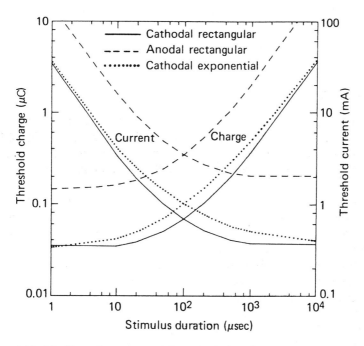

FIGURE 4.14. Myelinated nerve model strength-duration curves for monophasic stimuli. The left vertical axis indicates threshold charge for AP initiation. The right vertical axis indicates threshold current. The horizontal axis represents pulse duration for a rectangular stimulus or decay time constant for an exponential stimulus. (From Reilly et al., 1985.)

Fig. 4.14. This value differs from the simple product of membrane capacitance and resistance in the subthreshold region of linear response, for which the product $R_m C_m$ is only $66\,\mu s$. As suggested by Eq. (4.20), we can more simply estimate τ_e as the ratio Q_0/I_0, which results in $\tau_e = 92\,\mu s$. This estimate agrees quite well with the value obtained by a least-squares fit of the entire S-D curve.

The value of τ_e cited above applies to a point electrode whose radial distance from the axon is equal to an internodal space. As noted in Sect. 4.3, τ_e is expected to increase as the membrane current is distributed more gradually along the axon. We would therefore expect τ_e to be larger at increased electrode distances. This expectation is verified with the model results reported in Sect. 4.7. When the distance to a point electrode is varied from $\frac{1}{2}$ to 4 internodal spaces, the observed value of the τ_e ranges from 56 to $128\,\mu s$.

Values of τ_e obtained with the myelinated model agree quite well with in-vivo experimentation of animal nerve (Reilly et al., 1985), and is within the range of experimental time constants reported for electrocutaneous nerve stimulation (see Table 7.1). Nevertheless, experimental time constants vary over a large range, many being more than twice the values predicted with the nerve model. It is hypothesized that electrocutaneous stimulation may originate at neural end structures (receptors, free nerve endings, and motor neuron end plates). We will further consider models for neural end structures.

Polarity Sensitivity

The vertical separation between anodic and cathodic stimulation in Fig. 4.14 reflects the polarity sensitivity ratio (P) introduced in Fig. 4.13. In Fig. 4.14, P ranges from 4.2 at a pulse duration of $1\,\mu s$ to 5.6 at a pulse duration of $10\,ms$. As noted in Chapter 7, polarity ratios for electrocutaneous stimulation are closer to 1.3.

It is difficult to reconcile P values from sensory experiments with those obtained with the myelinated nerve model as long as the stimulating electrode is positioned far from the end of the model axon. As seen in Fig. 4.13, the minimum value of P is about 4.5 for a variety of electrode/neuron geometries when the electrode is distant from an axon terminus. But as the electrode moves closer to the terminus, the model predicts smaller polarity ratios, and even predicts ratios less than unity in some cases.

Ranck (1975) discusses a qualitative model for stimulation near the terminus of a CNS neuron and shows that it is possible for anodic excitation to occur at lower stimulus levels than with cathodic excitation. I have not encountered polarity ratios less than 1 in electrocutaneous experiments. Nevertheless, Ranck's analysis and Fig. 4.13 suggest that the modest polarity ratios found in electrocutaneous experiments may occur because the principal sites of stimulation are near terminal structures (receptors, free nerve endings). Figure 4.13 shows that P ranges from about 4.5 to less than unity, depending on the electrode/ neuron configuration. With cutaneous stimulation, there will be a mix of orientations of excitable structures in the skin beneath the electrode with a spectrum of polarity ratios. The modest ratios observed in sensory experiments may represent a macroscopic average of this spectrum. A further analysis of this issue is presented in Sect. 4.7.

Current Density and Electric Field Relationships

Recall from cable theory (Sect. 4.3) that a longitudinal electric field is necessary to support excitation. The electric field must also have a spatial gradient, as can be appreciated with reference to Eq. (4.31), which shows that second differences[4] of the external voltages are the driving forces for changes in membrane potential. If the electric field were uniform and the axon were infinitely long in both directions, there would in theory be zero net current transfer at every node. However, the field within the biological medium is never uniform. Furthermore, an effective field gradient will be realized if the orientation of the axon changes with respect to a locally uniform field, or if the axon is terminated in the field (as with receptors, free nerve endings, or nerve connections at muscle fibers).

Figure 4.15 illustrates three modes whereby a nerve fiber may be excited by an external electric field. We designate these as *end*, *bend*, and *spatial gradient modes*. End and bend modes of stimulation have been analyzed by several researchers (Reilly and Bauer, 1987; Coburn, 1989; Nagarajan et al., 1993; Rubinstein, 1993; Struijk et al., 1993; Abdeen and Stuchly, 1994).

The myelinated nerve model can be used to analyze these excitation modes. For a straight fiber terminated at a node and oriented along a uniform electric field of magnitude E, the nodal voltages are

$$V_{e,n} = V_{e,1} + ELn \tag{4.34}$$

where $V_{e,1}$ is a reference voltage at the first node, L is the internodal separation, and n is the node number. The membrane response is independent of $V_{e,1}$, which may be taken as zero for convenience. Equation (4.34) applies if the first node faces the cathode of the current source. Table 4.2 lists monophasic excitation thresholds for unbent fibers of 5, 10, and $20\,\mu m$—effectively covering the diameter spectrum of myelinated fibers. The thresholds are inversely proportional to fiber diameter for this mode of stimulation. For long stimulus pulses (≥ 1 ms), the threshold converges to a minimum E-field value; for short duration pulses ($t \leq 5\,\mu s$), the threshold converges to a minimum value of the product of Et (field times duration). These convergent measures, listed in Part A of Table 4.2, are analogous to

---

[4] A second difference is calculated as $(V_{n-1} - V_n) - (V_n - V_{n+1}) = V_{n-1} - 2V_n + V_{n+1}$, where V_n is the voltage at the nth node.

the rheobase current and charge noted in Fig. 4.14 for a discrete current stimulus. Although the in-situ electric field is the primary force governing stimulation, current density is perhaps a more frequently cited stimulus parameter. In Part B of the table, the thresholds are given as current density values, assuming a bulk conductivity $\sigma = 0.2\,\text{S/m}$. For the simulations represented in Table 4.2, the S-D time constant was found to be $\tau_e = 120\,\mu s$, as determined by a linear least square fit of the threshold data to Eq. 4.6. The thresholds listed in Table 4.2 bracket experimentally determined values if one properly accounts for waveform and geometric factors (see Table 9.7 and related discussion). In all cases, excitation was initiated at the terminus facing the cathode of the electric field source.

We also determined minimum thresholds of excitation for a bent fiber. For a 90° bend at node N, we applied a constant voltage to the first N nodes, and incremented the voltage as in Eq. (4.34) at each successive node above N. For a 180° bend at node N, we first decremented the voltage at successive nodes up to N, and then incremented the voltage for nodes above N. By this method, it was possible to simulate a sharply bent fiber within a uniform E-field. Table 4.3 lists threshold requirements associated with end and bend

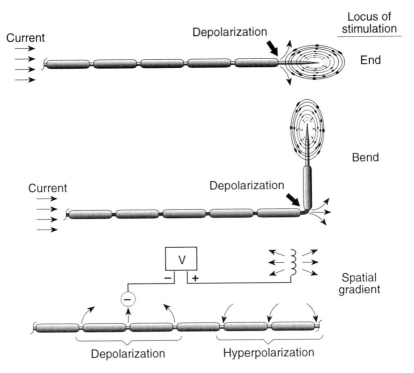

FIGURE 4.15. Modes of neural stimulation. Excitation is initiated at points of maximal current efflux across neural membrane. Potential excitation sites consist of fiber terminals, sharp bends, and maximal spatial gradient of E-field.

TABLE 4.2. Minimum stimulus thresholds with uniform field excitation; single monophasic stimuli.

	Fiber diameter (μm)		
	5	10	20
A. *Field strength criteria*			
1. E_{min} (V/m)	24.6	12.3	6.2
2. $(Et)_{min}$ (Vs/m)	2.98×10^{-3}	1.49×10^{-3}	0.75×10^{-3}
B. *Current density criteria*			
1. J_{min} (A/m^2)	4.92	2.46	1.23
2. q_{min} (C/m^2)	6.0×10^{-4}	3.0×10^{-4}	1.5×10^{-4}

Thresholds A.1 and B.1 apply to long pulses ($t \geq 1$ ms). Thresholds A.2 and B.2 apply to short pulses ($t \leq 5\mu$s). Current and charge density determined for conductivity $\sigma = 0.2$ S/m.
Source: From Reilly (1988).

modes of stimulation. The first column indicates the bend angle. The second column lists the node at which the bend occurs. The first listed case with a bend angle of 0° applies to a terminated fiber, which is the standard case used to evaluate MRI excitation thresholds (see Sect. 9.7). The third column lists the rheobase excitation threshold. The last column lists the strength duration time constant, as determined by the ratio $(Et)_{min}/E_{min}$ obtained with short- and long-duration stimuli.

For the cases listed in Table 4.3, the lowest thresholds apply to a fiber with a sharp 180° bend at a location distant from the terminus. However, such a condition would not be realistic for practical fiber trajectories. If we examine the remaining cases, we see that the straight, terminated fiber provides the lowest practical threshold. For the bent fiber cases, excitation was initiated at the bend node. Note that gradual bends would necessitate

TABLE 4.3. Excitation requirements for end and bend modes of stimulation.

Bend angle (deg.)	Bend node (#)	E threshold (V/m)	τ_e (μs)
0	1	6.21	128.2
90	2	8.55	126.0
90	4	9.84	114.3
90	6	9.96	112.9
90	8	9.96	112.9
180	2	6.56	101.4
180	4	5.45	105.0
180	6	5.10	110.3
180	8	5.04	111.6

Thresholds apply to 20-μm nerve fiber within constant E-field that is oriented parallel to the nerve beyond the bend point.

higher excitation thresholds, since the second derivative of voltage would necessarily be lower as compared with a sharp bend.

As will be apparent in succeeding chapters, the predicted rheobase for a 20-μm fiber in the terminated mode closely corresponds to thresholds obtained in many experiments with humans and animals. For instance, Havel and colleagues (1997) reported perception thresholds with magnetic stimulation through a coil encircling the arm of human subjects. At the threshold of perception, the rheobase E-field induced in the periphery of the arm was determined to be 5.9 V/m—which is quite close to the theoretical value of 6.2 V/m for a 20-μm myelinated fiber.

4.6   Response to Biphasic and Repetitive Stimuli

Strength-Duration Relationships

The current reversal of a biphasic pulse can reverse a developing AP that was excited by the initial phase. As a result, a biphasic pulse may have a higher threshold than a monophasic pulse. Figure 4.16 shows S-D curves from the myelinated nerve model for three types of stimuli: a monophasic constant current (rectangular) stimulus, a symmetric biphasic rectangular stimulus, and a sinusoidal stimulus. The data apply to stimulation via a point electrode 2 mm radially distant from a 20-μm-diameter fiber. Stimuli consist of a single biphasic current with an initial cathodic phase followed by an anodic phase of the same duration and equal magnitude. The phase duration indicated by the horizontal axis is that for the initial cathodic half-cycle. Stimulus magnitude is given in terms of peak current on the right vertical axis, and in terms of the charge in a single monophasic phase of the stimulus on the left vertical axis. The charge is computed by $Q = It_p$ for the rectangular waveforms and from $Q = (2/\pi) It_p$ or the sinusoidal waveforms (I is threshold current amplitude and t_p is phase duration).

The current reversal in the biphasic waveforms increases the threshold for a propagating AP. This situation can be seen in Fig. 4.16 by comparing the monophasic and biphasic rectangular stimulus thresholds. For long durations, the threshold current is the same for the two stimuli. However, as the stimulus duration becomes short relative to the S-D time constant (about 100 μs), the biphasic current has an elevated threshold. The degree of elevation is magnified as the stimulus duration as reduced.

The thresholds illustrated in Fig. 4.16 apply to stimuli with an initial cathodic phase. Thresholds are greater if the initial phase is anodic, but only if the phase duration is less than 100 μs. For a single cycle of a sine wave, the model shows that if the initial phase of the stimulus is anodic, thresholds are greater than initial cathodic thresholds by 5% at a phase duration of 100 μs, 10% at 50 μs, 45% at 10 μs, and 60% at 5 μs.

Figure 4.17 illustrates threshold multipliers (M) based on the myelinated nerve model for biphasic rectangular pulses with uniform field excitation

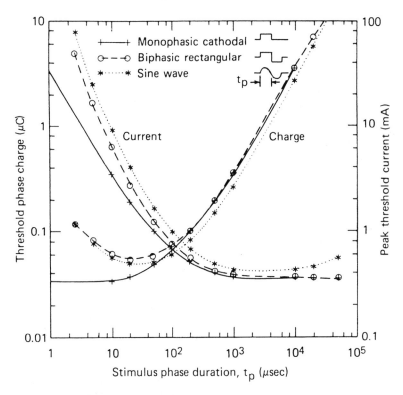

FIGURE 4.16. Strength-duration relationships derived from the myelinated nerve model: current thresholds and charge thresholds for single-pulse monophasic and for single-cycle biphasic stimuli with initial cathodic phase, point electrode 2 mm distant from 20 μm fiber. Threshold current refers to the peak of the stimulus waveform. Charge refers to a single phase for biphasic stimuli. (From Reilly et al., 1985.)

(Reilly, 1988). The vertical axis gives the threshold multiplier for a double pulse relative to a single pulse. The portion of the figure above $M = 1$ applies to a biphasic pulse doublet, where the current reversal has the same magnitude and duration as the initial pulse. The portion of Fig. 4.16 below $M = 1$ is for a monophasic pulse doublet. Figure 4.17 applies if the initial pulse is cathodic. Stimulation is also possible with an initial anodic pulse, but the thresholds are elevated.

According to Figure 4.17, biphasic thresholds are elevated by an amount that depends on the pulse duration and the time delay before current reversal. Thresholds are most elevated when the pulse is short and the current reversal immediately follows the initial pulse. If the phase reversal is delayed by 100 μs or more, there is little detectable effect on the threshold. An implication of the results shown in Figs. 4.16 and 4.17 is that the membrane integrates the stimulus over a duration roughly equal to the S-D

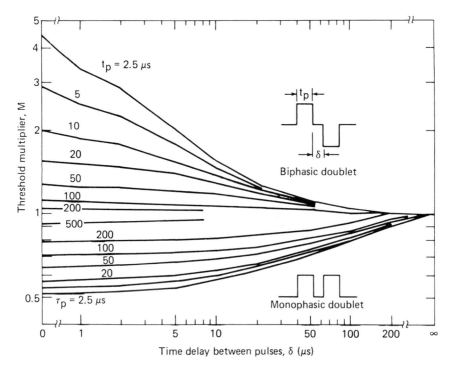

FIGURE 4.17. Threshold multipliers for biphasic and monophasic pulse doublets: Uniform field excitation of terminated axon. (From Reilly, 1988.)

time constant. This integration is nonlinear: the biphasic waveforms all inject zero net charge, but have finite threshold magnitudes. Various features observed with biphasic stimulation of the nerve model are also seen experimentally in cardiac (Chapter 6) and nerve excitation (Chapters 7 and 8), as long as appropriate adjustments are made for experimental strength-duration time constants.

Sinusoidal Stimuli

Figure 4.16 illustrates excitation thresholds for sinusoidal stimuli consisting of a single cycle. The excitation threshold also depends on the number of stimulus cycles. Figure 4.18 illustrates the response of the myelinated nerve model to threshold-level sinusoidal stimuli at a frequency of 5 kHz (point electrode 2 mm distant from 20-μm fiber). The stimulus is composed of a single cycle in example a, two cycles in b, and three cycles in c. The threshold current requirement is reduced according to the data above the figure as the number of stimulus cycles is varied from one to three.

Figure 4.19 illustrates the relationship between threshold and stimulus duration for a sinusoidal current with frequency from 25 to 400 kHz; the initial phase was cathodic in these examples. When evaluated at half-cycle multiples, there is an oscillating threshold with minima at odd numbers of half-cycles and maxima at even multiples. The broken lines in Fig. 4.19 show the duration of the stimuli. For the cases displayed, thresholds at stimulus durations beyond 1.28 ms converge to a minimum plateau that is within a few percent of the threshold that would apply to a cathodic monophasic square-wave pulse having a duration equal to the phase duration (one-half period) of the sinewave stimulus. This observation is an emphirical one. As noted in Sect. 9.7, threshold behavior similar to that in Fig. 4.19 has been experimentally observed in human subjects using variable length sinusoidal magnetic stimulation (Budinger et al., 1991).

The thresholds in Fig. 4.19 oscillate because a sinusoidal stimulus that has an even number of half-cycles is charge balanced (zero net charge), and one with an odd number of half-cycles will transfer the maximum net charge.

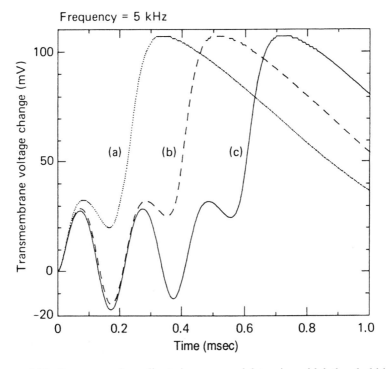

FIGURE 4.18. Response of myelinated nerve model to sinusoidal threshold-level stimuli at a frequency of 5 kHz. Stimulus durations: (a) one cycle, (b) two cycles, and (c) three cycles. Peak values of threshold currents are listed above figure. (From Reilly et al., 1985.)

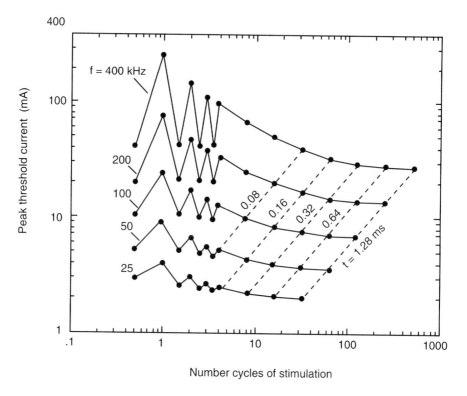

FIGURE 4.19. Excitation thresholds as a function of number of cycles of sinusoidal stimulation. Stimulus duration stepped in half-cycle increments out to four cycles, and full cycle increments beyond that point. Dashed lines indicate duration of stimulus. Point electrode 2 mm distant from 20 μm fiber.

The gradually falling aspect of the threshold is a consequence of the non-linear conductance of the membrane. Recall from Chapter 3 that as the membrane depolarization approaches the excited state, ionic conductance for inward- and outward-going current differs. As a consequence, an oscillating stimulus will build up a bias voltage on the membrane that increases with each successive cycle of stimulation, as experimentally verified with biphasic stimulation of both nerve and cardiac tissue (see Chapters 6–8).

Single monophasic pulses produce a single AP. An oscillating stimulus has the ability to produce a train of APs, which can greatly enhance the intensity of the electrical response in nerve (Chapter 7), cardiac (Chapter 6), and skeletal muscle (Chapter 8) responses. Repetitive responses can be predicted from the nonlinear models described in Chapter 3 for both myelinated and unmyelinated nerves.

Figure 4.20 illustrates the response of the myelinated nerve model to continuous sinusoidal stimulation. The axon response is shown two nodes distant from the excitation node in order to verify AP propagation. The stimulus levels indicated are multiples of the single-cycle threshold. The AP repetition rates are 100, 250, and 500 Hz for threshold multiples of 1.0, 1.2, and 1.5, respectively. In each case, APs are synchronized with the stimulus. Similar phase locking has been observed in the neural responses of sensory systems to periodic stimulation (Kiang, 1965). Responses were also studied with the myelinated nerve model for continuous sinusoidal stimulation at 5 kHz. The AP repetition rates were 320, 470, and 540 Hz at threshold multiples of 1.0, 1.2, and 1.5, respectively. At the 1.5 stimulus multiple, there was a 4.4-ms refractory period after the first AP during which excitation did not occur. After this initial refractory period, a steady AP rate of 540 Hz was produced. At both frequencies there is a reduction in magnitude of the AP spike as the AP repetition rate increases. The maximum AP rate and

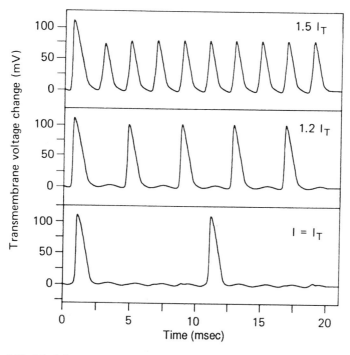

FIGURE 4.20. Model response to continuous sinusoidal stimulation at 500 Hz. The lower panel depicts the response to a stimulus current set at threshold level (I_T) for a single-cycle stimulus. Upper panels show response for stimulation 20% and 50% above the single-cycle threshold. (From Reilly et al., 1985.)

reduced depolarization voltage are consistent with the experimental
A-fiber response illustrated in Fig. 3.11.

Sinusoidal threshold response can be represented by strength-frequency
(S-F) curves as shown in Fig. 4.21. The horizontal axis in Fig. 4.21 is the
inverse of twice the phase duration in Fig. 4.16. Figure 4.21 also shows a
threshold curve for continuous sinusoidal stimulation.

Figure 4.21 includes experimental threshold curves for human perception
and muscle contraction (Dalziel, 1972; Anderson and Munson, 1951). The
experimental curves have been arbitrarily scaled on the vertical axis to
facilitate comparison of the curve shapes. The shapes of the experimental
data and continuous stimulation model results correspond reasonably well
considering that continuous simulation was used in the cited experimental
studies. Sinusoid thresholds rise below 40 Hz—at low frequencies, the slow

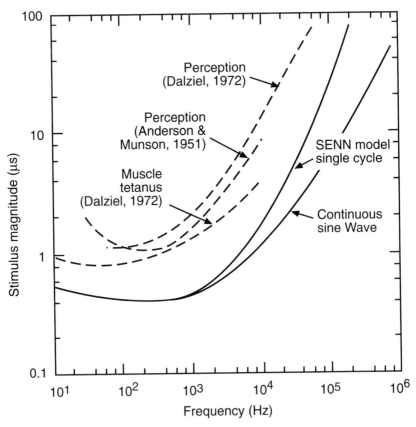

FIGURE 4.21. Strength-frequency curves for sinusoidal current stimuli. Dashed
curves are from experimental data. Solid curves apply to myelinated nerve model.
Experimental curves have been shifted vertically to facilitate comparisons. (From
Reilly, 1988.)

rate of change of the sinusoid prevents the membrane from building up a depolarizing voltage because membrane depolarization is counteracted by membrane leakage. Square-wave biphasic stimuli, in contrast, do not have a rate of change that depends on frequency. Consequently, there is no increase in thresholds at low square-wave frequencies.

To a first approximation, a functional S-F relationship might be obtained from a monophasic S-D curve by considering that a sinusoidal half-cycle corresponds to a single monophasic pulse. This simplification would, however, fail to account for the increased sinusoidal thresholds at both high and low frequencies relative to monophasic stimulation. An empirical fit to S-F data from the myelinated nerve model is given by

$$I_t = I_0 \left[1 - \exp\left(-\frac{f_e}{f} \right) \right]^{-a} \left[1 - \exp\left(-\frac{f}{f_0} \right) \right]^{-b} \tag{4.35}$$

where I_t is the threshold current, I_0 is the minimum threshold current, and f_e and f_0 are constants that determine the points of upturn in the S-F curve at high and low frequencies, respectively. An upper limit on the low frequency term (second bracket) of K_{DC} is assumed in Eq. (4.35) to account for the fact that excitation may be obtained with finite direct currents. Equation (4.35) has the asymptotic form

$$I_t \approx I_0 \left(\frac{f}{f_e} \right)^a \quad for f \gg f_e \tag{4.36}$$

and

$$I_t \approx I_0 \left(\frac{f_0}{f} \right)^b \quad for f \ll f_0 \tag{4.37}$$

Sinusoidal thresholds were obtained with the myelinated nerve model for stimulation by a point electrode positioned 2 mm radially distant from a 20-μm-diameter fiber. An empirical fit of Eq. (4.35) to the model thresholds indicates b = 0.8, and that below 80 kHz, a = 1.45 for single cycle stimulation, and a = 0.9 for continuous stimulation. From 80 to 400 kHz, a = 1.7 for single cycle, and a = 1.0 for continuous stimulation. Frequency constants for the nerve model were f_e = 5,400 Hz and f_0 = 10 Hz. Experimentally derived values of f_e and f_0 encompass wide range, as noted in Chapters 6 and 7. f_0 in the range 10 to 50 Hz is observed for nerve stimulation, and around 10 Hz for cardiac stimulation. Experimental geometric mean values of f_e are about 500 Hz for nerve excitation (Chap. 7), and about 115 Hz for cardiac excitation (Chap. 6). Differences in model an experimental values reflect the dynamic membrane responses of the different types of tissue being stimulated, as well as the distribution of stimulating current in an experimental situation.

Equation (4.35) has features in common with an S-F expression proposed over sixty ago by Hill and colleagues (1937), which can be equivalently stated as

$$\frac{I}{I_0} = \left[\left(1 + \frac{f_0^2}{f^2}\right)\left(1 + \frac{f^2}{f_e^2}\right)\right]^{1/2} \qquad (4.38)$$

The expression of Hill et al. involves upper and lower transition frequencies, as does Eq. (4.35), and has the asymptotic forms of Eqs. (4.36) and (4.37) with $a = b = 1$. The lower transition frequency was described by Hill et al. as arising from the neural property known as *accommodation*, that is, the adaptation of a nerve to a slowly varying or constant stimulus.

The frequency limits for which Eq (4.35) is valid are not known. While human sensory experiments with sinusoidal currents closely agree with the form of Eq. (4.35) up to at least 100 kHz (see Sect. 7.5), one could infer experimental verification up to perhaps 10 MHz based on sensory thresholds with pulsed stimuli—as noted in Chap. 7, the strength-duration law applies to human sensory data for capacitor discharges as brief as $0.1\,\mu$s. Experiments with rats show reasonable correspondence up to 1 MHz, as seen in Fig. 4.22 (LaCourse et al., 1985). The figure plots current thresholds (peak-to-peak) for excitation of rat's tibialis caudalis muscle, gastrocnemium muscle, or the innervating nerve of the gastrocnemius. Electrodes were wire loops placed directly on the muscle or nerve; excitation was determined by contraction of the affected muscle. The curve plots means for 7 to 10 subjects. Also plotted on the figure is a curve in which thresholds rise directly with frequency, as would be expected from the theoretical SENN model described previously. The curves taken together demonstrate a reasonable correspondence with the theoretical model. Deviations from the curve might be due to experimental difficulties, including the fact that the stimulus wire produced significant heating—the experiments were taken to the limits beyond which tissue destruction would have occurred as a result of thermal damage. The authors could have tested electrical stimulation without excessive tissue heating by limiting the duration of the sinusoidal current to a few milliseconds, although this apparently was not done.

Electrical stimulation of nerve has been observed during electrosurgery using devices that operate at frequencies in the vicinity of 500 kHz and above (LaCourse et al., 1985). For such high frequencies, one might expect that it would require unrealistically high current density based on the strength-frequency relationship for nerve excitation. A more likely explanation is that excitation during electrosurgery is due to a nonlinear sparking process at the cutting site. This process effectively rectifies the high frequency current, producing DC or low frequency current in the tissue that is effective for nerve excitation (Tucker et al., 1984; LaCourse et al., 1988, Slager et al., 1993). Others have suggested that nonlinear impedance of

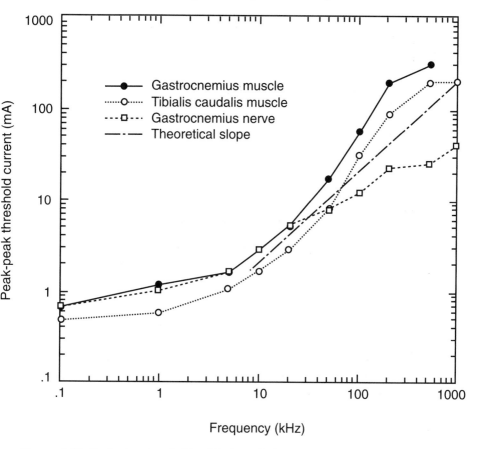

FIGURE 4.22. Excitation thresholds with sinusoidal current, 100 Hz—1 MHz. Stimulation via wire electrode contacting muscle or nerve. (Data from LaCourse et al., 1985).

tissue might provide a demodulation of low-frequency amplitude fluctuations on the otherwise high-frequency stimulus (Geddes et al., 1975).

Repetitive Stimuli

Repetitive stimuli can be more potent than a single stimulus through threshold reduction or through response enhancement due to multiple AP generation. In both cases an integration effect of the multiple pulses occurs. In the first case, the integration takes place at the membrane level. In the second case, response enhancement takes place at higher levels within the central nervous system for neurosensory effects, and at the muscle level for neuromuscular effects. The following considers membrane effects using the myelinated nerve model. Other integration effects are described in Chapters 7 and 8.

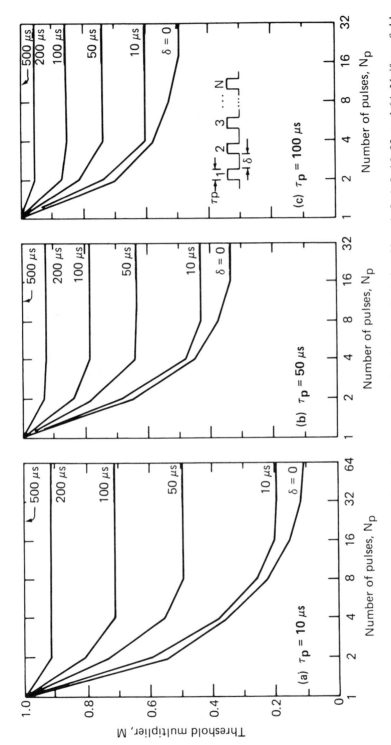

FIGURE 4.23. Threshold multiplier for repetitive pulse sequences—thresholds evaluated at $N_p = 1, 2, 4, 8, 16, 32,$ and 64. Uniform field excitation. (From Reilly, 1988).

The lower section of Fig. 4.17 ($M < 1$) illustrates multiple-pulse threshold effects for two pulses of the same polarity. The figure gives the threshold multiplier when there are two pulses of stimulation relative to that for a single pulse. The effects are most pronounced for short pulses and short interpulse delays. (A delay of zero is the same as a single pulse of twice the duration.) The nerve model was also exercised to evaluate the threshold modification for sequences of pulses, N_p, numbering 1, 2, 4, 8, 16, 32, 64, and 128. Figure 4.23 illustrates the results for $t_p = 10, 50$, and $100\,\mu s$ and for $\delta = 10, 50, 100, 200$, and $500\,\mu s$. As in Fig. 4.17 the vertical axis gives the threshold multiplier relative to a single pulse. The curve labeled $\delta d = 0$ applies to continuous stimulation over a time period corresponding to N_p. Pulse integration reduces thresholds increasingly so as the pulse duration and delay time are shortened. At a delay of $500\,\mu s$, the nerve model shows no measurable pulse integration effect.

Nonlinear Interaction of Multiple Waveforms

A linear system is one in which its response is proportional to the magnitude of an input stimulus. Clearly, the neural membrane is highly nonlinear when depolarized near its excitation threshold, where a slight change in polarization can trigger a propagating action potential. Even at potentials well below the action potential threshold where the membrane is only slightly nonlinear, one can observe nonlinear electrical response. For instance with an oscillatory stimulus, the membrane potential achieves a bias potential that asymptotically increases with each cycle of stimulation due to the differing membrane conductivity to inward and outward current. Nonlinear membrane properties are responsible for the variation of membrane potential and excitation thresholds with respect to the duration of a sinusoidal stimulus as seen in Figs. 4.18, 4.19, and 4.21.

As with all nonlinear systems, the response of the neural membrane to multiple stimuli will exhibit properties not observable in the response of the individual waveforms. That is why it is invalid to evaluate neural response to a multispectral stimulus waveform in terms of its response to the individual Fourier components of the waveform. While such "superposition" techniques are valid for linear systems, they are incapable of accounting for the response of a nonlinear system.

Nonlinear response of the neural membrane to multiple waveforms can best be studied with a nonlinear electrodynamic model, such as the SENN model described previously. As an example of such behavior, Figure 4.24 illustrates the threshold response to a dual stimulus consisting of a 0.2 ms square-wave "conditioning pulse" (CP), and a sinusoidal function. The simulation applies to a terminated 10-μm myelineated nerve in an E-field that is aligned with the fiber. The ordinate gives the peak threshold value of a sinusoidal E-field having a duration of 1 ms, and a frequency indicated by the abscissa. The CP is applied at a subthreshold level either preceding the

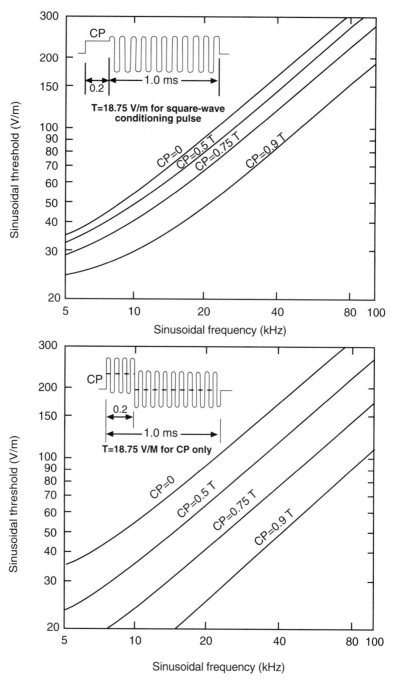

FIGURE 4.24. Excitation thresholds with dual-function stimulation of 10-μm fiber. CP = conditioning pulse at various fractions of excitation thresholds. Upper panel: sequential stimuli; lower panel: concurrent stimuli.

sinusoid (upper panel), or conjointly with it (lower panel). The parameter values indicate the strength of the CP as a fraction of its threshold value of 18.75 V/m if presented alone. For instance, CP = 0.5 means that the conditioning pulse attains an E-field of 9.37 V/m. The curve CP = 0 indicates the strength-frequency response of the neuron with a pure sinusoidal stimulus.

It can be seen that the CP can drastically lower the excitation threshold of the sinusoidal stimulus, especially if the two functions occur concurrently. For instance at frequencies above 10 kHz, a concurrent CP at 0.5, 0.75, and 0.9 will lower the excitation threshold of the sinusoidal stimulus by factors of 0.60, 0.47, and 0.29 respectively. These properties can be used to advantage in the design of focal magnetic stimulators, as described in Sect. 9.9.

Other aspects of nonlinear response have been demonstrated in experiments of human visual and auditory response to stimuli consisting of pairs of frequencies (Adrian, 1977). In these experiments, subject response depended on the difference frequency of two sinusoids mixed together, even though the individual frequencies had much higher thresholds if presented singly. These experiments are discussed in greater detail in Sect. 9.8. Sensitivity to difference frequencies are a characteristic of nonlinear systems, and that is a likely explanation for the observations in these experiments. The site of the nonlinear action was not explored in these experiments. It would be useful to examine whether the electrodynamics of the nerve membrane are responsible for such results.

4.7   Parameter Variation Effects

To further explore mechanisms for electrocutaneous stimulation, a parameter variation study was conducted with the myelinated nerve model (Reilly and Bauer, 1987). The membrane parameters varied were: fiber diameter (d), nodal gap width (W), internal axonal conductivity (g_a), membrane capacity (c_m), and membrane conductance (g_m). These were individually varied by factors of one half or two times the base-case value, while holding all other base-case parameters constant. For this study, the base-case fiber diameter was 10 μm; other parameters were as in Table 4.1. Geometric factors varied relative to the base case were radial distance from the central node and longitudinal distance between the electrode and the fiber terminus.

Four aspects of model response were evaluated in the parameter variation study:

1. Minimum charge (Q_0)—the AP initiation threshold in units of charge of a short duration pulse (1 μs). Charge is determined by the product of pulse width and current magnitude.

2. Minimum current (I_0)—the AP initiation threshold in units of peak current for a long pulse (2 ms).
3. S-D time constant (τ_e)—determined by a least-squares fit of the threshold currents for AP initiation to the theoretical S-D curve for an ideal linear membrane. (See Reilly and Larkin, 1983).
4. Polarity sensitivity ratio (P)—the absolute value of the AP initiation thresholds for anodic versus cathodic stimuli. P was determined separately for stimulus durations of 10 and 1,000 μs.

Table 4.4 summarizes results from the parameter variation study. The first row lists data for the base case in absolute units. The remaining entries in the table are expressed as multiples of the base-case values. Part A lists results for variations of fiber parameters that were set to one half or two times the base case value. In varying D, we kept the electrode distance equal to one internodal space. As a result, electrode separation, expressed in absolute distance units, varied with D. Part B lists results for variations in the radial distance (y_0) between the electrode (positioned over the center node) and the fiber; y_0 is expressed as a multiple of internodal units and is a unitless quantity. Part C lists results for variations in longitudinal distance with respect to the truncated end of the axon; $x_0 = -1$ means that the electrode was positioned one internodal unit beyond the terminus; $x_0 = 2$ and 5 means that the electrode was positioned above the second and fifth nodes, respectively, from the end.

In Part A of Table 4.4, we see that P is quite insensitive to the choice of membrane parameters. And except for c_m, the membrane parameter variations have only a modest effect on τ_e. In Part B, we see that the radial separation of the electrode from a central node has a significant effect on τ_e. The fact that τ_e falls with reduced distance is consistent with the theoretical expectations discussed in Sec. 4.3. In Part C, we see that the value of P can be significantly lowered when the electrode is placed near the truncated end of the axon. The degree of this reduction places the model values of P much more in conformance with experimental data from electrocutaneous sensory experiments.

Only passive membrane parameters were varied in the study. Bostock (1983) also examined the effects of variations in active membrane parameters. That study used a model for current injection at a single point on the model axon, rather than excitation through an external electrode as in the myelinated nerve model. As a result, it is difficult to compare Bostock's results with the myelinated nerve model. Despite differences in model assumptions, Bostock's results provide some guidance on the potential sensitivity of τ_e to active membrane parameter variations. Of the parameters examined by Bostock, the sodium activation rate constant had the greatest effect: for a factor of two reduction in the activation rate constant, there resulted at most a 21% increase in the value of τ_e.

TABLE 4.4. Summary of parameter variation effects.

Base case	Q_0 15.9 nC	I_0 0.18 mA	τ_e 92.3 μs	P for $t = 0.01$ ms 4.66	P for $t = 1.0$ ms 5.53
A. Variation of membrane parameters					
$D \times \frac{1}{2}$	0.50	0.50	1.00	1.00	1.00
$\times 2$	2.00	2.00	1.00	0.99	1.00
$G \times \frac{1}{2}$	0.76	0.82	0.93	0.98	1.10
$\times 2$	1.45	1.28	1.13	0.95	0.96
$r_i \times \frac{1}{2}$	0.70	0.99	0.71	1.00	1.00
$\times 2$	1.51	1.11	1.50	0.99	0.98
$c_m \times \frac{1}{2}$	0.70	0.99	0.71	1.00	1.00
$\times 2$	1.51	1.11	1.50	0.99	0.98
$g_m \times \frac{1}{2}$	1.27	1.12	1.13	0.98	0.98
$\times 2$	0.82	0.87	0.93	0.99	1.01
B. Variation of radial distance—electode centered along axon					
$y_0 = \frac{1}{4}$	0.12	0.14	0.61	1.06	1.24
$y_0 = \frac{1}{2}$	0.31	0.34	0.90	1.06	1.18
$y_0 = 1$	1.00	1.01	1.00	1.00	1.00
$y_0 = 2$	4.28	3.58	1.17	0.92	0.83
$y_0 = 4$	23.0	15.9	1.39	1.03	0.94
C. Variation of longitudinal distance—electrode near truncated end					
$x_0 = -1, y_0 = 1$	1.78	1.65	1.08	0.77	0.87
$x_0 = 2, y_0 = 1$	1.12	1.24	0.93	0.53	0.70
$x_0 = 5, y_0 = 1$	1.00	1.01	0.99	0.77	0.48
$x_0 = -1, y_0 = 2$	4.48	3.79	1.17	0.95	0.89
$x_0 = 2, y_0 = 2$	5.38	4.92	1.12	0.69	0.56
$x_0 = 5, y_0 = 2$	4.31	3.67	1.15	0.32	0.25

First row lists values in absolute units. Other table entries list values as a multiple of the base-case datum. x_0 and y_0 are dimensionless quantities (normalized to node spacing). $D = 10 \mu$m for base case.
Source: From Reilly and Bauer (1987).

We note that stimulation near the end of the truncated model axon results in values of P that are more nearly in line with experimental data. A possible hypothesis for experimental observations is that the principal site of transcutaneous sensory stimulation may be at or near neural end structures.

5
Electrical Properties of the Heart

HERMANN ANTONI

5.1 Cardiovascular System: General Anatomical and Functional Aspects

The heart provides the main driving force for the movement of blood through the vessels. It is composed of two hollow organs—its right and its left half—with muscular walls (Fig. 5.1a). Each half comprises an atrium (Ra, La) and a ventricle (Rv, Lv). The right half receives oxygen-depleted blood from the body and expels it via the pulmonary artery (Pa) to the lungs, where it is reoxygenated. Then the blood returns to the left half of the heart and is thence distributed via the aorta (Ao) to the organs of the body (Fig. 5.1b). The movement of the blood from the right to the left heart, by way of the lungs, is called the *pulmonary circulation*. Its movement to and from the rest of the body is the *systemic circulation*. Strictly speaking, the two constitute a single pathway of blood movement, with the propulsive force provided at two points by the two halves of the heart (Fig. 5.1b).

The pumping action of the heart consists of a rhythmic sequence of relaxations (*diastole*) and contractions (*systole*) of the chambers. During diastole, the ventricles fill with blood, and during systole, they expel it into the large vessels (aorta and pulmonary artery). Backflow from these arteries to the ventricles is prevented by the valves at their openings. Before entering the ventricles, the blood passes from the large veins into the associated atria, which act as booster pumps to help fill the ventricles. An additional pair of valves between the atria and the ventricles prevents backflow of blood during the ventricular systole. In the systemic circulation the arteries carry oxygenated blood, and in the pulmonary circulation the oxygenated blood is carried by the veins.

Because the demands made on the circulating blood vary with time, the heart must be able to adjust its activity over a wide range. For example, the volume of blood expelled per minute by one ventricle (cardiac output) is about 5 liters for an adult person at rest and rises to almost 30 liters during hard physical work. Because the two ventricles are arranged in series, their outputs must be nearly the same at each beat. This requires a mechanism for precise adjustment of the outputs of the two ventricles. Moreover, when the resistances to flow in the systemic or pulmonary circulation increase— for instance, because of extensive vasoconstriction—the ventricles quickly adapt to the changed conditions by contracting more strongly and raising the pressure sufficiently to propel the same volume of blood. Likewise, changes in venous return and diastolic filling are compensated by adjust- ment of the cardiac output. This astonishing adaptability of the heart arises from both intracardial (auto-), and extracardial regulation.

The former is brought about by intrinsic properties of the myocardium, mainly its ability to respond to increased extension (greater end-diastolic volume) with the development of a higher contractile force or of a higher stroke volume. The latter mechanism operates under the control of the endocrine and autonomic nervous systems. It enables the heart either to overcome a higher pressure or to eject a larger stroke volume without increased muscular extension, that is, without increase in the end-diastolic volume. Under normal conditions the autoregulation is brought into play when changes in filling occur without a general increase in physical activity, for instance because of changes in the position of the body that affect venous return.

The energy the heart requires for its mechanical work comes primarily from the oxidative decomposition of nutrients. In this regard cardiac and skeletal muscle differ fundamentally, for the latter can obtain a large part of the energy needed to meet short-term demands by anaerobic processes, and the "oxygen debt" that is built up can be repaid later. This is not possible with the heart, because it is exclusively dependent on oxidative energy supply. Although the weight of the heart of an adult is only about 0.5% of the total body weight, the O_2 demand of the heart is about 10% of the total resting O_2 consumption. When the body is performing hard work, the O_2 consumption of the heart can rise to four times the resting level.

The coronary vessels, which supply the heart, are part of the systemic circulation. There are two coronary arteries, both arising from the base of the aortic root (Fig. 5.2). The right coronary artery supplies most of the right ventricle; the rest of the heart is supplied by the left coronary artery. Venous drainage is mainly through the coronary sinus, which flows into the right atrium. At rest the total coronary blood flow amounts to about 250 to 300 ml/min; this is about 5% of the cardiac output. During normal resting activity the heart withdraws more oxygen from the blood than do the other organs. Of the 20 ml/dl of O_2 in the arterial blood, the heart extracts around 14 ml/dl. Therefore, when the load on the heart increases and more oxygen is required, it is essentially impossible to increase the rate of extraction. Increased O_2 requirement must be met primarily by increased blood flow, brought about by dilation of the vessels. The strongest stimulus to dilation of the coronary vessels is O_2 deficiency.

Because cardiac metabolism relies so heavily on oxidative reactions to provide energy, a sudden interruption of circulation (*ischemia*) results in

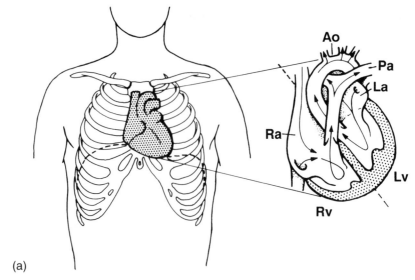

(a)

Pulmonary circulation

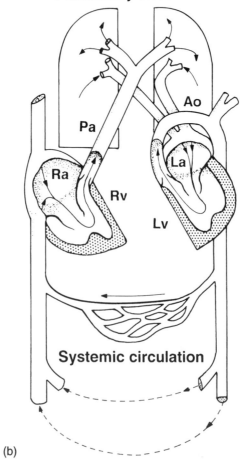

Systemic circulation

(b)

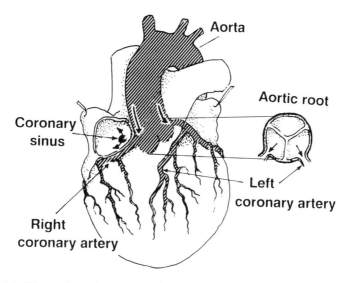

FIGURE 5.2. Human heart in a frontal view with the main coronary arteries originating from the aortic root. Venous return occurs via the coronary sinus to the right atrium.

extensive loss of function within a few minutes; contractions grow progressively weaker, a marked dilation develops, and after about 6 to 10 min the heart stops beating. Because coronary circulation is normally maintained by the pumping action of the heart itself, a strong diminution of the contractile force of the organ must also impair its blood supply, thus, in turn, weakening contraction still more. The same sequence of events takes place with ventricular fibrillation (see Sects. 5.8 and 5.10). When a circulatory breakdown affects the entire organism, the brain suffers irreversible damage after ischemia lasting only 8 to 10 min.

5.2   Origin and Spread of Excitation

Myocardial fibers, like nerve or skeletal muscle fibers, are excitable structures. The cell boundaries, which can be seen in the microscope as intercalated discs, offer no obstacle to the conduction of excitation. Because the

◀───────────────

FIGURE 5.1. (a) Simplified anatomy of the opened human heart in a frontal view. (b) Schematic diagram of the connections of the two halves of the heart with the pulmonary and systemic circulations. Ra = right atrium; La = left atrium; Rv = right ventricle; Lv = left ventricle; Ao = aorta; Pa = pulmonary artery.

musculature of the atria and ventricles forms a netlike structure, it behaves as a syncytium in which excitation arising anywhere in the atria or ventricles spreads out over all the unexcited fibers. This property provides the explanation of the all-or-none response of the heart; that is, when stimulated the heart either responds with excitation of all its fibers or gives no response. In a nerve or skeletal muscle, by contrast, each cell responds individually, so that only those fibers exposed to suprathreshold excitation discharge conducted impulses, while the others remain at rest.

Autorhythmicity and Geometry of Propagation

The rhythmic pulsation of the heart is maintained by excitatory signals generated within the heart itself. Under suitable conditions, therefore, a heart removed from the body will continue to beat at a constant frequency. This property is called *autorhythmicity*. Ordinarily, the spontaneous rhythmic triggering of excitation is performed exclusively by the specialized cells of the pacemaker and conducting system. The various elements in this system are diagrammed in Fig. 5.3.

Normally, the heartbeat is initiated in the sinoatrial (SA) node, in the wall of the right atrium near the superior vena cava. When the body is at rest, the SA node drives the heart at a rate of about 70 impulses/min. From the SA node the excitation first spreads over the working myocardium of both atria. The only pathway available for conduction to the ventricles is indicated in Fig. 5.3. All the rest of the atrioventricular boundary consists of nonexcitable connective tissue. As the excitation propagates through the conducting system it is briefly delayed in the atrioventricular (AV) node. Propagation velocity is high (about 2 m/s) through the remainder of the system—the common bundle, the left and right bundle

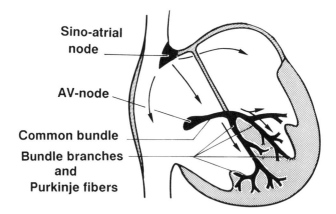

FIGURE 5.3. Arrangement of the pacemaker and conducting system of the human heart as seen in frontal section.

branches and their terminal network, the Purkinje fibers—so that the different ventricular regions are excited in rapid succession. From the Purkinje fibers, excitation spreads at a speed of about 1 m/s over the ventricular musculature.

Hierarchy of Pacemaker Activity and Artificial Pacemakers

The autorhythmicity of the heart is not entirely dependent on the operation of the SA node, since the other parts of the conduction system are also spontaneously excitable. But the intrinsic rhythm of these cells becomes considerably slower, the farther away from the SA node. Under normal conditions, therefore, these cells are triggered by the more rapid buildup of excitation in the higher centers, before they have a chance to trigger themselves. The SA node is the leading *primary pacemaker* of the heart, because it has the highest discharge rate.

If for any reason the SA node should fail to initiate the heartbeat, or if the excitation is not conducted to the atria (sinoatrial block), the AV node can substitute as a *secondary pacemaker* at a frequency of 40 to 60/min. If there should be a complete interruption of conduction from the atria to the ventricles (*complete heart block*), a tertiary center in the ventricular conducting system can take over as pacemaker for ventricular contraction.

In the case of the above-mentioned complete heart block, atria and ventricles beat entirely independently of one another, the atria at the frequency of the SA node and the ventricles at the considerably lower frequency of a tertiary center (30–40/min). When there is a sudden onset of total heart block, several seconds can elapse before the ventricular automaticity "wakes up." In this preautomatic pause an insufficient supply of blood to the brain may cause unconsciousness and convulsions (*Adams–Stokes syncope*). If the ventricular pacemakers fail altogether, the ventricular arrest leads to irreversible brain damage and eventually to death.

When conduction along the bundle branches is interrupted, the cardiac rhythm is not disturbed as long as at least one branch or subdivision of a branch remains functional. In this case the excitation spreads out from the terminals of the intact conduction system and eventually covers the whole ventricular myocardium; the time required for complete excitation is considerably longer than normal. In the absence of autorhythmicity it is therefore possible to keep the blood in circulation by artificial electrical stimulation of the ventricles. Electrical stimulation can sometimes be continued for years. The stimuli are generated by subcutaneously implanted, battery-driven miniature pacemakers and conducted to the heart by wire electrodes.

5.3   Elementary Processes of Excitation and Contraction

The action potential of the cardiac muscle cells, like that of neurons or skeletal muscle fibers, begins with a rapid reversal of the membrane potential, from the resting potential (approximately $-90\,\mathrm{mV}$) to the initial peak about $+30\,\mathrm{mV}$ see Fig. 5.4). This rising phase of the action potential (phase 0) lasts only a few milliseconds. During the subsequent period of repolarization, there are three phases that can be more or less clearly distinguished in different regions of the heart: Phase 1 is an initial short phase of repolarization, during which the membrane potential approaches zero. Phase 2 is a prolonged plateau following the initial peak, which is a very characteristic feature of cardiac muscle. Phase 3 is the terminal repolarization approaching the resting level (phase 4). The action potential of the cardiac musculature lasts about 200 to 400 ms—more than 100 times as long as that of a skeletal muscle or nerve fiber (see Chapter 3). The functional consequences, as we shall see, are considerable (see Sect. 5.4).

Ionic Mechanisms of Excitation

The action potential is generated by a complicated interplay of membrane potential changes, changes in ionic conductivity, and ion currents. Fundamentals of the ionic theory of excitation have been discussed in detail in Chapter 3. Here we shall give only a short recapitulation, with reference to the specific peculiarities of cardiac muscle (for details see Noble, 1984; Pelzer and Trautwein, 1987; Carmeliet, 1992; Antoni 1996). The resting potential of the myocardium is primarily a K^+ potential mainly determined by a specific K^+ channel (g_{K1}, see Table 5.1) and maintained by the electrogenic Na^+ pump. As in the neuron, the rapid upstroke phase of the action potential is brought about by a brief pronounced increase in Na^+ conductance, g_{Na}, which results in a massive Na^+ influx (see Fig. 5.4). This initial Na^+ influx, as in the neuron, is very rapidly inactivated. Hence, further mechanisms are required for the considerable delay in repolarization of the cardiac muscle tissue. These are (1) delayed activation and slow inactivation of L-type Ca^{2+} channels (g_{CaL}) which causes a depolarizing influx of calcium (slow inward current) and (2) a decrease in the resting (inwardly rectifying) K^+ conductance (g_{K1}) due to depolarization, which reduces the repolarizing K^+ outward current.

Repolarization of the myocardium results from a gradual decrease in g_{Ca} and delayed activation of another population of K^+ channels (g_K) which are activated by depolarization at the beginning of the action potential. The decrease in g_{Ca} diminishes the slow inward current, and the increase in g_K enhances the K^+ outward current. Yet another population of K^+ channels (g_{TO}) are responsible for the Phase 1 repolarization which is differently

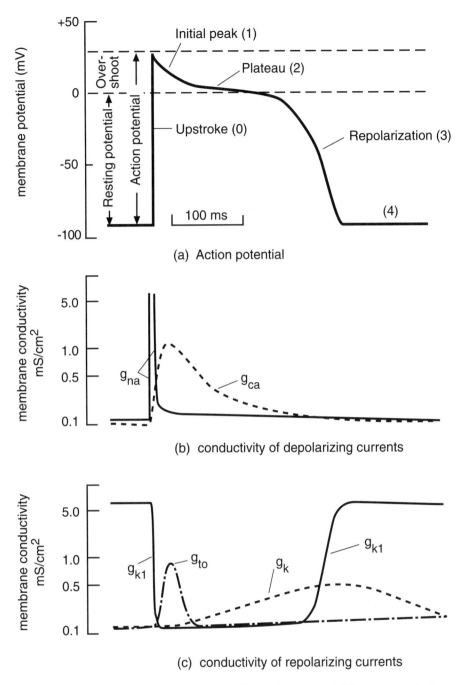

FIGURE 5.4. (a) General form of the cardiac action potential from a ventricular myocardial cell. (b) Changes in the conductivity of depolarizing ionic currents during the action potential. (c) Changes in the conductivity of repolarizing ionic currents. The magnitude of the corresponding ionic currents further depends on the difference between the membrane potential and the equilibrium potential of the corresponding ion.

TABLE 5.1. Ionic channels participating in excitation of mammalian cardiac cells. The symbol $> g <$ is used to characterize ionic channels and refers to their properties as conductances which carry ionic currents.

Notation	Characteristics and special functions
Ionic channels mainly carrying inward currents	
g_{Na} fast Na⁺ channel	Activated by depolarization. Inactivation is time- and voltage-dependent. Responsible for upstroke of the action potential in atrial and ventricular myocardium as well as in His–Purkinje system.
g_{CaL} L-type Ca²⁺ channel	Activation by depolarization. Inactivation depending on voltage and on internal Ca²⁺. Responsible for slow inward current (plateau phase of the action potential, upstroke in SA node) as well as for excitation contraction coupling.
g_f pacemaker channel	Nonspecific cationic channel. Carries mainly Na⁺ inward current when activated by polarization to high membrane potentials. Responsible for diastolic depolarization in His–Purkinje system and partly responsible for pacemaker activity of the SA node.
g_b background Na⁺ channel	Voltage independent channel in SA nodal cells carrying Na⁺ ions. The Na⁺ current is offset by an outward K⁺ current at the end of the action potential, but as the K⁺ current decays, it contributes to pacemaker behaviour.
Ionic channels mainly carrying outward currents	
g_{Kl} inward rectifier	K⁺ channel responsible for maintaining the resting potential near the K⁺ equilibrium potential in working myocardium and in His–Purkinje system. Shutting off during depolarization, and opening during repolarization, thus favoring the plateau phase and late repolarization.
g_K delayed rectifier	K⁺ channel slowly activating upon depolarization. Mainly responsible for repolarization of the action potential. Because of its slow inactivation partly responsible for pacemaker depolarization. g_K comprises two components g_{Kr} (rapidly activating) and g_{Ks} (slowly activating) which are differently developed in different regions of the heart
g_{TO}	K⁺ channel generating transient outward (=TO) current when activated by depolarization. Mainly present in atrial and Purkinje fibers as well as in subepicardial ventricular fibers. Responsible for the early phase of repolarization (phase 1) in these cells.

Data taken from Noble (1984), Pelzer & Trautwein (1987), Sanguinetti & Jurkiewicz (1990), Carmeliet (1992), Antoni (1996).

developed in different regions of the heart. When the membrane is at its resting potential, the depolarizing and repolarizing currents are in balance.

Mechanism of Contraction

While the excitatory processes take place at the surface membrane of the cardiac muscle cell, the contractile machinery is located in the interior. The main constituents of this machinery are the "contractile proteins": actin (molecular weight 42 kD) and myosin (approximately 500 kD). Actin and myosin form the thin and thick myofilaments of the myofibrils. These are contractile bundles about 1 μm in diameter, which are subdivided by the Z disks into compartments about 2.3 μm long, called *sarcomeres*. The structure of a sarcomere is illustrated in Fig. 5.5. There is a regular arrangement of the two filaments, with the myosin filaments forming the A band (isotropic band). On either side of the A bands are regions containing only thin filaments, which therefore appear light; these isotropic I bands extend to the Z lines.

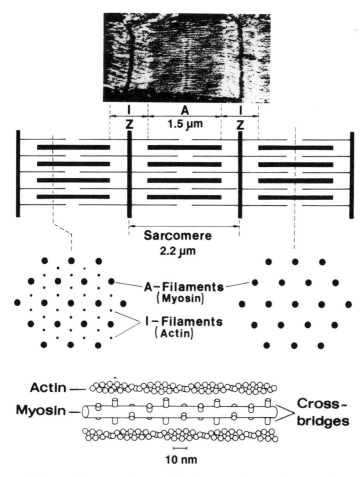

FIGURE 5.5. Top: Electron micrograph of mammalian cardiac muscle representing the functional unit of a myofibril. Middle: Banded structure of the myofibrils; arrangement of the myosin and the actin filaments. Bottom: Actin filaments and myosin filaments with cross-bridges.

The way in which they interact to bring about contraction is described by the sliding-filament theory (Huxley and Hanson, 1954). In the relaxed muscle the ends of the thick and thin filaments usually overlap only slightly. When the myofilaments contract, the thin actin filaments slide over the thick myosin filaments, moving between them toward the middle of the sarcomere. During the sliding process neither the myosin nor the actin filaments themselves shorten.

Excitation-Contraction Coupling

The transmission of an action potential from the excited cell membrane to the myofibrils in the depths of the cell requires several sequential processes in which calcium ions initiate contraction. This occurs when the concentration of intracellular free calcium rises above about 10^{-7} mol/l. Two intracellular tubular systems are involved in this process: the transverse tubular system (TTS) and the longitudinal system (*sarcoplasmic reticulum* (SPR), see Fig. 5.6). The T system is formed by invagination of the outer membrane and transmits the action potential to the interior of the cell. The

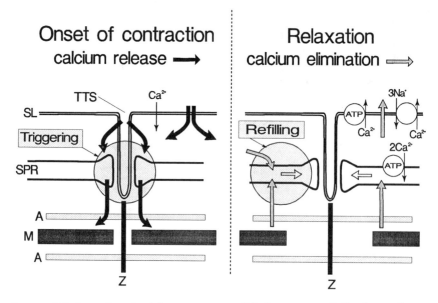

FIGURE. 5.6. Interplay of calcium movements (black arrows = Ca^{2+} release; shaded arrows = Ca^{2+} elimination) and contractile activation during the onset of contraction (left) and underlying relaxation (right). At the onset of contraction, transsarcolemmal Ca^{2+} influx induces a calcium-triggered Ca^{2+} release from the sarcoplasmic reticulum (Triggering). During relaxation, Ca^{2+} is partly eliminated from the cell and partly stored in the sarcoplasmic reticulum (Refilling). A = actin; M = myosin; SL = sarcolemma; SPR = sarcoplasmic reticulum; TTS = transverse tubular system; Z = Z-line of the sarcomere.

longitudinal system which represents an intracellular calcium store is in close contact with the T system through so-called lateral cisternae. When the action potential is propagated along the surface and into the T system, it causes an influx of calcium ions. The accumulation of calcium near the lateral cisternae triggers a further mobilization of calcium from the intracellular stores (see Fig. 5.6).

Upon repolarization of the membrane, the transmembrane influx of calcium ions ceases, and the calcium ions are removed by ATP-driven calcium pumps as well as by exchange mechanisms located in the longitudinal system and in the surface membrane (Fig. 5.6). Relaxation occurs when the myoplasmic concentration of activating calcium ions is reduced below about 10^{-7} mol/l. This reduction inhibits the interaction of actin and myosin cross bridges, which then detach.

5.4   Stimulation, Propagation, and Refractoriness

Elementary Mechanisms

As in other excitable tissues, action potentials of cardiac cells are elicited by depolarization of the resting membrane to threshold (Fig. 5.7). Except for the sinoatrial and the atrioventricular nodal cells, all myocardial fibers contain fast Na^+ channels, and the activation of these channels by depolarization is responsible for the initiation of the action potential. Normally, the spontaneous depolarization of pacemaker cells initiates excitation in the circumscribed pacemaker area as soon as the threshold is attained (see Sect. 5.5). Excitation is then propagated as a result of depolarization of still-resting fibers by their coupling to closely adjacent excited ones. Artificial stimulation works by the same mechanism.

To stimulate the heart as a whole, it is sufficient to stimulate only a fraction of about 50 closely coupled cardiac cells. Hence, the stimulus intensity required to do this is about the same for small, isolated myocardial preparations as for the whole organ. If a single isolated cardiac cell in culture is stimulated through intracellular electrodes, a current intensity of about 1 nA (with a duration of 5 ms) is required to elicit an action potential. If, on the other hand, a myocardial cell is in its normal connection within the tissue, suprathreshold intracellular stimulation, depending on the coupling resistance, requires more than 100 nA (Johna, 1989).

With extracellular stimulation, as it is applied in experiments on isolated cardiac tissue or by implanted pacemakers, threshold current strength for a given pulse duration depends strongly on the geometry of the electrodes and their position with respect to the cellular alignment in cardiac muscle. Under optimal conditions (electrodes diameter 0.7–1.0 mm; pulse duration 0.5 ms), pacemaker stimulation of the heart can be achieved with only 10–14 μA from an extracellular electrode (Irnich, 1973).

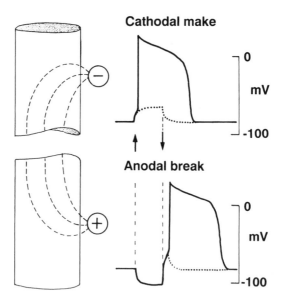

FIGURE 5.7. The two types of pulsed DC stimulation; changes in membrane potential during cathodic make and during anodic break. Action potentials are elicited when the membrane potential attains the threshold.

A sufficiently strong and long-lasting rectangular current pulse can stimulate twice: first, in the region of the cathode when the circuit is closed (*cathodic make*), and second, in the region of the anode when the circuit is opened (*anodic break*) (Fig. 5.7). This latter effect is in part the result of an increase in recovery from inactivation of the sodium channels. In part it is brought about by a decrease in potassium conductance during hyperpolarization. This declines slowly when the stimulus is switched off, and thus leads to a transient depolarization to threshold (dotted line in Fig. 5.7). For more details about cardiac stimulation, see Chapter 6.

Conduction of the Action Potential

Conduction in heart muscle occurs by local circulating currents, in the same fashion that conduction is brought about in nerve (see Fig. 3.10). These currents are driven by the sodium potential across the membrane of the excited region. The current displaces the charge stored on the adjacent membrane, depolarizing the membrane toward its threshold for increasing sodium conductance. This process will take place more quickly as more current flows to the resting regions and as the capacity that has to be discharged becomes smaller (Fozzard, 1979).

The structure and the functional properties of the ventricular conduction system allow rapid spread of excitation all over the ventricles, thus bringing about nearly synchronized contraction. This is especially pronounced in the hearts of large animals, where, as compared with the ordinary myocardium, the fibers of the conducting system exhibit considerably greater diameters with a correspondingly smaller internal resistance. The intercalated disks offer no obstacle to the conduction of cardiac excitation, because of special structures in the apposed portions of cardiac membrane, the so-called *gap junctions* or *nexuses*, which provide the electrical coupling between cardiac cells.

Disturbances of propagation (slowing or block) are usually attributed to a reduction of the excitatory inward currents: the fast sodium inward current in the atrial and ventricular myocardium or in the conducting system, and the slow calcium inward current in the AV node. The sodium inward current is reduced by depolarization due to inactivation of the voltage-dependent sodium channels. Depolarization itself may be the result of various factors, such as loss of internal or increase of external potassium or decrease of potassium conductance. However, the fast sodium inward current can also be reduced in the absence of depolarization by, for example, drugs with local anesthetic effects, which are frequently used as anti-arrhythmics. In the AV node, where fast sodium channels are absent, inhibition of the slow inward current by certain calcium antagonists (for instance, verapamil) can increase the normal AV delay.

If the excitatory inward currents are critically reduced, the ability of propagation will also depend on the geometry of the fibers: Normally, the wave of excitation will proceed without difficulty in a direction that branches off from the main direction at an angle of more than 90°. Under less favorable conditions, block of conduction can preferably occur at such branchings (Spach et al., 1982). Moreover, block of conduction may also be the result of an increase of the internal longitudinal resistance when conductance through the gap junctions becomes reduced. Such a change occurs when the intracellular concentration of free calcium rises above the maximal physiological concentration of about 10^{-5} mol/l. This can happen when calcium leaks into damaged cardiac cells. A similar effect is initiated by a rise of the intracellular concentration of hydrogen ions (acidosis) (Délèze, 1970; de Mello, 1972).

Definition and Mechanism of Refractoriness

When an action potential has been elicited in a cardiac muscle cell, this cell cannot be reexcited until its membrane potential has repolarized to a certain level. During the *absolute refractory period*, the cell is inexcitable; and during the subsequent *relative refractory period*, excitability gradually recovers as indicated in Fig. 5.8. Thus a new action potential can be elicited sooner with a stronger stimulus. Action potentials generated very early in the relative refractory period do not rise as sharply as normal action potentials and have a lower amplitude and a shorter duration (Fig. 5.8).

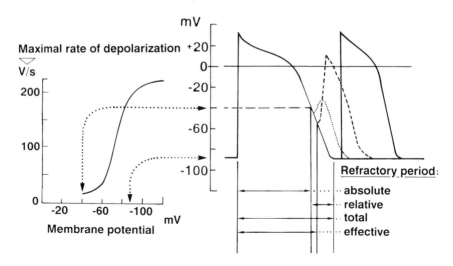

FIGURE 5.8. Left: Dependence of the maximal rate of depolarization of the action potential (as an indirect measure of the availability of the fast sodium channels) on the membrane potential prior to excitation. Right: Different types of refractoriness as related to the cardiac action potential.

The term *effective refractory period* is used to define the phase from the beginning of the excitatory cycle during which no propagated excitation can be elicited. This period lasts slightly longer than the absolute refractory period. When the threshold is not exactly determined, but instead stimuli of a constant suprathreshold diastolic strength (for instance, twice diastolic threshold) are used to determine the refractory period, its duration will comprise the absolute and part of the relative refractory period, during which the stimulus remains ineffective. This phase is usually called the *functional refractory period*.

The chief cause of refractory behavior is the inactivation of the fast Na^+ channels during prolonged depolarization. Not until the membrane has repolarized to approximately $-40\,mV$ do these channels begin to recover. The duration of the refractory period is therefore, as a rule, closely related to the duration of the action potential. When the action potential is short-ened or lengthened, the refractory period changes accordingly. But drugs that act as local anesthetics, inhibiting the initial Na^+ influx or retarding its recovery after inactivation, can prolong the refractory period without affecting action potential duration. In the absence of drugs, a similar behav-ior is observed in the AV nodal cells, which, after termination of an action potential, remain nonexcitable for a short time. The refractory period of these cells is thus not only voltage- but also time-dependent.

Another phenomenon that might be considered in this context is the *supernormal period*. This appears as a transient decrease in threshold for stimulation below its diastolic level immediately following the relative refractory period. This phase coincides with the terminal phase of repolarization, when the membrane potential has not yet fully recovered. It can explain the response to electrical stimuli at this instant which is otherwise ineffective. The decrease in threshold (increase in excitability) during the supernormal period can be attributed to the fact that the distance between actual membrane potential and threshold potential is reduced because the membrane retains moderate depolarization. This is more pronounced when the time course of the terminal phase of repolarization is slow, and it disappears if repolarization occurs abruptly.

The prolonged refractory period protects the musculature of the heart from too-rapid reexcitation, which could impair its function as a pump. At the same time, it prevents recycling of excitation in the muscular network of the heart, which would interfere with its rhythmic activity. Because the refractory period of the excited myocardial cells is normally longer than the time taken for spread of excitation over the atria or ventricles, a wave of excitation originating at the SA node or a heterotopic center can propagate over the heart only once and must then die out, for it encounters refractory tissue everywhere. Reentry (defined in Sect. 5.9) thus does not normally occur.

An action potential triggered immediately following the relative refractory period of the preceding impulse is normal, as Fig. 5.8 shows, in upstroke rate and amplitude. Its duration, however, is distinctly less than that of the preceding action potential. In fact, there is a close relationship between the duration of an action potential and the interval that preceded it, and thus between duration and repetition rate. The main cause of this phenomenon is an increase in the potassium conductance, which outlasts the repolarization phase of the action potential and returns only gradually to the basal level. When the interval between action potentials is short, the increased K^+ conductance accelerates repolarization of the next action potential.

When cardiac muscle is exposed to 50- or 60-Hz alternating current, it will respond in a way illustrated by Fig. 5.9. The first depolarizing AC half-wave triggers an action potential and then remains ineffective until the end of the corresponding (functional) refractory period, when a new action potential starts. For the reasons mentioned above, the duration of the second action potential is shorter, and so is its refractory period. Thus, 60-Hz AC, when applied to cardiac muscle, leads to a rhythmic response, but at a frequency increased in inverse proportion to the refractory period.

5.5 Regular and Ectopic Pacemakers

Elementary Events in Impulse Formation

The working myocardium of atria and ventricles is not automatically active: as outlined above, action potentials are generated by spread of excitation.

Stimulation with alternating current

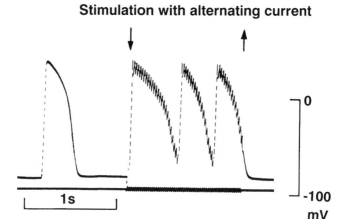

FIGURE 5.9. Electrical response of a single myocardial cell to stimulation with 50-Hz AC for 1.3s. Microelectrode recording from isolated myocardium of rhesus monkey.

In all cardiac muscle cells that are capable of autorhythmicity, depolarization toward the threshold occurs spontaneously. This elementary process of excitation can be observed directly by intracellular recording from a pacemaker cell. As shown in Fig. 5.10, the repolarization phase of such an action potential is followed—beginning at the *maximal diastolic potential*—by a slow depolarization which triggers a new action potential when the threshold is reached. The *slow diastolic depolarization* (pacemaker potential) is a local excitatory event, not propagated as is the action potential.

Actual and Potential Pacemakers

Normally, only a few cells in the SA node are responsible for timing the contraction of the heart (actual pacemakers). All the other fibers in the specialized tissue are excited in the same way as the working musculature, by conducted activity; that is, these potential pacemakers are rapidly depolarized by currents from activated sites before their intrinsic slow diastolic depolarization reaches threshold. Comparison of the two processes (Fig. 5.10) shows how a potential pacemaker can assume the leading role when the actual pacemaker ceases to function. Because the slow diastolic depolarization of the potential pacemaker takes longer to reach threshold, its discharge rate is lower. In the working myocardium, there is no automatic depolarization; the upstroke of the action potential triggered by the imposed current rises sharply from the resting-potential baseline.

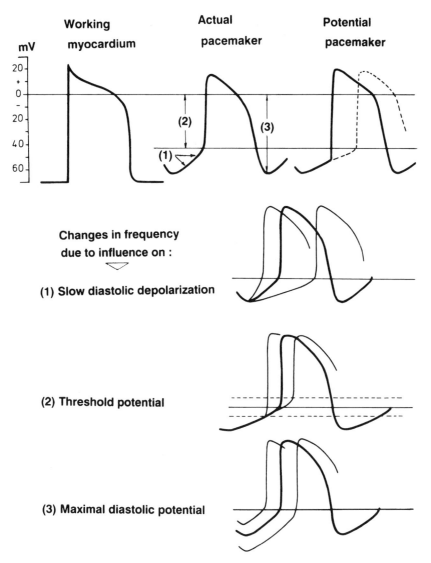

FIGURE 5.10. Top: Typical forms of action potentials in working myocardium as compared with actual and potential pacemakers. Bottom: Different mechanisms by which changes in frequency of pacemaker cells are brought about.

According to current opinion, the slow diastolic depolarizations of the SA node are mainly caused by the slow decline of the K^+ conductance that had increased during repolarization. This requires a relatively high background Na^+ conductance to produce depolarization. In the ventricular conducting system, the background Na^+ conductance is normally low.

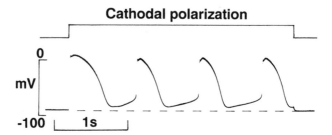

FIGURE 5.11. Induction of automatic activity in a cell of working myocardium (cat ventricle) by cathodic polarization.

Therefore, the membrane potential reaches relatively high levels just after the action potential, which permits extensive recovery of the rapid Na^+ system. The subsequent diastolic depolarizations involve a special ionic channel that does not operate in the SA node.

Ectopic Pacemakers

The capacity for spontaneous excitation is a primitive function of myocardial tissue. In the early embryonic stage, all the cells in the heart are spontaneously active. As fetal differentiation proceeds, the fibers of the prospective atrial and ventricular myocardium lose their autorhythmicity and develop a stable, high resting potential. But the stability of the resting potential can be lost under various conditions associated with partial depolarization of the membrane (catelectrotonus, stretching, hypokalemia, Ba^{2+} ions) (see Fig. 5.11). Then the affected fibers can develop diastolic depolarizations like those of natural pacemaker cells and, in some circumstances, can interfere with the rhythm of the heartbeat. On the other hand, depolarization due to elevated K^+ does not produce autorhythmicity, because a concomitant rise in K^+ conductance inhibits spontaneous activity. A center of autorhythmicity apart from the regular pacemaker tissue is called an *ectopic center* or *ectopic focus*.

5.6   Effects of Autonomic Nerves and of Changes in Electrolyte Composition

The cardioregulatory centers in the brain exert a direct influence on the activity of the heart, by way of sympathetic and parasympathetic nerves. This influence governs the rate of beat (chronotropic action), the systolic contractile force (inotropic action), and the velocity of atrioventricular conduction (dromotropic action). These actions of the autonomic nerves are mediated in the heart, as in all other organs, by chemical transmitters—

acetylcholine in the parasympathetic system and noradrenaline in the sympathetic system.

Parasympathetic and Sympathetic Innervation

The parasympathetic nerves supplying the heart branch off from the vagus nerves on both sides in the cervical region. The fibers on the right side pass primarily to the right atrium and are concentrated at the SA node. The AV node is reached chiefly by cardiac fibers from the left vagus nerve. Accordingly, the predominant effect of stimulation of the right vagus is on heart rate, and that of left vagus stimulation is on atrioventricular conduction. The parasympathetic innervation of the ventricles is sparse; its main influence is indirect, by inhibition of the sympathetic action.

The sympathetic nerve supply, unlike the parasympathetic, is nearly uniformly distributed to all parts of the heart. The sympathetic cardiac nerves come from the lateral horns of the upper thoracic segments of the spinal cord and make synaptic connections in the cervical and upper thoracic ganglia of the sympathetic trunk, in particular the stellate ganglion. The postganglionic fibers pass to the heart in several cardiac nerves. Sympathetic influences on the heart can also be exerted by catecholamines (mainly adrenaline) released from the adrenal medulla into the blood.

Chronotropy, Inotropy, and Dromotropy

Stimulation of the right vagus or direct application of acetylcholine to the SA node causes a decrease in heart rate (*negative chronotropy*); in the extreme case, cardiac arrest can result. Sympathetic stimulation or application of noradrenaline increases the heart rate (*positive chronotropy*). When vagus and sympathetic nerves are stimulated at the same time, the vagus action usually prevails and is followed by a delayed sympathetic influence when stimulation is terminated (Fig. 5.12). Modification of the autorhythmic activity of the SA node by these autonomic inputs occurs primarily by way of a change in the time course of the slow diastolic depolarization. Under the influence of the vagus, diastolic depolarization is retarded, so that it takes longer to reach threshold (Fig. 5.10). In the extreme case, diastolic depolarization is eliminated, and the membrane actually becomes hyperpolarized. The sympathetic fibers act to increase the rate of diastolic depolarization and thus shorten the time to threshold (see Table 5.2).

Because the positive chronotropic action of the sympathetic nerves extends to the entire conducting system of the heart, when a leading pacemaker center fails, the sympathetic input can determine when and to what

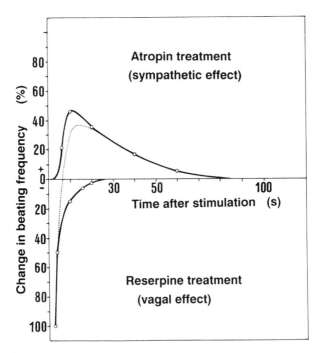

FIGURE 5.12. Changes in beating frequency of isolated right atrium (guinea pig) following 50-Hz field stimulation, 10 V/cm, 1 s. Dotted line: uninfluenced atria (controls). Vagal effects are derived from the difference between controls and atropine treatment, sympathetic effects from the difference between controls and reserpine treatment.

TABLE 5.2. Main action of several influences on the electrical and mechanical activity of the heart.

	SA node PA	AV node $\dot{V}_{max}$ and CV	Purkinje system $\dot{V}_{max}$ and CV	PA	Atrial and ventricular myocardium AP duration	Contraction Force
Warming	++	++	+	+	--	-
Cooling	--	--	-	-	++	+
Acidosis	-	0	0	-	+	-
Alkalosis	+	0	0	+	-	+
K_e increase	-	-	--	--	--	--
K_e decrease	0	0	0	++	0	+
Sympathetic (noradrenaline)	++	++	0	+	+	++
Parasympathetic (acetylcholine)	--	--	0	(-)	In atrium --	In atrium --
O_2 deficiency	-	-	-	0	-	-

$\dot{V}_{max}$ = maximal rate of depolarization of action potential. +, increase; ++, large increase; -, decrease; --, large decrease; 0, no prominent effect. PA = pacemaker activity; CV = conduction velocity; AP = action potential.

extent a subordinate center takes over as pacemaker. In the same way, however, an ectopic focus or rhythmicity can be stimulated to greater activity, so that the danger of arrhythmia increases.

An influence of the autonomic nerves on the conduction of excitation can normally be demonstrated only in the region of the AV node. The sympathetic fibers accelerate atrioventricular conduction and thus shorten the interval between the atrial and ventricular contractions (*positive dromotropic action*). The vagus—especially on the left side—retards atrioventricular conduction and in the extreme case can produce a transient complete AV block (*negative dromotropic action*). These effects of the autonomic transmitter substances are associated with a particular feature of the cells in the AV node. As discussed above, the fibers of the AV node closely resemble those of the SA node. Because there is no rapid inward Na^+ current, the upstroke is relatively slow and hence the conduction velocity is low. The vagus acts to decrease the rate of rise still further, whereas sympathetic activity increases it, with the corresponding effects on the velocity of atrioventricular conduction.

Vagal and Sympathetic Tone

The ventricles of most mammals, including humans, are influenced predominantly by the sympathetic system. By contrast, the atria can be shown to be subject to the continual antagonistic influence of both vagus and sympathetic nerves; this effect is most clearly evident in the activity of the SA node. It can be observed, for example, by intersecting or pharmacologically blocking one of the two sets of nerves; the action of the opponent then dominates. When the vagus input to the dog heart is removed, the rate of beating increases, from ≈100/min at rest to 150/min or higher; when the sympathetic input is removed, it falls to 60/min or less. This maintained activity of the autonomic nerves is called *vagal* and *sympathetic tone.* Because the rate of the completely denervated heart (the autonomic rate) is distinctly higher than the normal resting rate, it can be assumed that, under resting conditions, vagal tone predominates over sympathetic tone.

Mechanism of Autonomic Transmitter Actions

The effects of vagal stimulation and application of the parasympathetic transmitter, acetylcholine, are attributed to one fundamental action—an increase in the K^+ conductance of the excitable membrane. In general, such an influence is expressed in the tendency of the membrane potential to oppose depolarization, as is evident in both the retardation of the slow diastolic depolarization in the SA node, described above, and in a shortening of the action potential of the atrial myocardium. The reduction of the rate of rise of the action potential in the AV node can also be explained by a stronger outward K^+ current that counteracts the slow inward Ca^{2+} current. In the ventricular myocardium, by contrast, the sympathetic-antagonistic action dominates—that is, the main action is inhibition of noradrenaline release from the sympathetic nerve endings.

With respect to the mechanisms by which the sympathetic fibers (or their transmitters) act, there is convincing experimental evidence that they increase the slow inward Ca^{2+} current. That is, the contractile force becomes greater (positive inotropic action) because this effect has intensified the excitation-contraction coupling. The positive dromotropic action on the AV node is also likely, in view of the above considerations, to be related to enhancement of the slow inward Ca^{2+} current. As yet, there is no satisfactory explanation of the mechanism of the positive chronotropic sympathetic action. At the SA node, enhancement of the slow inward current is probably involved. In the case of the Purkinje fibers, however, an influence on a specific, hyperpolarization-activated pacemaker current is likely.

Autonomic transmitter substances are thought to bind to certain molecular configurations on the effector cell (the word "receptor" is used both for these subcellular structures and for sensory cells). The effects of noradrenaline and adrenaline on the heart, described above, are mediated by so-called β *receptors*. Sympathetic effects can be prevented by β *receptor blockers*, such as dichloroisoproterenol (DCI) and pronethalol. In the heart, as in other organs, the deadly nightshade poison atropine acts as an antagonist to the parasympathetic effects of acetylcholine.

Effects of the Ionic Environment and of Drugs

Of all the features of the extracellular solution that can affect the activity of the heart, the K^+ concentration is of the greatest practical importance. An increase in extracellular K^+ has two effects on the myocardium: (1) the resting potential is lowered because the gradient K_i^+/K_e^+ is less steep, and (2) the K^+ conductance of the excitable membrane is increased—as it is by acetylcholine in the atrial myocardium. Doubling of the K^+ concentration, from the normal 4 mmol/l to about 8 mmol/l, results in a slight depolarization accompanied by increased excitability (threshold for stimulation reduced) and increased conduction velocity. Moreover, this influence causes suppression of heterotopic centers of rhythmicity. A larger increase in K^+ (over 8 mmol/l) reduces excitability (threshold for stimulation increased) and reduces conduction velocity. When the extracellular K^+ concentration is lowered to less than 4 mmol/l, the stimulating influence on pacemaker activity in the ventricular conducting system dominates. The enhanced activity of heterotopic centers can lead to cardiac arrhythmias (Table 5.2).

The excitability-reducing action of large extracellular K^+ concentrations is turned to advantage during heart operations to immobilize the heart briefly for the surgical procedures (cardioplegic solutions). While the heart is inactive, circulation is maintained by an extracorporeal pump (heart–lung machine). Impairment of cardiac function due to increased blood K^+ during extreme muscular effort or in pathological conditions can be largely compensated by sympathetic activity.

Under normal conditions, changes in extracellular Ca^{2+} are of limited interest for the function of the heart, because the neuromuscular excitability is much more sensitive to such influences. Under experimental conditions, changes in the extracellular Ca^{2+} concentration rapidly affect the force of cardiac contraction. Complete excitation-contraction uncoupling can be achieved by the experimental withdrawal of extracellular Ca^{2+}.

Table 5.2 summarizes the most important physical and chemical influences on excitation and contraction of the heart; only the dominant effects are considered.

5.7 Electrocardiogram

As excitation spreads over the heart, an electric field is produced that can be sensed on the surface of the body. The changes in magnitude and direction of this field in time are reflected in alterations of potential differences measurable between various sites on the body surface. The electrocardiogram (ECG) represents such potential differences as a function of time. It is thus an indicator of cardiac excitation—not contraction!

Because the directly measured potentials usually amount to less than 1 mV, commercially available ECG recorders incorporate electronic amplifiers; these contain high-pass filters with a cutoff frequency near 0.1 Hz (a time constant of 2 s). Therefore, DC components and very slow changes of the potentials at the metal recording electrodes do not appear at the output. All electrocardiographs have a built-in means of monitoring amplitude in the form of a 1-mV calibration pulse set to cause a deflection of 1 cm.

ECG Form and Nomenclature; Relation to Cardiac Excitation

With electrodes attached to the right arm and left leg, the normal ECG looks like the curve shown in Fig. 5.13. There are both positive and negative deflections (waves), to which are assigned the letters P through T. By convention, within the QRS group, positive deflections are always designated as R and negative deflections as Q when they precede the R wave or as S when they follow it. By contrast, the P and T waves can be either positive or negative. The distance between two waves is called a segment. An interval comprises both wave and segments (for instance, the PQ interval, see Fig. 5.13). The RR interval, between the peaks of two successive R waves, corresponds to the period of the beat cycle.

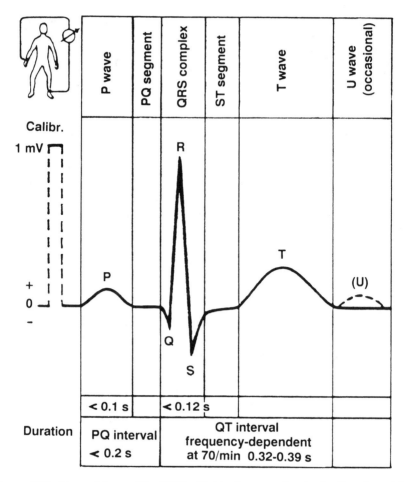

FIGURE 5.13. Normal form of the ECG with bipolar recording in the direction of the long axis of the heart. The times below the ECG curve are important limiting values for the duration of distinct parts of the curve.

An atrial part and a ventricular part can be distinguished in the ECG. The atrial part begins with the P wave, the expression of the spread of excitation over the atria. During the subsequent PQ segment, the atria as a whole are excited. The dying out of excitation in the atria coincides with the first deflection in the ventricular part of the curve, which extends from the beginning of Q to the end of T. The QRS complex is the expression of the spread of excitation over both ventricles, and the T wave reflects recovery from excitation in the ventricles. The intervening ST segment is analogous to the PQ segment in the atrial part, indicating total excitation of the ventricular myocardium. Occasionally, the T wave is followed by a so-called

U wave; this probably corresponds to the dying out of excitation in the terminal branches of the conducting system.

The PQ interval is the time elapsed from the onset of atrial excitation to the onset of ventricular excitation and is normally less than 0.2 s. A longer PQ interval indicates a disturbance in conduction in the region of the AV node or the bundle of His. When the QRS complex extends over more than 0.12 s, a disturbance of the spread of excitation over the ventricles is indicated. The overall duration of the QT interval depends on heart rate. When the heart rate increases from 40 to 180/min, for example, the QT duration falls from about 0.5 to 0.2 s.

Origin of ECG

As a wave of excitation passes over a cardiac muscle fiber, a potential gradient dV/dx is generated, the magnitude of which depends on the momentary phase of excitation. At the front of the wave, there is a steep gradient corresponding to the amplitude of the action potential. During the repolarization phase, there appear much smaller gradients in the opposite direction. To a first approximation, the excited myocardial fiber behaves in the physical sense as a *variable dipole*, the magnitude and direction of which are symbolized by an arrow (*vector*). By definition, the dipole vector points from minus to plus; that is, from the excited to the unexcited region; an excited site, as seen from the outside, is effectively electronegative as compared with an unexcited site. At every moment during the excitatory process, dipole fields across the surface of the heart sum to an integral vector. As this occurs, a large fraction of the vectors will neutralize one another, as observed from outside the system, because they exert equal effects in opposite directions.

When excitation spreads over the atria (P wave), the predominant direction of spread is from top to bottom; that is, most of the individual depolarization vectors generate an integral vector pointing toward the apex. When the atria are excited as a whole, the potential differences disappear transiently, for all the atrial fibers are in the plateau phase of the action potential (PQ segment). Only when the excitation moves into the ventricular myocardium do demonstrable potential gradients reappear. Spread of excitation over the ventricles begins on the left side of the ventricular septum and generates an integral vector pointing toward the base of the heart (beginning of QRS). Shortly thereafter, spread toward the apex predominates (largest QRS vector). During this phase, excitation moves through the ventricular wall from inside to outside. Spread through the ventricles is completed with the excitation of a basal region of the right ventricle, at which time the integral vector points toward the right and up (end of QRS). While the excitation was spreading over the ventricles (QRS), it died out in the atria. When the ventricles are totally excited (ST segment), the potential differences disappear briefly, as they did during atrial excitation (PQ segment).

If repolarization of the ventricles took place in the same sequence as depolarization and at the same rate, the behavior of the integral vector during recovery would be approximately the opposite of that during the spread of excitation. This is not the case because the process of repolarization is fundamentally slower than that of depolarization and rates of repolarization are not the same in the different parts of the ventricles. Repolarization occurs sooner at the apex than at the base and sooner in the subepicardial than in the subendocardial layers of the ventricles. Thus, during the ventricular recovery phase (T wave), the direction of the integral vector hardly changes; it points to the left. The different curve forms obtained with the arrangement of leads ordinarily used, on extremities and chest wall, are basically projections of the momentary integral vectors onto certain lead axes.

ECG Recording

In bipolar recordings, two recording electrodes are placed at defined sites on the body surface, and the potential difference between these electrodes is monitored. In unipolar recordings, only one recording electrode is placed at a defined site, and the potential is measured with respect to a reference electrode. This electrode can be thought of as positioned at the null point of the dipole, between positive and negative charge. In clinical practice, the following recording arrangements are the most commonly used today (Fig. 5.14).

Limb leads:
  Bipolar: standard Einthoven's triangle (leads I, II, III)
  Unipolar: Goldberger's augmented limb leads (aVR, aVL, aVF)
Chest leads:
  Bipolar: so-called small chest triangle of Nehb (D, A, I), not shown in Fig. 5.14
  Unipolar: Wilson's precordial leads (V1–V6)

Einthoven's Triangle

Because in bipolar recording from the limbs by the method of Einthoven the arms and legs act as extended electrodes, the actual recording sites are at the junction between limbs and trunk. These three points lie approximately on the corners of an equilateral triangle, and the sides of the triangle represent the lead axes. Figure 5.15 illustrates the way in which the relative amplitudes of the various ECG deflections in the three recordings are derived from the projection of the frontal plane vector onto the associated lead axes.

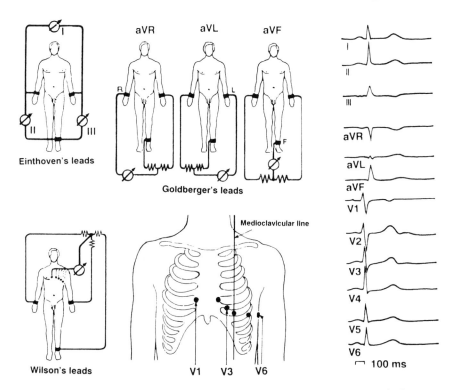

FIGURE 5.14. Arrangement of ECG leads in common use. Right: Typical curves recorded from a healthy subject.

Types of QRS Axis Orientation

The direction of the largest integral vector (the chief vector) during the spread of excitation is called the *electrical axis* of the heart. When the spread of excitation is normal, its direction in frontal projection agrees well with the anatomical long axis of the heart. Therefore, limb recordings can be used to infer the orientation of the heart. The various categories are based on the angle α between the electrical axis and the horizontal. In the normal range (shown at the top in Fig. 5.15), the angle to the horizontal varies from 0° to 90°. Angles above the horizontal are given a negative sign. The general categories of QRS axis orientation are: normal range ($0° < \alpha < +90°$); right axis deviation ($+90° < \alpha < +180°$); left axis deviation ($-120° < \alpha < 0°$).

For the construction of the electrical axis from the ECG by means of Einthoven's triangle (Fig. 5.15, top), two lead pairs suffice, for the third can be derived from the other two. At each instant during the excitatory cycle, it holds that: deflection in II = deflection in I + deflection in III (downward deflections having negative sign).

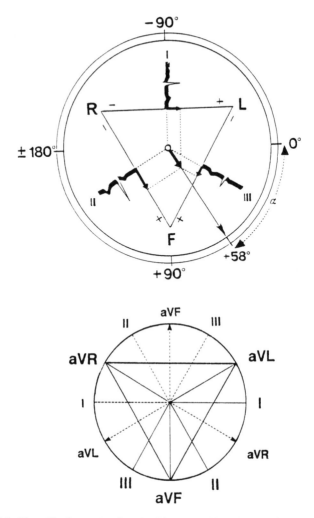

FIGURE 5.15. Top: Einthoven's triangle. The recording sites at the extremities are represented as the corners of an equilateral triangle, and the sites of the triangle correspond to the lead axes. Angle α of the QRS electrical axis indicated on the outer circle. The projection of an integral vector with angle α + 58° on the three axes is shown, and the magnitude of the various deflections in each axis is indicated by the customary curves. Bottom: Summary of axis orientations with the unipolar (Goldberger) and bipolar (Einthoven) limb leads.

In Goldberger's method, the voltage measured is that between one extremity—for example, the right arm (lead aVR)—and a reference electrode formed by voltage division between the two other limbs (see Fig. 5.14). With aVR recording, the lead axis on which the vector loop is projected is represented by the line bisecting the angle between I and II in the

Einthoven triangle (Fig. 5.15). The axes for aVL and aVF are found in an analogous way.

Unipolar Precordial Leads

Whereas the limb leads just described are fundamentally related to the frontal projection of the vector, the unipolar precordial leads of Wilson provide information chiefly about the horizontal vector projection. A reference electrode is produced by joining the three limb leads, and an exploring electrode records from specific points on the chest at the level of the heart (see Fig. 5.14). A positive deflection is seen when the instantaneous vector, projected onto the appropriate axis, points toward the recording site. If it points in the opposite direction, the deflection is negative.

Use of the ECG in Diagnosis

The ECG is an extremely useful tool in cardiological practice, for it reveals changes in the excitatory process that cause or result from impairment of the heart's activity. From routine ECG recordings, the physician can obtain information pertaining to the following:

Heart rate: Differentiation between the normal rate (60–90/min at rest), tachycardia (over 90/min), and bradycardia (below 60/min).

Origin of excitation: Decision whether the effective pacemaker is in the SA node or in the atria, in the AV node or in the right or left ventricle.

Abnormal rhythms: Distinction among the various kinds and sources (sinus arrhythmia, supraventricular and ventricular ectopic beats, flutter and fibrillation).

Abnormal conduction: Differentiation on the basis of degree and localization, delay or blockage of conduction (sinoatrial block, AV block, right or left bundle-branch block, fascicular block, or combinations of these).

QRS axis orientation: Indication of anatomical position of the heart; pathological types can indicate additional changes in the process of excitation (unilateral hypertrophy, bundle-branch block, etc.).

Extracardial influences: Evidence of autonomic effects, metabolic and endocrine abnormalities, electrolyte changes, poisoning, drug action (digitalis), etc.

Primary cardiac impairment: Indication of inadequate coronary circulation, myocardial O_2 deficiency, inflammation, influences of general pathological states, traumas, innate or acquired cardiac malfunctions, etc.

Myocardial infarction (complete interruption of coronary circulation in a circumscribed area). Evidence regarding localization, extent, and progress

It should, however, be absolutely clear that departures from the normal ECG except for a few typical modifications of rhythmicity or conduction as a rule give only tentative indications that a pathological state may exist. Whether an ECG is to be regarded as pathological or not can often be decided only on the basis of the total clinical picture. In no case, can one come to a final decision as to the cause of the observed deviations by examination of the ECG alone.

5.8   Abnormalities in Cardiac Rhythm as Reflected in the ECG

A few characteristic examples may indicate how disturbances of rhythmicity or conduction can be reflected in the ECG. The recordings, where not otherwise indicated, are from Einthoven's limb lead II (see Fig. 5.14). We first consider an analytical diagram as it is frequently used to illustrate the origin and spread of excitation in the heart. The bar SA in Fig. 5.16 symbolizes the rhythmic discharge of the SA node. The successive stages in the spread of excitation are shown from top to bottom, with the absolute refractory periods of the atria (A) and ventricles (V) indicated along the abscissa by rectangles. Figure 5.16a shows a normal ECG with the pacemaker in the SA node and the QRS complex preceded by a P wave of normal shape. Above the ECG trace, the process of excitation is diagrammed in the described way.

Rhythms Originating in the AV Junction

A source of rhythmicity in the AV junctional region (the AV node itself and the immediately adjacent conductile tissue) sends excitation back into the atria (including the SA node) as well as into the ventricles (Fig. 5.16b). Because excitation spreads through the atria in a direction opposite to normal, the P wave is negative. The QRS complex is unchanged, conduction occurring normally. Depending on the degree to which the retrograde atrial excitation is delayed with respect to the onset of ventricular excitation, the negative P wave can precede the QRS complex [Fig. 5.16b, (1)], disappear

FIGURE 5.16. Top: Illustration of the spread of excitation. The rhythmic discharge of the SA node is symbolized in the bar SA. The successive stages in the spread of excitation are shown from top to bottom, with the absolute refractory periods of the atria (A) and ventricles (V) indicated along the abscissa. AV summarizes the total atrioventricular conduction. Below: Normal time course of cardiac excitation (a) and of various disturbances (b–g): (b) Excitation generated at various parts of the AV junctional region. (c) Excitation originating in the ventricles spreads more slowly, and the QRS complex is severely deformed. (d) Interpolated ventricular extrasystoles of different origins. (e) Ventricular extrasystole with fully compensatory pause; S = normal SA interval. (f) Supraventricular extrasystole with incomplete compensatory pause. (g) Complete (third-degree) AV block.

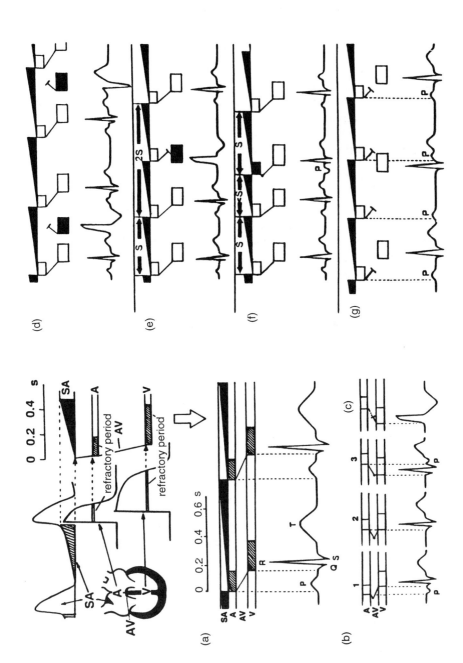

in it (2), or follow it (3). These variations are designated as upper, middle, and lower AV junctional rhythms.

Rhythms Originating in the Ventricles

Excitation arising at an ectopic focus in the ventricles spreads over various paths, depending on the source of the excitation (Fig. 5.16c). Because myocardial conduction is slower than conduction through the specialized system, the duration of spread through the myocardium is usually considerably extended. The differences in conduction path can cause pronounced deformation of the QRS complex.

Extrasystoles

Beats that fall outside the basic rhythm and temporarily change it are called extrasystoles. These may be *supraventricular* (SA node, atria, AV node) or *ventricular* in origin. In the simplest case, an extrasystole can be interpolated halfway between two normal beats and does not disturb the basic rhythm (Fig. 5.16d). Interpolated extrasystoles are rare, since the basic rhythm must be slow enough that the interval between excited phases is longer than an entire beat. When the basic heart rate is higher, a ventricular extrasystole is ordinarily followed by a so-called *compensatory pause*. As shown in Fig. 5.16e, the next regular excitation of the ventricles is prevented because they are still in the absolute refractory period of the extrasystole when the excitatory impulse from the SA node arrives. By the time the next impulse arrives, the ventricles have recovered, so that the first postextrasystolic beat occurs in the normal rhythm. But with supraventricular extrasystoles or ventricular extrasystoles that penetrate back to the SA node, the basic rhythm is shifted (Fig. 5.16f). The excitation conducted backward to the SA node interrupts the diastolic depolarization that has begun there, and a new cycle is initiated.

Atrioventricular Disturbances of Conduction

The ECG observed in cases of *complete AV block* is shown in Fig. 5.16g. As previously mentioned, the atria and ventricles beat independently of one another—the atria at the rate of the SA node and the ventricles at the lower rate of a tertiary pacemaker. *Incomplete AV block* is characterized by interruption of conduction at intervals, so that (for example) every second or third beat initiated by the SA node is conducted to the ventricles (2:1 or 3:1 block, respectively).

Atrial Flutter and Fibrillation

Atrial flutter and fibrillation are arrhythmias resulting from an uncoordinated spread of excitation, so that some atrial regions contract at the same time as others are relaxing (functional fragmentation). Atrial flutter is reflected in the ECG by waves with a regular sawtooth shape and a frequency of 220 to 350/min, which take the place of the P wave (Fig. 5.17a). Because of temporary AV block due to the refractory period of the

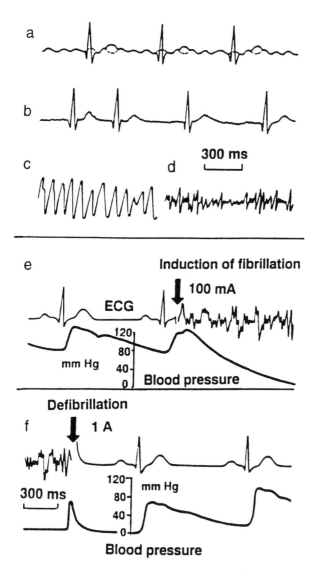

FIGURE 5.17. (a–d) ECG changes during flutter and fibrillation. (a) Atrial flutter with sawtooth-shaped flutter waves and 4:1 AV conduction. (b) Atrial fibrillation resulting in absolute arrhythmia of the ventricles. (c) Ventricular flutter. (d) Ventricular fibrillation. Time scale in f valid for all curves. (e, f) Induction and termination of ventricular fibrillation by electric current.

ventricular conducting system, normal QRS complexes appear at nearly regular intervals. In the ECG associated with atrial fibrillation (Fig. 5.17b), atrial activity appears as high-frequency (350–600/min) irregular intervals (absolute arrhythmia). There is a continuum of intermediate states between atrial flutter and fibrillation. In general, the hemodynamic effects of these atrial arrhythmias are only slight, and the patients are frequently quite unaware of the arrhythmia.

Ventricular Flutter and Fibrillation

With ventricular flutter and fibrillation electrical activity is uncoordinated, and the ventricles do not fill and expel the blood effectively. Circulation is arrested, and unconsciousness ensues; unless circulation is restored within minutes, death results. The ECG during ventricular flutter exhibits high-frequency, large-amplitude waves (Fig. 5.17c); whereas the fluctuations associated with ventricular fibrillation are very irregular, changing rapidly in frequency, shape, and amplitude (Fig. 5.17d). If flutter persists for several minutes, it changes more and more into fibrillation. Flutter and fibrillation can be set off by many kinds of heart damage: oxygen deficiency, coronary occlusion (infarction), cooling, overdoses of drugs or anesthetics, etc. Ventricular fibrillation is also the most common acute cause of death in electrical accidents.

5.9   Mechanism of Flutter and Fibrillation

There are two alternative concepts for the mechanisms underlying flutter and fibrillation. The first, originally proposed by Engelmann (1875) and later supported by Rothberger and Winterberg (1941), Scherf (1947), and others, assumes *ectopic automaticity* as the main cause of fibrillation. The second concept, proposed by Mines (1914), Garrey (1914), and Lewis, et al. (1920), postulates that *circulating excitation waves* (*reentry*) are responsible. The first case assumes that there are one or more ectopic foci that, by firing at high rates, overcome the regular formation and conduction of impulses in the heart. In the second case, fibrillation is attributed to a primary disturbance in the spread of excitation, in which excitatory waves circulate throughout extended regions of the heart.

It cannot be ruled out at the moment that either of the two mechanisms might operate under certain conditions. However, most of the phenomena that accompany the induction of fibrillation by any means can be explained convincingly only by the hypothesis of reentry. Direct evidence in favor of this hypothesis was presented by Allessie et al. (1973) for isolated rabbit atria and by Janse et al. (1980) for the left ventricles of isolated perfused porcine and canine hearts after coronary occlusion.

Conditions for Reentry; Anatomical and Functional Pathways

Reentry requires interconnected pathways along which excitation can proceed continuously. Since the days when Mines and Garrey developed the concept of reentry, many efforts have been made to identify anatomical structures in the heart that might represent reentry pathways. Rosenblueth and Garcia Ramos (1947) were able to show experimental evidence for such structures in the atrial wall surrounding the ostia of the great veins. A second pathway for reentry was shown by Wit et al. (1972) in meshes formed by Purkinje fibers (see Fig. 5.18), the excitability of which had been depressed. Moreover, Allessie et al. (1977) demonstrated reentry in a flat sheet of the atrial wall without macroscopically visible circuit pathways, showing that the development of reentry does not strictly depend on any preexisting ringlike anatomical structure, but may occur when the electrophysiological properties of myocardial tissue create a functional state that offers comparable conditions. Thus, a refractory zone may simulate a nonexcitable anatomical obstacle around which reentry can occur. In any case, reentry requires an excitation wave with a refractory zone shorter than the given interconnected pathway.

Length of the Excitation Wave

In a myocardial strand assumed to be infinite in length, excitation would proceed in a wavelike fashion. The distance between the excited and recovered parts is about 0.3 m, as can be calculated by multiplying the conduction velocity (≈ 1.0 m/s) by the duration of the myocardial action potential (≈ 0.3 s). Obviously, such a large excitation wave cannot bring about reentry in a human heart, because it is longer than any conceivable cardiac pathway. Thus, reentry requires the excitation wave to be shortened, either by decreasing the velocity of conduction, shortening the refractory period, or both.

The consistent relationship between heart size and probability of fibrillation was an early finding of Garrey (1914), who first emphasized that there must be a critical myocardial mass (or size) for reentry to occur. Furthermore, this relationship explains why sustained fibrillation is more likely in large hearts, whereas small hearts tend to defibrillate spontaneously.

Mechanisms of Abbreviated Refractory Periods

Several mechanisms can lead to an abbreviation of the cardiac excitatory process: (1) interval-dependent shortening of the cardiac action potential with increasing frequency of excitations, (2) shortening because of

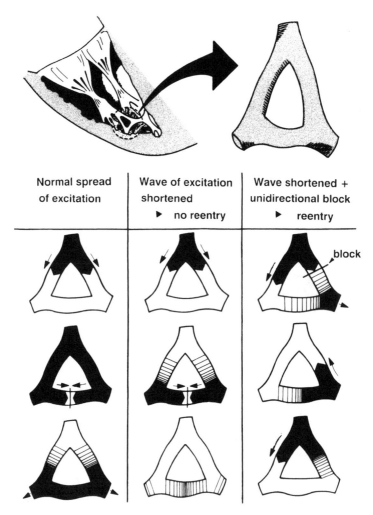

FIGURE 5.18. Representation of a loop-shaped piece of the ventricular conducting system (top). Absolute refractory zones of the excitation waves are represented in black, relative refractory zones are dashed. Left: The wave of excitation travels along the conducting system from top to bottom. Where the wavefronts meet, they extinguish each other because none of them can proceed into the absolute refractory zone of the other one. The same events will occur if the waves become shortened (middle). Right: Reentry is possible only if shortening of the waves of excitation is combined with transient block of conduction in one direction. This may occur when the heart is stimulated during its early relative refractory period.

metabolic disturbance caused by ischemia, and (3) shortening because of the electrotonic interaction between excited (depolarized) tissue proximal to a conduction block and unexcited tissue beyond the block.

1. An action potential initiated soon after a preceding one will be considerably shorter. The shorter the interval between the two action potentials, the shorter the duration of the second one. This influence depends on the rate of recovery from inactivation of the depolarizing ionic currents and on the rate of inactivation of the repolarizing currents (see Chapter 3). If an action potential is initiated before these processes are complete, its duration will be shorter than usual (Carmeliet, 1977).

2. In mild ischemia or in the early phase following coronary occlusion, there is a marked decrease in action potential duration accompanied by shortening of the refractory period (Elharrar et al., 1977). In more severe ischemia, the action potential duration decreases further, but the refractory period tends to lengthen, thus displaying the phenomenon of *postrepolarization refractoriness* (Lazzara et al., 1975).

3. If an action potential is propagated in excitable tissue, its shape depends not only on the active membrane properties of the excited tissue, but also on the passive membrane properties of the tissue still unexcited. During the fast upstroke of the action potential, the active membrane area supplies current to the neighbouring resting area, thus initiating excitation. This local circulating current, which depolarizes the resting membrane, simultaneously tends to repolarize the active membrane and to shorten the action potential. If propagation fails at a site of a conduction block, it can hasten the repolarization of the active membrane area proximal to the block. In this way, the duration of the action potential proximal to the site of a conduction block can be shortened by about 50% (Sasyniuk and Mendez, 1971).

Mechanisms of Slow Conduction

Very low conduction velocities of about 0.1 m/s or less can be obtained experimentally in myocardial preparations perfused with solutions containing both potassium in high concentrations and epinephrine (Cranefield et al., 1971). In such a medium, the fast inward current is inactivated by the potassium-induced depolarization, and simultaneously the slow inward current is enhanced by epinephrine. The result is an action potential, known as a *slow response*, which is slowly propagated with a conduction velocity of about 0.02 to 0.08 m/s. Under these conditions, it is possible to induce reentry within a loop of Purkinje fibers shorter than 15 mm.

Another possible mechanism of conduction delay occurs as a result of the electrotonic spread of excitation across a segment of inexcitable tissue (Jalife and Moe, 1981). An inexcitable segment—for instance, an ischemic zone—may be electrically well coupled with surrounding tissue but be unable to generate an all-or-nothing response. Thus, the spread of excitation across such tissue will be determined by the local circuit current that causes a depolarization in the excitable tissue beyond the inexcitable zone. A depolarization of sufficient amplitude may elicit an action potential in the distal tissue, although with marked delay. Such an electrotonically mediated conduction delay can result in conduction velocity less than 0.01 m/s and thus may allow reentry along circuit pathways only approximately 2 to 3 mm in length (Antzelevitch and Moe, 1981). Moreover, if the conduction delay across an inexcitable gap outlasts the refractory period of the tissue proximal to the gap, the excitation of the tissue beyond the gap can reexcite the proximal tissue by electrotonic transmission.

Conduction velocity also depends on the direction of propagation as related to the fiber orientation: propagation is slower in the direction transverse to the long cell axis than in the direction of this axis. Measurements of the axial resistance exhibit considerably higher values in the direction transverse to the long cell axis than in the longitudinal direction (Clerc, 1976; Spach et al., 1981). A possible explanation for the increased transverse axial resistance may be given by the fact that a substantial part of the resistance is located in the cell-to-cell junctions. Because the current in the transverse direction must pass through a junction every cell width, there are more cell junctions per unit length in the transverse direction, and hence the effective axial resistance will be greater.

Unidirectional Block and One-Way Conduction

To accomplish reentry, the excitation wave must travel in only one direction; that is, conduction must be blocked in one branch of the loop (Fig. 5.18, right). Furthermore, the blocked area must permit conduction in the opposite direction of the excitation wave arriving via the unblocked branch of the loop. If the effective refractory period of the area proximal to the block is shorter than the conduction time along the pathway, it will be able to reenter the loop. Thus, the occurrence of unidirectional conduction is an essential condition for reentry.

Probably the most important cause of unidirectional conduction block in cardiac tissue is given by nonuniformity in the recovery of excitability leading to spatial variations in refractoriness. As illustrated in Fig. 5.18 (right), the two pathways of the loop differ in their refractoriness. An impulse entering the system is therefore blocked in the right limb because of the still-existing refractory state of this region, whereas it can travel freely in the left limb where excitability has already recovered. After some delay (because of the travel time), the latter impulse can reenter the right limb, where the refractory period has meanwhile expired. Experimental evidence for the significance of nonuniformity in excitability was shown by Allessie et al. (1976), who found that reentrant tachycardias could be induced in a sheet of rabbit atrial tissue only if the excitation wave was elicited at a region of markedly inhomogeneous refractoriness.

Unidirectional conduction can also be associated with electrotonic propagation across a functional obstacle. Successful electrotonic transmission across an inexcitable gap in Purkinje fiber preparations depends on the mass of excited tissue proximal to the gap. If this tissue mass is too small to provide enough current for the excitation of a larger mass of tissue beyond the gap, conduction will fail. However, conduction in the opposite direction can succeed because the larger tissue mass is able to exert a sufficient depolarizing influence (Jalife and Moe, 1976).

Nonuniformity of active or passive cell properties leading to asymmetry in conduction are inherent features of cardiac tissue. Although not apparent under normal conditions where the safety factor of propagation is high, this nonuniformity can become obvious by the occurrence of unidirectional block if propagation is critically impaired, for instance, during propagation of a premature beat.

5.10   Vulnerable Period: Threshold for Fibrillation

Atrial or ventricular fibrillation can be induced by electrical stimulation coinciding with late atrial or ventricular systole—the so-called vulnerable period (Wiggers and Wégria, 1939). In the ventricles, the nonuniformity in recovery of excitability is maximal preceding the apex of the T wave of the ECG (Fig. 5.19). At this time, electrical stimulation elicits an excitation wave that encounters some regions fully recovered, others partially recovered, and others still absolutely refractory. Propagation of an electrically induced wavefront can thereby be initiated preferentially in certain directions, thus setting the stage for multiple reentry—the electrophysiological basis of ventricular fibrillation. The normal heart is susceptible to ventricular fibrillation only during this phase; at other times, stimulation is ineffective or results only in a single extrasystole.

In addition to the induction of fibrillation by electrical stimulation, a spontaneously generated extrasystole that occurs during a highly nonuniform phase of recovery may also induce fibrillation. This is the basis of the so-called *R-on-T phenomenon* in the ECG, which is considered a preliminary symptom of imminent fibrillation. Under certain conditions, the duration of the vulnerable period may span a longer time should marked nonuniformity in recovery persist, such as after a premature beat that exaggerates nonuniformity of recovery (Wégria et al., 1941). Furthermore, the vulnerable period may extend even beyond the T wave if the expiration of the refractory period of the myocardium lags behind the repolarization. This has been demonstrated during ischemia, in which postrepolarization refractoriness occurs (Williams et al., 1974). Under such conditions, late stimulation or a late coupled premature beat may induce tachycardia or ventricular fibrillation (El-Sherif et al., 1975).

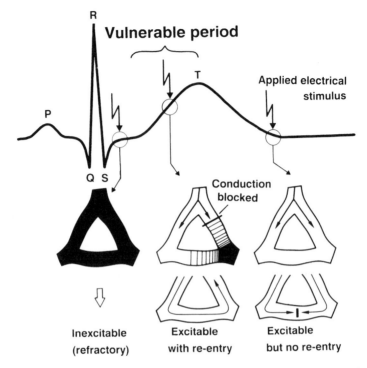

FIGURE 5.19. Diagram to explain the vulnerable period of the ventricles. The triangle-shaped figures below the ECG curve symbolize (as in Fig. 5.18) the branched network of the conducting system of the myocardium. In the vulnerable period, the conduction pathway is still partially refractory, so the wave of excitation generated by stimulation can propagate in only one direction and induce reentry in the same way as illustrated in Fig. 5.18. The ventricles would still be nonexcitable if they were stimulated earlier; at a later time, reentry is no longer possible because normal spread of excitation again takes place.

The Threshold for Fibrillation

The extra nonuniformity required for fibrillation is brought about in part by the direct influence of the current itself, which favors depolarization below the cathode and repolarization below the anode (see Chapter 4). If the inherent nonuniformity of excitability of the heart is low, the required increase in nonuniformity brought about by the applied current must be large, and thus the fibrillation threshold will be high. On the other hand, if the inherent nonuniformity is more pronounced, little additional extra nonuniformity will be required, and any agency that increases the degree of asynchrony of excitability recovery in heart muscle will facilitate the induction of ventricular fibrillation and lower the fibrillation threshold. Such conditions are given by localized or global myocardial ischemia, hypothermia, overdosage of quinidine or of ouabain, as well as by chloroform anesthesia (Han and Moe, 1964). The temporal dispersion of recovery of excitability and the degree of ventricular vulnerability are closely related. Likewise, metabolic acidosis predisposes the heart to ventricular fibrillation, whereas alkalosis exerts an opposite effect (Gerst et al., 1966). In canine experiments with coronary occlusion, the fibrillation threshold was also lowered if the site of stimulation remained unaffected by ischemia (Han, 1969). This means that, in such cases, the reduction of the fibrillation threshold is most probably the result of increased irregularity in the propagation of premature impulses rather than to a change of the excitability itself. Watson et al. (1973) determined ventricular fibrillation thresholds in human subjects in cardiopulmonary bypass using 50-Hz sinusoidal current applied both through epicardial electrodes and endocardial electrodes, alternatively. In the former case, the minimum fibrillating current was $735\,\mu A$ RMS; in the latter case, $67\,\mu A$. However, in one patient with grossly abnormal electrical activity of the heart, ventricular fibrillation could be induced with only $15\,\mu A$. This finding stresses the risk of micoelectrocution in hospitalized patients. In another study, performed in patients at the time of cardiac surgery, Horowitz et al., (1979) measured the fibrillation threshold using a 100-Hz train of rectangular pulses applied to the epicardial surface of the left ventricle. In 10 patients with normal left ventricles, the fibrillation threshold was $33.6 \pm 9.5\,\mathrm{mA}$. However, in 12 patients with 75% obstruction of the left anterior descending coronary artery, the ventricular fibrillation threshold was only $18.6 \pm 6.9\,\mathrm{mA}$. Hence, there is no doubt that a diseased heart is much more susceptible to fibrillation by electric shock.

Relation to the Threshold for Stimulation

The absolute value of the fibrillation threshold also depends on factors that determine the stimulation threshold, including the size and position of the electrodes, the current waveform, the duration of current flow, and the transition resistance. Only if these circumstances can be kept constant can relative changes in the fibrillation threshold be attributed to changes in the vulnerability of the heart. In Fig. 5.20, mean values for both thresholds are compared as the stimulus duration is increased from 0.5 ms to several milliseconds (Younossi et al., 1973). The values of the stimulation threshold reflect the *strength-duration curve* for the isolated whole heart. The curve drawn from the values for the fibrillation threshold shows an essentially similar shape, but it is shifted to a higher level by a factor of 10 to 20. Similar findings have been obtained by in-situ measurements from different species (Brooks et al., 1955).

The different levels of the two thresholds can be explained by the different conditions under which a cardiac response is elicited. To attain the stimulation threshold, a local density of current just high enough to stimulate a number of cells (approximately 50) with fully recovered excitability is

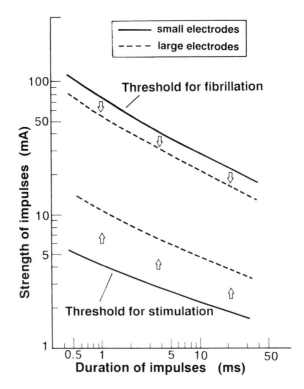

FIGURE 5.20. Relationship between stimulation threshold and fibrillation threshold in the isolated guinea pig heart ($n = 14$) for rectangular pulses (duration 0.5–50 ms). Note opposite effect of changes in size of stimulation electrodes on the two thresholds. Large electrodes were about $0.7\,cm^2$; small electrodes, $0.05\,cm^2$.

required. By contrast, to produce fibrillation, the cardiac response must be elicited during the early phase of recovery, and that requires a higher current intensity. Furthermore, to elicit premature responses at different sites of the heart, the current must spread throughout a certain mass of tissue, thus increasing the probability of multiple reentry. For the same reason, the two thresholds respond in an opposite way to changes in the electrode size (Fig. 5.20). Enlargement of the electrodes leads not only to a more extensive extrapolar current spread, but also to a decrease in the current density. The former effect is obviously responsible for the decrease in the fibrillation threshold; the latter increases the threshold for stimulation.

Effects of Alternating and Pulsed Direct Current

The fibrillation threshold cannot be predicted from the stimulation threshold when the current covers more than one cardiac cycle. For instance, the threshold for stimulation with pulsed DC is clearly below the value obtained with AC (Fig. 5.21). This result can be explained by the higher slope of the rising phase of the DC pulses compared with the sinusoidal AC waves. By contrast, the fibrillation threshold for DC is considerably higher than for AC (Hohnloser et al., 1982). Clearly, this result must be explained by causes other than the stimulatory efficacy of the two kinds of current. These causes can be derived from Fig. 5.9. The single fiber responded to 50-Hz AC with a series of action potentials at a frequency that depended on the duration of its refractory period and on the current strength. When the isolated heart is stimulated under comparable conditions, it shows a series of extrasystoles. A DC pulse exerts its stimulating effect only during the make-or-break phase of the current (Fig. 5.7). However, under the influence of long-lasting suprathreshold DC, the single fiber shows an additional effect; namely, the development of spontaneous activity (Fig. 5.11). The myocardial fiber then behaves like a pacemaker cell, exhibiting diastolic depolarizations and rhythmic firing. The frequency of these spontaneous action potentials, however, is much lower than during AC flow of comparable strength. Similarly, the frequency of extrasystoles of the whole heart

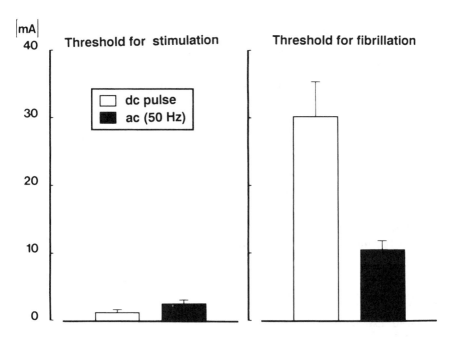

FIGURE 5.21. Right: Comparison of fibrillation thresholds for AC (50 Hz) and pulsed DC at a duration of 1 s. Mean values $\pm$ SD from guinea pig heart ($n = 5$). Left: Corresponding thresholds for stimulation. Size of electrodes was 0.05 cm^2 in all experiments.

is lower during DC flow than during AC even at a higher current intensity. It seems that the differences in the fibrillation thresholds for AC and DC can be related to the different numbers of repetitive extrasystoles that preceded the onset of fibrillation (see also Sect. 6.5).

Significance of Single and Repetitive Extrasystoles

Wégria et al. (1941) were the first to discuss a decrease of the fibrillation threshold after a premature response. They deduced this idea from the fact that the vulnerable period of a premature beat is prolonged and extends nearly to the end of the T wave, thus indicating a further state of nonuniform excitability. Han and Moe (1964) confirmed this conclusion by means of multiple-electrode recordings, which showed that the dispersion of the refractory period of nearby regions of the dog heart increased after a premature beat and thereby enhanced the intrinsic vulnerability; they measured a 35% decrease of the fibrillation threshold after a premature response (Han et al., 1966). This result was confirmed by Hauf et al. (1977), who found the decrease of the fibrillation threshold for a premature response to be 34% in the guinea pig heart.

A series of extrasystoles reduces the fibrillation threshold more so than does a single extrasystole (Roy et al., 1977). The dependence of the fibrillation threshold on the number of the repetitive extrasystoles that precede the onset of the arrhythmia explains why the fibrillation threshold decreases with increasing stimulus duration; that is, with an increasing number of prefibrillating extrasystoles. After the induction of a certain number of extrasystoles closely succeeding one another, the nonuniformity of excitability is at its maximum and cannot be increased further by inducing a greater number of extrasystoles. Hence, with increasing duration of AC flow, the fibrillation threshold reaches a minimum current plateau that is usually not much higher than the stimulation threshold (see Fig. 6.9).

5.11   Electrical Defibrillation

The first observations on electrical defibrillation date from about the same time as the induction of fibrillation by electric current was first described. Nevertheless, electrical defibrillation was not performed in humans before 1947 (Beck et al., 1947). Defibrillation requires short pulses of comparatively high intensity, applied through a large area. In case of success, fibrillation is terminated instantaneously, and normal cardiac activity is restored, usually without any sign of injury.

Defibrillation can be explained by synchronized stimulation of all excitable regions of the fibrillating heart by the electric shock. In this way, the heart as a whole is converted into an absolutely refractory state which prevents circus movement from going on. This state is comparable with the absolute refractory period at the end of the normal spread of excitation (Fig. 5.19). When interrupted in this way, reentry does not reappear, provided there are no premature excitations.

The large area of application of the defibrillation pulse and its comparatively high intensity of up to several amperes are required to achieve a suprathreshold stimulus strength in all regions of the fibrillating heart where excitable gaps are present at the instant of the shock. Since the excitable gaps perform circus movement too, defibrillation can also be effective if comparatively weak shocks are applied as a series of pulses in a circumscribed area via an intracardiac pacemaker electrode. Under these conditions, defibrillation occurs through the abolition of the excitable gaps, not instantaneously, but successively when they pass the region of stimulation.

Without doubt, electrical defibrillation is the most effective measure presently available against ventricular fibrillation. However, its application will lead to full success only when the circulatory arrest due to fibrillation does not exceed several minutes. Otherwise, irreversible damage of the brain must be expected. When the circulatory arrest lasts only a few seconds, unconsciousness may ensue as the first still-reversible disturbance of the central nervous function. After about 2 min, spontaneous respiration stops, and, increasing with time, the heart itself becomes more and more depressed, because its blood supply is interrupted too.

6
Cardiac Sensitivity to Electrical Stimulation

6.1 Introduction

It is important to understand the relationships between cardiac excitability and stimulus features such as duration, frequency, repetition pattern, wave-shape, electrode configuration, current density, and current pathway. Much of our knowledge of these effects is derived from studies directed toward therapeutic uses of cardiac stimulation, such as cardiac pacing and defi-brillation. We can better evaluate cardiac hazards if we understand the principles of cardiac electrophysiology presented in Chapter 5. The reviews in Chapters 3 and 4, concerning principles of excitable membranes, will also be referred to in this chapter.

6.2 Threshold Sensitivity with Respect to Cardiac Cycle

The sensitivity of the heart to electrical stimulation depends critically on the timing of the stimulus with respect to the cardiac cycle. Figure 6.1 (adapted from Jones and Geddes, 1977, and Geddes, 1985) shows the threshold sensitivity for both excitation and fibrillation as a function of the initiation time of brief (0.5–5 ms) rectangular current stimuli. The data were obtained in experiments with 120 dogs, with stimulation of the myocardium via a helical bipolar electrode. The membrane waveform illustrates the action potential (AP) that might be measured across the membrane of a single myocardial cell. The lower waveform shows a corresponding ECG that might be simultaneously measured with cutaneous electrodes (see Sect. 5.7). The horizontal axis of the figure applies to the initiation time of the stimulus with respect to an ongoing cardiac cycle, measured as a fraction of the QT interval of the ECG waveform. The QT interval is equivalent to the AP duration, as suggested by the drawings below the graph.

The period labeled "ARP" is the *absolute refractory period*, during which time the cellular membrane is nearly fully depolarized, and it is impossible to initiate a new AP process, no matter how strong the stimulus. The period labeled "RRP" is the *relative refractory period*—this is the repolarization phase of the AP process, during which the membrane may be reexcited by a sufficiently strong stimulus. Following the relative refractory period is the *diastolic period,* during which the tissue is most sensitive to excitation. The refractory behavior of cardiac tissue is largely a result of the *deactivation variable* of the cellular membrane during the AP process (refer to Sect. 3.2).

The excitation curves shown in Fig. 6.1 reflect the refractory condition of the excitable cardiac membrane, showing a threshold that declines sharply during the RRP to a minimum plateau that is sustained until the next cardiac cycle. Ventricular fibrillation (VF), however, can be induced only in

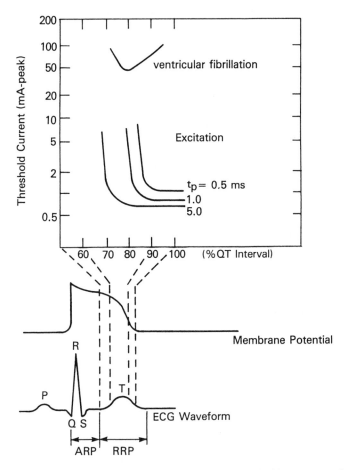

FIGURE 6.1. Excitation and ventricular fibrillation thresholds: upper, stimulation thresholds; middle, membrane potential; lower, ECG waveform. (Adapted from Geddes, 1985.)

TABLE 6.1. Ratio of VF to excitation thresholds for monophasic current pulse.

Current duration (ms)	Ratio I_F/I_{Ex}
0.5	145 ± 55
1	100 ± 27
2	112 ± 37
3	87 ± 25
4	81 ± 12
5	63 ± 37
10	52 ± ?

I_F determined during vulnerable period; I_{Ex} determined during diastolic period. Experiments with dogs (n = 120). Stimulation via bipolar helical electrode sutured to myocardium. *Source:* Data from Jones and Geddes (1977).

a narrow window during the repolarization phase of the heart—this is the so-called vulnerable period of the heart (see Sect. 5.1). The temporal point in the heart cycle when an extrasystolic excitation is most likely to cause VF displays a bell-shaped distribution. The parameters of that distribution (mean ± S.D.) were found to be 85 ± 9% of the heart's QT duration in sheep (Ferris et al., 1936), and 72 ± 9% in dogs (Jones and Geddes, 1977). Table 6.1 lists for the Jones and Geddes experiments the ratio of the minimum fibrillation current during the vulnerable period to the minimum excitation current during the diastolic period. The indicated ratio will depend on the conditions of stimulation, including the stimulus waveform, the electrode size, and the locus of excitation. The tabulated values apply to stimulation of dog hearts through a bipolar helical electrode sutured to the myocardium.

Only a few cardiac cells need to be excited to produce an extrasystolic contraction (Geddes, 1985), whereas ventricular fibrillation typically requires several conditions, namely: (1) excitation, (2) during the repolarization phase, and (3) involving a critical mass of the heart. Ventricular fibrillation can be initiated by exciting a site that is partially refractory as the normal AP propagates over the surface of the heart [see Sect. 5.9, and Antoni (1979)]. Reexcitation at that stage can result in an extraneous AP that fails to extinguish naturally, but rather circulates in an uncoordinated fashion called "reentry," that fails to produce pumping. It should not be assumed that excitation outside the vulnerable period is inconsequential to cardiac safety considerations. As pointed out in Sect. 6.4, extrasystolic excitation can significantly increase the susceptibility of the heart to fibrillation by a subsequent excitatory stimulus.

The heart is more sensitive during the diastolic interval to stimulation at the cathode electrode than at the anode, as are neurons (see Sect. 4.4). Nevertheless, cardiac tissue can be preferentially excited with an anodic stimulus during the relative refractory period (Brooks et al., 1955; Cranfied et al., 1957). Figure 6.2 (from Antoni, 1979) illustrates anodic and cathodic thresholds. An anodic pulse applied during the relative refractory period repolarizes the heart to a level that depends on the strength of the stimulus. If the repolarization is sufficient, the membrane may spontaneously revert to a depolarized state if the anodic stimulus is terminated within the AP time frame—the result can be the initiation of an extraneous AP (Hoffman and Cranfield, 1960). Cranfield and colleagues (1957) also demonstrated that, in the relative refractory period, anodic break may cause excitation at some minimum threshold, but may fail to produce excitation above a higher stimulus value. This phenomenon produces a "dead zone" between the anodic and cathodic thresholds where excitation does not occur.

The phenomenon of excitation by anodic and cathodic make and break currents was studied with two-dimensional cardiac simulation models (Roth, 1995). Although in all cases, excitation was initiated through cellular depolarization—a process most efficiently enacted beneath a cathode—it was found that a "virtual cathode" exists somewhat distant from an anode, much like that illustrated in Figure 4.12 for one-dimensional nerve models.

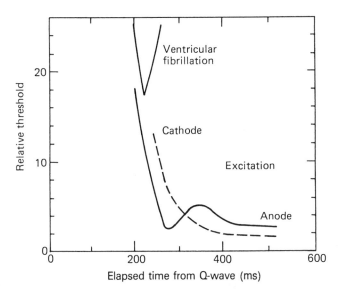

FIGURE 6.2. Excitation and ventricular fibrillation thresholds. [Adapted from Antoni (1979). Reprinted by permission of VCH Publishers, Inc., 220 East 23rd St., New York, NY 10010 from: *Progress in Pharmacology* (Stuttgart, Gustav Fisher Verlag), vol. 2/4, pp. 5–12, fig. 3 (page 6).]

Diastolic rheobase thresholds of excitation with this model for an electrode length of 1 mm and a surface area of $1.5\,mm^2$ were: 0.038, 0.41, 0.49, and $5.3\,mA$, respectively, for cathode make, anode make, cathode break, anode break.

When the stimulus sufficiently exceeds the VF threshold, a fibrillating heart may be defibrillated by wide-scale excitation that places the heart in a homogeneous state (see Sect. 5.11). There exists a threshold window within which fibrillation is possible. Below the lower limit of the window, VF is not possible because the stimulus is too feeble to cause excitation; above an upper limit, VF is avoided by wide-scale excitation. Consequently, exposure to high voltage or high current can result in massive injuries, without death from VF (see Sect. 10.2).

6.3 Strength-Duration Relations for Unidirectional Currents

An experimental relationship between the threshold intensity of a rectangular current stimulus and its duration defines a *strength-duration* (S-D) relationship. We have seen in Chapter 4 that experimental S-D data are frequently fitted with a hyperbolic (Eq. 4.15) or exponential (Eq. 4.17) expression. The parameter τ_e in either formulation may be determined by the ratio of minimum charge to minimum current:

$$\tau_e = \frac{Q_0}{I_0} \tag{6.1}$$

Equation (6.1) represents a simple and practical relationship for determining the parameter τ_e in both the hyperbolic and exponential relationships.

Although the S-D formulas were originally applied to neural excitation thresholds, these curves have been used to fit empirical data for cardiac excitation.[1] For example, the exponential relationship provides an excellent fit to ventricular excitation thresholds in dog and turtle hearts (Pearce et al., 1982) as well as excitation thresholds of sheep cardiac Purkinje fibers (Dominguez and Fozzard, 1970). Equally good fits have been obtained with the hyperbolic relation for excitation thresholds of dog hearts (Jones and Geddes, 1977). Comparison of experimental S-D curves for cardiac excitation in dogs reveals that both formulations give a good fit to experimental data, with no clear preference between the two (Mouchawar et al., 1989). Table 6.2 summarizes strength-duration data derived from a variety of sources. The table is organized by experimental end points consisting of excitation, fibrillation, and defibrillation. The column labeled "τ_e" lists em-

---

[1] As Used here, excitation refers to the production and propagation of an extrasystolic action potential—usually during the diastolic period of a previous beat.

pirical time constants, which in most cases were determined by Equation (6.1) or as specified in the cited work.

The medium value of τ_e in Table 6.2 is approximately 2.6 ms—a value that is roughly a factor of 10 above S-D time constants typically observed for neural excitation (see Sect. 4.5 and Table 7.1). Some general observations are pertinent to the information presented in Table 6.2.

Electrode Size

Irnich (1980) reports that τ_e increases monotonically with electrode size, as in Fig. 6.3. The Irnich data can be fitted by a relationship of the form $\tau_e = KA^{0.56}$, where A is the electrode area and K is a scaling constant. The direction of this relationship follows that observed with cardiac models (Roth, 1995) and for neural excitation (Pfeiffer, 1968), although it has not been observed by others (Talen, 1975). Smyth et al.'s data (1976) correspond closely to the Irnich relationship. For instance, for an electrode area of either 32 or 8 mm^2, the values of τ_e indicated by Smyth et al. are 2.3 and 1.7 ms, respectively—values that correspond very closely to the Irnich data.

Theoretical considerations (Jack et al., 1983, pp. 275, 418) predict that the S-D time constant should increase with the size of the area stimulated. With a small electrode, charge is applied rapidly to the membrane in a concentrated area, but more slowly to distant parts of the membrane. Consequently, with a small electrode, the membrane becomes more efficiently depolarized, and also acquires a local voltage gradient. These factors contribute to lower excitation thresholds and to shorter S-D time constants. On the other hand, if a larger area of membrane is excited, the time required for depolarization increases, and, because the depolarization is more diffuse, larger stimulus intensities are required for excitation. Jack et al., (1983, p. 418) reported that the size of the stimulated area affects the time constants much more in muscle than in nerve tissue.

Polarity

With cathodic stimulation, the site of depolarization is concentrated near the electrode; with anodic stimulation, it is spread more gradually over distant areas [see Sect. 4.4 and Reilly et al., (1985)]. As explained above, both excitation thresholds and S-D time constants are expected to be reduced as the area of stimulation is reduced. Consequently, cathodic stimulation should result in shorter time constants and lower excitation thresholds than anodic stimulation. Thalen et al. (1975) found that cathodic excitation thresholds were indeed below anodic thresholds by a factor of 2.6, in agreement with theoretical expectations; however, the observed S-D time constants with anodic stimulation were smaller than with cathodic stimulation by a factor of 2.5, a result not in conformance with the above arguments. Further work is needed to reconcile these experimental observations with theoretical expectations.

TABLE 6.2. Strength-duration data for cardiac stimulation by square-wave current waveforms.

τ_e (ms)	Species	Preparation	Stimulus locus	Electrode	Notes	Reference
A. Thresholds of excitation						
3.6	Rabbit	Isolated tissue	Papillary muscle	Large (E-field)		Knisley et al. (1992)
3.8	Guinea pig	Isolated tissue	Papillary muscle	Small		Weirich et al. (1985)
1.75	Rabbit	Isolated heart	Left ventricle	1 cm^2		Roy et al. (1985)
2.86 ± 0.99	Dog	In vivo	Trans-chest	8.1 cm diam.		Voorhes et al. (1983)
2.14 ± 0.97	Dog	In vivo	Ventricle	Small		Pearce et al. (1982)
7.31 ± 2.22	Turtle	In vivo	Ventricle	Small		Pearce et al. (1982)
0.68 to 3.94	*	*	*	4–94 mm^2		Irnich (1980)
0.206	Dog	In vivo	Myocardium	See note	(a)	Jones and Geddes (1977)
1.1	Human	In vivo	*	12 mm^2	(b)	Fozzard and Schoenberg (1972)
2.1	Dog	In vivo	Ventricle	Loop-and-catheter cathode	(c)	Thalen et al. (1975)
0.73	Dog	In vivo	Ventricle	Loop-and-catheter anode	(c)	Thalen et al. (1975)
3.75 ± 0.50	Sheep	Long (>8mm) Purkinje fiber	Near end	Pointlike	(d)	Fozzard and Schoenberg (1972)
29.3	Sheep	Short Purkinje fiber	Near end	Pointlike	(d)	Fozzard and Schoenberg (1972)
2.63 ± 0.91	Dog	In vivo	Trans-thoraz	4.1 cm dia.		Voorhees et al. (1992)
2.6 ± 0.88	Sheep	Purkinje fiber	Near end	Pipette	(d)	Dominguez and Fozzard (1970)
2.3	Human	In vivo	*	0.32 cm^2	(e)	Smyth et al. (1976)
1.7	Human	In vivo	*	0.18 cm^2	(e)	Smyth et al. (1976)
1.49	Dog	In vivo	Myocardium	7.9 mm^2		Mouchawar et al. (1989)

B. *Thresholds of fibrillation*

7.7	Guinea pig	Isolated heart	Whole heart	*		Weirich et al. (1985)
(~36.0)	Guinea pig	Isolated heart	*	*	(f)	Hohnloser et al. (1982)
1.70	Dog	In vivo	Myocardium	See note	(b)	Jones and Geddes (1977)

C. *Thresholds of defibrillation*

3.90	Dog	Isolated heart	Whole heart	Uniform field	(g)	Geddes et al. (1985)
2.76	Dog	Isolated heart	Right ventricle	Catheter	(g)	Wessale et al. (1980)
4.01	Dog	In vivo	Trans-chest		(g)	Bourland et al. (1978)
2.98	Pony	In vivo	Trans-chest		(g)	Bourland et al. (1978)
6.50	Dog	In vivo	Ventricle	*		Koning et al. (1975)

* Information sketchy or not available.

(a) τ_e monotonically increasing with electrode size.

(b) Electrode pair (1-cm separation) wrapped around 2-mm-OD tube and sutured to myocardium.

(c) Average of loop and catheter electrode time constants.

(d) τ_e strongly dependent on length of excised fiber, intracellular electrode.

(e) Data from 21-month electrode implant.

(f) Fibrillation induced by anodic break during vulnerable period of extrasystole; stimulus duration 30 to 100 ms. Time constant not indicative of single excitation (see text).

(g) Trapezoidal pulses with slopes not exceeding 15%.

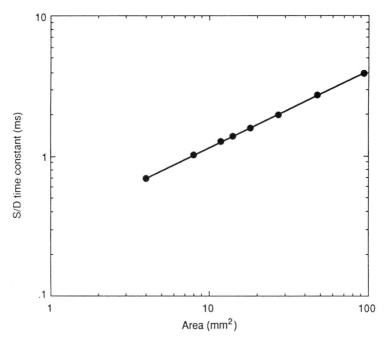

FIGURE 6.3. Relationship between S-D time constant and electrode area for cardiac excitation. (Data from Irnich, 1980).

Size of Tissue Preparation

When the stimulating current is introduced through an intracellular micro-electrode, the values of τ_e for small cardiac tissue samples depend critically on the size of the preparation. In the experiments of Fozzard and Schoenberg (1972), unusually large (29.3-ms) values of τ_e were seen with small preparations. However, long samples (>8 mm) of Purkinje fibers yielded values of τ_e that appeared typical of whole-heart preparations (3.75 ms). The direction of this effect follows theoretical expectations for intracellular excitation (Jack et al., 1983). See Chapter 4 for additional discussion of relationship between S-D time constants and the size of the stimulated tissue.

Fibrillation by Prolonged Currents

The S-D curves discussed above apply to single cardiac responses, in which case we expect thresbolds to follow Eqs. (4.15) and (4.17). However, when the duration of the stimulating current is prolonged beyond the S-D time constant, the fibrillation threshold may continue to drop due to the effects of multiple excitations (see Sect. 6.4). If we attempt to fit such data to the S-D formula, we may obtain time constants that appear to exceed that of mycardial excitation. The data presented by Hohnloser et al. (1982), for example, display an unusually long value of τ_e. The S.D. curve for these experiments declined steadily with stimulus duration, out to 60 ms, after which an increase in threshold was seen. According to the authors, the fact that fibrillation could be initiated either by cathodic make or anodic break accounted for the observed S-D relationship. The timing of these events with respect to the cardiac cycle is of major importance, according to the authors. Knickerbocker (1973) also observed fibrillation thresholds that declined steadily with prolonged stimulation out to 2 s— the maximum period of stimulation in his experiment (refer to Sect. 6.8 and Fig. 6.11).

Consistency of Data

There is a wide spread of reported τ_e values, with a medium of about 2.6 ms. The thresholds of fibrillation listed in Table 6.2 demonstrate a particularly wide variance in comparison with the other categories, although it is difficult to judge the extent to which this is simply a consequence of relatively few data points. In general, it is more difficult to establish minimum fibrillation thresholds than excitation thresholds, because the former depend more critically on the on and off times of the stimulus current with respect to the cardiac cycle, as well as on the number of prior extrasystolic excitations (refer to Sect. 6.5).

6.4   Biphasic and Sinusoidal Stimulation

Before considering how oscillatory currents produce cardiac excitation, it is helpful to make comparisons with corresponding properties of neural excitation. Figure 4.16 illustrates S-D curves from a neuroelectric model for monophasic and biphasic waveforms. Compared with the S-D curve for a monophasic pulse, thresholds for sinusoidal stimuli rise at durations below the S-D time constant. This upturn occurs because the current reversal of the sinusoidal cycle tends to reverse an action potential that was started by the initial phase of the stimulus. To compensate for this reversal, the biphasic wave requires a higher current to achieve threshold. This cancellation effect is most effective as the phase duration t_p is reduced. For long t_p, thresholds depend more on the rate of rise and peak value of the current. It should not be construed that thresholds for all oscillatory stimuli are necessarily elevated above monophasic stimuli of the same phase duration. Indeed, if the sinusoidal cycle is repeated as an oscillatory waveform, thresholds may decrease with successive oscillations, as noted in Fig. 4.19.

In Chapter 4, we presented a functional relationship for strength frequency (S-F) curves obtained from monophasic S-D curves by considering that a sinusoidal half-cycle corresponds to a single monophasic pulse. This simplification would, however, fail to account for high- and low-frequency upturns in S-F curves relative to S-D curves. A better empirical fit to S-F data from the myelinated nerve model is described by Eq. (4.35), which involves high- and low-frequency transition parameters f_e and f_o, respectively. While the equation was derived from a neuroelectric model, it provides a reasonably good fit to cardiac stimulation data if the equation parameters are chosen correctly.

Table 6.3 includes f_e values determined from S-D curves published by several sources listed in the right-hand column. The table is organized by beat rate effects, excitation, and fibrillation. Data apply to continuous cycle stimuli except for rows 2 and 4, which apply to single-cycle stimulus waveforms. The data listed under the heading f_e were determined by a rough curve fit of the experimental data to Eq. (4.36). Excluding item 1, the geometric mean of the f_e values in Table 6.3 is 115 Hz. Item 1 was excluded from this average because it was felt that the mechanism affecting the beating rate in this experiment is fundamentally different from the other two categories by virtue of the unusually small value f_e and the sharply defined frequency sensitivity of the threshold current. The average value $1/(2f_e) = 4.3$ ms compares favorably with the S-D time constant $\tau_e = 2.6$ ms derived from the data in Table 6.2. This comparison appears adequately consistent considering the great diversity of experimental methods represented in Tables 6.2 and 6.3.

Figure 6.4 plots a number of experimental threshold data points taken from S-D curves published by several of the sources listed in Table 6.3. The plotted data have been normalized by the minimum threshold current in the various experiments. Figure 6.4 also shows a plot of the empirical formulas defined by Eq. (4.35), in which the following relationships have been used for cardiac excitation: $f_e = 115$ Hz, $f_0 = 10$ Hz, $a = 1.45$ for single-cycle stimulus and 0.9 for continuous-cycle stimulus, and $b = 0.8$. The value of f_e has been taken from the geometric mean of the data listed in Table 6.3. The other values have been taken from the myelinated nerve model for neural excitation (refer to Chapter 4). Of the data plotted in Fig. 6.4, that of Weirich et al. (1985) apply to single-cycle stimuli; the other data apply to continuous-cycle stimulation. The empirical curves plotted in Fig. 6.4 provide a reasonably good fit to the combined experimental data; better fits could be obtained if the parameters f_e and f_0 were determined separately for each individual experiment. The low-frequency term, K_L, should have an upper limit, K_{DC}, that accounts for the finite threshold observed with DC stimulation. Fibrillation thresholds for DC currents have been reported by several authors. The data of Ferris et al. (1936) indicate K_{DC} values of about 5.2; Biegelmeier (1987) reported values of K_{DC} in the range of 3 to 4 for durations of several seconds and a value of about 1.0 for durations of a fraction of a second.

TABLE 6.3. Strength-frequency data for cardiac stimulation by sinusoidal current waveforms.

f_c (Hz)	Species	Preparation	Stimulus locus	Electrode	Notes	References
A. Thresholds of beat rate change						
3	Frog	Isolated heart	Whole heart	Uniform field	(a)	Kloss and Carstenson (1982)
B. Thresholds of excitation						
278	Guinea pig	Isolated tissue	Papillary muscle	AgAgCl	(b)	Weirich et al. (1985)
50	Guinea pig	Isolated heart	Aorta/apex	AgAgCl	(c)	Weirich et al. (1983)
C. Thresholds of fibrillation						
250	Guinea pig	Isolated heart	Whole heart	AgAgCl	(b)	Weirich et al. (1985)
72	Guinea pig	Isolated heart	Aorta/apex	AgAgCl	(c)	Weirich et al. (1983)
79	Dog	In vivo	Ventricle	3.1 cm^2		Kugelberg (1976)
148	Human	In vivo	*	4-mm diam.		Kugelberg (1976)
100	Dog	In vivo	Various	Various		Geddes et al. (1971)
100	Dog	In vivo	Trans-chest	8 × 10 cm	(d)	Geddes and Baker (1971)
120	Dog	In vivo	Trans-chest	1/2-in. wire	(e)	Geddes and Baker (1971)

* Information sketchy or not available.
(a) Thresholds for changing beat rate by 30%.
(b) Single-cycle stimulus.
(c) Current duration = 1 s.
(d) Current flow transverse to long axis of animal.
(e) Current flow longitudinal to long axis of animal.

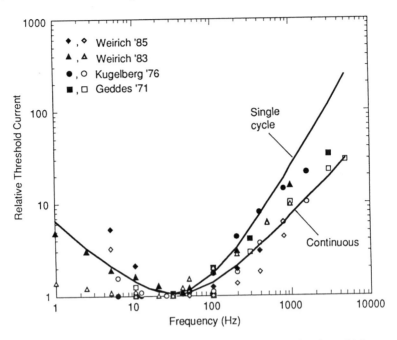

FIGURE 6.4. Strength-frequency data for cardiac stimulation by sinusoidal currents. Points are examples from published experimental data, solid curves represent analytic expression for single-cycle (upper) and continuous-cycle (lower) stimuli. Two examples are provided for each source, as indicated by open and filled symbols.

6.5   Duration Sensitivity for Oscillatory Stimuli

Cardiac sensitivity to oscillating currents is affected by a complex of factors not present with single DC pulses, including: (1) A current reversal can reverse an excitation process begun on an initial stimulus phase; (2) the excitation threshold resulting from a prolonged AC current steadily drops with the current duration; (3) an AC current may promote a train of extrasystolic excitations, which successively make the heart more susceptible to ventricular fibrillation. The first two of these effects was demonstrated in Chapter 4 for the neural membrane—Fig. 4.19 shows a threshold that oscillates with successive current phases, but nevertheless steadily declines on the average. The third effect has been developed in sect. 5.10.

Excitation Sensitivity

The cancellation effect of a biphasic reversal has been observed for cardiac stimulation. Figure 6.5 (adapted from Green et al., 1985) shows the ventricular fibrillation threshold versus the duration of a 50-Hz sine wave initiated at its peak. A threshold minimum occurs at approximately the point of current reversal (5 ms). Although the current dip is small in comparison with the variance of the measurements, it was consistently reproduced in several experiments. Furthermore, when the stimulus was initiated at the zero crossing of the current waveform, the minimum was shifted to 10 ms. The effects of current reversal on the fibrillation threshold shown in Fig. 6.5 is relatively weak. But, by analogy with neural stimulation, a stronger effect is expected at higher stimulus frequencies. Indeed, biphasic current reversal of a 1-kHz waveform has been shown to require current levels that are significantly elevated relative to a monophasic waveform (refer to discussion concerning Fig. 6.18).

Stimulation by an alternating current involves a nonlinear integration of the current that crosses the excitable membrane. The membrane conductance is approximately constant well below the excitation threshold. However, as the depolarization voltage approaches the value needed for excitation, the conductance for membrane current influx and efflux becomes increasingly asymmetric. As a result, membrane depolarization caused by one phase of cyclic stimulus may be only partially offset by a current reversal on a succeeding half-cycle. This asymmetric process can result in a buildup of the membrane depolarization on successive cycles of an oscillatory stimulus. This process of membrane depolarization was demonstrated in the cardiac system with stimulation of the papillary muscle of rhesus monkeys by 50-Hz alternating currents (Antoni et al., 1969). In that study, AC depolarization effects were observed experimentally and were

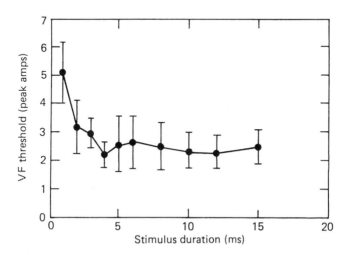

FIGURE 6.5. Ventricular fibrillation threshold versus duration of 50-Hz sinusoidal stimulus initiated at peak. Electrodes right fore limb to left hind limb of dogs ($n = 12$). Means and S.D. shown. (From Green et al., 1985.)

also verified with computer simulation of the electrical response of Purkinje fibers.

Fibrillation Sensitivity

As with nerve stimulation (Fig 4.20), oscillatory stimuli can produce a sequence of excitations in the heart—an effect that can significantly enhance the biological response that might otherwise result from a single excitation (see Sect. 5.10). This phenomenon has important safety consequences, because the fibrillation threshold falls steadily with increasing number of extrasystolic excitations. Figure 6.6 illustrates this effect with experimental data from in-vivo excitation of dog hearts (from Sugimoto et al., 1967)—the vertical axis gives the VF threshold to a 60-Hz current; the horizontal axis gives the number of prior ventricular responses produced by a conditioning 60-Hz current with variable duration. The duration of current for six responses was about 1 s. Over the range of one to six ventricular responses, the average VF threshold drops by a factor of 30. When the conditioning current was changed to a train of unidirectional rectangular pulses and the VF stimulus was a single DC pulse, a response curve nearly identical to that shown in Fig. 6.6 was obtained. Clearly, the ability of a

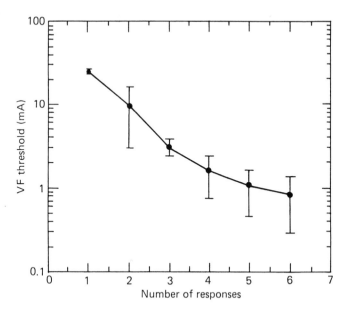

FIGURE 6.6. Ventricular fibrillation thresholds as a function of number of excitations from 60-Hz stimulation. Experiments with dogs ($n = 6$); mean ± standard deviation plotted. Needle electrodes inserted into epicardial surface. (Data from Sugimoto et al., 1967.)

TABLE 6.4. Comparison of AC and DC excitation
and fibrillation thresholds—experiments with dogs.

	DC (t = 1 ms) (mA-peak)	AC (50 Hz, t = 5 s) (mA-peak)
Ex. thresholds	0.10 + 0.05	0.11 ± 0.03
VF thresholds	15 ± 4.3	0.26 ± 0.08
Ratio VF/Ex.	150	2.4

VF stimuli applied during vulnerable period to in-vivo dog
hearts (n = 14); 25-gauge bipolar needle electrodes spaced
5 mm apart, inserted 2 mm into epicardial surface. ± values
indicate standard deviation.
Source: Data from Sugimoto et al. (1967).

stimulus to produce multiple cardiac excitations significantly affects its
potency for causing fibrillation.

Table 6.4 compares excitation and ventricular fibrillation thresholds for
in-vivo stimulation of dog hearts. The VF threshold exceeds the excitation
threshold by a multiple of 150 when the stimulating current is a single DC
pulse; with 60-Hz stimulation of 5 s duration, the multiple is only 2.4. Nevertheless, the 60-Hz excitation threshold is nearly the same as that for a
single DC pulse. Table 6.5 summarizes AC and pulsed DC thresholds from
the data of Hohnloser et al. (1982). With DC stimulation of 1 s duration, the
ratio of the VF threshold to the excitation threshold is 23:1; with AC
stimulation, the ratio is only 4.2:1.

Both alternating and direct currents can disrupt the normal rhythm of the
heart and promote multiple excitations (c.f. Fig. 3.15). The greater fibrillation efficacy of 60 Hz relative to a DC pulse arises because the AC stimulus
is a more efficient promoter of multiple ventricular responses. The lowering
of the fibrillation threshold by an extrasystolic excitation arises because
extra beats create a higher degree of nonhomogeneity within the heart,
making it more susceptible to fibrillation (see Sect. 5.10). A burst of closely

TABLE 6.5. Comparison of AC and DC excitation
and fibrillation thresholds—experiments with
guinea pig hearts.

	DC (t = 1 s) (mA-peak)	AC (50 Hz, t = 1 s) (mA-peak)
Ex. thresholds	1.3 ± 0.6	2.6 ± 0.8
VF thresholds	30 ± 8.5	11 ± 2.2
Ratio VF/Ex.	23.1	4.2

Isolated perfused guinea pig hearts (n = 3); ring-shaped
electrode attached to apex of heart.
Source: Data from Hohnloser et al. (1982).

spaced extrasystoles especially enhances this effect (Hohnloser et al., 1982). The maximum achievable rate of extra-systolic excitation is enhanced by the fact that the AP duration of an extrasystole can be significantly shorter than the normal AP duration. This phenomenon was demonstrated by simulation of Purkinje fiber response such as seen in Fig. 3.14.

Figure 6.7 illustrates the relationship between current magnitude and the number of extrasystolic excitations induced by a 3-s burst of 50-Hz alternating (solid dots) or direct (open dots) current introduced into isolated guinea pig hearts (Hohnloser et al., 1982). The terminal values of each curve show the fibrillation threshold. It is evident that both AC and DC stimulation are capable of producing multiple excitations, but the AC stimulus is much more effective.

Figure 6.8, from Weirich et al. (1983), demonstrates the relationship between the extrasystolic rate and the fibrillation threshold for sinusoidal

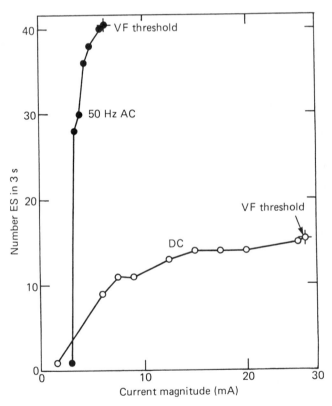

FIGURE 6.7. Relationship between current strength and number of extrasystoles for DC and 50-Hz AC stimulation of 3-s duration. Terminal values represent ventricular fibrillation thresholds. Perfused guinea pig hearts; ring electrode at apex of heart. (From Hohnloser et al., 1982.)

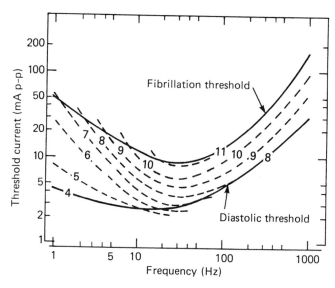

FIGURE 6.8. Threshold cardiac stimulation by sinusoidal current of 1-s duration. Ring electrode at heart apex. Isolated, perfused guinea pig hearts (n = 4). Upper solid curve shows VF threshold; lower solid curve shows excitation threshold during diastolic interval; broken curves show contours of constant estrasystole number. Mean thresholds shown. (From Weirich et al., 1983.)

currents of 1 s duration with frequency ranging from 1 to 1,000 Hz. The upper solid line represents the fibrillation threshold; the lower solid line represents the threshold for inducing an excitation during the diastolic interval of the heart; the broken lines represent current levels at which the extrasystolic rate is constant. The thresholds in Fig. 6.8 have a minimum around 30 Hz. Above that frequency, fibrillation and diastolic thresholds follow constant excitation rates of 11 and 8 per second, respectively. Below 30 Hz, fibrillation and diastolic thresholds also rise, but are associated with decreasing excitation rates. The upturn in thresholds above 30 Hz may be explained by the fact that as the AC frequency is increased, the alternate phases of alternating current have an increasing tendency for mutual cancellation. This explanation is strengthened by the observation that thresholds do not rise when the sinusoidal stimulus is replaced by a train of 1-ms unidirectional pulses (Weirich et al., 1983). Figure 6.8 also shows a low-frequency upturn that is associated with decreasing excitation rate as the frequency is lowered. Although a low-frequency upturn is expected for an excitation curve (see Fig. 6.4), it is not easy to explain the association of VF thresholds with decreasing excitation rates because it becomes less probable that an excitation will coincide with the vulnerable period of the heart.

The Z Relationship for Fibrillation by AC Currents

Various experiments have demonstrated that when VF thresholds are plotted versus duration of an AC stimulus, the result is a sigmoidal-shaped curve with an asymptotic maximum at short durations (a few cycles of stimulation) and an asymptotic minimum at long durations (several seconds). This so-called Z relationship was evident in early experiments of Ferris et al. (1936) and was later substantiated by others.

Figure 6.9 illustrates the Z relationship for fibrillation by AC stimulation as determined by Roy et al. (1977). In these experiments, stimulation was via catheter electrodes of various areas, introduced into the right ventricle of dog hearts. The authors obtained an empirical fit of thresholds with the following equation:

$$I_T = I_0\left[1 - \exp\left(-K_n N^b\right)\right]^{-1} \quad \left(\text{for } N \geq 3\right) \quad (6.2)$$

where I_O is the minimum current corresponding to the longest duration (1,000 cycles of 60 Hz); K_n and b are empirical constants given in Table 6.6. The value of K_n listed for an area of 500 mm^2 was obtained by this writer using logarithmic extrapolation of the K_n values at the three areas tested (A = 0.22, 14, and 90 mm^2). This extrapolated value represents a rough estimate of the threshold dependence for cardiac excitation over a large area. Equation (6.2) provides an excellent fit to Roy et al.'s experimental data, provided that $N \geq 3$; for $N \leq 3$, measured thresholds reach an asymptotic maximum, whereas Eq. (6.2) continues to rise, thereby overstating the experimental thresholds. These experiments show that the threshold dependence on electrode area is a strong function of the duration of the AC stimulus. At 1,000 cycles of 60-Hz stimulation ($t = 17$ s), there is an 18:1 ratio of thresholds between the largest and smallest electrodes tested (90 and 0.22 mm^2); with 1 cycle of current, the ratio of thresholds is perhaps 2:1. The final column of Table 6.6 lists the ratio of the upper and lower plateaus of the Z curves, based on the data from Roy et al. The upper plateau has been determined from Eq. (6.2) at $N = 3$. It can be seen that the plateau ratio diminishes as the area is increased. The value listed for $A = 500$ mm^2 is an estimate for large-area contacts or possibly for whole-heart stimulation.

Figure 6.10 illustrates VF thresholds in guinea pigs versus the duration of AC stimulation having frequencies of 5, 50, and 200 Hz. The curves for 50 and 200 Hz differ by a nearly constant multiple that reflects the frequency sensitivity of the excitation process (refer to Sect. 6.4). The curve at 5 Hz, in contrast, converges to the 50-Hz thresholds at short durations and to the 200-Hz thresholds at long durations. The indicated VF thresholds depend in a complex way on the individual extrasystolic excitations, their rate, and their timing with respect to the cardiac cycle. At 1 s stimulation, the difference in thresholds for the 50- and 200-Hz curves closely corre-

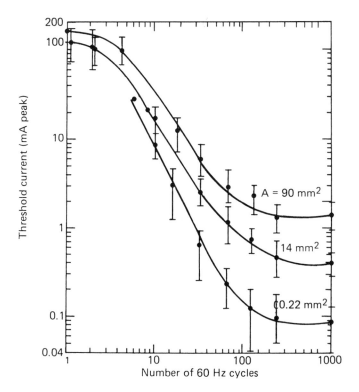

FIGURE 6.9. Relationship between VF threshold and number of 60-Hz cycles. Separate curves apply to area of catheter electrodes inserted into right ventricle. Mean and S.D. values shown for 50 dogs. (From Roy et al., 1977, © 1977 IEEE.)

sponds to the differential sensitivity of diasystolic stimulation (see Fig. 6.8). The 5-Hz VF threshold is, however, elevated relative to that at 50 Hz because of its lower rate of extrasystolic production. As the duration of stimulation is shortened, the extrasystolic production rate for the three

TABLE 6.6. Empirical constants for Eq. (6.2).

Area(mm^2)	K_n	b	I_{max}/I_{min}
(500)	(3.7×10^{-3})	(1.5)	(50)
90	1.8×10^{-3}	1.5	90
14	8.3×10^{-4}	1.5	200
0.22	1.1×10^{-4}	2.0	1,000

Empirical constants for $A = 90$, 14, and 0.22 mm^2 were determined by Roy et al. (1977). Values for $A = 500$ mm^2 have been extrapolated from Roy et al.'s data. Values listed for I_{max}/I_{min} represent ratio of upper and lower plateaus of fibrillation thresholds.

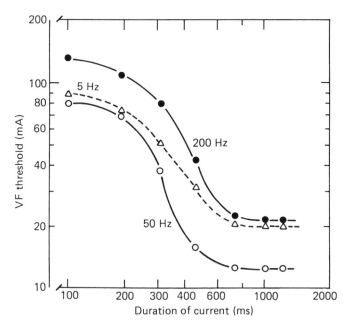

FIGURE 6.10. Fibrillation threshold versus duration of current for three frequencies. Data from isolated perfused guinea pig hearts. (Adapted from Antoni, 1985.)

curves becomes less disparate. Consequently, for short stimulus durations, VF thresholds more closely reflect diastolic excitation sensitivity at all three stimulation frequencies.

Figure 6.11 illustrates VF thresholds for 20-Hz AC and DC currents applied to dogs (left foreleg to right hind leg). Both AC and DC stimuli result in steadily declining thresholds for increasing stimulus duration, although the rate of decline in AC thresholds is much greater. As explained above, the differential rate of decline arises because the extrasystolic rate associated with the AC waveform is greater than that for the DC waveform. (see also Fig. 5.21).

Considering the mechanisms discussed above, we can represent the duration sensitivity of 50/60-Hz fibrillating currents as shown in Fig. 6.12. It is assumed that short-duration stimuli coincide with the vulnerable period of the cardiac cycle. The thresholds in regions (1) and (2) follow a strength-duration curve for pulsed stimuli—the assumed S-D time constant is 3 ms, as suggested in Sect. 6.3. The lower curve in region (1) is drawn under the assumption that the stimulus current initiation time results in a unidirectional (monophasic) pulse, in which case the threshold converges to a t^{-1} relationship. If the initiation time is near a current reversal such that the current pulse is biphasic, the threshold will be elevated as indicated by the upper curve in region (1). For a balanced biphasic waveform, the relationship will be approximately $t^{-1.5}$, as suggested by the discussion in Sect. 6.4

and by analogy with neural membrane properties (see Chapter 4). For intermediate cases of biphasic asymmetry, the threshold would fall in between the two curves. In region (2), the stimulus duration is longer than the S-D time constant (indicated as t_o in Fig. 6.12), and the threshold curve converges to a constant-current relationship. In region (3), the stimulus consists of more than one AC cycle, leading to multiple extrasystoles. The VF threshold drops with each successive excitation, reaching a minimum plateau in region (4).

The ratio, R, of the plateau thresholds in regions (2) and (4) and the transition duration, t_1 and t_2, vary with the experimental conditions, including the tested species. Table 6.7 summarizes empirical values of R, t_1, and t_2 applying to 50/60-Hz exposure to the whole heart or through a large-area cardiac contact electrode.

In applying animal data to humans, Biegelmeier and Lee (1980) argue that differences in heart beat rate between humans and animals should be taken into account, such that the ratio R and upper transition duration t_2 would be increased in a human relative to an animal with a much faster beat rate. This idea is bolstered by the data in Table 6.7 for guinea pigs—the transition time t_2 and the ratio R are both quite short with respect to the data for larger animals having longer cardiac cycles. Biegelmeier proposes human VF thresholds as indicated in the lower section of Table 6.7. The

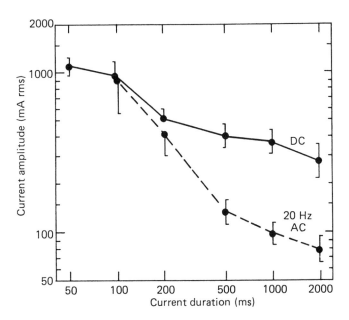

FIGURE 6.11. VF thresholds versus duration for DC and 20-Hz AC currents; in-vivo stimulation of dogs (n = 10–25). Error bars show 95% confidence limits. (From Knickerbocker, 1973, © 1973 IEEE.)

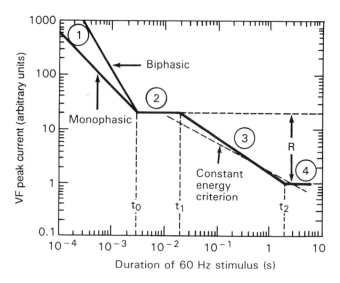

FIGURE 6.12. VF threshold versus 60-Hz stimulus duration. Numbered portions of the curve represent regions of differing sensitivity (see text). Short-duration stimuli are assumed to coincide with vulnerable period.

TABLE 6.7. Z-curve parameters for 50/60-Hz currents.

Species	Current path	Threshold ratio R	Transition time (s) t_1	t_2	Reference	Notes
A. *Measured relationships*						
Sheep	Right fore/ left hind limb	10	0.27	1.4	Ferris et al. (1936)	(a)
Pig		10	0.03	4.0	Jacobsen et al. (1975)	
Dog	Ventricle (5 cm² elect.)	50	0.05	2.0	Roy et al. (1977)	(a)
Dog	Fore/hind limb	25	0.02	2.0	Biegelmeier and Lee (1980)	(b)
Guinea pig	*	6	0.20	0.80	Antoni (1985)	
B. *Postulated relationships for humans*						
Human	Hand/foot	20	0.20	5.0	Beigelmeier and Lee (1980)	(c)
Human	Hand/foot	20	0.10	2.0	Beigelmeier (1987)	(c)

*Information sketchy or not available.
(a) Area relationship extrapolated to 5 cm² (see text).
(b) Derived from data of Kouwenhoven et al. (1959), Scott et al. (1973), and Kiselev (1963).
(c) Postulated from animal data; adjustment for beat rate; 50% probability curves.

bottom entry represents the recommendations of the International Electrotechnical Commission (IEC) for human safety criteria (see Fig. 11.10).

The extent to which Fig. 6.12 would apply to frequencies other than 50/60 Hz has not been established, although its general features are expected to be applicable to a much wider frequency range. The data of Antoni (1985) (Fig. 6.10) indicate that over the time interval of 0.1 s to 1 s, the Z curve for 200 Hz is shifted above the curve for 50 Hz by an approximately constant multiplicative factor, consistent with the frequency sensitivity discussed in Sect. 6.4. If the duration of stimulation were to be reduced below 0.1 s to a fraction of a 200-Hz cycle, we would expect the 50-Hz and 200-Hz curves to converge to the same curve indicated in regions (1) and (2).

6.6   Energy Criteria and Impulse Currents

Engineers in North America have traditionally evaluated safe current thresholds for 60-Hz exposure using the so-called *electrocution equation* of Dalziel (1960, 1968):

$$I_f = Kt^{-1/2} \qquad (6.3)$$

Where I_f is the fibrillation threshold, t is the duration of exposure, and K is a scaling constant that depends on body weight and the assumed percentile rank of sensitivity. The limits on t are from 8.3 ms to 5 s. Equation (6.3) is fundamentally an energy criterion in which $I^2 t$ is a constant. Values of K for various body weights and percentile ranks are given in Table 6.8, based on Daiziel's criteria (see Fig. 6.15). The values listed under the category "MNF" were said to apply to a "maximum nonfibrillating" current. The 50 percentile rank was also represented as an *average* in some of Daiziel's writings. Daiziel concluded that VF thresholds are linearly related to body weight, as discussed in Sect. 6.7.

TABLE 6.8. Values of K for electrocution equation.

Body weight (kg)	I_F (mA) for percentile rank		
	50%	0.5%	MNF[a]
20	177	78	61
50	368	185	116
70	496	260	156

[a]MNF indicates "maximum nonfibrillating current."
Source: Based on data from Dalziel (1968).

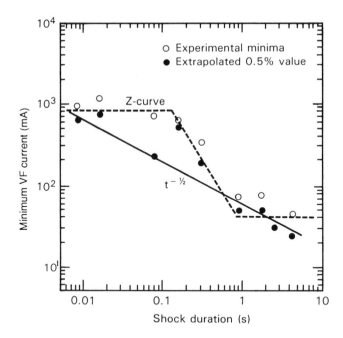

FIGURE 6.13. Minimum VF thresholds obtained in experiments with 45 dogs using data of Ferris and Kiselev. (Adapted from Dalziel, 1968, © 1968 IEEE.)

Daiziel based the time relationship in Eq. (6.3) on his analysis of the 50/60-Hz fibrillation data from dogs developed by Ferris et al. (1936) and Kiselev (1963). Figure 6.13, for example, shows one of Daiziel's curve fits to Ferris's and Kiseiev's combined data involving 45 dogs. The open points represent the minimum fibrillation thresholds obtained at each tested duration. The closed points represent Dalziel's estimates of the 0.5 percentile rank values, obtained by extrapolating the statistical distribution of VF thresholds pertaining to each tested value of t. The solid line shows Daiziel's energy fit to the extrapolated points; the broken indicates a Z-curve interpretation of the same data. Except for one extrapolated point at $0.083 \, \text{s}$, the minimum experimental points and the extrapolated points both appear to fit a Z curve.

It was not unreasonable for Dalziel to have postulated a constant energy fit to the empirical data available to him. For one thing, the $t^{-1/2}$ relationship does provide a reasonable fit to the data when viewed over the range of durations from 0.08 to 5 s, as indicated in Figs. 6.12 and 6.13. Also, an energy criterion has a certain intuitive appeal to many engineers. Furthermore, the role of extrasystolic excitations in lowering VF thresholds was not widely known at the time that Dalziel published his work.

Although the electrocution equation may not correspond strictly to theoretical expectations, it does give a reasonable fit to the VF threshold/time relationship for 60-Hz currents with durations exceeding 1 cycle. A difficulty, however, is that some analysts have attempted to ascribe an energy criterion to a wide variety of waveforms, including short-duration impulse currents. This view has probably been reinforced by other work of Dalziel (1953), in which he suggests an energy criterion for short-duration impulse currents.

It should be clear from preceding discussions that the excitation or fibrillation capability of a stimulus is not predicted by its energy content. We have seen in Sect. 6.3 that short-duration unidirectional currents instead produce a well-defined strength-duration curve that is unrelated to constant energy. For durations well below the strength-duration time constant, the unidirectional stimulus threshold converges to a constant charge criterion (see Sect. 6.3). If the transient is biphasic, the stimulus becomes less effective (see Sect. 6.4).

Short-duration charge thresholds for electrodes placed on the heart have been reported for a variety of open-heart experiments and procedures. For cutaneous electrodes, direct experimental data on transient thresholds are lacking, and we need to draw inferences from data applying to longer stimulus durations. Referring to Fig. 6.12, the charge threshold for a monophasic current may be estimated as

$$Q_0 = I_L R t_0 \qquad (6.4)$$

where Q_0 is the minimum charge threshold for short durations ($t < 3\,\mathrm{ms}$), and I_L is the peak current threshold for a long-duration ($t = 3\,\mathrm{s}$) stimulus; R and t_0 are defined in Fig. 6.12. As an example, consider the median threshold for hand-to-foot exposure of an individual weighing 20 kg. From Table 6.10, Dalziel's peak fibrillation threshold is $0.102\sqrt{2} = 0.144$ A. If we use $R = 20$ and $t_0 = 3\,\mathrm{ms}$, the mean fibrillation charge threshold from Eq. (6.4) is calculated as 8.6 mC. That value compares favorably with the minimum fibrilliating charge of 5 mC observed experimentally by Pelaska (1963) using capacitor discharges across the thorax of dogs. Cardiac arrhythmia's were observed by Pelaska with capacitor discharge as low as 0.25 mC. It is probable that the later value does not represent a minimum threshold for arrhythmia's, because stimuli below 0.25 mC were apparently not tested by Pelaska.

When a capacitor discharge is lead through an inductance, a decaying oscillatory stimulus is produced, with a net charge displacement still given by $Q = CV$. The excitation potency of an oscillatory discharge can, under some circumstances, be greater than a purely capacitive discharge, since the amount of charge initially displaced can exceed the stored charge on the capacitor (Reilly and Larkin, 1985b). Indeed, fibrillation thresholds have been observed to be lower when inductance is added to a capacitive discharge circuit (Dalziel, 1953), although this is not always the case (Pelaska, 1965). Neural excitation thresholds, expressed in terms of net capacitor charge, have also been observed to be lower when inductance is added to the discharge circuit (Reilly and Larkin, 1985b). To account for the overall effect of added inductance, we would have to consider the initial displaced charge and the frequency and decay time constant of the oscillatory current waveform. The interrelationship among these factors has not presently been defined.

6.7   Body-Size Scaling

Much of our empirical knowledge of electrical hazards has been derived from animal experimentation. In interpreting such data for human applications, we must make some judgments concerning scaling from animal data. Differences in physiology, anatomy, and size may all have a bearing on the scaling method. Body size is one variable that has been studied systematically.

Experimental Body-Size Data

Much of our knowledge of electrical safety is derived from animal experimentation. Early experiments made use of the dog and rat for application to human safety (Kouwenhoven and Langworthy, 1931; Hooker et al., 1932; Kouwenhoven, 1949). Several studies have suggested that average thresholds increase with the size of the experimental animal. In early studies, Ferris and colleagues (1936) plotted the VF thresholds for several species of animal (guinea pig, rabbit, cat, sheep, pig, calf, and dog) and postulated a linear relationship between body weight and the VF threshold, as shown in Fig. 6.14. The thresholds were also correlated with the weight of the heart itself, although the relationship appeared to be a nonlinear one. Dalziel later used Ferris's data, along with Kiselev's (1963) data on dogs to obtain a linear regression fit:

$$I_f = 3.68 \ W + 28.5 \qquad (6.5)$$

where W is the body weight (kg), and I_f is the average VF threshold (mA-rms) applied between fore and hind limbs for a duration of 3 s. The correlation coefficient for Eq. (6.5) was reported as $r = 0.74$. Figure 6.15 plots Dalziel's regression line, along with the average thresholds and weights of the larger animals tested by Ferris and Kiselev; the middle line shows Dalziel's 0.5 percentile rank, which he obtained by extrapolating the statistical distributions of the VF thresholds. Dalziel proposed a nonfibrillating current safety criterion for a hypothetical 50-kg person as

$$I_f = 116t^{-1/2} \ \left(\text{mA-rms}\right) \qquad (6.6)$$

In this formula, the time relationship follows Eq. (6.3), and the body-weight relationship follows the lower curve in Fig. 6.15. The minimum non-

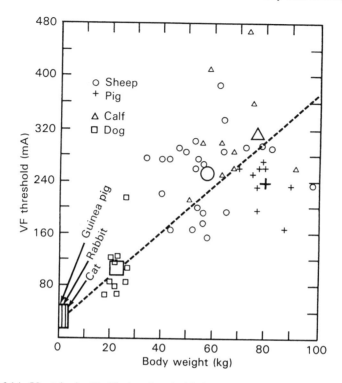

FIGURE 6.14. Ventricular fibrillation thresholds for several species of animal, plotted against body weight. Large symbols indicate average values. Current path, left fore limb to right hind limb; duration, 3 s. (From Ferris et al., 1936, © 1936 AIEE.)

fibrillating criterion of Dalziel has been widely used for safety analysis in North America (see Sect. 11.8 for additional details).

The body-weight relationship was also examined by Geddes and colleagues (1973). Using data from 104 animals of several species (rabbits, puppies, one monkey, dogs, goats, and ponies), regression fits were obtained as a power law of the form:

$$I_f = KW^a \tag{6.7}$$

Table 6.9 lists Geddes' regression coefficients for several electrode arrangements. Although these coefficients were obtained with a stimulus duration of 5 s, it would not be unreasonable to apply the same values over the range 2 to 5 s, since little threshold variation was observed over that range. Variations in the coefficient K indicate a sensitivity to the current pathway, which is discussed in some detail in Sect. 6.10.

Table 6.9 indicates that the VF threshold is nearly proportional to the square root of body weight, rather than the linear relationship given by Eq.

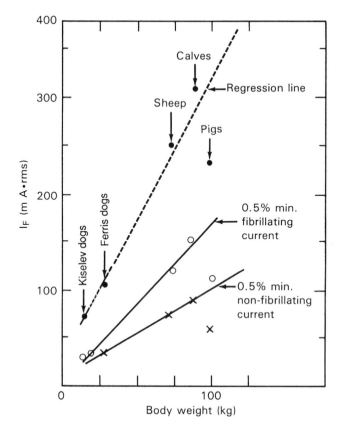

FIGURE 6.15. Relationship between fibrillating current and body weight for 3-s shocks of 60 Hz (from Dalziel, 1968, © 1968 IEEE).

TABLE   6.9. Body-weight   regression parameters.

Electrodes	K	a	r
Fore/fore	69.4	0.533	0.887
Left fore/right hind	29.7	0.510	0.929
Left fore/left hind	33.6	0.437	0.749

Five species of animal (n = 104); 60-Hz current; applicable durations, 2–5 s.
Source: Data from Geddes et al. (1973).

(6.7). Nevertheless, the two regression formulas are in reasonable agreement over a fairly wide range of body weights, as seen in Table 6.10, especially for the front-to-hind limb thresholds. For small human body weights (20 kg), the Dalziel formula gives a more conservative estimate of VF thresholds; at large body weights (100 kg), the Geddes relationship is more conservative.

A body-weight relationship is not universally accepted for the VF threshold in human safety applications. Points of view on the subject were presented in a specialists' panel meeting (Geddes and Antoni, 1985). In that discussion, Biegelmeier pointed out that when the VF thresholds of several species are plotted against body weight (as in Fig. 6.14), one indeed observes an apparent body-weight relationship. But when the thresholds of a single species are plotted against body weight, the statistical correlation largely disappears. Rather, the regression line may represent an interspecies dependency that is dominated by factors other than body weight per se.

As a consequence of the arguments stated above, the IEC does not include a body-weight law in its safety criteria. Rather, the IEC bases its criteria on experimental data from dogs (Biegelmejer, 1986, 1987), which it considers to be conservative in view of the demonstratedly lower VF thresholds of dogs as compared with humans. Especially pertinent are experimental thresholds obtained for both dogs and humans using the same arrangements of electrodes placed in the heart or on the myocardium (Green et al., 1972; Rafferty et al., 1975a, 1975b). Considering several end points (ECG irregularities, pump failure, and ventricular fibrillation), median thresholds for humans were observed to be a factor of two or more above corresponding thresholds for dogs.

A number of researchers have considered the dog heart to provide a reasonably conservative model for human safety applications. In the experiments of Green and Rafferty cited above, the smallest current having an observable effect—that of heart rhythm disturbance—occurred at $60\,\mu$A in dogs, and at $80\,\mu$A in humans. Consequently, the authors recommended for

TABLE 6.10. Median fibrillation thresholds (mA) for 3-s shocks using regression formulas of Dalziel and Geddes.

W	Geddes			Dalziel
(kg)	F/F	r.F./l.H.	l.F./l.H.	F/H
20	343	137	124	102
50	558	218	185	213
70	668	259	215	286
100	808	311	251	397

Column headings indicate current pathway: F, H indicate fore, hind limb; r, l indicate right, left.
Source: Data are based on regression formulas of Dalziel (1968) (Eq. 6.5), and Geddes et al. (1973) (Eq. 6.7).

open-heart surgery a maximum 50-Hz leakage current of 60μA for a wire placed in the right ventricle. That value, however, would not represent an absolute lower limit on the VF threshold for open-heart exposure. Roy and colleagues (1976) found the minimum VF current threshold to be inversely related to the size of the contact electrode—a minimum value of 18μA-rms was observed for right-ventricle exposure of dog hearts to 60-Hz currents through a contact electrode having an area of 0.224 mm^2. The threshold voltage of the contact electrode relative to the surrounding tissue was, however, relatively insensitive to electrode area—the minimum voltage necessary to produce VF was found to be in the vicinity of 100 mV-rms, almost without regard to electrode size.

Scaling Arguments

As shown in Sect. 4.3, the electric field aligned with the long axis of an excitable cell is responsible for the depolarizing force on the cell's membrane. Current density is, however, a more commonly cited variable in analyses of electric shock. For a given quantity of current injected through the limbs, current density in the heart will generally diminish as the size of the animal is increased. The direction of this effect corresponds to the body-weight formulas mentioned above. On the other hand, fibrillation requires not just excitation at some locus, but a critical mass of excited tissue. One might argue that in the larger heart, the mass of the heart would tend to counterbalance its reduced current density. If the fibrillation threshold were related to the fraction of body current intercepted by the heart, as suggested by Bridges (1985), then we would conclude that the fibrillation threshold ought to be about the same for animals of nearly similar geometries but of different sizes and weights.

Under the reentry theory for electrically introduced fibrillation (see Chapter 5), below some critical size, a small heart ought to be more difficult to fibrillate than a large heart. This could occur if the reentry paths were too small to simultaneously support states that were excited, partially refractory, and resting. Experimental evidence indeed shows that the heart of a small animal tends to revert spontaneously from a fibrillated to a normal state (Zipes, 1975). This observation, however, says little about the *threshold* of fibrillation.

Gross anatomical features can complicate the scaling of data from animal to human. Most experimental data have been derived from quadrupeds and applied to a bipedal human. The geometry of the heart and thorax is considerably different between a quadruped and a human, and it is likely that considerations other than body weight are important to the scaling question (Lee, 1966). A further complication concerns the position of the animal during experimentation, because the heart shifts significantly within the thorax, depending on the position of the animal (Bridges, 1985). Such shifts could cause the heart to alter its position with respect to a region of high current density when current is introduced through the limbs.

Although there are uncertainties associated with body-weight scaling, it is clear that interspecies differences do exist and that smaller animals generally have lower VF thresholds than larger animals. Less clear is the appropriate scaling method within a single species, especially humans.

6.8   Statistical Distribution of Thresholds

Various experiments on both human and animals indicate the log-normal distribution for fibrillating currents applied between the front and hind limbs. An example is shown in Fig. 6.16 for fibrillating currents in dogs, using combined data from the experiments of Ferris and Kiselev. The log-normal distribution plotted as a straight line in the figure provides a much better fit than the linear-normal distribution suggested by early investigators (Dalziel, 1968).

Table 6.11 summarizes distribution parameters extracted from the data of several investigators who applied currents across the limbs of experimental animals. The experimental data were provided in terms of VF thresholds for current durations ranging from 8 ms to 5 s and with several different waveform types. The thresholds listed in Table 6.11 give the observed 10 and 90 percentile ranks, normalized by the median value. The ± values indicate the standard deviation of the normalized percentile ranks across the various experimental distributions. In all, 29 distributions were represented in the cited reports. There was no recognizable sensitivity in the normalized distributions with respect to the duration of current, the waveform type, or the tested species. In view of this invariance, it is reasonable to represent an overall statistic as listed at the bottom of the table.

Table 6.12 provides a summary of log-normal rank values corresponding to a distribution having a 10 percentile rank of 0.66, consistent with the combined experiments of Table 6.11. The data are listed to the 0.5 percentile rank—a value frequently used in safety analysis. The data probably do not support projections below that value. The log-normal model indicates that the 0.5 percentile rank lies below the median by the multiplicative factor 0.43. This value is not substantially different from the projections of Dalziel, who used a linear-normal model to fit experimental data below the 50 percentile rank.

Compared with the animal data discussed above, a much broader distribution has been reported for human patients undergoing open-heart surgery for valve replacement. Watson and colleagues (1973) measured VF thresholds in 56 patients exposed to 50-Hz alternating current for several electrode arrangements, including 2.5-cm^2 disk electrodes placed on the epicardium, needle electrodes inserted into the myocardium, and bipolar

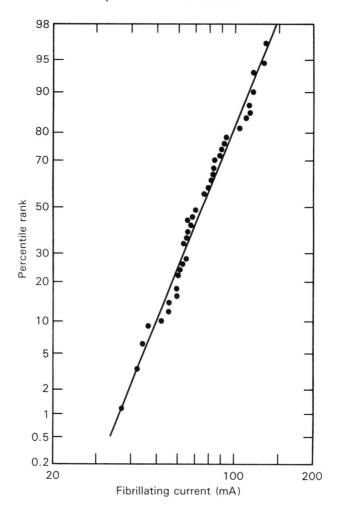

FIGURE 6.16. Statistical distribution of VF threshold for dogs; 3-s shocks applied front-to-hind limb. (Combined data from Ferris et al., 1936 and Kiselev, 1963, as summarized by Dalziel, 1968).

pacing electrodes placed on the apex of the right ventricle. Figure 6.17 illustrates the distribution of fibrillating current. The original data of Watson et al. have been replotted on coordinates on which a straight line would conform to a log-normal distribution. The ratio of the 10 percentile rank to the median VF current for Watson et al.'s data averaged 0.42 for the three electrode arrangements; the 90 percentile rank was above the median value by an average multiple of 2.54. These values indicate a significantly broader statistical spread than with the animal data summarized in Table 6.11. It is difficult to determine whether the greater spread in human open-

TABLE 6.11. Experimental statistical distribution parameters for fibrillating currents (front/hind limb paths).

Species	Current type	Duration(s)	Normalized I_F(mA)[a] 10%	90%	Data source[b]
Sheep	60 Hz	0.03 − 3.0	0.59 + 0.16	1.90 + 0.76	Ferris et al. (1936)
Dogs	60 Hz	0.008 − 5.0	0.65 + 0.10	1.84 + 0.57	Kouwenhoven et all. (1959)
Dogs	50 Hz	3.0	0.67	1.57	Kiselev (1963)
Dogs	50 Hz; pulsed[c]	$1.15t_c$	0.70 + 0.08	1.48 + 0.17	Jacobsen et all. (1975)
Dogs	20 Hz	0.1 − 2.0	0.64 + 0.06	1.59 + 0.18	Knickerbocker (1973)
Dogs	DC	0.05 − 2.0	0.71 + 0.07	1.49 + 0.12	Knickerbocker (1973)
Overall data (n = 29)			0.66 + 0.10	1.69 + 0.47	

[a] I_F values are normalized by median.
[b] Data of Ferris, Kouwenhoven, and Kiselev are summarized by Dalziel (1968).
[c] Duration was 150% of cardiac period.

heart measurements is a consequence of variations in the placement of the electrodes with respect to the most sensitive excitable regions of the heart, a reflection of the pathological state of the patient sample, or some other mechanism not yet identified.

6.9   Combined AC and DC Stimuli

As noted in Sects. 5.10 and 6.3 to 6.5, cardiac sensitivity to AC and to monophasic DC pulses can differ significantly. It follows that excitation and fibrillation thresholds for AC can be modified when a DC component is

TABLE 6.12. Log-normal percentiles for fibrillating currents (front/hind limb paths).

Percentile rank (%)	Normalized I_F
99	2.14
95	1.67
90	1.51
75	1.24
50	1.00
25	0.80
10	0.66
5	0.60
1	0.47
0.5	0.43

I_F has been normalized by median value.

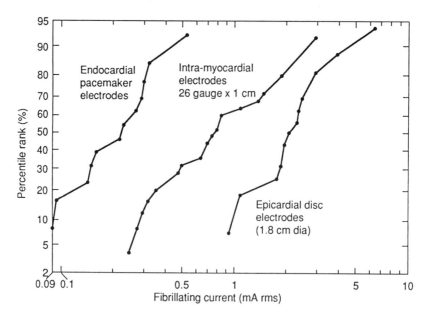

FIGURE 6.17. Distribution of 50-Hz fibrillating current in patients undergoing surgery for valve replacement. Straight lines on this format represent log-normal distributions. (Data from Watson et al., 1973.)

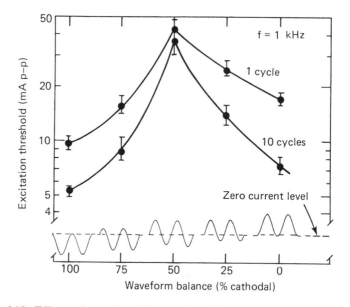

FIGURE 6.18. Effects of waveform bias on stimulation by 1-kHz sine wave. Isolated rabbit hearts, 1-cm^2 active electrode on left ventricle. Horizontal axis indicates degree of DC offset. Brackets indicate $\pm$ S.D. (Adapted from Roy et al., 1985.)

included. The effects of DC bias are illustrated in Fig. 6.18 for a sinusoidal stimulus at 1 kHz, with a duration of either 1 or 10 cycles (Roy et al., 1985). For these data, stimuli were provided to isolated rabbit hearts with a 1-cm electrode attached to the left ventricle and an indifferent electrode at a distant location. The vertical axis in Fig. 6.18 indicates the excitation threshold for cardiac pacing; the horizontal axis indicates the degree of DC bias as a percentage of the peak of the AC component. At the extremes of the horizontal axis, stimulus current was either fully cathodic or fully anodic; in between are intermediate cases of DC bias, with a symmetric biphasic wave represented at the center. Thresholds for symmetric sine waves are significantly above the purely anodic or purely cathodic waveforms, reflecting the cancellation effects of current reversal (see Sects. 6.4 and 6.5). Thresholds for the monophasic currents are considerably below those for the balanced biphasic current. The cathodic threshold is lower than the anodic threshold by factors of 1.8 and 1.4 for the 1-cycle and 10-cycle stimuli, respectively.

Figure 6.19 illustrates cardiac excitation thresholds for various monophasic waveforms applying to the previously described experiments of Roy

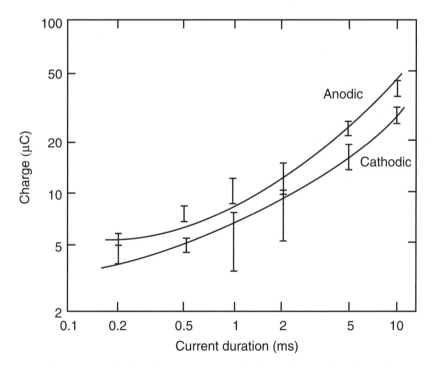

FIGURE 6.19. Strength-duration curves for stimulation by rectangular pulses and DC biased sine waves of frequency 0.5, 1, 10, 50, 500, and 1,000 kHz. Rabbit hearts ($n = 5$); 1-cm^2 electrode at left ventricle. Brackets indicate range of mean thresholds for all waveforms. (Adapted from Roy et al., 1985.)

et al. In this case, the DC bias was such that the sinusoidal current was unidirectional. Stimuli consisted of biased sine waves with frequencies of 0.5, 1, 10, 50, 500, and 1,000 kHz and with durations ranging from 0.2 to 10 ms, as indicated on the horizontal axis of the figure. An additional stimulus waveform consisted of a rectangular pulse of the indicated duration. Thresholds are expressed on the vertical axis in units of transferred charge. The brackets shown on the curves encompass the mean excitation thresholds over the entire set of stimulus waveforms. When measured in charge units, thresholds among the various waveforms all followed the strength-duration curve of a rectangular monophasic pulse. Cathodic thresholds in Fig. 6.19 are consistently lower than anodic thresholds by an average ratio of 1.8 ± 0.2. This polarity sensitivity difference agrees with data reported by Thalen et al. (1975) and Cranfield et al. (1975). Similar polarity differences have also been seen with peripheral nerve stimulation (Sect. 4.4).

Effects of waveform bias illustrated in Figs. 6.18 and 6.19 apply to excitation thresholds with waveforms of limited duration (up to 10 ms). It is much more complex to determine VF thresholds for biased waveforms, particularly when the stimulus duration is sufficiently prolonged that multiple excitations may be produced. The experiments of Knickerbocker (1973) are valuable in assessing VF thresholds for mixed waveforms of prolonged duration. These experiments were carried out on anesthetized dogs, with electrodes attached from the left foreleg to the right hind leg. The stimuli consisted of DC or 20-Hz AC administered either singly or in combination; durations ranged from 0.2 to 2 s. Figure 6.20 gives a rough summary of the mean VF thresholds. The mix of AC and DC components is plotted on Cartesian coordinates, with the points on the axes corresponding to pure DC (vertical axis) or pure AC (horizontal axis). The single-component thresholds correspond to the data of Fig. 6.11.

A biased AC waveform will be purely monophasic when the DC component is at least $\sqrt{2}I_{AC}$, where I_{AC} is the rms AC current. This relationship is represented by the diagonal line in Fig. 6.20, above which the waveforms are purely monophasic and below which they are biphasic with varying degrees of asymmetry. At all the durations studied, the pure AC thresholds were lower than the pure DC thresholds. The threshold contours for the mixed waveforms form patterns that depend not only on the excitatory power of the stimuli but also on the associated excitation rate.

6.10 Electrodes and Current Density

Electrode Area

For a given current, the average current density beneath an electrode increases inversely with the electrode area. At the same time, the smaller

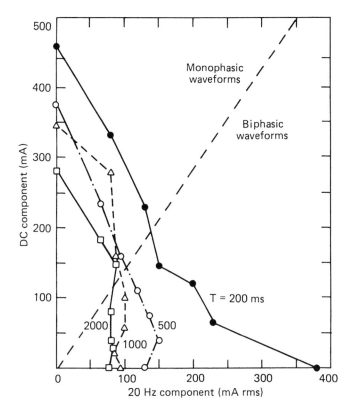

FIGURE 6.20. VF threshold contours for combined 20-Hz AC and DC currents of prolonged durations. Limb-to-limb electrodes with dogs. (Data from Knickerbocker, 1973.)

electrode results in a reduced amount of tissue that may be brought to excitation. A unified theory does not exist to predict the effect of electrode area on VF thresholds, although there is considerable empirical data on this issue (see Sect. 5.10).

Figure 6.21 summarizes VF thresholds for 60-Hz currents from electrodes in contact with the heart (adapted from Roy, 1980). The individual points are average and minimum thresholds from a variety of published data applying to 60-Hz stimuli of at least 2 s duration. Considering the diversity of experimental conditions and tested species, the consistency of the data is remarkable. The vertical axis of Fig. 6.21 expresses the ratio of threshold current to electrode contact area. Threshold current density determined this way drops as the electrode area is increased to about 5 cm², after which very little further drop is encountered. The average thresholds converge to about 0.5 mA/cm² for large electrodes, a value that is supported by data from both human and dog heart experiments (Starmer and Whalen, 1973).

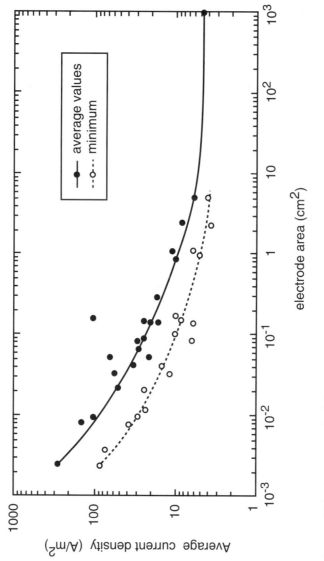

FIGURE 6.21. Average and minimum current density for VF thresholds with 60-Hz stimulating current of at least 2-s duration. Data points apply to a variety of different experiments. (Adapted from Roy, 1980.)

Minimum current density thresholds were typically about a factor of 2 below the averages for the various experiments evaluated by Roy. Using the log-normal model of Table 6.12, it is estimated that the "minimum" values observed by Roy might represent 1 percentile values of a statistical distribution. Additional current density thresholds (Ruiz et al., 1985) are consistent with Fig. 6.21.

Roy et al. (1986) suggest the following electrode-size relationship for VF thresholds based on experiments with dogs in which a disk electrode was applied to the epicardial surface of the right ventricle:

$$I_f = Kd^c \tag{6.8}$$

where $c = 0$ for $d \leq 2$ mm, $c = 1.5$ for $4 \leq d \leq 15.9$ mm, $c = 2.0$ for $d \geq 15.9$ mm, I_f is the VF threshold, d is the electrode diameter area, and K is a scaling constant. Roy's relationship states that, below a lower critical size (2-mm diam. = 0.03-cm^2 area), the VF threshold current is directly proportional to electrode area (square of diameter). In between those critical sizes, current thresholds are proportional to the 1.5 power of electrode diameter (0.75 power of area). Analogous relationships were reported for excitation thresholds (Lindermans and Van der Gon, 1978), in which threshold current was found to be independent of electrode area for diameters below 0.4 mm and proportional to the 1.5 power of diameter between 0.8 mm and 1.8 cm. Lindermans and colleagues justified their experimental results with theoretical calculations of current density beneath a contact electrode and the assumption of a *liminal length* of 0.3 mm for myocardial excitation; i.e., a minimum size of depolarized tissue necessary to initiate excitation. That liminal length is consistent with the value suggested in studies of individual Purkinje fibers (Fozzard and Schoenberg, 1972).

The relationships between electrode area and threshold current density will be different for excitation and fibrillation. With isolated guinea pig hearts stimulated by DC pulses, Weirich and colleagues (1985) showed that VF thresholds expressed in units of current are lowered if the electrode size is increased, but the opposite effect is seen with thresholds of excitation (see Fig. 5.20). Current thresholds of excitation go up with electrode size because a minimum current density is required for stimulation. To attain fibrillation, however, it is not sufficient merely to stimulate a few cells— rather, a critical mass of heart muscle must be excited to cause the degree of nonhomogeneity that is required for fibrillation.

The observations of Weirich et al. are supported by Table 6.13, which lists current and current density thresholds for excitation and ventricular fibrillation of dog hearts by 2-ms DC pulses (from Chen et al. 1975). When expressed in units of current magnitude, thresholds of excitation drop steadily as the electrode size is made smaller, whereas fibrillation thresholds do not differ greatly over the range of areas tested and actually rise at the smallest tested area. The trends seen in Table 6.13 demonstrate that, as far as excitation is concerned, the heart is more easily excited by a given

TABLE 6.13. Current density thresholds for stimulation of dog hearts by DC pulses ($t = 2$ ms).

Electrode area (cm²)	VF (mA/cm²)	threshold (mA)	EX. (mA/cm²)	threshold (mA)	VF/Ex. ratio
0.005	5,000	25.0	14.0	0.07	357
0.08	275	22.0	2.0	0.16	140
0.28	104	29.1	1.8	0.50	58

Pulse duration = 2 ms. In vivo stimulation of dog hearts ($n = 20$). Active electrode implanted into myocardium.
Source: Data from Chen et al. (1975).

amount of current into a small electrode. With fibrillation, on the other hand, there exists a trade-off between the excitation threshold and the critical mass of excited tissue required to support fibrillation (Antoni, 1979). Table 6.14 gives additional data applying to the pacing of human hearts by 2-ms pulses (from Furman et al., 1967). The current densities listed may be compared with the data for dog hearts given in Table 6.13. At an electrode area of 0.28 cm², the thresholds in the experiments with dog hearts are given as 1.8 mA/cm², and in the experiments with human hearts, as 3.0 mA/cm². Considering the different experimental conditions and tested species, these data are reasonably consistent. The data of Tables 6.13 and 6.14 provide further support for the proposition that experimental thresholds obtained from dogs provide conservative estimates for human applications.

Current Density

The foregoing discussion has dealt with current density thresholds inferred from experiments in which an electrode was held in contact with cardiac tissue. Current may also be introduced less directly, such as with electrodes

TABLE 6.14. Current density thresholds for stimulation of human hearts by DC pulses ($t = 2$ ms).

Electrode area (cm²)	Ex. Threshold (mA/cm²)	(mA)
0.12	3.27	0.39
0.28	3.00	0.84
0.49	3.14	1.54
0.87	2.30	2.00

In vivo stimulation of human hearts. Active catheter electrode inserted into right ventricular apex.
Source: Data from Furman et al. (1967).

applied to the body at locations distant from the heart, or with current induced by exposure to time-varying magnetic fields. In these cases, current may be applied more or less uniformly to the whole heart rather than in a concentrated locus beneath a contact electrode.

Stimulation thresholds for whole-heart exposure are traditionally evaluated in terms of current density. Nevertheless, the electric field within the biological medium may be a more relevant parameter, as suggested by Starmer and Whalen (1973), who observed that VF thresholds are better correlated with spatial potential differences than with current density. This observation is consistent with theoretical and experimental studies with nerve fibers (Sect. 4.4), which show that the relevant excitation parameter is the electric field longitudinal to the fiber rather than the current density per se. The electric field (E) and the current density (J) are simply related by the conductivity (σ) of the medium by $J = E\sigma$. Although current density is often cited as an experimental parameter, conductivity is not always given in published descriptions of experimental data. For that reason, the electric field may be difficult to ascertain.

Current density is most commonly determined by dividing stimulus current by the area of the electrode, as with the data in Fig. 6.21 and Tables 6.13 and 6.14. Current density determined this way will correctly give the *average* value beneath the electrode, but may understate the maximum density because current tends to concentrate at the edges of a contact electrode (see Fig. 2.9).

Considering the possibility of nonuniform current distribution at the interface between an electrode and biological tissue, the values plotted in Fig. 6.21 probably provide conservative estimates of the maximum current density. With this caveat in mind, the current density of $0.5\,\text{mA/cm}^2$ shown for the largest exlectrodes in Fig. 6.21 estimates the average current density threshold for VF by whole-heart exposure to prolonged ($>2\,\text{s}$) 60-Hz currents; $0.25\,\text{mA/cm}$ is a more conservative estimate of a 1 percentile experimental value. Table 6.4 indicates that excitation thresholds are about 40% of the 60-Hz fibrillation value for prolonged ($t > 2\,\text{s}$) 60-Hz and also for monophasic pulse stimulation ($t \approx 10\,\text{ms}$). Applying this factor to the VF current density results in an estimated average threshold of $0.2\,\text{mA/cm}^2$ for excitation by either prolonged 60-Hz or single DC pulses and a 1 percentile threshold of $0.1\,\text{mA/cm}^2$. That minimum value is quite close to the value of $0.12\,\text{mA/cm}^2$ calculated as a median threshold for excitation of large ($20\,\mu\text{m}$) peripheral nerves (Table 4.2). The current densities cited above have been obtained in experiments with healthy animals. It is not known whether these values are representative of the human heart when it is in a pathological state (see Sect. 6.8).

The current density required for defibrillation is much greater than that required for excitation, or fibrillation. To achieve defibrillation, one must ensure that a critical mass of cardiac tissue distant from a defibrillation electrode is uniformly excited. To achieve defibrillation, it is estimated that

80% of the heart must be subject to a current density of at least $35\,mA/cm^2$ (Sepulveda et al., 1990).

Locus and Direction of Stimulus Current

Thresholds of fibrillation will vary with the location of electrodes, whether they are contacting the heart or are applied externally. Table 6.15 (adapted from Roy et al., 1987) lists VF thresholds for a 1-cm² electrode applied to five different regions of a pig's heart. An indifferent flexible electrode (15 × 56 cm) was wrapped around the back and sides of the animal. Seven animals were tested using a 60-Hz stimulus of 5 s duration. The data in the table are listed in the order of increasing average current density; the stimulated regions are identified in Fig. 6.22. The thresholds span a range of 2.2:1 from the least to the most sensitive area. The most sensitive area is the apex; regions of the right ventricle have higher thresholds than those on the left ventricle. At the heart apex, the average and minimum thresholds listed in Table 6.15 compare favorably with the VF thresholds noted in the previous section if electrode area is properly considered. The authors pointed out that their results are consistent with the concept that a critical mass of cardiac muscle is required for sustained fibrillation—the apex, having the greatest muscle thickness, is associated with the lowest threshold. Excitation via external electrodes would be expected to have the lowest thresholds if the electrodes were oriented so as to provide the greatest exposure to the apex of the heart.

Table 6.16 indicates the variation of VF thresholds with respect to the pathway of 50/60-Hz currents delivered for long duration ($t > 3$ s) through electrodes placed at various locations on the body. The tabulated data, derived from the references listed at the bottom of the table, indicate relative thresholds with a single column. In column 1, for example, the value $I_f = 1.56$ for a hand-to-hand path indicates a finding that 56% more current was needed to induce VF than with a right hand-to-foot path having $I_f = 1.0$. A listing of 1.0 in one column does not necessarily indicate the same VF

TABLE 6.15. VF thresholds for 60-Hz currents applied through a 1-cm² active electrode (data listed in order of increasing average threshold).

Electrode position	n	min	avg	rms
		VF threshold (mA)		
1	16	0.66	0.98	0.23
2	14	1.06	1.42	0.52
3	12	1.19	1.82	0.42
4	15	1.09	1.83	0.51
5	14	1.35	2.17	0.71

Source: Data from Roy et al. (1987).

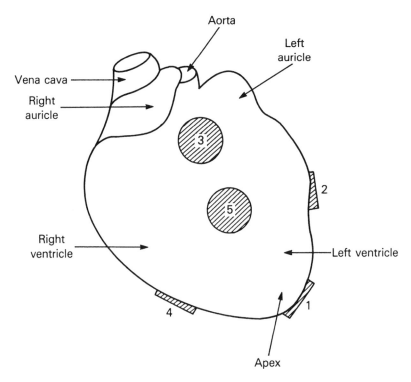

FIGURE 6.22. Diagram of pig's heart showing position of 1-cm^2 electrodes numbered in order of increasing average VF thresholds. (From Roy et al., 1987.)

threshold as a listing of 1.0 in another column. The I_f data in the first three columns apply to experiments with animals (Ferris et al., 1936; Geddes et al., 1973; Roy et al., 1986); the values in the last column are criteria recommended in Europe by the IEC for evaluation of electrical hazards (IEC, 1982).

The relative current thresholds listed in the table are a result of several factors, including the following:

1. *Current density.* The current density in the vicinity of the heart will vary with the placement of external electrodes. In experiments with dogs, it has been reported that 7% of the applied current flows through the heart with a forelimb-to-hind limb path, whereas only 3% is intercepted by the heart for current applied across the forelimbs (Kouwenhoven et al., 1932).

2. *Excitable fiber orientation.* We have seen in Chapter 4 that current aligned with the long axis of an excitable fiber is much more effective in producing excitation than current flow in a transverse direction.

3. *Exposure of sensitive region.* As suggested by Table 6.15, we may expect a 2:1 difference in sensitivity for stimulation to different regions of

TABLE 6.16. Relative VF sensitivity with respect to path way of 50/60-Hz current (large area, $t \geq 3\,s$).

Electrode positions	Relative I_F			
A. *Hand-to-feet paths*				
Left hand-right foot (left fore-right hind)		1.0		1.0
Left hand-left foot (left fore-left hind)		1.13		1.0
Right hand-left foot (right fore-left hind)	1.0			1.25
B. *Hand-to-hand paths*				
Right hand-left hand (right fore-left fore)	1.56	2.3		2.5
C. *Foot-to-foot paths*				
Left foot-right foot (left hind-right hind)	>50			
D. *Misc. hand paths*				
Right hand-back (right fore-back)				3.33
Left hand-back (left fore-back)				1.43
Right hand-chest (right fore-chest)				0.77
Left hand-chest (left fore-chest)				0.67
Hand-seat				1.43
E. *Other paths*				
Left chest-right chest	1.04		1.4	
Left foot-head	1.20			
Chest-back			1.0	

References: (1) Ferris et al. (1936); (2) Geddes et al. (1973); (3) Roy et al. (1986); (4) IEC (1982).

the heart. Variation in the placement of external electrodes may provide more or less exposure to sensitive regions of the heart.

4. *Anisotropic impedance properties.* The conductivity of muscle tissue is greatest when the current is aligned with the direction of the muscle fibers (Chilbert et al., 1983). The conductivity parallel and perpendicular to the fibers of ventricular muscle demonstrate a ratio of about 2:1 (Ruch et al., 1963). Accordingly, the direction of current would affect its distribution within the heart, and the mass of excited tissue.

5. *Polarity of DC stimulation.* We have seen in Chapter 4 that an excitable fiber will be depolarized most effectively at end structures that are oriented toward the cathode of a current source.

While the factors mentioned above may all play a role in the heart's sensitivity to electrode placement and the current path, it is not possible to separate out their relative importance on the basis of available data. However, a number of generalizations can be made, as noted below.

Longitudinal versus Transverse Current

Current flow along the long axis of the body (longitudinal flow) is associated with significantly lower thresholds than with current flowing in an orthogonal (transverse) direction. Hand-to-foot pathways are observed to have significantly lower thresholds than hand-to-hand paths—the ratio of thresholds for the two orientations is about 2:1. Table 6.10 lists average thresholds for longitudinal paths, attributed to subjects of different body weights.

Foot-to-Foot Paths

Available data indicate that current applied from one foot to another is very unlikely to result in VF. Ferris et al. (1936) found that current applied from foot to foot failed to produce VF at a current level that was 50 times greater than the VF threshold for hand-to-foot contacts. The relatively high thresholds for foot-to-foot contact results because relatively little current reaches the heart when it is introduced through the legs or hind limbs.

Chest Electrodes

When one of the electrodes is placed on the chest, the VF threshold depends critically on its precise placement with respect to anatomical features of the heart. The lowest threshold for a chest electrode is obtained when it is placed directly over the apex of the heart (Geddes et al., 1973)—a finding consistent with open-heart tests (Roy et al., 1987). With large-area ($300 \, cm^2$) electrodes on the chest, VF thresholds have been found to be 1.4 times lower when the current path was from the chest to the back of the animal than with a transverse (side-to side) orientation (Roy et al., 1986). Thresholds of stimulation with a large area chest electrode were measured typically between 40 and 70 mA with a minimum value of 20 mA in tests of humans (Zoll et al., 1985). The 60-Hz fibrillation thresholds for 200-mm^2 chest electrodes on dogs averaged 68 mA (Roy et al., 1986).

DC Stimulation

As with AC stimulation, a longitudinal orientation of DC is associated with significantly lower thresholds than with transverse current flow. Fatal accidents in Europe involving DC exposure have been reported to occur only with a longitudinal current path (Biegelmeier, 1987). Additionally, the polarity of the DC electrodes significantly affects VF thresholds, with positive foot electrodes being the most hazardous. In early experiments (Ferris et al., 1936), it was found that the VF threshold was about 36% lower when the feet were at a positive potential relative to a hand electrode as compared with the opposite polarity. Biegelmeier (1987) reports that VF thresholds with positive foot and negative hand electrodes are about one-half the thresholds pertaining to the opposite direction of current flow.

7
Sensory Responses to Electrical Stimulation

7.1 Introduction

Sensory sensitivity to electrical stimulation depends on a host of factors associated with the stimulus waveform, its method of delivery, and subjective variables. In most situations involving electrical safety or acceptability, current is applied to the body by cutaneous electrodes. There are also practical applications in which electric current may be applied subcutaneously or induced internally by external electromagnetic fields. Although the emphasis in this chapter is on electrocutaneous stimulation, many of the principles discussed may be applied to other modes of stimulation. The reader is directed to Chapter 9 for additional discussion of peripheral nerve stimulation by time-varying magnetic field effects or by induced shock within intense electric field environments. In addition to sensory effects described in this chapter, stimulation by electric current and electromagnetic fields can also elicit visual and auditory sensations. These will be treated in Sect. 9.8.

7.2 Mechanisms of Electrical Transduction

Current of a fraction of a microampere can be detected when the finger is gently drawn across a surface charged with small AC potentials (Grimnes, 1983b). Such levels are roughly 100 times less than commonly tested electrical thresholds. Detection of such small current results from electromechanical forces arising from electrostatic compression across the *stratum corneum* (the outermost layer of dead skin cells). As analyzed by Grimnes, the electrostatic force K is

$$K = \frac{A\varepsilon v^2}{2d^2} \tag{7.1}$$

where A is the contact area, ε is the dielectric constant of the corneum, d is its thickness, and v is the instantaneous voltage. The compression of the corneum would not normally be sensed. But when the skin is moved along the charged surface, there is a vibratory frictional force on the finger that is maximized on each half-cycle of the alternating voltage. This vibrational force stimulates mechanoreceptors and is responsible for the detection of microampere currents. Grimnes estimates that the minimum voltage contributing to a detectable vibration is about 1.5 V at 50 Hz.

The detection of microampere currents through mechanical vibration is, for most purposes, of passing interest, although it may be important for a researcher to know about it when designing perception tests. Of greater significance is the mode of detection when the current level is raised to roughly 0.1 mA or above. At that point, perception can be initiated by the electrical excitation of neural structures, according to the mechanisms discussed in Chapters 3 and 4.

Exactly what is excited with electrocutaneous stimulation, and what is the specific site of initiation? At the lowest levels of stimulation, it is likely that peripheral structures are involved, because these are closest to the surface electrode. Among fiber classes, the larger-diameter myelinated fibers have the lowest electrical thresholds, and circumstantial evidence presented in this chapter points to the involvement of one or another class of mechanoreceptor. The precise site of cutaneous electrical stimulation is unknown; whether stimulation is initiated at the axon proper, at the site of the generator potential of sensory receptors, or along free nerve endings has not been demonstrated. Some evidence, however, exists, as noted in Chapter 4, that the site of initiation is near neural end structures, including receptors or free nerve endings. Electrocutaneous perception is a local phenomenon; subjects typically report sensation occurring locally at the electrode site rather than remotely as might be supposed if the excitation occurred on the axons of deeper-lying nerves. It is only when the current is raised substantially above the perception level that distributed sensations are felt.

If the current is raised sufficiently above the threshold of perception, excitation of unmyelinated nociceptors becomes possible. Because of their higher electrical thresholds and generally deeper sites, these structures are not likely to be involved at perception threshold levels. At still higher current levels (some tens of milliamperes for long-duration stimuli), thermal detection due to tissue heating becomes possible. Neuroelectric thresholds may exceed thermal thresholds if the waveform of the electric current is inefficient for electrical stimulation, such as with sinusoidal currents of very high frequency ($>10^5$ Hz).

The electrical stimulus is nonspecific as to the receptor class that might be stimulated. Nevertheless, there is some degree of selectivity on the basis of electrical properties—thresholds tend to be lower as the neuron is closer to the corneum, as the fiber diameter is increased, and as its orientation is aligned with the internally generated electric field (i.e., direction of current flow). At suprathreshold levels, an array of different fiber types will be stimulated, including deeper-lying structures such as small-diameter nociceptors and motor neurons.

One might question whether the electrical stimulus would feel like some natural modality corresponding to the receptor type that is being excited. Research, subjects have preferentially selected punctuate pressure descriptors from a list of 10 choices for capacitor discharge stimulation applied cutaneously to the forearm, as shown in Fig. 7.1 (Reilly and Larkin, 1985a). Similar results have been obtained for stimulation by 300-ms square-wave pulses: At levels just above perception, subjects used descriptors of distributed pressure; but as the stimulus level was increased, they used punctuate pressure descriptors (Tashiro and Higashiyama, 1981).

Electrical stimulation by pulse trains evokes sensations closer to natural stimuli when the nerve is stimulated directly by a percutaneous microelectrode. In the experiments of Vallbo and colleagues (1984), a microelectrode was inserted above the elbow into the median nerve, which has a receptive field in the hand. When stimulated by a train of pulses, subjects reported localized sensations in the hand akin to steady pressure, transient pressure, or vibration, depending on the particular fiber being stimulated and the electrical stimulation parameters. The authors reported that stimulation of a single nerve fiber could elicit perceptible sensations for some classes of fibers.

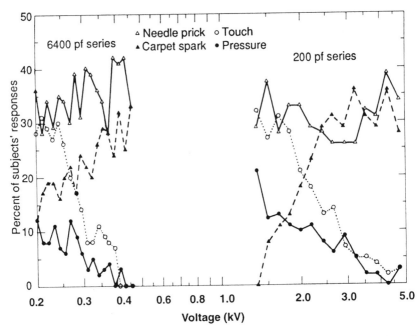

FIGURE 7.1. Qualitative ratings by subjects receiving suprathreshold capacitor discharges. (From Reilly and Larkin, 1984.)

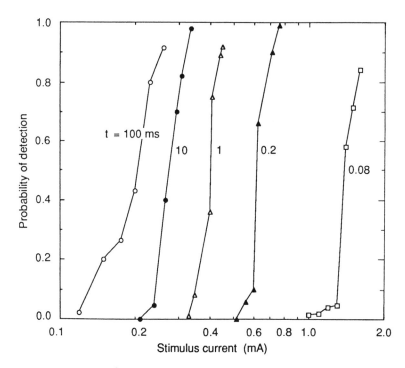

FIGURE 7.2. Psychometric functions for detection of square-wave pulses applied cutaneously; 0.5-cm^2 saline-treated electrode on forearm. (Data from Rollman, 1974.)

Some understanding of the neural pathways for electrical stimulation can be inferred from experiments in which A or C fibers are selectively blocked in human subjects (Torebjörk and Hallin, 1973). When mainly A-fiber response is present, weak electric shocks are felt as tactile sensations, and strong shocks are perceived as a short, sharp blow without prolonged pain. When A-fiber response is blocked, there is an impaired discrimination of weak stimuli, and strong stimuli evoke painful sensations because of C-fiber activity.

The electrocutaneous stimulus is a very unnatural one in terms of its spatial distribution of stimulation, lack of receptor specificity, abnormal neural recruitment properties, abnormal temporal patterns of excited action potentials, and concurrent recruitment of sensory and motor neurons. These factors all contribute to a sensory quality that is unlike any for which nature has equipped us and probably contribute to the highly aversive quality that most people attribute to electrical stimulation only somewhat above their perception.

The relatively small dynamic range of electrical stimulation is reflected in extremely steep *psychometric functions* (i.e., the functional relationship between the stimulus level and the probability of detection). Figure 7.2

illustrates psychometric functions applying to a single individual detecting cutaneously applied square-wave pulses of various durations (Rollman, 1974). The coefficient of variation (standard deviation to mean ratio) was typically 0.08, a value that is significantly smaller than that observed for other sensory modalities. Because of the steep psychometric functions, a change of stimulus magnitude of only 20% typically elevates the detection probability from 10% to 90% for each curve shown in Fig. 7.2. Figure 7.2 also demonstrates that thresholds are inversely related to the duration of simulation, as discussed below.

7.3 Perception of Transient Monophasic Currents

Strength–Duration Relationships

Sensory sensitivity varies greatly with stimulus parameters, including waveshape, duration, repetition pattern, polarity, and whether the stimulus is monophasic or biphasic. The relationship between the intensity of a monophasic current and the threshold of reaction has been the most systematically studied aspect of waveform dependency. The S–D relationships for neural excitation have a long history of exploration, beginning with the early studies of Weiss (1901) and Lapicque (1907). In Chapter 4, the early formulations are described in terms of an exponential [Eq. (4.15)] or hyperbolic [Eq. (4.17)] relationship. Although their mathematical expressions are different, the two formulas express similar results (Fig. 4.3).

Perception thresholds converge to a minimum current (I_0) as the current duration is made very long and to a minimum charge value (Q_0) as it is made very short. The minimum current has traditionally been called *rheobase*. Whether the exponential or hyperbolic formulations are used, the S–D curve may be described in terms of two parameters: the equivalent S–D time constant (τ_e) and the minimum (rheobase) threshold. The time constant is a measure of the temporal responsiveness of the excitable system. In both exponential and parabolic formulations, τ_e may be obtained by the ratio Q_0/I_0 [Eq. (4.20)].

The S–D relationship applies to both perception and suprathreshold values. Figure 7.3 illustrates neurosensory and neuromotor S–D curves from the experiments of Alon and colleagues (1983). The upper set of curves represents thresholds as the peak current of a monophasic square-wave stimulus. The lower set shows stimulus charge (product of amplitude and duration). The broken line indicates the exponential expressions [Eqs. (4.6) and (4.9)], adjusted such that the values Q_0 and I_0 correspond to the experimental perception threshold values. With this adjustment, the experimental and analytic curves agree remarkably well.

Table 7.1 summarizes S–D data from various sensory stimulation experiments. The column headed τ_e refers to the S–D time constant and the

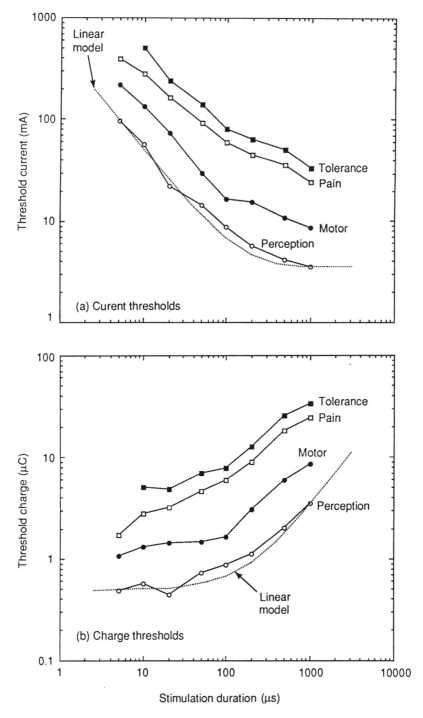

FIGURE 7.3. Strength-duration curves for sensory and motor reactions to square-wave pulses; forearm stimulation, 4-cm² electrode. (Data from Alon et al., 1983.)

TABLE 7.1. Strength-duration data for nerve stimulation.

Subject	N	Stimulus locus	Electrode	Reaction	τ_e (μs)	I_0 (mA)	Q_0 (μC)	Reference
A. Human subjects								
Human	6	Forearm	4-cm² carbon sponge	Perception	138	3.5	0.48	Alon et al. (1983)
Human	6	Forearm		Motor	127	8.5	1.08	Alon et al. (1983)
Human	6	Forearm		Pain	70	24.5	1.72	Alon et al. (1983)
Human	6	Forearm		Tolerance	142	33.8	4.80	Alon et al. (1983)
Human	4	*	2-mm diam.	Pain	760	0.5	0.38	Notermans (1966)
Human	3	Finger (Cap. discharge)	1.1-mm diam.	Perception	200–900	—	0.12	Larkin and Reilly (1984)
Human	*	Ulnar nerve, forearm	0.5-cm² elect. paste	Perception	480	0.25	0.12	Rollman (1975)
Human	3	Finger pad	0.4-cm²	Perception	217	0.69	0.15	Hahn (1958)
Human	*	Forehead; forearm	0.1 mm²	Perception	189	1.8	0.34	Girvin et al. (1982)
Human	6	Ulnar nerve	*	AP	<260	2.0	<0.52	Heckmann (1972)
Human	6	Ulnar nerve	*	Muscle twitch	275	3.2	0.88	Heckmann (1972)
Human	*	Denervated muscle	*	Muscle twitch	8333	6.0	50	Heckmann (1972)
Human	*	Enervated muscle	*	Muscle twitch	30–700	*	*	Harris (1971)
Human	*	Denervated muscle	*	Muscle twitch	>4300	*	*	Harris (1971)
Human	2	Forearm	3.9-cm² elect. paste	Muscle twitch	140	9.9	1.39	Crago et al. (1974)
Human	12	Foot	8mm dia.	Perception	295	1.1	0.32	Friedli & Meyer (1984)
Human	8	Finger	8mm dia.	Perception	383	0.5	0.19	Friedli & Meyer (1984)
B. Animal, in-vitro nerve preparation								
Rat, cat	12	Tibialis ant. musc.	10mm, intra. musc.	Muscle twitch	85	5.2	0.44	Crago et al. (1974)
Rat, cat	12	Peroneal nerve	*	AP	55	0.38	0.021	Crago et al. (1974)
Rabbit	2	Tibial nerve (4–13 μm diam.)	0.1-mm diam., in vivo (subcutaneous)	AP	80–100	—	—	Reilly et al. (1985)
Cat	*	Aβ fiber	In virto	AP	30	—	—	Li and Bak (1976)
Cat	*	Aδ fiber	In vitro	AP	650	—	—	Li and Bak (1974)
Toad	*	Myelinated nerve	In vitro	AP	148	—	—	Tasaki and Sato (1951)
Dog	4	Back	Magnetic coil	Muscle twitch	148	—	—	Bourland et al. (1990)
Dog	4	Chest	2.5cm dia.	Inspiration	245	49.4	—	Voorhees et al. (1992)
C. Theoretical model								
SENN model	—	Myelinated nerve	Point elect.	AP	92–128	—	—	Reilly and Bauer (1987)
SENN model	—	Myelinated nerve	Uniform field	AP	120	—	—	Reilly and Bauer (1987)

*Information not available.
—Information not applicable for table summary.

columns headed I_0 and Q_0 to the minimum current and charge thresholds. In most cases, the tabulated value of τ_e has been determined by Q_0/I_0. In a few cases, where the minimum thresholds were not available, τ_e was determined by chronaxy/0.693 [see Eq. (4.19b)] or by a simple curve fit of Eq. (4.6). The table entries are subdivided according to categories of human perception, animal or nerve preparations and a theoretical model for myelinated nerve. The "human" category includes both neurosensory and neuromuscular responses. The values of Q_0 and I_0 depend on a number of factors such as electrode size, the locus of stimulation, the reaction threshold being tested, and whether the current is supplied transcutaneously or subcutaneously. These factors are discussed in detail in later sections of this chapter. The ratio Q_0/I_0 is much less sensitive to these factors.

Empirical time constants for neuromuscular reactions do not differ significantly from neurosensory values, although the absolute threshold for motor responses is typically higher. Although motor neurons typically have larger diameters than sensory neurons, and hence would be expected to have lower electrical thresholds, the fact that they are situated in deeper strata below the skin accounts for their higher thresholds. When the muscle is denervated, motor stimulation takes place by direct excitation of muscle fibers rather than motor neurons (see Sect. 8.3). In this case, motor thresholds rise significantly, and the time constants increase dramatically (Bauwens, 1971; Harris, 1971). The fact that the parameters of the S–D curves vary with conditions of neuropathy (Friedli and Meyer, 1984) suggests a potential diagnostic tool.

Although sensory time constants encompass a range similar to that noted for neural stimulation (Parts B and C of Table 7.1), there is a greater representation toward the larger values with sensory data. The geometric mean of τ_e values given in Table 7.1 for human perception is about 270μs; when the data of Parts B and C are included, the geometric mean is closer to 170μs. The myelinated nerve SENN model indicates a range of τ_e depending on the distance between a point electrode and the nerve fiber. For uniform E-field excitation, the model indicates $\tau_e = 121\,\mu s$; in comparison, a value of 140μs was obtained under experimental conditions closely approximating uniform field stimulation, in which the internal E-field was induced by magnetic field exposure over a large area of the torso (Schaefer et al., 1991).

Experimental S–D time constants are not simply a property of the excitable tissue, but depend on the method of stimulation as well. For instance, time constants have been found to increase monotonically with electrode size in nerve and neuromuscular preparations (Pfeiffer, 1968; Davis, 1923) as well as in cardiac tissue (see Sect. 6.2); this dependency is consistent with theoretical arguments, although the effect is expected to be much stronger in muscle than in nerve tissue (Jack et al., 1983).

For the data described in Table 7.1, absolute perception thresholds for short-duration stimuli and small electrodes converge to a constant charge value of 0.1 μC, and for long-duration stimuli to a constant amplitude value of about 0.25 mA. These values must be qualified by a variety of factors affecting sensitivity such as body locus, electrode size, skin temperature, tactile masking, electrode contact, and stimulus waveform features. Additionally, individual differences in sensitivity will result in a distribution of threshold values.

Capacitor Discharges[1]

Exposure to transient electric shock is a common occurrence. We have all experienced shocks when we walked across a carpet on a dry day and then touched a grounded object. In such cases, our body acts as a capacitor that stores electric charges at levels of several thousand volts. Then, when we come sufficiently close to a grounded object, the stored charge is suddenly discharged at some discrete body location through a spark that may be felt, seen, and heard. The peak current of a carpet spark can be very large— typically more than an ampere—a level that could be lethal if sustained. Fortunately, the event is very brief, in the submicrosecond range. As a result, the shock is well below lethal intensity, but nonetheless can be annoying to many people.

Capacitor discharges may be used beneficially in biomedical applications. One research application takes advantage of the fact that a capacitor discharge has a sudden current onset, but a gradual decay (Accornero et al, 1977). This produces a stimulus that is effective on its leading edge, while avoiding the possibility of excitation on the current break as with a square-wave stimulus.

Various cutaneous sensations are possible with capacitor discharges. Near threshold, the sensations commonly resemble touch or mild pinprick. Above threshold, the pinprick may be followed by a burning sensation akin to the delayed pain related to C-fiber activity. Bishop (1943) reported other sense qualities as well and suggested that a punctuate "spark" from a capacitor could be used selectively on the skin to excite discrete neural channels. The capacitor discharge stimulus is monophasic (current flows in only one direction) and can be made brief in relation to the depolarization process in neural tissue. In these two respects, capacitor discharges resemble the short-duration, constant-current pulses often used in sensory and electrophysiological research.

In two other respects, capacitor discharge stimuli may differ from the current pulses commonly in use. First, the discharge waveform can have both a very high initial voltage and a very high current peak, even at threshold levels of stimulation. Furthermore, while constant-current stimu-

---

[1] Portions of this section have been adapted from Larkin and Reilly (1984) and Reilly and Larkin (1984).

lation requires an electrode in good contact with the skin, a high-voltage capacitor discharge can occur without contact if the electrode is simply brought close enough to sustain an electric arc between it and the skin. These characteristics are also present in electrostatic "carpet sparks" of the sort familiar to all who live in dry, upholstered environments (Chakravarti & Pontrelli, 1976).

The capacitor discharge stimulus is typically associated with a high-peak current density at a relatively high voltage. The electrical properties of the skin are much less significant with high-voltage capacitor discharge stimuli than with traditional constant-voltage or constant-current devices because the skin impedance breaks down in response to high-voltage and high-current densities of a capacitor discharge (see Sect. 2.5).

When a capacitance (C) discharges through a resistance (R), both current and voltage have simple exponential waveforms given by:

$$I(t) = I_0 e^{-t/\tau} \tag{7.2}$$

$$V(t) = I_0 \left(1 - e^{-t/\tau}\right) \tag{7.3}$$

where V_0 is the initial voltage on the capacitor, $I_0 = V_0/R$, and τ is the discharge time constant given by the product RC. The amount of stored charge that may be released in the stimulus is

$$Q = CV_0 \tag{7.4}$$

Although Eqs. (7.2) and (7.3) represent an ideal case, discharges to biological materials, including human skin, exhibit additional complexities such as nonlinear impedance (see Chapter 2). Despite these complexities, the capacitor discharge waveform is typically very close to Eqs. (7.2) and (7.3), so that an exponential approximation is sufficient. While the time course of the discharge current may be expected to follow the exponential form [Eq. (7.2)], discharges to intact skin are highly dependent on the initial voltage as well as the capacitance, as indicated by Figs. 2.29 and 2.35.

We have systematically explored the sensory and biophysical aspects of capacitor stimuli. As an example of this work, Fig. 7.4 illustrates mean perception thresholds for anodic capacitor discharges (Reilly and Larkin, 1987). Thresholds are shown for delivery of the stimulus by: (a) the finger tapping an energized electrode; (b) discharge to a large (1.27-cm-diameter) circular electrode held against the fingertip; (c) discharge to a small (0.11-cm-diameter) electrode held against the fingertip; and (d) discharge to a needle piercing the corneum of the forearm. The curve shapes and their relative displacements are based on intensive study of six expert subjects (Reilly and Larkin, 1985a), but their absolute positions have been slightly adjusted along the vertical axis to reflect measurements with a sample of 124 subjects (see Sect. 7.10). In Fig. 7.4, threshold charge at 100 pF is 0.25 μC for procedure (a); 0.19 μC for procedure (b); 0.11 μC for procedure (c); and 0.07 μC for procedure (d).

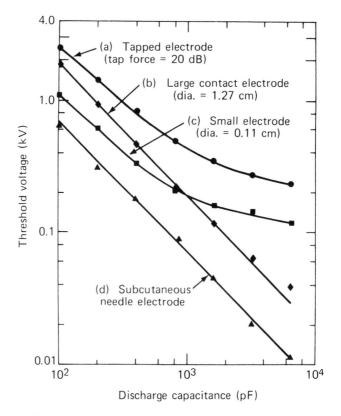

FIGURE 7.4. Threshold sensitivity contours for four methods of stimulation using capacitor discharges of positive polarity. (Measured data slightly adjusted to conform to large-sample results—see text.) (From Reilly and Larkin, 1987.)

Two of the curves in Fig. 7.4 follow straight lines representing constant charge; that is, CV = constant. The other two curves depart significantly from a constant-charge contour at the larger capacitances. Besides differences in curve shapes, there are significant vertical displacements among the curves. The contour shapes of Fig. 7.4 can be understood with reference to S–D relationships for capacitor discharges, as seen in theoretical models [Fig. 4.14 and Eq. (4.14)], as well as experimental perception data (Wessale et al., 1992). Figure 7.5 illustrates S–D data from the author's experiments for capacitor discharge stimuli (Larkin and Reilly, 1984), plotted in terms of normalized charge units against the measured discharge time constant, τ. The vertical axis has been normalized by the minimum threshold charge corresponding to the smallest values of τ (obtained with a 100 pF capacitor discharge). The curve in Fig. 7.5 is an analytic S–D expression according to Eq. (4.14), with the parameter $\tau_e = 0.6$ ms. The figure demonstrates that, for monophasic current pulses that are brief relative to τ_e, thresholds converge to a minimum charge. This minimum charge is not sensitive to the fine structure of the stimulus waveform as long as the stimulus remains monophasic. For biphasic (oscillating) waveforms, the waveshape can be critical (see Sects. 4.6, 7.5).

It is possible to explain the contour shapes of Fig. 7.4 in the light of S–D relationships. For the range of parameters represented in the figure, stimulus time constants range from about $0.1\,\mu s$ to more than $1,000\,\mu s$ (see Fig. 2.35). For procedures (b) and (d), discharge time constants were in all cases below $3\,\mu s$, a value much smaller than the time constants typical of excitable membranes. These brief time constants result from the relatively low impedance at the electrode/skin interface associated with the capacitive discharge. In curve (b), the low impedance results from the large contact area of the electrode; in curve (d), the electrode is small, but it bypasses the high-impedance corneal layer of skin. In contrast, discharge impedance associated with curves (a) and (c) becomes very large as the stimulus voltage is reduced. At small capacitance values, the time constants associated with all four curves in Fig. 7.4 are sufficiently short to be within the charge-dependent region of the S-D curve. In this region, the vertical displacements of the threshold curves can be explained by other factors discussed below.

Capacitor discharge thresholds developed by Swiss and Austrian groups form the basis of acceptability criteria published by the IEC (1987). The

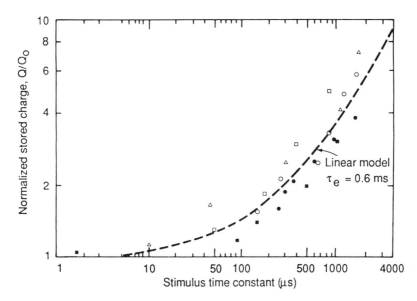

FIGURE 7.5. Strength-duration data—perception of cutaneous capacitor discharges for various procedures. (Thresholds normalized by minimum perception charge.) (Adapted from Larkin and Reilly, 1984 in *Perception and Psychophysics*, vol. 36, pp. 68–78, reprinted by permission of Psychonomic Society, Inc.)

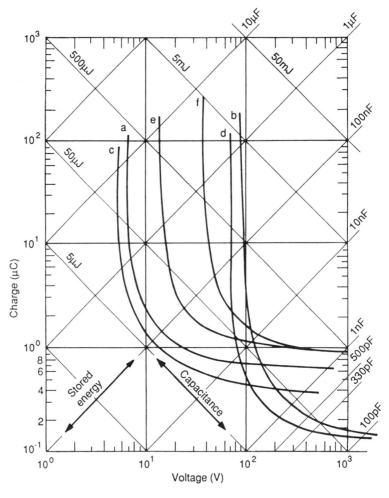

(a) 80 X 150 mm electrodes, adults
(b) 5 mm rod electrodes, adults
(c) As in (a), children age 4-14 yrs.
(d) As in (b), children age 4-14 yrs.
(e) 80 X 100 mm electrodes, adults
(f) 4 mm dia. wire electrodes

FIGURE 7.6. Perception thresholds for capacitor discharges. Curves *a-d* from Swiss measurement; curves *e* and *f* from Austrian measurements. (From Biegelmeier, 1986.)

European measurements are shown in Fig. 7.6, as summarized by Biegelmeier (1986). Curves *a* and *c* apply to Swiss experiments in which subjects grasped with both hands cylinders of 80-mm diameter and length 100 mm. Capacitor discharges were applied to the cylinder electrodes. Curve *a* is for adults, and curve *c* is for children. Curves *b* and *d,* for adults and children, respectively, apply to a 5-mm-diameter rod electrode placed on the fingertip. For the Austrian data, curve *e* applies to an 80-mm cylinder grasped in the hands.

There are notable similarities in the contact electrode thresholds of Fig. 7.4 (curves *b* and *c*) and European data if one allows for differences in plotting formats. In both sets of data, thresholds are defined by constant charge at small capacitances and then approach constant voltage at large capacitances. Also, in both sets of data, thresholds for large and small electrodes cross over when plotted against capacitance: at small capacitances, thresholds are lower for the smaller electrodes; at larger capacitances, the reverse is true. The curve shapes, and their relative displacements, can be understood if we account for the effects of discharge time constant and electrode size (see Sect. 7.6).

7.4   Suprathreshold Responses

Magnitude Scaling

One method of characterizing subjective intensity is through magnitude-scaling experiments. For instance, a subject may be given a reference stimulus that is assigned some arbitrary numerical value and asked to rate other stimuli as a proportion or multiple of the reference. The absolute numbers themselves are meaningless, but their rate of growth with respect to the stimulus intensity reveals the subject's internal sensory interpretation.

In many cases, experimental data on subjective magnitude (M) versus stimulus intensity (I) is fitted to a simple power function of the form (Stevens, 1975).

$$M = \alpha I^{\beta} \tag{7.5}$$

where α is a magnitude-scaling constant, and β indicates the rate of growth of sensory magnitude. Alternatively, the relationship is sometimes expressed in a form requiring three experimental parameters (Kaczmarek et al., 1992):

$$M = \alpha\left(I - I_0\right)^{\beta} \tag{7.6}$$

where I_0 is the threshold-level stimulus intensity. The advantages of the two forms have been debated over decades in a broad literature. The simpler form of Eq. (7.5) is preferred by many for electrical stimulation; it has been noted to give similar residual variance as compared with the more complex form of Eq. (7.6) (Reilly et al., 1982) and is argued to more closely convey the subjective impression of observers (Rollman, 1974). At any rate, the form of the two equations is similar for stimulus intensities sufficiently above the perception threshold. A variety of values for the exponent β has been reported for electrical stimulation (Tashiro and Higashyama, 1972); to

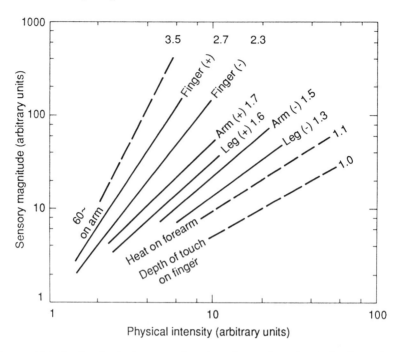

FIGURE 7.7. Psychophysical power function slopes for electrical, thermal, and mechanical stimulation. Solid lines represent spark discharges at negative ($-$) or positive ($+$) polarity. (Dashed lines from Stevens et al., 1958, 1974.) The relative positions of the functions are arbitrary on these coordinates. Slope values are given at upper right for each function. (From Reilly et al., 1982.)

some extent, these variations may be traced to features of the stimulus waveform or to its locus or method of application.

Slopes of psychophysical power functions for electrical stimulation and two other modes of natural stimulation are illustrated in Fig. 7.7 (Reilly et al., 1982). The numerical values assigned to each curve represent the exponent β from Eq. (7.5). Values of β for capacitor discharge stimulation are significantly greater than that for heat or pressure, but not as great as with 60-Hz stimulation. Stimulation of the finger results in uniformly higher slopes (larger exponents) than for stimulation on the arm or leg. Negative polarity discharges produce lower exponents than positive-polarity discharges; on the average, the difference is 18%.

Categorical Scaling

Suprathreshold reactions may also be tested using *categorical* ratings in which subjects choose from a list of affective (how unpleasant) or intensive (how strong) descriptors. Although it might be thought that affective and intensive scales would be unfailingly linked, this is not always the case. Indeed, with the administration of narcotic analgesia, it is possible to reduce significantly the intensive aspects of painful electrical stimulation without affecting its unpleasantness (Gracely et al., 1979).

Figure 7.8 provides an example of suprathreshold measurements including both subjective magnitude scaling and intensive descriptors (Reilly and Larkin, 1984). In these experiments, subjects were remarkably consistent in their use of the two scales even though they were derived in separate procedures. As a result, it is possible to display both on a single graph. These data show that the growth of sensation magnitude is much greater than that of stimulus magnitude. When fitted by a power function, perceived magnitude grows at about the 2.5 power of stimulus magnitude for stimulation of the finger, the 1.6 power for the arm, and the 1.4 power for the leg. This range of exponents corresponds well with values reported in past studies using pulsed electrocutaneous stimulation (Sternbach and

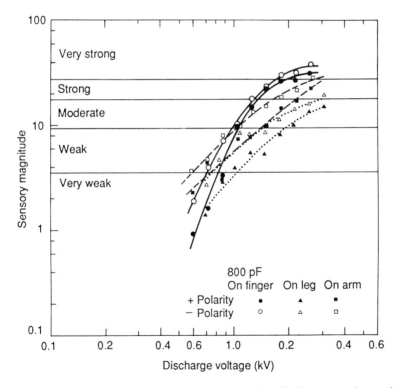

FIGURE 7.8. Growth of sensory magnitude for capacitor dischargers to three stimulation sites on the body. Vertical coordinate shows numerical magnitude judgments and ranges of adjectival rating categories. Composite data for eight subjects. (From Reilly and Larkin, 1984.)

Tursky, 1965; Rollman, 1974; Sachs et al., 1980; Higashiyama and Rollman, 1991).

The faster growth of sensation magnitude for the fingertip may be a consequence of its small volume relative to the arm or leg. Because of the volume constraint, current density becomes uniform along the finger beyond the stimulation point. This appears to result in a more spatially extensive sensory excitation: at suprathreshold levels, subjects report extended sensations along the finger.

In Fig. 7.8, sensation magnitude tends to progress along a reduced slope at the larger stimulus levels. A similar description of a two-limbed power curve has been noted by others (Rosner & Goff, 1967). The reason for the deceleration may be explained by a general theory of interaction between excitatory and inhibitory neural processes, in which inhibition increases faster with stimulus level than does excitation (Atkinson, 1982). It is equally plausible, however, that the change in slope reflects a shift among populations of excitable neurons. The group of polymodal nociceptive fibers, for example, would respond only when their relatively high excitation thresholds are reached. Other groups of neurons may respond at various thresholds of the electrical stimulus. Because electrical stimulation has no specialized transduction mechanism, it is likely that the psychophysical function reflects a mixture of neural populations. The finding that both A and C fibers can participate in the process of pain transduction (Meyer and Campbell, 1981; Campbell et al., 1979) tends to support this hypothesis.

Stimulation levels corresponding to specific affective or intensive adjectives may be conveniently reported as multiples of perception thresholds. Suprathreshold multiples, if properly determined, are relatively consistent, despite wide variations in absolute thresholds due to factors such as electrode size or intersubjective sensitivity variations. Table 7.2 lists suprathreshold responses as multiples of the mean perception threshold as reported by investigators who used single monophasic pulse stimuli. In the data of Larkin and Reilly, for example, "tolerance thresholds" were determined by presenting the subjects with pairs of stimuli in an ascending sequence. The tolerance limit was reached when subjects indicated an unwillingness to accept a second stimulus or to proceed to the next higher level. Tolerance limits determined in this manner are highly dependent on the context of the experimental procedure.

AC or Repetitive Stimuli

A brief electrical stimulus will elicit a single action potential (AP) on excited sensory neurons. The growth of sensory magnitude with increasing strength of a brief stimulus can be accounted for by the corresponding increase in recruitment of sensory neurons, including less-sensitive nociceptive fibers residing in deep subcutaneous strata. An additional factor, however, is involved in the growth of sensation for continued AC or repeti-

TABLE 7.2. Average suprathreshold multipliers for single monophasic pulses.

Body locus	N	Electrode	Stimulus[a]	Multiple above perception			Reference
				Unpleasant	Pain	Tolerance	
Finger	8	0.8-mm diam.	CD, 800 pF	2.3	3.5	7.1	Reilly and Larkin (1984)
Forearm	8	0.8-mm diam.	CD, 800 pF	3.5	5.5	11.0	Reilly and Larkin (1984)
Leg	8	0.8-mm diam.	CD, 800 pF	3.1	—	9.1	Reilly and Larkin (1984)
Finger	124	Tapped electrode	CD, 200 pF	2.6	—	—	Larkin and Reilly (1986)
Triceps	6	4-cm² carbon sponge	Pulse, 5–10 μs	—	4.3	8.9	Alon et al. (1983)
Triceps	6	4-cm² carbon sponge	Pulse, 20–1000 μs	—	7.4	10.5	Alon et al. (1983)

[a] CD, capacitor discharge stimulus.

tive pulse stimulation: sensory magnitude will also grow because of the higher CNS responses to repetitive APs (see Chapter 3). As the magnitude of an alternating current or pulse train is increased above the threshold level, there is a corresponding increase in the rate of AP production. As a result, continued AC or repetitively pulsed stimulation may be judged to be more intense than a single brief electrical stimulus of equal magnitude.

Table 7.3 lists suprathreshold multiples for AC or repetitive stimuli based on reported experimental data. Although it is clear that AC or repetitive stimuli produce a dynamic range from perception to pain that is much less than with single pulsed stimuli (Table 7.2), there is much more variance in the multiples for the reported data. The pain multiples of Budinger and of Bourland are small in comparison to the other experiments. In these experiments, stimulation was induced by exposure of the torso to bursts of sinusoidal magnetic fields, and excitation was typically felt at the buttocks, lower back, or flank. It is not clear whether the relatively small multiples applying to these experiments were the result of the relatively extended spatial distribution of excitation or to some other unique feature associated with magnetic stimulation. The multiples of Higgins appear to be particularly great. In this case, stimulation was provided by a small concentric electrode arrangement on the forearm. The relatively large multiples in Higgins' experiments might be explained by the restricted depth and lateral extent of excitation associated with the particular electrode arrangement. From these observations, one might suspect that painful sensations would be enhanced as the spatial extent of electrical excitation is made greater, although experimental verification of this conjecture is lacking.

TABLE 7.3. Average suprathreshold multipliers for AC or repetitive pulse stimulation.

Body locus	N	Electrode	Stimulus[a]	Multiple above perception			Reference
				Unpleasant	Pain	Tolerance	
Fingertip	6	Tapped electrode	60-Hz AC	1.8	2.4	3.5	Reilly and Larkin (1987)
Forearm	40	1-cm diam.	Pulse train, 100 Hz	—	2.0	5.7	Rollman and Harris (1987)
Fingertip	2	*	100–10 kHz AC	—	2.9	4.0	Hawkes and Warm (1960)
Fingertip	367	Tapped electrode	10-kHz AC	—	2.3[a]	—	Chatterjee et al. (1986)
Forearm	12	Concentric ring	60-Hz AC	4.2	6.6	11.8	Higgins et al. (1971)
Fingertip	12	Saline bath	60-Hz AC	1.3	—	—	Currence et al. (1987)
Torso	2	Magnetic field	600–1950 Hz AC	—	1.3	—	Budinger et al. (1994)
Torso	52	Magnetic field	128 pulses	—	1.6[b]	2.0	Bourland et al. (1997)

*Information sketchy or not available.
[a] Chatterjee threshold described as "pain," but may be functionally equivalent to "tolerance".
[b] Bourland category of "uncomfortable" attributed to "pain".
Note: N = number of subjects tested.

Figure 7.9 illustrates sensory power functions for stimulation by 0.1-ms pulses, delivered at a 60-Hz rate (from Rollman, 1974). The parameter n indicates the number of pulses in the train; β indicates the sensory magnitude power law, according to Eq. (7.5). It is seen that the power exponent grows with increasing n. More striking is the increase in sensory magnitude with increasing n for pulse trains at the same stimulus intensity. At 3.0-mA current level, for example, perceived sensory magnitude increases sixfold as the number of pulses increases from 1 to 30.

Dynamic Range of Electrical Stimulation

The data in Table 7.2 and 7.3 show that the dynamic range for electrocutaneous stimulation is very small. For stimulation by single pulses, the ratio of pain to perception threshold is only 3.5 to 7.4, depending on the locus of stimulation; for AC or repetitive stimulation, the dynamic range is even less. Contrast the measured dynamic range of electrical stimulation with that for hearing or pressure sensitivity, both of which have dynamic ranges of about 100,000 to 1. We can thus appreciate the importance of perception threshold measurements in electrical sensitivity studies. If we can measure the perception threshold accurately, we know that a relatively small increase will result in a strong perceptual effect.

The ordering of tolerance values on the finger, arm, and leg has been found to be consistent among subjects who may differ widely in absolute threshold levels (Reilly et al., 1982). Overall, the fingertip is more sensitive than the leg by nearly a factor of 2, and the arm takes a midway position. This ordering contrasts with measurements taken at the perception threshold (see Sect. 7.7), in which electrical stimulation is more easily detected on the forearm than on the fingertip. One factor that may account for this difference is that current flow through the finger is more volume limited than in the arm or leg. Consequently, at points distant from the introduction of current, current density may be higher in the finger than at equivalent distances on the arm or leg. Subjects experience suprathreshold stimulation

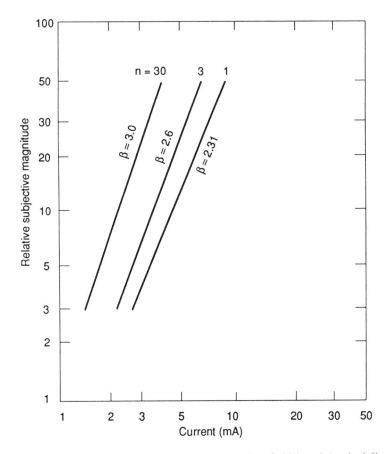

FIGURE 7.9. Sensory power functions for train of pulses (width = 0.1 ms), delivered at 60-Hz rate; n indicates number of pulses in train; β indicates power function exponent. (Adapted from Rollman, 1974, in *Conference on Cutaneous Communication Systems and Devices*, pp. 38–51, reprinted by permission of Psychonomic Society, Inc.)

as "radiating" through the finger and hand. This indicates that current density remains relatively high over a broad region and that more than one population of afferent neurons may be involved. On the leg or arm, in contrast, the sensation appears to be more limited spatially.

There is a wide variation among individual tolerance limits. Although part of this variation may reflect individual differences in sensitivity, it is clear that "tolerance" is not simply a sensory limitation. Subjects with high tolerance limits may have a detached attitude about the possibility of injury, whereas those with low limits may be more cautious or fearful. This variation does not seem tied to an individual's experience or familiarity with electrical shocks. Indeed, the author has encountered subjects who had no experience, but were willing to accept levels substantially higher than those set by the highly experienced investigators. The acceptance of a large stimulus does not necessarily imply a lack of cutaneous sensitivity. Individuals who tend to have very nearly the same thresholds and give numerical magnitude estimates in the same range may have markedly different tolerance levels.

In view of the preceding comments, the tolerance limits in Tables 7.2 and 7.3 cannot be taken as absolute limits on an individual's capacity for painful stimulation. The concept of tolerance has empirical application only as a relative, context-dependent measure. Implied demand, or social expectations, influence a person's willingness to cooperate in a scientific endeavor. For some individuals, neither pain nor the potential for skin injury deters their voluntary exposure to electric shock if they believe that there is some benefit to be gained.

7.5   Stimulus Waveform Factors[2]

Sect. 7.3 showed that neural excitation thresholds are markedly sensitive to the duration of a monophasic stimulus current. Other features of the stimulus waveform, such as biphasic properties and pulse repetition patterns, can be equally important. A general theoretical framework for evaluating such effects can be provided by appropriate neural excitation models, such as the SENN model discussed in Chapter 4.

Biphasic Stimuli

As noted in Sect. 4.6, the current reversal of a biphasic pulse can suppress a developing AP that was elicited by the initial phase. To compensate for the reversal, a biphasic pulse must present a higher initial current. When thresholds are measured in terms of the initial current, the biphasic pulse

---

[2]Portions of this section have been adapted from Reilly (1989).

TABLE 7.4. Relative thresholds of biphasic stimuli: comparison of SENN model and experiment.

Waveform parameters		Threshold ratio (Q_1/Q_2)	
t_p (Q_1/Q_2) (μs)	δ (Q_1/Q_2) (μs)	Model	Experiment
20, 20	10, 50	1.20	1.19
50, 50	10, 50	1.10	1.12
50, 20	50, 50	1.17	1.23

Notes: t_p = phase duration; δ = interpulse delay.
Source: Experimental data from Bütikofer and Lawrence (1978). SENN model as in Chapter 4.

may therefore have a higher threshold than a monophasic pulse. According to Fig. 4.17, the threshold elevation depends on the phase duration and the time delay for current reversal. Thresholds are elevated more when the pulse is short and when the current reversal immediately follows the initial pulse.

Experimental results with biphasic stimuli generally confirm the modeled threshold relationships of Fig. 4.17. One study tested suprathreshold sensory sensitivity of the hand to biphasic pulses having durations of 20 to 50 μs, and with delays from 10 to 50 μs (Bütikoffer and Lawrence, 1978). Although it is difficult to compare absolute thresholds with model results, we can compare the relative sensitivity for different waveforms. Table 7.4 lists the relative thresholds of biphasic pulse doublets having different phase durations (t_p) and interpulse delays (δ). The experimental and model results agree within a few percentage points. Other experimental data confirm that a biphasic stimulus has a reduced efficacy for neuromuscular stimulation as compared with a single monophasic pulse of the same phase duration (Gorman and Mortimer, 1983; van den Honert and Mortimer, 1979a).

The finding that thresholds are elevated for a single biphasic pulse should not be interpreted as implying that thresholds for all oscillatory stimuli are necessarily elevated above monophasic stimuli of the same phase duration. Indeed, if a biphasic stimulus is repeated as an oscillatory waveform, thresholds may decrease with successive oscillations to the point that they eventually reach the single pulse monophasic threshold (see Sect. 4.6).

Repetitive Stimuli

We have seen in Chapter 4 that the threshold for exciting a single AP by a sequence of pulses may be lower than the threshold for a single stimulus pulse. Repetitive stimuli can also enhance sensory (and motor) response if multiple APs are generated. In either case, there is an integration effect of the multiple pulses. In the first case, the integration takes place at the membrane level. In the second case, response enhancement takes place at the CNS level for neurosensory effects and at the muscle level for neuro-muscular effects (see Chapters 3 and 8).

Multiple pulse threshold effects predicted by a neuroelectric model are illustrated for two pulses in the lower section ($M < 1$) of Fig. 4.17. The effects are most pronounced for short pulses and short interpulse delays. The neuroelectric model also predicts that thresholds consistently fall as the pulse number is increased until a minimum plateau is reached. Figure 4.23 suggests a temporal summation process that effectively sums repetitive pulses, with an integration time that is roughly four times the S-D time constant. At a delay of $500\,\mu$s, the neuroelectric model shows no measurable integration effect. Similar results with neuroelectric models have been reported by others (Bütikofer and Lawrence, 1979).

Tasaki and Sato (1951) experimented with myelinated toad nerve to examine the shift in threshold of a stimulus pulse caused by a previous conditioning pulse of either the same or the opposite polarity. The duration of the two pulses was $10\,\mu$s, and the interpulse delay varied from 0 to $200\,\mu$s. Results from these experiments are shown in Table 7.5; tabulated values express the threshold of a pulse doublet as a multiple of a single-pulse threshold. Also shown are theoretical predictions taken from Fig. 4.17; the experimental data and theoretical prediction agree quite well. Even for pulse delays as long as $5\,$ms, cutaneous perception threshold reductions on the order of 7 to 8% have been reported for continuous pulse trains with pulse durations from 100 to $400\,\mu$s (Hahn, 1958).

The thresholds of perception as well as suprathreshold reactions both decline with increasing number of pulses in a stimulus train. Figure 7.10 illustrates data of Gibson (1968) obtained with monophasic pulses individually having durations of $0.5\,$ms and delivered at a rate of $100\,$Hz. The vertical axis indicates the threshold normalized by the value obtained with $n=1$. The plotted curves represent averages of thresholds obtained for stimulation at two different groups of body loci, which appeared to have different functional dependencies with pulse number. Group I applies to the

TABLE 7.5. Threshold multiples for monophasic and biphasic pulse doublets ($t_p = 10\,\mu$s).

	Theory		Experiment	
Delay (μs)	Monophasic	Biphasic	Monophasic	Biphasic
0	0.54	1.99	0.60	1.40
20	0.65	1.21	0.76	1.36
50	0.72	1.07	0.70	1.30
100	0.80	1.02	0.82	1.15
200	0.91	1.00	0.95	1.04

Source: Theoretical data derived from SENN model (see Chapter 4). Tabulated values are the threshold of a two-pulse doublet expressed as a ratio of the single-pulse threshold. Experimental data derived from Tasaki and Sato (1951).

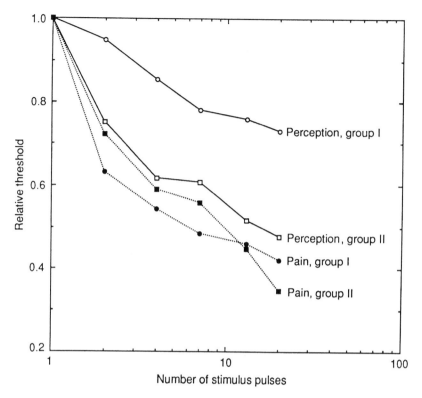

FIGURE 7.10. Relative thresholds versus number of stimulus pulses at 100-Hz rate; pulse width = 0.5 ms. Designations I and II refer to averages for two different groups of body loci. (Data extracted from Gibson, 1968.)

fingertip, lip, and ear; group II applies to the forearm and back. The coefficient of variation of the data represented in the figure was typically 0.1. Notermans (1966) related the threshold of pain to the number of pulses, each having a width of 5 ms and delivered at a rate of 100 Hz. His results, when plotted on a format similar to that in Fig. 7.10, nearly coincide with the pain data of Gibson.

We have seen, in connection with Fig. 4.22, that a theoretical neuron model effectively sums pulsed stimuli over a duration roughly four times the S–D time constant. The data of both Gibson and Notermans demonstrate thresholds that continue to decline with increasing n, out to the maximum duration of the pulse train (about 200 ms). This observation indicates a temporal integration time constant of roughly 50 ms. It is difficult to justify a time constant that long solely from the properties of a single neuron, whose S–D time constants are typically a fraction of a millisecond (see Table 7.1). The precise mechanism of temporal integration in electrical perception remains poorly understood.

Besides lowering thresholds relative to a single pulse, multiple pulses that are individually near or above threshold can generate a train of APs. The consequences can be elevated sensory magnitudes for afferent stimulation or enhanced muscle contraction for efferent stimulation. Sensory magnitude and unpleasantness in both A and C fibers are related to several factors, including the AP frequency and burst duration (Campbell et al., 1979; Gybels et al., 1979). Figure 7.11 illustrates the effects of sensory enhancement because of multiple pulses that are individually above threshold. The stimulus was designed to study 60-Hz electric field induction effects, in which a train of capacitor discharges is produced on individual half-cycles of a 60-Hz waveform (i.e., 120-Hz repetition rate) (Reilly and Larkin, 1987). Additional details concerning this figure are given in Sect.

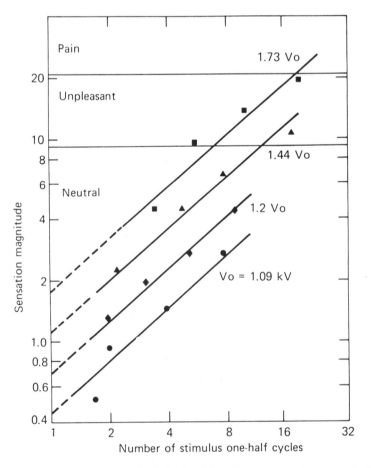

FIGURE 7.11. Effect of the number of stimulus cycles on sensation magnitude for 60-Hz AC electrical field-inducted stimuli, discharge capacitance = 100 pF. (From Reilly and Larkin, 1987.)

9.4. The horizontal axis gives the number of stimuli, and the vertical axis gives the perceived magnitude (an arbitrary, but consistent, scale assigned by subjects). Also shown are regions of subjects' qualitative descriptors. In this experiment, sensory magnitude grows at about the 0.8 power of the number of stimuli. Notice that a stimulus that is perceptible, but qualitatively "neutral," can become highly unpleasant when presented as a repetitive train.

Rollman (1974) found that sensory magnitude scaled approximately as the 0.5 power of the number of pulses up to $N = 30$ (pulse width = 0.1 ms, repetition interval = 16 ms; see Fig. 7.9). Experiments with biphasic pulse trains (Sachs et al., 1980) found that sensory magnitude scaled with N to the 0.8 power; these results were likely a combination of membrane integration and multiple AP effects, because the experimental repetition interval (50 μs) was much too small to support an AP on each pulse.

Sinusoidal Stimuli

The sinusoidal stimulus is a special case biphasic waveform, which has been discussed in Chapter 4 and also in Sect. 7.4. We have seen that the current reversal of a biphasic waveform can reverse the development of membrane excitation caused by the initial phase of the stimulus and elevate excitation thresholds. The degree of desensitization caused by the biphasic current reversal is increased as the duration of the stimulus is reduced. This phenomenon partially accounts for the high-frequency upturn noted in strength–frequency (S–F) curves for neural excitation (Fig. 4.21). The S–F curve reaches a minimum plateau as the frequency is reduced. If the frequency is reduced further, the threshold begins to rise again as the time rate of change of the sinusoidal stimulus becomes small.

The overall shape of the frequency sensitivity curve has been developed in Chapter 4 using a neuroelectric model for myelinated nerve. A mathematical fit to the thresholds determined from the model expressed by Eq. (4.35) is based on an analogy to the exponential S–D expression for monophasic stimulation. The expression includes terms to account for the threshold rise at high and low frequencies. By analogy with the S–D expression, a high-frequency parameter f_e specifies the frequency at which the threshold upturn occurs; a similar term f_0 applies to the low-frequency upturn. The low-frequency term is constrained to a maximum value to account for the fact that stimulation is possible with direct current. A maximum value of $K_L \leq 4.6$ may be inferred from the ratio of the perception threshold for continuous DC to that for 60-Hz AC (Dalziel and Mansfield, 1950b).

Although we can suggest a functional form of an S–F curve from theoretical principles, the parameters of this functional relationship need to be determined experimentally. Table 7.6 summarizes experimental S–F data determined from neurosensory and neuromuscular responses to sinusoidal

TABLE 7.6. Empirical strength frequency curve parameters.

Species	Locus of stimuli	N	Electrode	Response	I_0 (mA)	f_e (Hz)	f_0 (Hz)	Ref.
Human (m)	Hand	115	Hand-held wire	Perception	1.1	500	<60	a
Human	Arm	2	0.27 cm²; elect. paste (concentric)	Perception	0.11	1,000	50	b
Human	Thorax	8	1.0 cm²	Perception	0.2	100	<20	c
Human	Neck	8	25 cm²	Perception	0.5	150	<20	c
Human	Fingertip	2	*	Perception	0.4	500	<100	d
				Tolerance	1.0	400	<100	d
				Pain	1.8	460	<100	d
Human	Hand		Large-area grip	Tetanus	15.0	600	8	a
Frog	Sciatic nerve	*	In vitro, elect. spac. = 2 mm	AP		1,500	~30	e
			elect. spac. = 25 mm	AP		533	~30	e
Rat	Gast. tib. musc.	8	Subcutaneous	Muscle twitch		200	<100	f
Toad	Mye. nerve	*	In vitro	AP		200	<100	g
Frog	Mye. nerve	*	In vitro	AP		500	28	h

*Information incomplete or not available.

References: (a) Dalziel (1954); (b) Anderson and Munson (1951); (c) Geddes et al. (1969); (d) Hawkes and Warms (1960); (e) Hill et al. (1937); (f) LaCourse et al. (1985); (g) Tasaki and Sato (1951); (h) Wyss (1963).

stimulation. The table lists experimental estimates of f_e and f_0 from Eq. (4.35). The table also lists the minimum threshold for electrocutaneous stimulation. For proper comparison of the threshold, one should take into account the electrode size (see Sect. 7.6).

The values of f_e in Table 7.6 encompass a rather wide range, with a median (and also geometric mean) of 500 Hz. Part of the variation might be due to the electrode size and its configuration. In the experiments of Hill et al. (1937), f_e changed from 1,500 to 533 Hz when a bipolar electrode spacing was changed from 2 to 25 mm. For the electrocutaneous sensory data reported in Table 7.6, there appears to be a tendency toward large f_e values with smaller electrodes, but there are insufficient data to establish a clear relationship. The direction of the effect of electrode size on f_e is consistent with the corresponding effect on the S–D time constant, τ_e, which increases monotonically with electrode size in neural stimulation (Pfeiffer, 1968; Davis, 1923) and also with cardiac stimulation (see Sect. 6.3).

The form of the S–F curve suggests that the perception threshold current ought to increase indefinitely as its frequency is increased. As a practical matter, however, if the electrical threshold is raised high enough, sensory thresholds will be dominated by thermal perception. Because heating effects are largely independent of frequency, perception thresholds cease to rise at that point. Experimental data with continuous sinusoidal stimuli show that, at a frequency of about 100 kHz, perception thresholds reach a maximum plateau as a result of thermal perception (Chatterjee et al., 1986; Dalziel and Mansfield, 1950b) as seen in Fig. 7.12. However, for pulsed stimuli, the frequency above which thermal perception will dominate electrical excitation depends greatly on the fraction of on-time (duty factor), as the heating capacity of electrical current (i.e., its rms value) is proportional to the square root of the duty factor (see Sect. 11.3). Consequently, one can extend the frequency where thermal perception dominates electrical perception by using pulsed stimuli of low duty factor.

Thermal perception of contact current was tested by Rogers (1981) in the frequency range 2 to 20 MHz. Fifty adult subjects touched an electrode energized by a radio frequency source; the return path was through a ground plane on which the subjects stood. Perception and discomfort thresholds rose with increasing frequency. The median perception

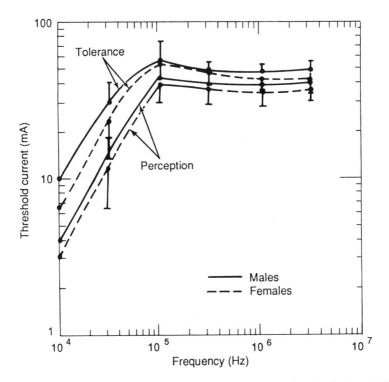

FIGURE 7.12. Perception and tolerance thresholds for AC stimulation of high frequency. Finger contact on 25-mm² plate. (Adapted from Chatterjee et al., 1986, © 1986 IEEE.)

thresholds were 45, 50, 80, 86, and 97 mA at frequencies of 2, 5, 10, 15, and 20 MHz, respectively, for stimulation on the back of the finger. Discomfort thresholds at those frequencies were 150, 173, 183, 190, and 206 mA. The ratio of discomfort to pain ranged from 3.3 at 2 MHz, to 2.1 at 20 MHz; these ratios are not unlike those for electrical stimulation at much lower frequencies (Table 7.3). Fingertip thresholds were approximately a factor of 2 higher than those on the back of the finger. With a grasping contact, current of 500 mA or more could be conducted for short periods without discomfort.

The reason for the gradually rising thresholds with frequency was not addressed by Rogers but might be explained by the impedance properties of the skin. As the frequency is increased, capacitive coupling of the skin would increasingly bypass resistive elements (see Sect. 2.2). Accordingly, I^2R energy would tend to be dissipated in deeper tissue rather than in a concentrated locus within the skin as the frequency is increased.

Figure 7.13 illustrates S–F curves from several of the references listed in Table 7.6. Also drawn is a theoretical curve defined by Eq. (4.35), using the empirical constants $f_e = 500\,\text{Hz}$ and $f_0 = 30\,\text{Hz}$. In general, the analytic expression provides a good fit to the functional form of the S–F curves, provided that the constants f_e and f_0 are properly selected.

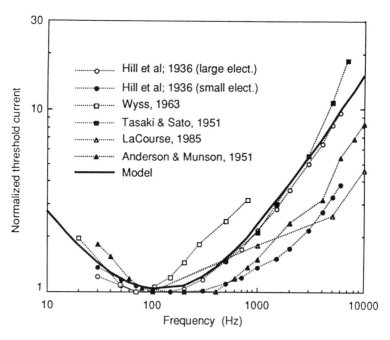

FIGURE 7.13. Strength-frequency curves for perception of sinusoidal currents applied cutaneously. Symbols represent experimental data from several investigators. Solid curve is analytic expression with $f_e = 500\,\text{Hz}$ and $f_0 = 30\,\text{Hz}$.

The analytic S–F expression provides a good fit to experimental data in most cases. However, that particular representation may not apply universally. In patients with diabetic neuropathy, frequency sensitivity reported by Katims et al. (1987) deviates significantly from the analytic form. In Katims' patients, thresholds were typically elevated above the values noted for normal individuals. It was especially notable that perception thresholds on the hands and feet of the diabetic patients were significantly smaller at 2,000 Hz than at 250 Hz; the direction of this frequency sensitivity is opposite to that found in normal people. The electrophysiological basis for this finding is presently unknown.

The S–F expression of Eq. (4.35) has been verified for human perception thresholds up to about 100 kHz. However, one would expect the relationship to apply to perception thresholds at much greater frequencies. As noted in Sect. 4.6, neuromuscular stimulation has been observed to approximately follow frequency-proportional thresholds up to 1 MHz. Also, one could infer experimental correspondence up to several MHz based on the fact that the strength-duration law can be verified for capacitor discharges as brief as $0.1 \mu s$.

Polarity Effects

Measurements taken under a wide variety of experimental conditions demonstrate lower thresholds for monophasic stimuli of negative polarity than with positive polarity. In our experiments perception thresholds for negative polarity stimuli averaged 23% lower than those for positive polarity (Reilly and Larkin, 1983). We found similar polarity differences (about 30%) when the discharges were applied to corneum-piercing needles, demonstrating that polarity-sensitive mechanisms do not depend on the properties of the corneum.

Similar polarity sensitivity has been found by others (Higashiyama and Rollman, 1991). Gibson (1968), for example, reported cathodic perception thresholds to be approximately 50 to 75% of those for anodic pulses. Girvin and colleagues (1982) tested perception thresholds for monophasic pulses having durations ranging from 63 to $1,000 \mu s$. At the shorter durations, the ratio of cathodic to anodic thresholds was 0.77; at the longer durations, the ratio was 0.63. Polarity sensitivity similar to that found for sensory excitation has also been noted for simulation of cardiac tissue (refer to Chapter 6).

We have seen in Chapter 4 that polarity sensitivity is predicted from theoretical considerations of neural excitation. With anodic stimulation, the distribution of current influx and efflux along the axon dictates that the nerve may be excited by current from an external electrode of either polarity, but that a cathodic electrode should be significantly more effective (Fig. 4.12). A model for excitation near the terminal structures of nerves also demonstrates that cathodic thresholds are lower than anodic thresholds and that the magnitude of the polarity effect is similar to that noted in sensory experiments (Fig. 4.13).

Biphasic stimuli may also exhibit a sensitivity to the polarity of the initial phase, but the polarity effects are generally less than with single monophasic pulses. The neuroelectric model described in Chapter 4, for example, shows that with short-duration biphasic pulse doublets, the threshold of excitation is lower if the initial phase is cathodic rather than anodic.

7.6   Electrodes and Current Density

In Chapter 4, we showed that the relevant force contributing to electrical excitation is the electric field within the biological medium. The field has the maximum effect when it is aligned along the long axis of an excitable cell (neuron, nerve ending, or muscle fiber). The electric field, E, is related to the current density, J, by $E = J/\sigma$, where σ is the conductivity of the medium. It can be rather difficult to determine the spatial distribution of E or J in a typical experimental situation involving electrodes within or contacting biological materials. Frequently, one has only a gross estimate of the average current density at the surface of the stimulating electrode.

Cutaneous Electrodes

The relationship between electrode area and perception sensitivity was determined for capacitor discharges to contacts of different sizes (Reilly and Larkin, 1984; Reilly et al., 1983). To avoid artifacts of electrode placement, the precise point of contact was varied from one trial to the next, within a perimeter defined by the largest electrode. A low capacitance (100 pF) was used in these tests so that stimulus discharges would have time constants within the charge-dependent region of the S–D relationship regardless of the electrode size. The electrode was either dry or treated with electrode paste.

Results for untreated skin (Fig. 7.14) show that electrode size is a critical parameter only for diameters greater than about 1 mm. Below this point, sensitivity is nearly constant. We hypothesize that for dry skin, current is conducted through discrete channels beneath a contact electrode, so that effective current density depends on the number and size of these channels and not simply on the electrode size. This hypothesis is consistent with the observation that current concentration is not homogeneous in the corneal layer of skin (Mueller et al., 1953; Panescu et al., 1993). As discussed in Sect. 2.2, tests with electroplating on the skin surface suggest a density of about one channel per square millimeter (Saunders, 1974). This estimate corresponds very well with the plateau in Fig. 7.14. For electrodes smaller than 1 mm^2, a single channel of excitation may be produced in dry skin. For larger

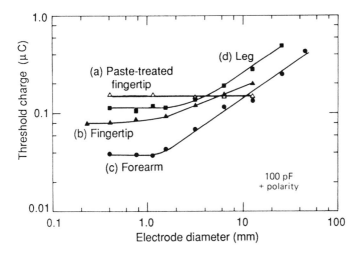

FIGURE 7.14. Effect of electrode contact size on perception of discharge from 100-pF capacitor. Curve (*a*) applies to paste-treated skin; the others apply to dry skin. (From Reilly and Larkin, 1985a.)

electrodes, the discharge current may pass through the dry epidermis in more than one place. If so, current density would be constant for any electrode smaller than about 1mm^2 and would decrease only when the electrode is made to cover at least two current channels.

If current density is a critical parameter for human sensitivity, then thresholds eventually must rise as electrode area is increased. This effect is expected whether the current is uniformly distributed or is concentrated in small current channels. For the data shown in Fig. 7.14, thresholds above 1-mm diameter increase as the one-third power of area on the forearm and leg and as the one-sixth power on the fingertip. Slopes falling between the one-sixth and one-third powers are evident in the data of previous investigators who used 60-Hz stimulation, as shown in Fig. 7.15 (Jackson and Riess, 1934; Forbes and Bernstein, 1935; Nethken and Bulot, 1967; Dalziel, 1954).

Although cutaneous sensory thresholds rise as the electrode size is increased, the rate of rise is much less than would be expected if thresholds were simply inversely proportional to electrode area. The lack of area proportionally is most likely related to the nonuniform conduction of current beneath a cutaneous electrode. While electrode paste treatment or skin hydration would be expected to reduce this effect, the current distribution beneath even a treated electrode is not uniformly distributed (refer to Sect. 2.2). Furthermore, the presence of conduction "hot spots" in treated skin cannot be ruled out.

When the skin is treated with electrode paste, the area relationship is markedly altered. In Fig. 7.14, for example, thresholds on the paste-treated

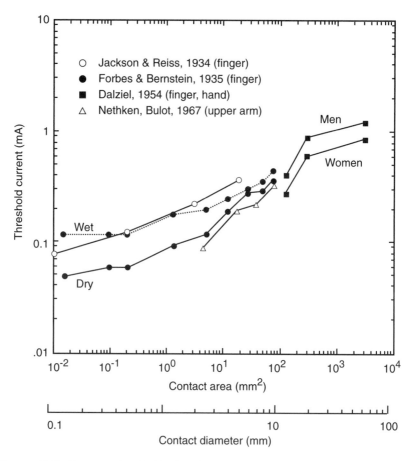

FIGURE 7.15. Sensory thresholds for 60-Hz stimulation versus area of contact. Data of Jackson and Reiss apply to pain threshold; other data apply to perception threshold.

fingertip are elevated above the dry-skin thresholds by a factor of about 2 for small-diameter contacts. As the electrode size is increased, the thresholds of dry and paste-treated skin eventually cross each other. The data of Forbes and Bernstein shown in Fig. 7.15 also demonstrate that paste treatment elevates thresholds by a factor of 2 above that for a small, dry electrode and that the thresholds of dry and treated electrodes eventually converge as the electrode size is increased. The elevation of thresholds for hydrated skin can be explained by the fact that treatment would tend to result in a more even distribution of current beneath the contact electrode, thereby reducing (but not eliminating) current-density "hot-spots" and electrode-edge current enhancement. Furthermore, current is more readily conducted beyond the edges of the electrode if the skin is hydrated. The crossover of thresholds at larger contact areas is more difficult to explain.

The role of skin impedance breakdown in electrical sensation should not be discounted. It has been noted that, with rising applied voltage, the sensation of electric shock on the dry, untreated skin is accompanied by a dramatic reduction in impedance (Nute, 1985; Gibson, 1968; Mason and Mackay, 1976; Mueller et al., 1953; Paneseu et al., 1993). Such a decrease is likely to be caused by skin breakdown in discrete locations, leading to a sudden increase in current density in those locations.

Based on the preceding, we can postulate thresholds for brief monophasic stimuli delivered to the finger or hand as given in Table 7.7. The smallest listed "perception" threshold of $0.11 \mu C$ applies to a 0.01-cm^2 contact electrode; that value has been determined from the capacitor discharge measurements of Reilly and Larkin (1987), as indicated by curve (c) at 100-pF capacitance in Fig. 7.4. The corresponding value on the forearm has been taken from the body-location sensitivity data, as discussed in Sect. 7.7. Values at other contact areas have been calculated by assuming that thresholds rise as the one-sixth power on the fingertip or hand and as the one-third power on the forearm, as discussed previously. The thresholds for unpleasantness, pain, and tolerance have been determined from Table 7.2. One postulate used to derive Table 7.7 is that the area power law determined for perception currents also applies to the suprathreshold categories of annoyance, pain, and tolerance. A second postulate is that the suprathreshold multiples of Table 7.2 are invariant with respect to electrode size. The perception threshold for the finger/hand in Table 7.7 compares favorably with the European data using capacitive discharges (Fig. 7.6). Note that in the Swiss large-electrode data, the contact area is probably on the order of 100 cm^2.

TABLE 7.7. Calculated sensory thresholds for single monophasic stimuli to finger or hand (threshold in μC for short-duration currents).

Electrode area (cm^2)	Finger, hand threshold (μC)				Forearm threshold (μC) area			
	Percept.	Annoy.	Pain.	Tol.	Percept.	Annoy.	Pain.	Tol.
0.01	0.11	0.25	0.39	0.78	0.09	0.31	0.48	0.97
0.1	0.16	0.37	0.56	1.14	0.19	0.66	1.04	2.08
1.0	0.23	0.53	0.81	1.63	0.41	1.43	2.24	4.49
10.0	0.34	0.78	1.19	2.41	0.88	3.08	4.84	9.68
100.0	0.51	1.17	1.79	3.62	1.89	6.62	10.40	20.79

Notes: Median threshold for adults. Contact to finger or hand. Brief-duration ($<20\mu s$) currents. Area power law = 1/6 for finger/hand; 1/3 for arm. Suprathreshold multiples from Table 7.2.

Current Density Considerations

In many instances, researchers report current density thresholds for peripheral nerve stimulation. Because of the aforementioned difficulties, thresholds calculated from the average current density beneath a cutaneous electrode may be unrealistically low. Current density thresholds have been calculated from a theoretical myelinated nerve model (Sect. 4.4). For fiber diameters of 5, 10, and 20 μm, the current density thresholds were calculated as 0.492, 0.246, and 0.123 mA/cm^2, respectively, for medium conductivity of 0.2 S/m (Table 4.2). In Chapter 9, these theoretical thresholds are compared with experimental values applying to stimulation by externally applied magnetic fields or by in-vitro nerve stimulation by uniform electric fields. In these experiments, difficulties due to nonuniform electrode conduction are not present. If allowances are made for variations in stimulation parameters, the experimental values are found to be within the range predicted by the theoretical model. Accordingly, the value of about 0.1 mA/cm^2 appears to represent a reasonable estimate of the current density threshold for stimulation of the most sensitive peripheral nerves.

7.7   Body Location Sensitivity

It is well known that different parts of the human body have different sensitivities to tactile and thermal stimulation. Figure 7.16 displays the rank-orders of body-part sensitivity for three measures of tactile sensitivity: pressure, two-point discrimination, and point localization (Weinstein, 1968). The rank-ordering applies to a mixed population of male and female subjects. Females were found to have consistently lower pressure thresholds on all body parts than males, as illustrated in Fig. 7.17; the vertical scale expresses the logarithm of the sensitivity relative to 0.1 mg. The gender difference is appreciable, reaching 10:1, for example, on the back, forearm, and leg. No such difference appears in Weinstein's data for spatial discrimination. We shall return to a discussion of male/female differences in sensitivity in Sect. 7.10, where we show that females tend to be more sensitive to electrical stimulation as well, and we advance a hypothesis for these differences. Relative pressure sensitivity in absolute units varies in Fig. 7.17 by nearly 10:1 from the most sensitive regions (the face) to the least sensitive (the foot).

Electrical sensitivity also varies considerably across different body parts, as demonstrated in our experiments (Reilly et al., 1982; Reilly and Larkin, 1984). In these experiments, sensitivity thresholds were measured for five individuals at each of seven body loci: forehead, cheek (both right and left), tip of third finger, thenar eminence (portion of the palm at the base of the thumb), underside of forearm, back of the midcalf, and a point 3 cm above the ankle bone. The stimuli were discharges from a 200-pF capacitor. To

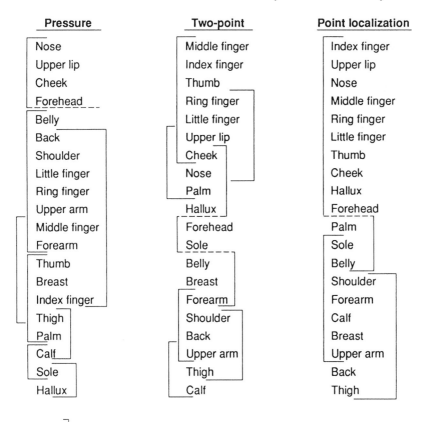

Pressure	Two-point	Point localization
Nose	Middle finger	Index finger
Upper lip	Index finger	Upper lip
Cheek	Thumb	Nose
Forehead	Ring finger	Middle finger
Belly	Little finger	Ring finger
Back	Upper lip	Little finger
Shoulder	Cheek	Thumb
Little finger	Nose	Cheek
Ring finger	Palm	Hallux
Upper arm	Hallux	Forehead
Middle finger	Forehead	Palm
Forearm	Sole	Sole
Thumb	Belly	Belly
Breast	Breast	Shoulder
Index finger	Forearm	Forearm
Thigh	Shoulder	Calf
Palm	Back	Breast
Calf	Upper arm	Upper arm
Sole	Thigh	Back
Hallux	Calf	Thigh

⎤ Items grouped within a bracket not significantly different.

– – – – All upper items significantly different from all lower items.

FIGURE 7.16. Rank-order of tactile sensitivity across various body parts, for three measures of sensitivity. (From Weinstein, in D. R. Kenshalo, ed., *The Skin Senses* 1968. Courtesy of Charles C Thomas, Springfield, Il.)

minimize the influence of concomitant tactile stimulation, a 1.6-mm-diameter contact electrode was used, flush mounted in a large plastic holder. At each body locus, the electrode contacted a slightly different spot to avoid an average measurement that might have been unduly influenced by spots of unusually high or low sensitivity.

Figure 7.18 illustrates the results, ordered in decreasing sensitivity from left to right. Results have been normalized in relation to the fingertip threshold for each subject. The fingertip was chosen for this purpose because it showed smaller variability than the other points, both among and within subjects. Electrical sensitivity across body loci is seen to vary by about 4:1, with points on the face being the most sensitive, the lower leg the least sensitive, and the finger about in midrange.

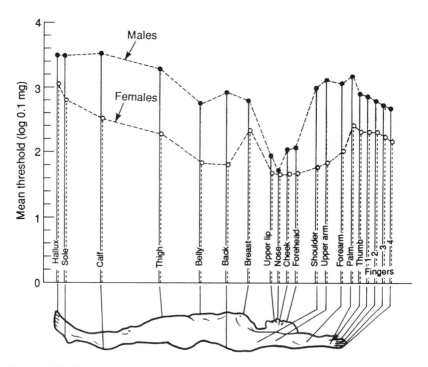

FIGURE 7.17. Pressure sensitivity for males and females (Adapted from Weinstein, in D. R. Kenshalo, ed., *The Skin Senses,* 1968. Courtesy of Charles C Thomas, Springfield Il.)

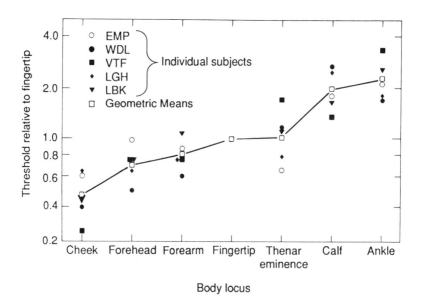

FIGURE 7.18. Threshold data for seven body loci, normalized to the threshold for the fingertip, flush-tip contact electrode, 200-pF capacitance. (From Reilly and Larkin, 1984.)

Five of the seven locations tested in this experiment were included in Weinstein's study of cutaneous tactile sensitivity; a sixth, the thenar eminence, can be compared with Weinstein's measurement on the palm. By comparing Figs. 7.17 and 7.18, it can be seen that the rank ordering of pressure and that of electrical sensitivity are similar. On the basis of these results, it is reasonable to expect that electrical sensitivity at other points of stimulation will correlate highly with tactile pressure sensitivity.

Although the ankle is about one-half as sensitive to electrical stimulation as the fingertip, this difference should not be construed to mean that the lower extremities are in all conditions protected from accidental exposure that cannot be felt on the fingertip. There may be situations in which it is natural to use some force in touching an electrical source with the fingertip, but not with the ankle. In such cases, the force of the finger tip can mask an electrical sensation (see Sect. 7.9).

We observe that the further the site of stimulation is from the brain, the higher the threshold. Whether this observation implies some connection with the synchronous timing of AP volleys, or differences in receptor type or density at various body locations, can only be guessed. Lacking sufficient histological and experimental data, any explanation for electrical sensitivity differences at different body locations remains speculative.

7.8 Skin Temperature[3]

It is not surprising that human sensitivity to electrical stimulation depends on skin temperature, as the responsiveness of the underlying neural fibers depends on their temperature. But there is a major difficulty in extrapolating from the temperature-dependent behavior of a single neuron to the temperature-dependent behavior of human skin sensitivity: The skin, together with its blood supply, is a very efficient thermoregulator that protects underlying tissues from the stress of changing temperatures at the body surface. It is important, therefore, to measure human sensitivity under controlled conditions of ambient temperature. Larkin and Reilly (1986) performed experiments for this purpose, using an air-immersion technique that mimics exposure to environmental heat and cold. The experiments tested both perception thresholds and suprathreshold responses for capacitor discharges on the fingertip and on the forearm. An electrode-tapping procedure was used for the fingertip, and a contact electrode was used for the forearm; the stimulated arm was placed inside an insulated, temperature-controlled chamber.

Perception threshold measurements, illustrated in Fig. 7.19, demonstrate that the effects of skin temperature are similar on the fingertip and forearm.

---

[3] Portions at this section have been adapted from Larkin and Reilly (1986) and from Reilly and Larkin (1985a).

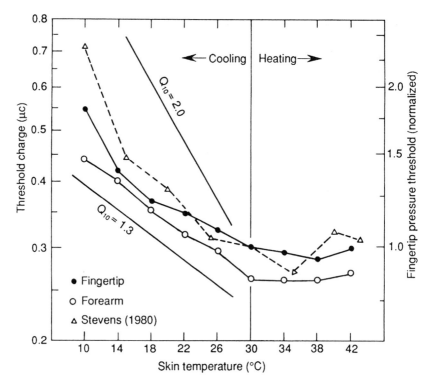

FIGURE 7.19. Effect of skin temperature on cutaneous thresholds. Solid lines with circles show detectability of electrical pulse on fingertip and forearm. Broken line shows punctuate-pressure sensitivity on fingertip. (From Stevens, 1980, Fig. 1.) Reference lines of constant slope are shown for $Q_{10} = 1.3$ and 2.0. (From Larkin and Reilly, 1986.)

Warming the skin affects sensitivity very little in either case, but there is a drastic effect with cooling. The somewhat higher rate of climb for the fingertip at low temperatures is consistent with its relatively lower priority in thermoregulation compared with the forearm. Figure 7.19 also shows normalized punctuate-pressure thresholds adapted from a study of temperature effects on pressure thresholds (Stevens et al., 1977; Stevens, 1980). The temperature dependence of pressure and electrical thresholds is similar: Pressure sensitivity degrades if the skin is cooled, but shows very little dependence on warming, up to 43 °C.

In a suprathreshold procedure, subjects adjusted a stimulus on the cooled or heated right hand to match the perceived intensity of a capacitor discharge stimulus delivered to the left hand that was maintained at room temperature. The stimuli presented to the left hand ranged from just perceptible levels to the threshold of pain. Results of the sensation-matching procedure are given in Fig. 7.20. The horizontal axis plots the charge delivered to the left fingertip. The vertical axis shows the stimulus needed on thermally treated skin to achieve an equivalent level of sensation. The most striking aspect of these data is that the matching charge falls along straight lines parallel to the diagonal. Thus, the change in sensitivity is uniform over the range of stimuli tested and can be represented by a multiplicative constant for each temperature. The multipliers for the matching experiment are listed on Fig. 7.20. They indicate that, for example, charge must be raised by about 60% to reach the same sensation magnitude on the skin at 10 °C, compared with skin at normal temperature (30 °C). These multipliers fall very close to similar numbers derived from the threshold data in Fig. 7.19. Thus, skin temperature affects perception sensitivity and suprathreshold sensitivity nearly identically.

The fact that heating or cooling affects electrical sensitivity differently is not surprising in view of the body's more efficient thermoregulation against heat than against cold. The tolerance of pain to skin temperature occurs around 45 °C, at which point tissue damage begins to occur with prolonged

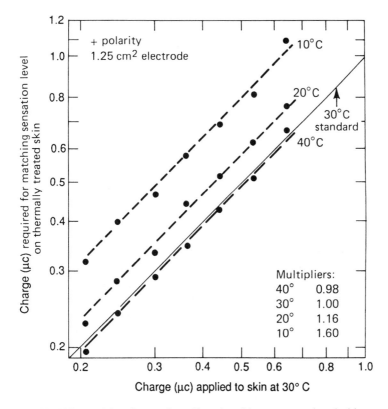

FIGURE 7.20. Effect of heating and cooling the skin on suprathreshold sensation levels. Measurements for capacitor discharges at 200 pF to the fingertip. (From Larkin and Reilly, 1986.)

heating (Chapter 10; Fig. 11.12), and nerve conduction is irreversibly blocked (Klump and Zimmerman, 1980). On the other hand, the body can withstand cooling well below normal temperatures. Below 30°C, the skin loses heat faster than it can be restored metabolically (Tregear, 1966). It can therefore be expected that neural responsiveness would be affected by cooling of the skin much more so than with heating.

The temperature dependence of neural activity is frequently reported in terms of a Q_{10} coefficient, which is the ratio of a given index of activity at one temperature to that when the temperature is shifted by 10°C. For a variety of indices of neural responsiveness, Q_{10} coefficients for isolated neural preparations typically range from about 1.3 to 3.0 (Larkin and Reilly, 1986). The data of Fig. 7.19 indicate a Q_{10} value of 1.3 for cooling on the arm. On the fingertip, a Q_{10} value around 1.3 applies to temperatures down to about 14°C, below which a value around 2.0 is indicated.

7.9 Tactile Masking

The mechanical stimulation of touch can have a masking effect on electrical sensation. Most of us learn, for example, that a carpet spark is less bothersome if we grasp the grounded object firmly, rather than tentatively. The attenuating effects of contact force are well known among electricians and others who are called upon to touch "hot wires." The explanation of this phenomenon is not fully understood, but is likely to include the fact that electrical and mechanical stimulation excite the same neural pathways in the skin. At low voltages, people commonly report that capacitive discharges feel very much like tactile pressure, mild pricks with a needle, or pinpoint touches. Furthermore, we have seen that both tactile perception and electrical perception share several features in common, including similar rank-ordering with respect to body-part sensitivity, temperature sensitivity, and male/female differential sensitivity.

We studied the relationship between tactile force and electrical threshold by measuring perception thresholds when subjects tapped an energized electrode with calibrated force (Reilly and Larkin, 1983). Subjects were trained to tap with a force within 2 dB of a target position on an accelerometer scale. The scale ranged from 0 dB (the lightest force for repeated tapping) to 50 dB. Repeated tapping at a force of 60 dB was painful. Steps of 10 dB on this scale represent approximately equal intervals of subjective sensation, as well as of physical intensity. Thus, 50 dB is a "strong" tap, 40 dB is "firm," 30 dB is "moderate," 20 dB is "light," and 10 dB is "very light." Tap force was calibrated in a mechanical arrangement, in which a force of 20 dB was registered by dropping a steel ball with 11 μJ of kinetic energy (Reilly et al., 1982).

Figure 7.21 illustrates results from an experiment in which capacitance was held fixed while tap force was changed randomly from one trial to

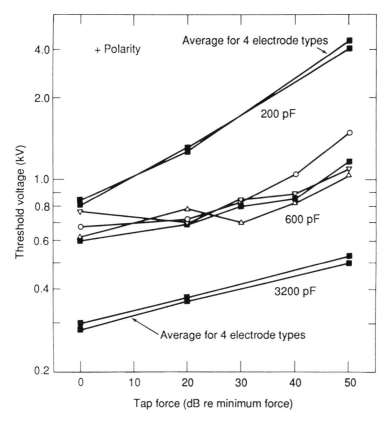

FIGURE 7.21. Effect of tap force on detection threshold for single-spark discharges into the fingertip. Each symbol represents an individual subject. Except where noted, a 1-mm raised-tip electrode was used. (From Reilly and Larkin, 1985a.)

another. The middle set of curves represents results for four subjects using a tip electrode (1-mm-diameter tip, elevated 0.5 mm above an insulating holder) at 600 pF. The threshold voltage at the highest tap force, 50 dB, is approximately double the voltage at the minimum tap force. A second set of experiments was performed with capacitor discharges at 200 and 3,200 pF. Results for one individual are plotted in the top and bottom of Fig. 7.21. The data represent trials using a variety of electrodes of different shapes and sizes. Two general conclusions were made from these experiments. First, the electrode size and shape have almost no effect in the tapping procedure. Second, the masking effect of tap force is greater for electrical stimuli with shorter durations. The masking of electrical thresholds by touch can be large and should not be discounted in any assessment of human sensitivity. Accordingly, tap force should be monitored or controlled in any test involving active touching.

Rollman (1974) tested the masking effects of a conditioning pressure stimulus on electrical thresholds and the reverse (electrical masking of pressure stimulus). The degree of masking depended on the time sequence of two stimuli. For coincident pressure and electrical stimuli matched to the same subjective intensity (determined by cross-modal matching to an audible tone), Rollman reported a 1- to 3-dB threshold shift in the electrical stimulus if the mechanical stimulus is the conditioning event; for the reverse situation (detection of pressure, masking by electrical), the pressure threshold was shifted by 12 dB.

A concurrent pressure stimulus has been observed to raise thresholds of painful 60-Hz electrical stimulation (Higgins et al., 1971). This observation was offered as evidence supporting Melzack and Wall's (1965) so-called "gate theory" in which stimulation of myelinated fibers is said to inhibit pain perception from nearby unmyelinated nociceptors.

7.10   Individual Differences in Electrical Sensitivity[4]

It is not uncommon for one person's detection threshold to be as much as twice that of another person, although the measurement procedures are in all respects identical. These individual differences persist from day to day and from week to week; they are not temporary artifacts of the measuring procedure. In spite of these differences, electrical sensitivity data exhibit a high degree of parametric invariance, in the sense that curves of sensitivity (for example, psychometric functions or equal-sensation contours) commonly have the same shape regardless of who is picked as the test subject. Curves for different test subjects can frequently be related to one another by a simple multiple. Consequently, basic studies of sensitivity need not require a large number of participants. However, none can serve as a basis for estimating the distribution of sensitivity variation in the general population. For that purpose, a large sample of test participants is needed.

Potential Sources of Variability

Variations in sensory measurements fall into three general categories: trial-to-trial variations, long-term sensitivity drifts, and subject-to-subject differences. Trial-to-trial variations in thresholds are observed when an individual is tested repeatedly in a single session. For capacitor discharge stimuli, these variations depend on the method of applying the stimulus (Larkin and Reilly, 1984). For stimuli delivered by actively tapping an energized electrode with the fingertip, perception thresholds exhibit a coefficient of variation (ratio of standard deviation to mean) of about 0.12; for

---

[4] Portions of this section have been adapted from Larkin and Reilly (1984).

capacitor discharges delivered to an electrode held in contact with the skin, the observed coefficient of variation is about 0.05. In the contact-electrode case, the coefficient of variation is comparable to that observed in other studies using constant-current pulses (Rollman, 1974). In the tapped electrode case, the electrical sensation can be masked by tactile stimulation, raising the threshold of detectability and increasing its variance. Gradual threshold drifts reflect a second source of variability: Even highly practiced subjects do not give precisely the same results from day to day. We examined individual sensitivity drifts in a longitudinal study in which four subjects were tested once weekly over a period of 8 weeks (Reilly et al., 1984). Treating the session means for each subject as a statistical variable, the coefficient of variation across sessions averaged 0.11 for the four subjects. These long-term drifts cannot be separated from intersubjective comparisons without enormous experimental effort. Such an effort would not be repaid by any substantial increase in the precision with which two individuals can be compared, because the drifts appear to account for only a small fraction of the total intersubjective variance. Individual differences in sensitivity are a much greater source of variability in sensory measurements. These differences persist over time and greatly outweigh the fluctuations observed from trial to trial or from one week to the next.

Elevated sensory thresholds exist in individuals with neuropathological conditions (Friedli and Meyer, 1984; Katims et al., 1986; Notermans, 1967; Pruna et al., 1989), as well as psychopathological conditions (Hare, 1968). Neuromotor excitation threshold shifts also exist in individuals with neuropathy (Parry, 1971; Harris, 1971). Significant variations from normal populations have been observed in the thresholds of electrocutaneous perception in uremic (Katims et al., 1991) and HIV-positive individuals (Taylor et al., 1992). Among individuals who appear to be normal, wide variations in electrical sensory thresholds are also found. The basis for these variations is not well understood.

Distribution of Sensitivity

We studied electrical sensitivity of 124 subjects (Larkin et al., 1986). Rather than make an extensive study of people chosen completely at random—a practical impossibility in any case—the study focused on specific subgroups whose sensitivity might be expected to be either high or low. Subgroups consisted of college students (25 women and 24 men), 25 female office workers, and 50 men engaged in skilled trades (electricians, carpenters, plumbers, and sheet-metal workers). The sampling strategy allowed direct comparisons between men and women and between occupational groups. Among the groups, there was a wide distribution of ages and body sizes, so that a statistical estimate could be made of their relationship to electrical sensitivity. A variety of physical attributes was quantified; namely, age, height, weight, temperature of fingertip and forearm, circumference of fingertip and forearm, index of habitual physical activity, and index of prior shock experience.

Statistical distributions of sensitivity were determined separately for the subgroups. Geometric means are summarized in Table 7.8. One indication is that there are sensitivity differences related to sex and occupation: males appear to have higher thresholds than females, and maintenance men have higher thresholds than college men. As shown below, the apparent group differences appear to be artifacts of a body-size dependency.

Table 7.8 also provides the geometric means of ratios among individuals' sensitivity values. The column labeled B/A gives the ratios of perception for finger versus arm stimulation. This ratio not only reflects the greater sensitivity of the arm relative to the finger (see Sect. 7.7), it also arises because of tactile masking and electrode size effects. The effect of tactile masking is to elevate thresholds for fingertip stimulation relative to a contact electrode condition (see Sect. 7.9). The combined effects of these factors result in an average fingertip-to-arm threshold ratio (B/A) of 2.9. The column labeled C/B indicates the multiple above the perception threshold needed to produce the rating "definitely annoying." These "perception-to-annoyance" ratios agree well with results from other experiments designed specifically to study suprathreshold reactions (Sect. 7.4). The column labeled C/D indicates the relative sensitivity of 200- versus 6,400-pF discharges from a

TABLE 7.8. Geometric averages and ratios of thresholds from the large sample study.

		Perception thresholds		Annoyance levels				
		(A)	(B)	(C)	(D)			
		Arm-contact electrode	Fingertip tapping	Fingertip tapping	Fingertip tapping		Ratios	
Group	N	200 pF	200 pF	200 pF	6,400 pF	B/A	C/B	C/D
Female students	25	0.083	0.252	0.576	1.91	3.0	2.3	9.7
Male students	24	0.107	0.277	0.664	2.14	2.6	2.4	9.9
Female office workers	25	0.074	0.266	0.648	1.95	3.6	2.4	10.7
Male maintenance workers	50	0.113	0.314	0.924	2.94	2.8	2.9	10.0
Total sample	124	0.097	0.284	0.734	2.33	2.9	2.6	10.1

Notes: Data given in μC of charge for positive-polarity stimulation; values of capacitance are listed at heads of columns.

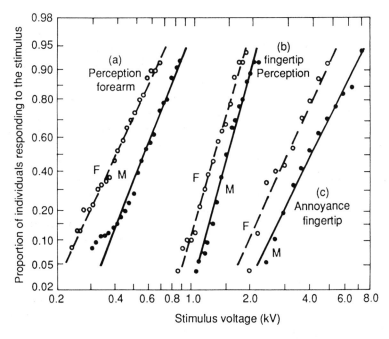

FIGURE 7.22. Cumulative probability that a capacitive discharge (+polarity, 200 pF) exceeds perception or annoyance thresholds; samples of 74 men (M) and 50 women (F). Discharges on the fingertip via a tapped electrode; discharges on the arm via a 0.5-cm-diameter contact electrode. (From Larkin et al., 1986.)

tapped electrode: On the average, 3.2 times as much charge is required at 6,400 pF to produce the same annoyance rating obtained at 200 pF. This capacitance effect stems from differences in discharge time constants at the different capacitance values (see Sect. 7.3).

Threshold distributions are well represented by the log-normal distribution, both for subgroups and for pooled data. Figure 7.22 gives an example of sensitivity distributions for 200-pF discharges, separated into male and female subgroups. Voltage levels are indicated on the horizontal axis. The vertical axis plots the relative number of individuals reporting perception [curves (a) and (b)] or annoyance [curves (c)]. The horizontal axis is logarithmic, and the vertical axis is subdivided according to the standard Gaussian probability distribution. The data are nearly linear on these plots, indicating the log-normal distribution. The straight lines through the data are minimum chi-square fits to the log-normal model. This depiction omits some outliers and therefore slightly underrepresents the total variability in the sample. The voltage ratio spanning the middle 90% of the fingertip perception thresholds (that is, from the 5th to the 95th percentile) is approximately 2:1. The ratio is closer to 3:1 in the case of forearm thresholds and the annoyance ratings.

As explained below, electrical thresholds are correlated with measures of body size. Consequently, subgroups that are correlated with body size will exhibit a threshold distribution that is less dispersed than a more heterogeneous group, as illustrated in Table 7.9 by the data of Larkin and Reilly. For comparison, statistical parameters from other experimenters are also shown. In the first subgroup of the table, adult males and females are combined in presumed equal numbers. The second subgroup, applies to either males or females. Although these subgroups are associated with significantly different thresholds, the distributions are virtually indistinguishable when their thresholds are normalized with respect to their respective medians (see Fig. 7.22). The third subgroup, assumed to be of homogeneous body weight, has weight-adjusted thresholds in accordance with a regression formula presented below.

For analysis of electrical acceptability, we would like to know the distribution of thresholds across a population of exposed individuals. For ex-

TABLE 7.9. Statistical distributions of sensory thresholds.

Body locus	Subgroup	N	Stimulus	Measurement	Normalized threshold 10%	Normalized threshold 90%	Ref.
Forearm	M + F	124	200-pF cap. discharge	Perception	0.54	1.67	a
Fingertip	M + F	124	200-pF cap. discharge	Perception	0.71	1.43	a
Fingertip	M + F	124	200-pF cap. discharge	Annoyance	0.70	1.81	a
Forearm	M or F	74, 50	200-pF cap. discharge	Perception	0.61	1.52	a
Fingertip	M or F	74, 50	200-pF cap. discharge	Perception	0.78	1.32	a
Fingertip	M or F	74, 50	200-pF cap. discharge	Annoyance	0.74	1.55	a
Forearm	Weight-adjusted	124	200-pF cap. discharge	Perception	0.73	1.36	a
Fingertip	Weight-adjusted	124	200-pF cap. discharge	Perception	0.79	1.26	a
Forearm	M + F	40	1-ms pulses @ 10 pps	Perception	0.34	1.69	b
Forearm	M + F	40	1-ms pulses @ 10 pps	Pain	0.43	1.57	b
Hand (grip)	M	115	Continuous DC	Perception	0.63	1.50	c
Hand (grip)	M	115	60-Hz AC	Perception	0.73	1.27	c

Thresholds normalized by median value.
References: (a) Larkin and Reilly (1984); (b) Rollman and Harris (1987); (c) Dalziel and Mansfield (1950b).

TABLE 7.10. Normalized perception threshold distribution for log-normal model of adult population.

Percentile rank (%)	M + F		Weight adjusted	
	Forearm	Finger	Forearm	Finger
99	3.11	1.91	1.78	1.51
95	2.24	1.58	1.50	1.33
90	1.85	1.41	1.36	1.25
75	1.40	1.21	1.18	1.12
50	1.00	1.00	1.00	1.00
25	0.72	0.83	0.85	0.89
10	0.54	0.71	0.73	0.80
5	0.45	0.63	0.67	0.75
1	0.32	0.52	0.56	0.66
0.5	0.29	0.49	0.53	0.63

Note: Data express multiples of median threshold.

ample, one may wish to select a criterion of exposure that protects some percentile rank of sensitive individuals. Table 7.10 indicates thresholds for a representative log-normal model, based on the author's measurements. The data listed under (M + F) apply to a group of adults, with males and females in equal proportion. The "weight-adjusted" data apply to calculated thresholds, using body-weight regression formula. The data in this table have been normalized by the median values of the respective subgroups.

Correlates of Sensitivity

The cumulative distributions plotted in Fig. 7.22 show an unmistakable difference between the two sexes. For a homogeneous group of students, for example, the forearm thresholds for males is 1.3 times that for females (column A of Table 7.8). The direction of this difference is not surprising in view of the research literature on sex differences in cutaneous sensitivity. Women reportedly are more sensitive than men to pressure (see Sect. 7.7), to vibratory stimulation (Verillo, 1979), and to pain (Notermans and Tophoff, 1975). Women are also reported to be more sensitive to a variety of electrical stimuli, including pulse trains (Rollman and Harris, 1987), 60-Hz currents (Dalziel, 1972), and sinusoidal currents of frequency from 10^4 to 10^7 Hz (Chatterjee et al., 1986). The reported ratio of the thresholds for males versus females is typically consistent with our findings (about 1.3:1).

To my knowledge, no physiological hypothesis has ever been advanced to account for these differences, although there have been suggestions that cultural and social factors may account for them—especially in the case of pain tolerance. Our data do not support a cultural explanation, since detection thresholds were determined in a manner that effectively eliminates all sources of bias in response tendencies, and the ratio between annoyance level and threshold level is remarkably constant across the two sexes (see the column labeled C/B in Table 7.8).

The data in Table 7.8 also suggest that higher electrical thresholds might be associated with occupational groups involved in strenuous labor. There is a common belief that those who lead more strenuous lives acquire a more stoic or tolerant attitude toward physical discomfort and pain. The physical demands of daily life may contribute to this effect if skin sensitivity is altered by injury to the epidermis or by the development of hard callused layers. Whether pain tolerance is culturally acquired, as many have suggested (Zborowski, 1952; Sternbach and Tursky, 1965; Craig and Weiss, 1971; Wolff and Langley, 1975; Clark and Clark, 1980), or reflects an underlying physiological adaptation, is not at issue in the present study. We might have expected college students and office workers to differ greatly from electricians, carpenters, and sheet-metal workers, who are more frequently exposed to strenuous physical labor and harsh weather. If the threshold and annoyance levels of the maintenance workers are compared with those of the other groups (Table 7.8), there is a difference in the expected direction. However, the difference is nearly as large for forearm sensitivity, where calluses do not develop, as it is for the fingertip. This observation suggests that skin thickness alone does not account for the result.

Much the same conclusion applies in the case of physical exercise and the subjects' reported levels of previous exposure to electric shock. Neither of these factors is well correlated with sensitivity; they do not explain why maintenance workers have higher thresholds and annoyance levels than the other subgroups.

It might seem that a factor related to sex and occupation could partially account for individual sensitivity differences. But multiple regression analysis shows that the apparent sex and occupation differences appear to be artifacts of an underlying body-size dependency: Woman in our sample group tended to be smaller than the men, and maintenance workers were larger than college men (Larkin and Reilly, 1986). When these size differences were taken into account, no statistically significant differences in sensitivity remained. A regression equation based on body weight was determined:

$$V = ae^{bW} \qquad (7.7)$$

where V is the predicted median perception or annoyance voltage, W is the subject's weight in kilograms, and a and b are empirical parameters. Best-fit values for a and b that apply to discharges to the fingertip, from a 200-pF capacitor were, for perception and annoyance, respectively, $a = 971, 1904$; and $b = 0.0053, 0.0093$. Statistical variations about the median indicated by Eq. (7.7) follow the log-normal distribution, with a 90th percentile value at a factor of approximately 1.5 times the median and a 10th percentile at approximately 0.7 times the median.

Equation (7.7) provided a better overall fit (greatest r^2) to the data than the other two-parameter functions. However, from theoretical considerations (Schmidt–Nielsen, 1984), one might expect a power function of body weight. In this case we can represent the data by

$$V = cW^d \tag{7.8}$$

where c and d are empirical parameters. The least-squares fit of Eq. (7.7) was only marginally better than that of Eq. (7.8). The estimated parameter values that apply to discharges to the fingertip, from a tapped 200-pF capacitor, were for perception and annoyance, respectively: $c = 204, 121$; and $d = 0.386, 0.678$. With 200-pF discharges to a 0.5-cm-diameter electrode held against the arm, perception parameter values were $c = 13.8$ and $d = 0.702$.

Table 7.11 illustrates weight-dependent fingertip thresholds for discharges of 200 pF based on Eqs. (7.7) and (7.8). The two formulations closely agree over the weight range of the test subjects (41 to 128 kg). Note that the dynamic range of sensitivity, represented by the perception-to-annoyance multiple, increases from low to high body weights.

Downward extrapolation of Eqs. (7.7) and (7.8) yields rather different estimates of sensitivity at the lower body weights. A representative weight for children (23 kg) is included in Table 7.11. In this case, the power function gives a value 16% lower for perception and 26% lower for annoyance, compared with the exponential function. Without further testing, it is not possible to make a secure choice between these alternatives nor to be confident that extrapolation to children is valid.

The regression formulas presented above used body weight as a matter of convenience. Other relationships with body size (e.g., forearm diameter) yield equally valid correlates of sensitivity, since various measures of body size are strongly correlated with one another. Based on these findings, it

TABLE 7.11. Calculated thresholds for various body-weight classes, discharge to the fingertip from a 200-pF capacitor, tapped electrode delivery.

Weight		Perception		Annoyance		Annoyance multiple	
(kg)	(lb)	(7.7)	(7.8)	(7.7)	(7.8)	(7.7)	(7.8)
23	50	0.22	0.18	0.47	0.35	2.15	1.87
45	100	0.25	0.24	0.58	0.55	2.45	2.29
68	150	0.28	0.28	0.72	0.73	2.57	2.58
91	200	0.31	0.31	0.88	0.88	2.81	2.81
113	250	0.35	0.34	1.09	1.03	3.11	3.00

Thresholds are expressed in μC of charge. Corresponding voltages are calculated according to Eq. (7.7) or (7.8) as indicated in column head.

Source: From Reilly and Larkin (1987).

would be interesting to examine whether male/female differences in tactile pressure sensitivity might also reflect an underlying body-size effect. The fact that pressure sensitivity of women's breasts is more closely related to the inverse of breast size, rather than to the overall size of the women (Weinstein, 1963), suggests that the size of the stimulated body part is relevant.

An intriguing question remains: why is electrical sensitivity related to body size? Conceivably, the relationship might be explainable if cutaneous receptor density were inversely related to the size of the individual. However, until further histological and experimental evidence becomes available, the etiology of the body-size role in electrical sensitivity can only be guessed.

7.11   Startle Reactions[5]

Muscle movement that is initiated through reflex activity involves both sensory and motor responses (see Sect. 3.6). Reflex activity operates outside conscious control to regulate many vital body functions. When this activity is in response to a surprise sensory insult, the reaction is frequently called a "startle reaction." A startle reaction to electrical stimulation could potentially pose a hazard to an individual who was engaged in certain inherently dangerous activities, such as climbing a ladder, handling hot liquids, or handling power cutting tools.

It is very difficult to design a meaningful experiment to define startle reactions. One difficulty arises from the fact that it is necessary to introduce the element of surprise in the context of informed consent on the part of experimental subjects. Expectations in an experimental setting are likely to introduce a bias in the measured reaction threshold or the extent of the reaction.

Published data on startle reactions to electrical stimulation were obtained from experiments carried out at the Underwriters Laboratories (UL) in connection with the development of ANSI standards on leakage current (Smoot and Stevenson, 1968a and 1968b; Stevenson, 1969). These studies were designed to help define levels of current that could pose a potential hazard because of a startle reaction. Men and women volunteers were asked to transfer rice from one container to another using a small metal cup held in the fingers. Random levels of current were applied at unexpected times to various parts of the hand, wrist, or lower arm, including the small cup itself. The return electrode was through the other arm immersed in salt water. Subjects' reactions were given numerical ratings according to the distance that the hand moved during a startle reaction, the speed of movement, and the amount of rice spilled. A panel consisting of

---

[5] This section has been adapted from Reilly (1978b).

approximately 240 people viewed the volunteers' reactions by videotape and made judgments as to what reaction might pose a hazard in the home, such as because of spillage of hot grease or falling from a ladder. Most panel members were engineers and scientists concerned with electrical safety, although 11 homemakers were also represented on the panel. A borderline hazardous reaction was typically judged as one corresponding to between 3 and 12 in. of motion at a moderately fast rate, with some rice spillage (usually less than 10%).

Initial tests established that women were somewhat more sensitive than men. As a result, succeeding tests were carried out with women to establish a conservative estimate of hazardous thresholds. Based on 20 women tested, the average reaction judged as borderline hazardous was elicited at a conducted current[6] of approximately 2.2 mA when applied to the lower arm and 3.2 mA when applied to the cup held between the fingers. These average reaction currents may be compared with the median perception levels for women of 0.73 mA for gripped conductors, 0.59 mA for pinched contacts (Thompson, 1933), and 0.24 mA for lightly touched contacts (Dalziel, 1954). Thus, it appears that average reaction currents are at least a factor of 3 above average perception currents. Such a threshold multiple would place the startle threshold stimulus in the painful affective category, as can be seen from the data of Table 7.3.

Based on the UL reaction tests, a limit of 0.5 mA was adopted as the ANSI standard for leakage current from portable appliances (see Chapter 11). We would like to determine the probability that the specified leakage might cause a reaction fitting into the "borderline hazardous" category defined by the UL observers. Lacking the statistical distribution of reaction thresholds, we assume that the distribution of reaction thresholds has the same coefficient of variation (standard deviation to mean ratio) as do perception thresholds. The coefficient of variation of the 60-Hz perception threshold for touched or gripped contacts is about 0.37 (Dalziel, 1954). Applying that value to a normal distribution for reaction currents along with the mean reaction values mentioned above, it is estimated that an unexpected current of 0.5 mA applied to a woman's lower arm will have a 2% probability of causing a reaction fitting into the UL "borderline hazardous" category. If the contact is pinched between the thumb and fingers, the probability is 1%.

7.12   Electrical Stimulation of Domestic Animals

Farmers are becoming increasingly aware of the phenomenon of *stray voltage*, a term used to describe unintended electrical potentials between contact points that may be accessed by a human or farm animal. Stray

---

[6] Because of body impedance, conducted current was on the average 0.72 times the available short-circuit current.

voltage of sufficient magnitude can cause unpleasant shocks to farm animals, which can ultimately lead to animal handling problems. Much of the attention to stray voltage problems has focused on dairy cows, for which potentials of only a few volts at 60 Hz can be unpleasant. With consideration of such low potentials, it is not surprising that stray voltage can be a real problem on some farms. Consequently, much research has been devoted to understanding the conditions under which dairy cows may be disturbed by low-level electrical stimuli. A useful summary of some of this research has been given by Lefcourt (1991).

Researchers typically apply 60-Hz currents to electrodes contacting various body surfaces of the animal. Locations of contact electrodes may vary considerably in given situations. The electrical impedance model of Fig. 2.41 can be used to determine median cow impedance for a variety of exposure conditions.

It is not possible for a cow to directly tell us when she can perceive a current as can humans. Instead, one must infer perception from the animal's behavior. Often an experimenter tests the minimum current that elicits some reaction. If the experiment is carefully designed, the minimum reaction current can approach the animal's threshold of perception.

Reactions to 60-Hz Stimulation

Table 7.12 summarizes experimental data on cow perception of 50- or 60-Hz current (Reilly, 1994). The upper section of the table lists minimum reaction thresholds (i.e., the just-noticeable response of the animal, such as lifting a foot, vocalization, sudden movement, or opening the mouth). Animals tested by Norrell, Currence, and Gustafson were trained to react to break the current. It is possible that these trained animals may have responded to just-perceptible current. The other reaction thresholds may lie somewhat above the minimum perception thresholds.

The lower part of Table 7.12 applies to avoidance reactions. In these tests, cows were given the opportunity to receive a food treat after 25 to 30 presses of a lever with the muzzle, each press resulting in an electric shock. The listed current is the minimum value that suppressed the plate-pressing activity.

In Table 7.12, median minimal behavior thresholds for muzzle contact range from 2.5 to 4.8 mA with an average of 3.9 mA; for contact by other body regions, generally greater thresholds are observed. For the 10% most sensitive cows, reported thresholds are in the range 1.0 to 3.2 mA. It is clear from the avoidance threshold data that the minimum listed currents are not particularly aversive. Norell's cows, for instance, were initially not willing to endure 30 shocks of 1.53 mA to the muzzle to receive a food treat. But after only two sessions, the same cows were willing to accept 30 shocks of 3.63 mA. In this context, current less than 3.63 mA to the muzzle is seen as only a mild deterrent. Whittleson's cows were willing to endure

TABLE 7.12. Bovine Electrical Thresholds: 60/50-Hz Current.

N(a)	Current Path	Reaction	Threshold (mA) @ given percentile		Ref.	Notes
			10%	50%(b)		
Minimal Behavior Thresholds						
7H	mouth/4 hooves	Open mouth to break current	1.2	2.7	[a]	c
	2 front/2 rear hooves	Lift foot to break current	2.4	3.7		
6H	mouth/4 hooves	Open mouth to break current	1.1	4.2	[b]	c
	2 front/2 rear hooves	Raise hoof to break current	1.0	3.3		f
	shoulder/4 hooves	Raise hoof to break current	2.3	9.0		g
18H, 3G	1 front/2 rear hooves	Any (twitch, lift foot, hump back, sudden movement)	2.5	3.5	[c]	d
5H	rt front knee/rt rear hock	Mild reaction (flinch, vocalization)	—	2.5	[d]	e
		distinct reaction (startle)	—	3.5		
120	nose clip/4 hooves	Mild reaction (twitch, leg move)	3.2	4.8	[f]	j
Avoidance Thresholds						
6H	muzzle/4 feet	Suppress 30 plate presses			[a]	h
		session 1	—	1.53		
		session 2	—	2.85		
		session 3	—	3.63		
7J	rt front teat/4 hooves	Suppress 25 plate presses	—	7.1	[e]	h, i
	4 teats/4 hooves	Suppress 25 plate presses	—	5.6		
	rump/4 hooves	Suppress 25 plate presses	—	6.1		
	chest/4 hooves	Suppress 25 plate presses	—	4.0		

Notes:
(a) Number of subjects; H = Holstein; G = Guernsey; J = Jersey.
(b) 10 and 50 percentiles given where available; otherwise, mean substituted for 50%.
(c) Experimental percentile adjusted for null response.
(d) Current duration t = 0.17, 1 s.
(e) EKG electrodes + electrode gel on shaved leg.
(f) Shoulder electrode = 60 cm^2 with EKG gel.
(g) 50% estimated by extrapolation.
(h) 30 (or 25) plate presses required for food treat—each plate press causes shock.
(i) 50-Hz current used.
(j) Transient thresholds adjusted for continuous 60 Hz, rms.

References:
[a] Norell et al. (1983); [b] Gustafson et al. (1985); [c] Currence et al. (1990)
[d] Lefcourt (1982); [e] Whittleson et al. (1975); [f] Reinemann et al. (1997).

even greater current levels to other body parts for the privilege of a food treat.

Reactions to Transient Stimulation

Despite the fact that most cow research has used 60-Hz stimulus currents, a detailed examination often reveals a highly complex voltage or current waveform consisting of a distorted 60-Hz sinusoidal variation, on which is superimposed short-duration transients. These transients may be produced when on- or off-farm equipment is switched on or off. Transients may be brief in duration, but can exceed the amplitude of the more persistent 60-Hz background and display a wide variety of waveform characteristics with various patterns of oscillation and repetition.

A number of researchers have tested bovine reactions to transient or intermittent currents (Gustafaon, 1988; Currence et al., 1990, Reinemann et al., 1994; Reinemann et al., 1996; Reinemann et al., 1998). In the experiments of Reinemann and colleagues, Holstein cows received electrical stimuli through nonpiercing ball-end clips attached to the cow's nose; the return path was through the four feet. Electrical thresholds were evinced

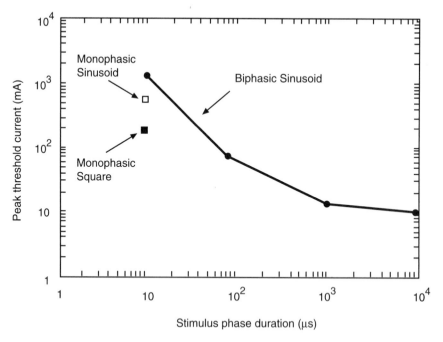

FIGURE 7.23. Strength-duration data for reaction thresholds of dairy cows. Curve applies to single sinusoidal cycle; points apply to monophasic stimuli. (From Reinemann et al., 1996).

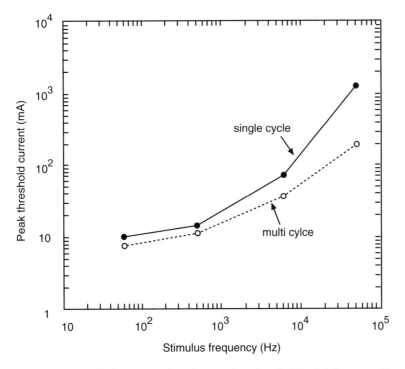

FIGURE 7.24. Strength-frequency data for reaction thresholds of dairy cows. Curves apply to single or multiple cycle sinusoidal currents. (From Reinemann et al., 1996.)

through changes in behavior (leg motion, facial twitch, or tail switch) or through animal movement recorded by a load cell within the floor of the stall. Electrical transients consisted of monophasic square wave stimuli or a single cycle of a sinusoidal current.

Figure 7.23 shows average strength-duration data for 120 cows for a monophasic square wave (single point), a monophasic half-cycle sine wave (single point), and single biphasic cycle of a sine (curve). The horizontal axis gives the "phase duration," which is the duration of a single mono-phasic phase of the stimulus, as indicated in Fig. 4.16. Using Eq. (4.20) along with the charge threshold of the monophasic square wave and the rheobase current (at $t = 10 \, \text{ms}$), we infer from these data a strength-dura-tion time constant $\tau_e = 200 \, \mu\text{s}$. This value is similar to that found in human sensory experiments (Table 7.1). The experimental data agree well with the theoretical SENN model (Fig. 4.16), except that the time scale of the experimental curve is shifted to a longer strength-duration time constant—an observation that is often found with human sensory data.

Figure 7.24 shows biphasic thresholds on a strength-frequency format. Thresholds are shown for a single cycle of a sine wave and for a continuous sine wave. These data are similar to the theoretical SENN model data shown in Fig. 4.21, except that the experimental curve is shifted to lower frequency values as compared with the theoretical model. This frequency shift is a consequence of the temporal shift in the strength-duration curve as noted above.

Statistical Distribution of Bovine Reaction Thresholds

Reaction thresholds were obtained for 120 cows with stimulation by a single-cycle 60-Hz current (Reinemann et al., 1998). The cumulative frequency distribution is shown in Fig. 7.25. On this plotting format, the log-normal distribution would appear as a straight line. The data conform well to the log-normal distribution between the 5- and 99-percentile ranks. It would require a larger sample to determine whether conformance to this statistical model exists over a wider percentile range. The range in cow thresholds from the 90- to 10-percentile ranks encompasses a ratio of 2.37. Similar results have been obtained in tests of human perception of transient currents, for which a log-normal model also applies, with the range from 90- to 10-percentile thresholds encompassing a ratio of 2.0 for stimulation of the fingertip and 3.4 for stimulation of the forearm (Table 7.10).

In comparing our results with other published data, note that Figure 7.25 represents peak currents for transient stimulation, whereas most other researchers report rms values for continuous 60-Hz current. The median reaction threshold in Figure 7.25 is 9.0mA-peak. If we adjust that value to an rms metric (multiplier = 0.707) and further adjust from single cycle to continuous stimulation as indicated in Fig. 7.24 (multiplier = 0.75), we would infer a median threshold of 4.8mA-rms for continuous 60-Hz stimulation.

Magnetic Field Stimulation of Cows

One might ask: If stray voltage of only a few volts can cause animal behavior problems, might not low-level electromagnetic fields on the farmstead also cause reactions in farm animals and result in behavior problems? Indeed, public agencies including Public Service Commissions, governmental groups, and the courts have confronted such questions because of law suits brought on behalf of farmers (primarily dairy) who have alleged disturbing perception of extremely low-frequency fields by their farm animals. To address these concerns, a theoretical study was conducted on the conditions necessary to excite peripheral nerves of cows via time varying magnetic fields (Reilly, 1995). Using models of magnetic excitation that are developed in Chapter 9, this study showed that stimulation of peripheral nerves of a 1300-lb cow would require a magnetic flux density of 54mT at a frequency of 60Hz over the entire body of the animal. Such a field is several

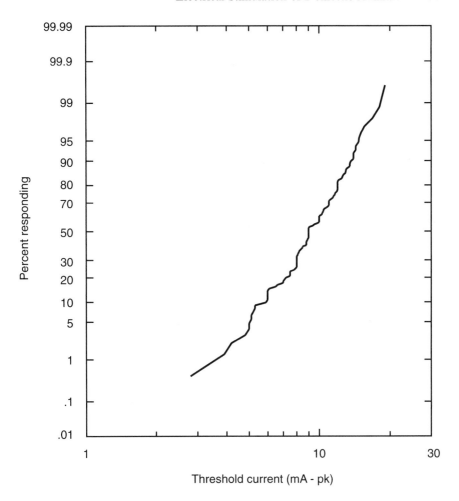

F<small>IGURE</small> 7.25. Cumulative frequency distribution for reaction threshold of 120 diary cows for single-cycle, 60-Hz transient current. Threshold current refers to the peak of the stimulus waveform where an animal responded.

orders of magnitude greater than what can be encountered in the home or on the farmstead.

Cow magnetic thresholds can be approximately scaled to other animals using the body weight adjustment of Eq. (9.22). In the cow example presented here, the body weight is 1300 lb. For a 180-lb human, for instance, the body weight scaling factor relative to a cow is 1.93. Using that factor as a multiple of the cow threshold, we arrive at an extrapolated threshold of 104 mT for the human; that estimate does not differ greatly from the value obtained with a more exact analysis based on the human body shape (Fig. 9.17).

7.13   Visual and Auditory Effects

Other sensory responses to electrical stimulation include visual and auditory effects, which have been studied using both contact electrodes and electromagnetic exposure. These effects, which are treated in detail in Sect. 9.8, will be mentioned in brief here.

It is possible to stimulate auditory channels intentionally for application to cochlear prostheses or unintentionally in the case of accidental exposure to electric current or electromagnetic fields.

Auditory effects from pulsed microwave stimuli involve the absorption of microwave energy within the brain when exposed by an electromagnetic field (Fig. 9.22). The absorbed energy launches an acoustic wave through thermoelastic expansion of brain tissue, and the acoustic wave is conveyed to the inner ear through bone conduction within the skull.

Auditory sensations can be elicited with audio-frequency currents through electrodes placed on the head (Fig. 9.21). Auditory thresholds obtained in this manner are near neuromuscular thresholds. A plausible explanation for these auditory effects is that excitation of muscle units convey vibratory stimuli to the inner ear through bone conduction.

Visual sensations called *phosphenes* can be produced by magnetic fields or electric currents applied to the head (Fig. 9.20). The rheobase E-field threshold for phosphenes is approximately a factor of 100 below that for action potential excitation; S–D time constants for phosphenes are approximately 100 times greater than corresponding values for nerve excitation and 10 times greater than those for excitation of muscle tissue (Sect. 9.8). The responsible mechanism is thought to involve the polarization of neurosynaptic processes through electric fields within the retina (Sect. 3.7).

Electrophosphenes are of interest in the development of acceptability criteria for electrical exposure because the underlying neurosynaptic interactions are also found within the central nervous system (Sect. 11.2).

8
Skeletal Muscle Response to Electrical Stimulation

JAMES D. SWEENEY

8.1 Introduction

In the eighteenth and nineteenth centuries, many seminal investigations into the basic properties of electricity were electrophysiological in nature and focused on neuromuscular phenomena. In Galvani's commentary (1791), a frog nerve-muscle preparation was used to propose a theory of "animal electricity" in which muscle fibers were thought to store charge. Volta (1800) later demonstrated that the bimetallic electrodes used by Galvani were a source of electrical stimulation current rather than a pathway for discharge of animal electricity. In modern society, electrical stimulation of skeletal muscle may occur unintentionally as one possible effect of an accidental electric shock or intentionally in medical devices for artificially induced muscle exercise or control.

This chapter presents the fundamental principles and effects of electrical stimulation of skeletal muscle. We first introduce the principles of skeletal muscle electrical stimulation in general and then focus on electrically induced skeletal muscle responses that may occur during intentional stimulation or electric shock. To achieve an understanding of the skeletal muscle responses that may be elicited by any form of electrical stimulation, it is necessary to consider the basic structure and function of muscle and the nerves that innervate muscle.

8.2 Neuromuscular Structure and Function

Skeletal Muscle Innervation

The myelinated nerve fibers that innervate skeletal muscle fibers derive from α-motor neurons (see Fig. 3.1), whose cell bodies lie within the ventral horn of the spinal cord. These efferent axons (approximately 9 to 20μm in diameter) leave the spinal cord through spinal-nerve ventral roots and project out to muscles through peripheral nerve trunks. Skeletal muscles may contain thousands of individual muscle fibers. However, a one-to-one relation does not generally exist between α-motor neurons and muscle fibers. Rather, as an α-motor neuron fiber nears and enters a skeletal muscle, it branches into a number of collateral axons, each of which innervates a single skeletal muscle fiber. An α-motor neuron and the skeletal muscle fibers it innervates are known as a *motor unit*. These skeletal muscle fibers, which upon contracting contribute to total muscle force, are known as *extrafusal* muscle fibers. *Intrafusal* muscle fibers produce only a minute fraction of total muscle force output. They are innervated by efferent myelinated γ-motor neuron fibers (averaging 5 μm in diameter) and afferent sensory myelinated Ia and II fibers (averaging about 17 and 8 μm in diameter, respectively). The sensory receptors within skeletal muscle, the *muscle spindles*, incorporate a small number of intrafusal fibers. These receptors sense muscle length and the rate of change of muscle length and are responsible for the well-known *stretch reflex* of skeletal muscle—that is, when a skeletal muscle is stretched, excitation of muscle spindles elicits reflex activation of the muscle. The second type of skeletal muscle receptors, the *Golgi tendon organs,* lie in series with muscle fibers as they enter tendons. These receptors are innervated by afferent myelinated Ib nerve fibers (averaging 16 μm in diameter) and report muscle tension information to the central nervous system.

Peripheral Nerve Trunk Structure

Peripheral nerve trunks in general carry motor, sensory, and sympathetic nervous system information between the central nervous system and the periphery (or vice versa). Axons are not simply "packed" into nerve trunks. Peripheral nerve trunks usually contain several fascicles (i.e., bundles, also called "funiculi"), each of which contains many axons as well as endoneurial connective tissue (see Fig. 8.1). A thin perineurial connective tissue sheath encloses each fascicle. The surface of the nerve trunk is covered by epineurial connective tissue. The fascicular structure of a nerve trunk is not constant throughout its course. On the contrary, regrouping and redistribution of fascicles (and the axons they contain) occur throughout the nerve trunk. For example, in the far periphery, some fascicles tend to exit from main nerve trunks and provide innervation to one or two skeletal muscles. They therefore must, at this point, contain all of the appropriate muscular efferent and afferent axons. As peripheral nerve trunks near the spinal roots, however, axons that innervate particular muscles must regroup into a multisegmental arrangement of fascicles. Sunderland (1978) provides an excellent review of this and related material. An implication of this aspect of trunk structure is that more central electrical stimulation of nerve trunks tends to elicit more diffuse muscular responses. Electrical stimulation of nerve trunks or branches in the far periphery can give quite specific responses.

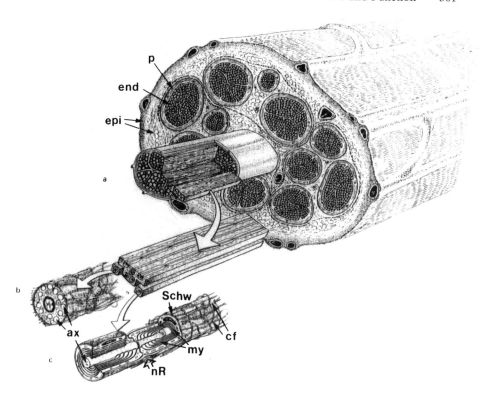

FIGURE 8.1. Microanatomy of a peripheral nerve trunk and its components. (a) Fascicles surrounded by a multilaminated perineurium (p) are embedded in a loose connective tissue, the epineurium (epi). The outer layers of the epineurium are condensed into a sheath. *(b)* and *(c)* illustrate the appearance of unmyelinated axons and myelinated nerve fibers, respectively. Schw, Schwann cell; my, myelin sheath; ax, axon; nR, node of Ranvier (components not to scale). (From Lundborg, 1988, p. 33.)

Muscle Action Potentials and Excitation-Contraction Coupling

Action potentials of α-motor neurons are generated by the cell body given sufficient dendritic input. Each action potential then propagates along a myelinated nerve fiber until it reaches a synapse at the neuromuscular junction (or motor end plate) (see Fig. 3.1). At the neuromuscular junction, neurotransmitter (acetylcholine or ACH) is released into the synaptic cleft in response to the presynaptic action potential. The postsynaptic response on the muscle fiber membrane (sarcolemma) is, in general, a muscle action potential (MAP), which propagates away bidirectionally given that the motor end plate tends to be located toward the middle of the fiber. Muscle action potentials penetrate each fiber through the transverse-tubule

(T-tubule) system. Propagation of an MAP into the T-tubule network of a fiber elicits release of calcium ions by the sarcoplasmic reticulum system into the proximity of the myofilaments. These calcium ions, upon binding with troponin C, initiate muscle fiber contraction (see below). This process, whereby muscle action potentials lead to contraction, is known as *excitation-contraction coupling*.

Skeletal Muscle Force Production

The skeletal muscle filaments (i.e., myofilaments) of Fig. 3.23 consist of relatively thick myosin filaments and thinner actin filaments (see Fig. 5.5). Each myosin filament contains many individual myosin molecules, each of which possesses a "head" and a "tail." The myosin tails are wrapped together to form a myosin filament, with the myosin heads protruding outward. (Each head and the "arm" of myosin attaching the head to the myosin filament are known as a *cross-bridge*.) Each myosin head possesses strong ATPase activity. That is, ATP is readily bound and cleaved (into ADP and phosphate). An actin filament consists of a double-helix F-actin molecule as well as tropomyosin strands, which in resting muscle act to cover "active" sites on the actin, and troponin complexes, which are attached to the tropomyosin molecules.

During muscle contraction, the excitation-contraction coupling process described above results in the release of calcium ions near the myofilaments. Calcium binding to troponin is thought to elicit movement of tropomyosin so that actin "active" sites are uncovered. Myosin heads, upon binding ATP, produce ADP and phosphate (which remain bound to the head) and achieve a conformational state wherein the heads protrude toward surrounding actin filaments. When actin "active" sites become exposed, myosin heads attach to these sites. This binding is thought to elicit a new conformational change in the myosin cross-bridge, which "pulls" the actin relative to the myosin—providing the force of muscle contraction (Murphy et al., 1996). The myosin head then releases from the actin, frees the bound ADP and phosphate, and can bind a new molecule of ATP. At this point, the cycle can repeat if sufficient free calcium and ATP are present at the myofilaments. A calcium pump in the sarcoplasmic reticulum acts continuously to decrease the calcium concentration near the myofibrils. (Thus, in response to a single action potential, a brief surge in myofibrillar calcium concentration, and therefore a brief contractile response, occurs.) To prevent rapid depletion of ATP concentrations within a contracting muscle, ADP molecules must be rephosphorylated. Some energy for this process is immediately available from phosphocreatine. However, anaerobic and aerobic metabolic processes must provide the energy for contractions that last longer than a few seconds. Glycolysis of stored carbohydrates (such as glycogen) is a relatively fast, but metabolically inefficient, anaerobic process by which new ATP can be formed. Oxidative (aerobic) metabolic processing of carbohydrates, fats, or proteins can provide slower, lower, but steadier levels of energy for ATP production.

The contractile output of a single motor unit receiving a single action potential input is a muscle twitch (actually, such a twitch represents a summation of responses from the individual muscle fibers within the motor unit). As we have seen in Chapter 3, two fundamental physiological means exist by which the nervous system can modulate force production within a skeletal muscle: temporal summation (also known as rate modulation) and recruitment. In temporal summation (Fig. 3.24), the rate of action potential input to the fiber is controlled so that twitch summation results in a desired force level. Note that this summation process is in general nonlinear and can be quite time dependent. Recruitment in this context refers to the activation of greater or lesser numbers of motor units within the skeletal muscle (i.e., α-motor neuron fibers).

Differentiation of Skeletal Muscle Fiber Types

The muscle fibers of motor units that innervate a given skeletal muscle can be categorized with respect to physiological, ultrastructural, and metabolic characteristics. Table 8.1 presents a highly simplified, but useful, differentiation among fiber types based on three common classification schemes (see Burke, 1981, for a detailed review). In general, muscle fiber and motor unit properties lie within an overlapping continuum—and classification schemes serve mainly as aids in relating structure-function properties of the muscle. The classification method of Brooke and Kaiser (1970) places muscle fibers into three categories, I, IIA, and IIB (based on histochemical measurements of ATPase reactivities). Peter et al. (1972) differentiated fiber types based on a combination of physiological and metabolic properties (SO, slow oxidative; FOG, fast oxidative-glycolytic; FG, fast glycolytic). The method of Burke et al. (1973) differentiates motor units on the basis of their twitch speed and fatigue resistance (S, slow; FR, fast-resistant; FF, fast-fatigable). An "intermediate" type of fast muscle fiber

TABLE 8.1. Correlation of skeletal muscle major fiber type physiological, ultrastructural, and metabolic properties.

Fiber type	IIB, FG	IIAB	IIA, FOG	I, SO
Corresponding motor unit type	FF	FI	FR	S
Myofibrillar ATPase, pH 9.4	High	High	High	Low
Glycogen	High	High	High	Low
Phosphorylase	High	High	High	Low
Neutral fat	Low		Medium	High
Capillary supply	Sparse		Rich	Rich
Aerobic metabolism capacity	Low	Medium	Medium-High	High
Anaerobic metabolism capacity	High	High	High	Low

Source: Adapted from Burke, 1981, p. 349 with permission from Oxford University Press.

also exists in some muscles (type IIAB or FI, fast intermediate resistance to fatigue).

Slow-twitch skeletal muscle fibers are optimized for maintenance of relatively low force levels, at relatively low speeds, but for prolonged periods of time. Fast-twitch fibers are optimized for quick production of relatively high force levels for briefer time intervals. We see from Table 8.1 that the physiological, ultrastructural, and metabolic properties of the major skeletal muscle fiber types reflect a trade-off between the ability to produce force quickly and powerfully, or slowly and steadily. Type I fibers generate force relatively slowly (myofibrillar ATPase activity at high pH is relatively low), yet can maintain force well through a relatively high aerobic metabolic capacity (high neutral fat content, rich capillary supply, etc.). In the other extreme case, type IIB fibers can generate force quickly (myofibrillar ATPase activity is relatively high) via the use of anaerobic metabolic mechanisms (high glycogen storage, high phosphorylase). However, muscle force fatigue occurs relatively rapidly in these fibers. Type IIAB and IIA fibers represent varying degrees of compromise between the design requirements of force generation, speed/power, and fatigue resistance.

Recruitment and Firing Patterns

In normal physiological use, motor unit recruitment and firing patterns also appear to be optimized for performing desired motor tasks (see Burke, 1981, 1986). In 1965, Henneman and colleagues advanced a theory known as the "size principle" of motor neuron recruitment (Henneman et al., 1965a, 1965b), in which it was proposed that small α-motor neurons innervating slow motor units are recruited at lower "functional thresholds" than large α-motor neurons, which innervate fast motor units. The term "functional threshold" is analogous to "excitability level" in this context and essentially reflects the level of synaptic input necessary to excite the motor neuron (Burke, 1981).

The logical implication of this theory is that, in "mixed" skeletal muscles containing a distribution of muscle fiber types, low-force tasks requiring fatigue resistance (such as force generation to maintain posture) are met by recruiting slower motor units. Only when greater and/or faster force levels are desired are the more readily fatigued fast motor units brought to bear. Although this theory has, to a great extent, stood the test of time and further experimentation (see Henneman and Mendell, 1981; Burke, 1981, 1986), it is clear that some exceptions do occur in normal physiological use and that motor neuron size alone does not completely determine functional excitability. For example, in rapid, powerful "ballistic" movements, relatively synchronous activation of both slow and fast motor units, and even preferential activation of fast units, can occur.

In general, α-motor neurons innervating slow-twitch motor units tend to have average firing rates that are lower than those that innervate fast-twitch units (see Burke, 1968, 1981). This seems sensible in that twitches generated by slow motor units start to fuse at approximately 5 to 10 Hz and reach a tetanic state by 25 to 30 Hz, while fast motor units may require 80 to 100 Hz to reach total fusion (see Burke, 1981; Wuerker et al., 1965). In typical physiological use, however, motor units rarely exhibit sustained firing frequencies that even approach these tetanic fusion limits. For example, in the human extensor digitorum communis muscle, Monster and Chan (1977) observed that virtually all motor units began firing at about 8 Hz regardless of functional threshold (i.e., motor unit type) and increased their firing rates to a maximum of 16 to 24 Hz at maximal voluntary force levels.

Burke (1986) has suggested that recruitment patterns that fit the size principle probably meet the normal functional demands placed on muscles well while maintaining an acceptable metabolic cost. In instances where recruitment deviates from the "orderly" pattern dictated by the size principle, metabolic cost is sacrificed to achieve less frequently needed fast and/or powerful motor demands.

Fatigue in Normal Physiological Use

Muscle fatigue is often defined as an inability to generate the required or expected force (Edwards, 1981). Bigland-Ritchie et al. (1986) have pointed out that neuromuscular fatigue might more generally be defined as "any reduction in the maximum force generating *capacity*, regardless of what type of work is being done." It should also be emphasized that fatigue should be regarded as a normal protective measure for survival rather than as a deficit or abnormal phenomenon. If fatigue mechanisms did not exist, it would be possible to exercise muscles to the point where ATP levels were completely exhausted and muscle rigor (i.e., irreversible binding of cross-bridges to actin) occurred.

We have seen that voluntary skeletal muscle contractions depend upon a sequence of events involving the motor neuron (dendritic input, cell body excitability, action potential propagation, motor end-plate presynaptic transmission), excitation-contraction coupling (motor end-plate postsynaptic transmission, muscle action potential propagation along the sarcolemma and into the T-tubules, calcium release from the sarcoplasmic reticulum), and the actual contractile machinery (ATP dependent cross-bridge cycling, ATP rephosphorylation through metabolic processes). Fatigue may be induced due to impairment of any of these processes (Bigland-Ritchie et al., 1986; Fitts, 1994). It is generally accepted that both short-lasting and long-lasting components exist in physiological fatigue. Quickly reversible effects include sarcoplasmic electrolyte shifts and pH changes that may impair MAP propagation through the T-tubule system, sarcoplasmic reticulum calcium ion release, and/or force development at the myofilaments. Recent evidence implicates a particularly important role in the fatigue process for alterations in intracellular calcium exchange. Such changes may occur secondary to reductions in the rates of calcium uptake and release via the sarcoplasmic reticulum (see Williams and Klug, 1995). Force development is obviously also dependent upon metabolic processes and products. At the motor unit level, a clear correlation between aerobic (oxidative) metabolic capacity and resistance to fatigue exists. As mentioned above, the fatigue-resistant fiber types S (SO) and FR (FOG) can be associated with the high aerobic capacity I and IIA categories; while easily fatigued type FF (FG) units are associated with type IIB low aerobic capacity (high anaerobic capacity). Longer-lasting fatigue may be due to depletion of metabolic energy stores (such as glycogen) or excitation-contraction coupling impairment (i.e., "low frequency fatigue" (Edwards et al., 1977)).

In normal physiological use, motor unit firing patterns appear to be adaptable to match fatigue levels. Bigland-Ritchie et al. (1983) have reported that, in a series of brief maximal voluntary contractions of the human pollicis muscle, the average firing rate of more than 200 motor units (from five subjects) dropped from 29.8 Hz (± 6.4) to 18.8 Hz (± 4.6) between 30 and 60 s after contraction onset, and to 14.3 Hz (± 4.4) after 60 to 90 s. In the same study, twitch contraction times before and immediately after a 60 s maximal voluntary contraction were not significantly different. However, twitch relaxation times were prolonged by approximately 50%. This has the effect of lowering the firing frequency needed to achieve fused tetanic contractions. Thus, firing frequency decreases during fatigue may in part occur because higher rates are unnecessary to achieve fused force output.

Dynamic Nature of Skeletal Muscle Fiber Types

A great deal of evidence exists that the skeletal muscle fiber "type" of an individual motor unit is not fixed, but is instead a dynamic quantity that depends on the neural input to and motor usage of the unit (see reviews of Pette and Vrobova, 1985; Burke, 1981). Prolonged endurance training, or constant electric stimulation at low frequencies (10 Hz), can result in a transformation of fiber type populations within mixed skeletal muscles. This transformation is in the direction of faster to slower twitch and increased aerobic metabolism capacity. Increases in fatty acid metabolic capacity, fiber cross-sectional areas, whole-muscle fatigue resistance, and capillary density can be elicited within weeks to months after training onset. At moderate levels of exercise demand, oxidative capacity increases are most dramatic as the result of apparent conversion of type FF fibers to type FR. With constant electrical stimulation, actual conversion of type II (fast) myosin to type I (slow) eventually can be brought about in addition to the metabolic and ultrastructural changes already noted. Conversely, extended disuse of skeletal muscle tends to result in fiber atrophy, decreases in tetanic force output per fiber, and aerobic capacity loss in fatigue-resistant fibers.

The properties of muscle therefore appear to be optimized for normal use, as determined by overall activity level, and to some extent activity and force production patterns. Fast-twitch fibers, which normally see relatively low levels of phasic activity, appear capable of adapting to new high, tonic levels of activity. Conversely, muscle fiber inactivity appears to result in an opposite shift to metabolic profiles more appropriate for infrequent, phasic levels of activation.

8.3   Fundamental Principles of Skeletal Muscle Electrical Stimulation

Strength–Duration (S–D) Relations for Neuromuscular Excitation

The most basic requirement for electrical stimulation of a skeletal muscle fiber is that an action potential(s) must be elicited on the α-motor neuron innervating the fiber or within the membrane of the muscle fiber itself. As has been discussed in Sect. 3.5, the electric excitability of myelinated nerve is generally much greater than that of skeletal muscle (see Parry, 1971; Walthard and Tchicaloff, 1971; Mortimer, 1981). Figure 8.2 depicts S–D curves (see Sect. 4.2) for transcutaneous stimulation of human skeletal muscles by regulated-voltage waveforms. We see that, in this example, direct electrical stimulation of denervated muscle requires substantially higher rheobase (i.e., the stimulus strength needed for large pulsewidths) and chronaxy (i.e., the stimulus pulsewidth at twice rheobase) values than for normal stimulation, where α-motor neuron axons can be activated.

Strength–duration curves for cat skeletal muscle stimulation with an implanted intramuscular electrode (see Sect. 8.4) and regulated-current monophasic stimuli are shown in Fig. 8.3. In these experiments on cat skeletal muscle (Mortimer, 1981), evoked isometric muscle response was held constant at a small fraction of total possible force. Once again, we see that the curve for direct muscle stimulation (in this case, following administration of the motor end-plate blocking-agent curare) shifts up and to the right in comparison to the curve for neural stimulation. Crago et al. (1974) conducted similar experiments on intramuscular stimulation of curarized and noncurarized cat skeletal muscle. They found that, over a range of stimulus pulse durations and amplitudes, direct (curarized) muscle stimulation accounted for only 3 to 7% of isometric muscle force evoked by neural (uncurarized) stimulation with identical parameters.

An S–D curve time constant τ_e can be determined for both neural and direct muscle stimulation by performing a mathematical fit of experimental data to Lapicque's form of the S–D curve as expressed by Eq. (4.17). Such data indicate τ_e for denervated skeletal muscle of 4.3 to 57.6 ms, and for intact muscle of 20 to 700 μs (Oester and Licht, 1971). In that just-suprathreshold intact muscle stimulation is accomplished indirectly through excitation of α-motor neuron axons, we would expect that the S–D time

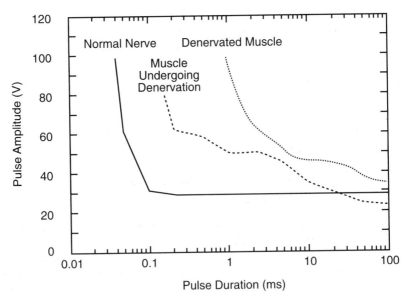

FIGURE 8.2. Strength–duration curves for transcutaneous stimulation of human Peroneus Longus skeletal muscles by regulated voltage waveforms. The left-hand curve is typical of normally innervated muscle. The middle curve was obtained following a left lateral popliteal nerve lesion when the Peroneus Longus was undergoing denervation atrophy. The right-hand curve, measured six weeks after the middle record, is typical of denervated muscle. (Adapted from Sunderland, 1978, p. 259.) (Reprinted from S. Sunderland, Nerves and Nerve Injuries, Churchill Livingstone, 1978, p. 259.)

constant values for intact muscle and large myelinated nerve stimulation would be similar. Indeed, for mammalian peripheral myelinated nerve fibers, experimentally obtained values from $29\mu s$ to over $800\mu s$ can be determined (Ranck, 1975; Li and Bak, 1976).

We have seen in Chapter 4 that τ_e is a property not just of the excitable membrane, but also of the method of excitation. In general, τ_e becomes greater as the stimulus current crossing the cell membrane becomes more gradually distributed along the longitudinal dimension. The absolute minimum values of τ_e for neural and direct muscle monophasic rectangular stimulation are the actual *membrane* time constants (τ_m) for α-motor neuron nodes of Ranvier and skeletal muscle sarcolemma.[1] Based on data obtained from rabbit peripheral myelinated axons by Chiu and colleagues (Chiu et al., 1979; Chiu and Ritchie, 1981), the author has estimated that

---

[1] This situation occurs, for example, when a regulated current stimulus is applied directly across a small patch of excitable membrane. The S–D time constant τ_e then becomes equal to the membrane time constant τ_m [i.e., Eq. (4.4) simplifies to Eq. (4.5)].

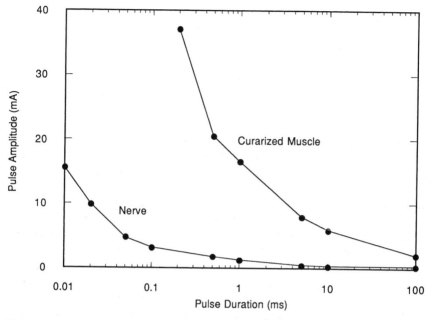

FIGURE 8.3. Strength–duration curves for cat skeletal muscle stimulation with an intramuscular electrode. In these experiments, evoked muscle response was held constant at a small fraction of total possible muscle force. The left-hand curve represents data for direct nerve stimulation. The right-hand curve represents data for direct muscle fiber stimulation (following administration of curare). (Adapted from Mortimer, 1981, p. 170 with permission from Oxford University Press.)

node of Ranvier τ_m values would lie in the range of 10 to $20\,\mu s$. Albuquerque and Thesleff (1968) found that, for innervated rat skeletal muscles, sarcolemmal τ_m was approximately 1.7 ms. While interspecies differences are to be expected, these values indeed fall just below the lowest experimentally obtained measurements of S–D time constants mentioned above.

Regulated-Current versus Regulated-Voltage Stimulation

Many typical stimulation waveforms are rectangular in shape. A regulated-current stimulus (of magnitude I) with pulsewidth t will deliver a fixed total charge Q per stimulus, such that $Q = It$. Regardless of electrode impedance and potential shifts (either during a stimulus, or between stimuli), because the current of the stimulus is regulated, a reproducible applied electric field can be created within the tissue(s) to be stimulated (see Sect. 2.1). For this reason, many recently designed implanted neuromuscular stimulation systems utilize regulated-current stimulus output stages. Transcutaneous neuromuscular stimulation systems (i.e., systems that use surface electrodes) often produce regulated-voltage waveforms. Despite the drawback that stimulation results may be less reproducible, in this case regulating the voltage of a stimulus is safer—a regulated-current stimulus passed through a dislodged or broken surface electrode could produce skin irritation or even burns due to the resultant high current density.

Biphasic Stimulation

Balanced-charge biphasic (BCB) regulated-current stimuli are used with many implantable electrode neuromuscular stimulation systems (see Sect. 8.4 below). With such waveforms, a cathodic rectangular pulse is followed by an anodic phase of equal charge (either rectangular in shape or with the form of a capacitive discharge) (see Robblee and Rose, 1990). While the "primary" cathodic phase serves to stimulate excitable tissue(s), the "secondary" anodic phase performs charge balancing mainly for electrochemical reasons. Without some form of electrochemical reversal, deleterious cathodic reactions such as alkaline pH swings, hydrogen gas evolution, and oxidizing agent(s) formation can occur at the electrode-tissue interface [see Sect. 8.4 and Robblee and Rose, (1990)]. However, although the intent of BCB stimulation is to minimize adverse electrochemical reactions, we have seen in Sect. 4.6 that the anodic current reversal of a biphasic stimulus can act to abolish an action potential developing in response to the cathodic phase. This physiological effect has been studied in vitro on isolated Xenopus Laevis myelinated nerve fibers and in vivo using a cat nerve-muscle preparation (van den Honert and Mortimer, 1979a). It was found that a delay between the primary and secondary phases of approximately $100 \mu s$ effectively prevented this effect from occurring. Modeling of myelinated nerve BCB electrical stimulation also predicts this result (see Fig. 4.16). Gorman and Mortimer (1983) found that the effect could be used potentially to advantage for separating the stimulation thresholds of differing diameter α-motor neuron axons.

Fatigue and Conduction Failure in Response to Electrical Stimulation

Trains of electrical pulses at rates of 10 to 20 Hz and at magnitudes that activate α-motor neuron axons can elicit a long-lasting depression in skeletal muscle force output in addition to metabolic deficit fatigue. This "low-frequency fatigue" phenomenon (Edwards et al., 1977), which also occurs in normal physiological use (see Sect. 8.2), is thought to occur (at least in part) because of a deficit in excitation-contraction coupling. That is, in response to each muscle action potential, less calcium is released by the sarcoplasmic reticulum. With increased stimulation frequencies of 20 to 100 Hz, "high-frequency fatigue" may be elicited. Motor neuron propagation failure at axon branch points, neurotransmitter depletion at the motor endplate, and MAP propagation failure all may contribute to this effect. Bigland-Ritchie et al. (1986) feel that high-frequency fatigue is avoided in normal physiological use because motor neuron firing rates remain below the frequencies needed for propagation failure.

Bowman and McNeal (1986) studied in vivo the response characteristics of single α-motor neuron axons of cat sciatic nerves when very high-frequency trains (100–10,000 Hz) of balanced-charge biphasic stimuli were applied. At frequencies of 100 to 1,000 Hz, axonal firing rates were generally equal to or were subharmonics of the stimulation frequency. Over a 3-min stimulation period, an axon often typically initially fired at the same rate as the stimulus train for a very brief time, shifted to firing with every other stimulus, gradually shifted to firing with every third stimulus, and sometimes then shifted to every fourth. Increased stimulus amplitudes tended to delay the downward shift in firing frequency with time. Axonal firing frequencies in this range tend to rapidly deplete supplies of ACH at the neuromuscular junction. This type of "junction fatigue," first studied by Wedensky (1884) and Waller in 1885 (see Bowman and McNeal, 1986), rapidly results in contractile force loss (see McNeal et al., 1973; Solomonow et al., 1983).

At stimulation frequencies of 2,000 to 10,000 Hz, Bowman and MeNeal found that axons would briefly respond with firing rates of several hundred Hertz, but stopped firing completely within seconds after stimulation initiation. This type of "electrical conduction block" could be maintained for periods up to 20 min. with recovery following stimulation termination occurring within 1 s. The mechanism involved in this type of very high-frequency conduction block, which has also been studied by a number of other investigators using rectangular stimuli or AC sinusoidal trains (e.g., Cattell and Gerard, 1935; Katz, 1939; Tasaki and Sato, 1951; Tanner, 1962), is not completely known. Polarization buildup at nodes of Ranvier due to differing conductivities for inward and outward currents (see Sect. 4.6) may act to depolarize excitable membranes to the point where strong sodium channel inactivation results.

Sinusoidal Stimulation Strength–Frequency (S–F) Effects

Nonlinear membrane polarization buildup is also thought to account for the upward deflection in AC sinusoidal stimulation S–F curves above a minimum level that typically occurs at about 150 Hz for myelinated axons (see Fig. 4.20, right). At stimulation frequencies below about 100 to 150 Hz, rectangular stimulation thresholds for α-motor neurons are independent of stimulation frequency, and each stimulus tends to produce one action potential. AC sinusoidal trains, however, exhibit an upturn in their strength-frequency behavior (see Fig. 4.20, left). That is, higher-amplitude sinusoidal trains are needed to excite the nerve. The well-known "accommodation" of excitable membrane thresholds (see Sect. 4.6) in response to slowly increasing stimuli is thought to be responsible for this effect. For example, Sato and Ushiyama (1950) demonstrated in an in-vitro toad motor nerve preparation that accurate mathematical prediction of the form of S–F curves below 40 Hz could be performed through consideration of membrane accommodation properties [see Tasaki (1982) for an excellent review of this and related topics].

As sinusoidal frequencies approach DC, it becomes quite difficult to stimulate peripheral nerves. Stimulation may be elicited at the onset of direct current as with any rectangular pulse. However, following the onset of direct current, additional stimulation of α-motor neurons or skeletal muscle is not easily achieved except at very high stimulus amplitudes. Extremely high-amplitude stimuli, at very high frequencies (2,000 Hz) or at DC, have been found to produce "tetanoid" responses directly in isolated (curarized) frog skeletal muscles (Koniarek, 1989). After initiation of this type of response, a portion of each muscle fiber appears to reach a constant depolarization level.

8.4   Functional Neuromuscular Stimulation Systems

Electrically induced stimulation of skeletal muscle for artificially controlled exercise and restoration of motor function (lost because of injury or disease) is generally the objective of functional neuromuscular stimulation (FNS) systems. In addition, FNS electrode and stimulation technologies recently have also been adapted for use in other applications, such as dynamic myoplasty. A number of excellent reviews of and texts on FNS, and the general area of functional electrical stimulation (FES), already exist (e.g., Vodovnik, 1981; Mortimer, 1981; Peckham, 1987; Agnew and McCreery, 1990; Rattay, 1990; Peckham and Gray, 1996). In this section, we will briefly present the current state of FNS and related applications, methods, models, and electropathology and will introduce basic principles that can provide greater understanding of accidental stimulation in electrical accidents.

Objectives and Applications of FNS Systems and Technologies

It has been estimated that there are over 200,000 spinal cord injured people in the United States. Spinal cord injuries occur mostly in younger people, with the most common age at injury being 19 years and the mean age at injury being 29 years (Young et al., 1982). A conservative estimate of the annual cost of care for spinal cord injured persons is one and a half billion dollars. In the paraplegic person (spinal cord damage in the thoracic or lumbar region), the freedom to walk, run, or simply stand from a chair to reach a shelf is restricted or lost. Quadriplegic individuals (spinal cord damage in the cervical region) will have lost, in addition, some or all of the ability to manipulate and control their environment. Functional neuromuscular stimulation of α-motor neurons that remain intact below the level of spinal cord damage provides a means by which some control of lost motor function can be reasserted. At this time, a number of research centers around the world are developing systems designed to provide upper extremity control (e.g., Peckham, 1983; Hoshimiya et al., 1989; Smith et al., 1996; Keith et al., 1996; Triolo et al., 1996) and lower extremity control (e.g., Waters et al., 1985; Holle et al., 1984; Kralj et al., 1986; Marsolais and Kobetic, 1987; Graupe, 1989; Jaeger et al., 1989; Shimada et al., 1996). FNS can also be used for motor restoration or assistance in other disabled populations where α-motor neurons remain intact, but central control has been lost or impaired (e.g., stroke, cerebral palsy).

The term "myoplasty" refers to any clinical procedure wherein skeletal muscle is transferred within the body (e.g., in plastic surgery). In "dynamic myoplasty," electrical stimulation is used to activate muscle that has been surgically relocated (for a recent review, see Grandjean et al., 1996). In "dynamic cardiomyoplasty," a skeletal muscle (usually the left latissimus dorsi) is mobilized except for its neurovascular pedicle, internalized through a window created by partial resection of the second or third rib, and secured to the heart. Leriche and Fontaine proposed as early as 1933 that skeletal muscle might be used to replace damaged myocardial tissue. Early efforts in this area were frustrated primarily by the skeletal muscle fatigue that resulted when the muscle was subjected to cardiovascular work loads (see Kantrowitz, 1990). With the discovery (see Sect. 8.2) that chronic electrical stimulation can be used to improved skeletal muscle fatigue resistance through conversion of type II (fast) fibers to type I (slow), this application of FNS technologies has seen renewed investigation (Carpentier and Chachques, 1985). In the dynamic cardiomyoplasty surgical procedure, skeletal muscle may be wrapped around the heart, used to replace resected myocardium, or both (see Magovern et al., 1990; Carpentier et al., 1991). Electrical stimulation of the latissimus dorsi muscle in synchrony with the heart is accomplished using an implanted neuromuscular stimulator and electrodes (Grandjean et al., 1991). Worldwide clinical evaluation of the dynamic cardiomyoplasty approach to cardiac assistance is now underway (e.g., Moreira et al., 1996; Furnary et al., 1996; Oh et al., 1996). Alternative approaches to the use of electrically stimulated skeletal muscle for cardiac assistance, which are also under experimental study, include creation of "skeletal muscle ventricles" (by wrapping muscle into or around a pumping chamber much like a ventricle of the heart) and "aortomyplasty" (where muscle is wrapped around the aorta) (see e.g., Chachques et al., 1996; Niinami et al., 1996). Dynamic myoplasty for treatment of fecal incontinence is also under investigation (e.g., Williams et al., 1991; Baeten et al., 1995).

Electrical Excitation of Skeletal Muscle by Functional Neuromuscular Stimulation

Excitation of a given motor nerve fiber can be achieved, in general, if an electric field is introduced (by electrodes) that involves a focalized change in the extracellular voltage gradient. Equations such as (4.31) imply that a second-difference function involving extracellular potentials at nodes of Ranvier [see Sect. 4.4 and Rattay (1986)] characterizes the "driving force" of electric field distributions. In Eq. (4.31), the term $(V_{e,n+1} - 2V_{e,n} + V_{e,n+1})$ represents a second-difference sampling of the longitudinal extracellular potential distribution along the fiber. Focal changes in the extracellular electric field can therefore cause depolarization of the axonal membrane at nodes of Ranvier and generation of action potentials that will propagate toward a muscle.

In spinal cord injured people, most motor neurons in the spinal cord will remain intact if they are below the level of the injury. Electrical stimulation of their axons can therefore be used to cause muscle contraction. The restoration of functional movement in general requires the fine gradation of force so that adequate control can be achieved and damage to the body or to objects in the environment can be avoided. Just as skeletal-muscle force output is controlled in normal physiological use by recruitment of α-motor neurons and by rate modulation of each motor neuron's firing, electrical stimulation can be used to control the same variables artificially. FNS system designers often intuitively strive to replace the body's normal control strategies; however, major limitations still exist in the present ability to achieve such control.

Electrodes for Functional Neuromuscular Stimulation

An excitatory electric field can be generated by an FNS system using a variety of electrode types (for reviews, see Mortimer et al., 1995; Sweeney et al., 1996). Stimulation using electrodes located on the skin surface is the most straightforward possible approach. It was determined as early as 1833 by Duchenne de Boulogne that skeletal muscles could be stimulated by passing current percutaneously via cloth-covered electrodes (see Licht, 1971; Walthard and Tchicaloff, 1971). Duchenne found that contractions were most easily obtained by stimulating at a discrete location on the skin overlying each muscle. He called such areas "points of election" and theorized that the muscle beneath each point was hyperexcitable. Remak and von Ziemssen subsequently showed that such locations, now known as "motor points," actually overlie sites where nerve trunks penetrate and branch out into the deep face of muscles. Coers (1955) refined this essentially correct viewpoint. He found that the greatest concentration of motor end-plates in a superficial muscle is located in the superficial part of the muscle beneath the skin surface motor point. "Motor lines" reflect linear projections onto the skin surface of underlying superficially located motor nerve trunks. Walthard and Tchicaloff (1971) provide extensive maps and tabulations of approximate motor point and motor line locations.

Stimulus parameters for surface-electrode motor point stimulation vary from one muscle to another and depend on muscle/nerve anatomy; whether voltage or current is regulated; the waveform; and the electrode(s) type, size, placement, and configuration. Vodovnik et al. (1967) found that for regulated-voltage rectangular stimuli, while threshold parameters for stimulation of upper extremity muscles were quite variable, typical threshold values (for 200-μs pulses at 50 Hz) were on the order of 20 V. Voltages in excess of 100 V may sometimes be necessary to achieve high activation levels (Vodovnik et al., 1981). Figure 8.4 presents typical S–D curves for surface-electrode regulated-current stimulation of four upper-extremity muscles. These curves were obtained by placing an electrode over the motor point of each muscle and measuring, for each pulse duration, the just-threshold pulse amplitude. Crago et al. (1974) have reported similar results (e.g., for stimulation of the flexor digitorum sublimis muscle in a typical subject: 9.9 mA threshold, for 0.5-ms-duration stimuli at 50 Hz). Bajzek and Jaeger (1987) found that, for triceps surae stimulation, similar or somewhat

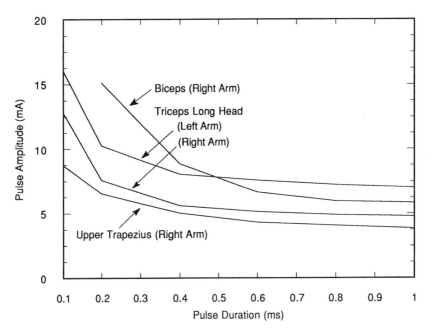

FIGURE 8.4. Strength–duration curves for surface-electrode threshold stimulation of four upper-extremity muscles. (Frequency = 50 Hz; C-3 level spinal cord injured patient "H.K.") (Adapted from Vodovnik et al., 1967.)

higher charge parameters were necessary (e.g., on the order of 50 to $100\mu s$ pulse duration thresholds for 60 to 80mA stimuli at 25Hz).

An increase in surface-electrode stimulus intensity above the threshold level tends to activate a larger volume of the underlying muscle and therefore increase force output until maximal muscle activation is achieved. Equation (4.31) implies that, for a given electric field distribution beneath a surface electrode, a "shell" will exist under the electrode that separates (for each size nerve fiber) regions where excitation will occur. Figure 8.5 presents a family of S–D curves for varying levels of torque about the elbow given surface-electrode stimulation of the biceps muscle. Each curve essentially represents a constant degree of biceps muscle activation (i.e., a constant level of neural excitation penetration into the muscle body). Prediction of the relationship between surface-electrode stimulation parameters and underlying muscle activation level is nontrivial. Rattay (1988) has modeled the excitation of nerve fibers under disk-shaped surface electrodes using the "activating function" approach. He found, as have others, that the tendency for high current densities to arise at the edge of a surface electrode will ultimately limit the depth to which excitation can be elicited (i.e., at sufficiently high edge current densities, burns can occur).

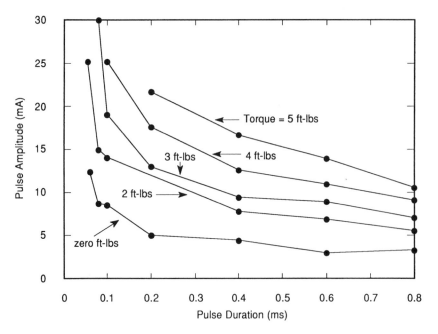

FIGURE 8.5. Strength–duration curves for surface-electrode tetanic stimulation of the biceps muscle at five levels of isometric torque about the elbow. Curve for zero ft-lbs reflects the just-threshold response. (Frequency = 50Hz; elbow angle = 75°; subject "P.C.") (Adapted from Vodovnik et al., 1967.)

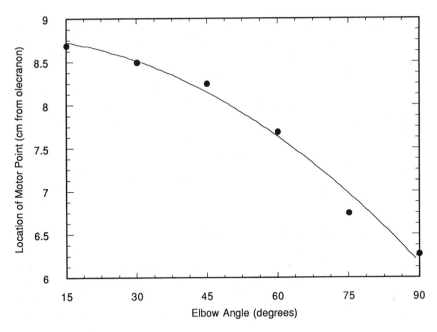

FIGURE 8.6. Location of the motor point of the medial head of the triceps muscle versus elbow angle for surface stimulation using a 1-in.-diameter electrode (pulse amplitude = 15 mA; pulse duration = 0.2 ms; frequency = 50 Hz; subject "W.C.") (Adapted from Crochetiere et al., 1967.)

A fundamental difficulty with surface motor point stimulation is that, while electrodes are applied to the *skin,* the motor point is inherently dependent on anatomical properties of the underlying *muscle.* Thus, as shortening or lengthening of a muscle occurs during a motor task, and as limb positions or joint angles change, the location of a motor point will shift with respect to fixed locations on the skin surface. Figure 8.6 illustrates this effect for surface electrode stimulation of the medial head of the triceps. In these experiments, Crochetiere et al. (1967) estimated that the 2.5 cm shift in the motor point skin surface location (over an elbow angle range of 15–90°) corresponded well with a theoretical expected excursion of 2.5 cm in the underlying muscle.

In a recent study of joint angle and electrode position effects on surface stimulation characteristics, McNeal and Baker (1988) found that, in activation of the hamstring muscles, a large intersubject variability existed in predicting surface regions of greatest excitability. Changes in regions of excitability also generally occurred as the knee was flexed. Excitability regions for surface stimulation of the quadriceps muscles showed less dependence on joint angle. Surface electrode stimulation also tends to be relatively non-selective because the electric field created is necessarily widespread. For example, stimulating the innervation of deep muscles without also stimulating overlying superficial muscles is impossible. Despite these drawbacks, however, the non-invasiveness of surface stimulation has great appeal; and a number of surface electrode based FNS systems have been reported. Patients using surface electrode systems for lower-extremity FNS have been able to stand, to walk at speeds approximately one-quarter of normal and to climb stairs (Kralj et al., 1983; Petrofsky and Phillips, 1983). In stimulation systems intended purely for artificially generated muscle exercise (as opposed to function restoration), surface stimulation is clearly preferred.

Intramuscular electrodes (Caldwell and Reswick, 1975; Crago et al., 1980; Handa et al., 1989; Memberg et al., 1994), which can be percutaneously placed or fully implanted into skeletal muscles, and implanted epimysial electrodes that are placed on skeletal muscle surfaces (Grandjean and Mortimer, 1986) enable selective activation of individual muscles (or even muscle regions). FNS research systems that use intramuscular or epimysial electrodes are presently limited by the need to implant, maintain, and use a large number of electrodes if many muscles are to be controlled (e.g., see Marsolais and Kobetic, 1987). An exciting recent development in intramuscular electrode design is the concept of "injectable" microstimulators that receive power and command signals by inductive coupling (Loeb et al., 1991).

So-called "neural" electrode systems place electrodes directly into or adjacent to peripheral nerve trunks (Naples et al., 1990) and offer the possibility of controlling multiple muscles with a single implant. As presented in Sect. 8.2, in the distal segments of peripheral nerve trunks, individual fascicles (bundles) tend to contain the innervation for one or two skeletal muscles (Sunderland, 1978). Horch and colleagues have developed and demonstrated the feasibility of an implantable intrafascicular electrode stimulation system that uses Teflon-insulated 25-μm diameter Pt-Ir wires. Single pulse charge thresholds for muscle activation using such intrafascicular electrodes can be remarkably low—typically about 1 nC (Nannini and Horch, 1991; Yoshida and Horch, 1993). Tyler and Durand (1993, 1994) have studied an interfascicular stimulation electrode design (the "slowly penetrating interfascicular nerve electrode", or "SPINE") that allows placement of electrodes between but not within fascicles. Epineural stimulation (i.e., surgical implantation of electrodes onto the epineurium of a nerve trunk) has been used by one research group (Holle, 1984). All of these intraneural and epineural electrode systems are potentially suitable for selective activation of individual nerve trunk fascicles and, because they are not introduced into or onto muscles, may be more reliable and in some instances easier to implant than intramuscular or epimysial electrodes. Nerve cuff electrodes, which involve placement of electrodes (within insulating tubes or sheaths) near but not within the peripheral nerve may offer similar advantages as intra- or epineural electrodes but without invasive penetration of the nerve trunk. A peroneal nerve cuff electrode, developed by Medtronic, Inc., and McNeal and colleagues (Waters et al., 1975, 1985; McNeal et al., 1977), has been used for correction of footdrop in a large patient population; with some implants functioning well after many years. McNeal and Bowman (1985) tested in animals multiple-electrode cuffs intended to provide greater selectivity of muscle activation. Electrode position and good contact (i.e., close apposition of the electrodes to the nerve trunk) were found to be important factors in successfully achieving selectivity of stimulation with this design. Agnew and coworkers (e.g., Agnew et al., 1989) have developed a helical cuff electrode intended to reduce possible mechanical damage to nerve trunks. Spiral nerve cuffs (Naples et al., 1988) suitable for fascicle selective stimulation have been developed by several groups. These cuff designs enable direct apposition of electrodes to a nerve trunk surface without eliciting the mechanical damage normally caused by tightly fitting cuff systems (Naples et al., 1990; Sweeney et al., 1990; Veraart et al., 1993; Rozman et al., 1993; Sweeney et al., 1995). Most recently, Walter et al. (1997) have demonstrated the feasibility of producing selective hand and wrist movements in a raccoon animal model using multielectrode nerve cuff stimulation.

Recruitment and Rate Modulation in Functional Neuromuscular Stimulation

With regulated-current BCB rectangular stimuli, recruitment levels are in general controlled by altering the charge within the primary stimulus (i.e., by changing the amplitude or the pulse width). Figure 8.7 illustrates an example of isometric force recruitment using intramuscular electrodes implanted in cat soleus muscle. Recruitment by pulse width and pulse amplitude modulation is compared. Recall from Sect. 4.5 that a minimum charge threshold is approached for very small values of pulse width. This property explains the slight shift toward lower charge injection values across most of the range of recruitment for pulse width modulation in comparison to amplitude modulation for this example.

Recruitment selectivity in FNS involves two separate aspects—spatial selectivity and nerve fiber diameter selectivity. In the above discussion, we introduced the concept that implanted electrodes offer greater selectivity in activating different muscles than does surface stimulation. This enhanced "spatial selectivity" occurs because the electric field introduced can be focused closer to the α motor neuron fibers of interest. An aspect of FNS not yet discussed, fiber diameter selectivity, relates to the tendency to stimulate subpopulations of nerve fibers (at the same spatial location) based on their size. Consider again the form of the spatially extended nonlinear nodal (SENN) myelinated nerve model equation (4.31). In Eqs. (4.27) to (4.29) and (4.32a), the terms C_m, G_a, and $I_{i,n}$ are all proportional to fiber

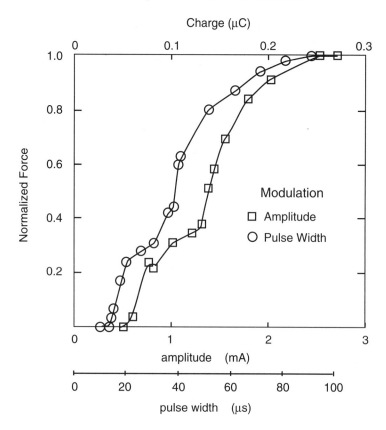

FIGURE 8.7. A comparison of isometric force recruitment by pulse width and amplitude modulation. Circles: normalized force as a function of pulsewidth (zero to $100\mu s$) at a fixed stimulus amplitude (2.7 mA); squares: normalized force as a function of stimulus amplitude (zero to 2.7 mA) at a fixed pulse width ($100\mu s$). Abscissa also gives the total charge per stimulus (top scale) for either modulation scheme. Data are for 10-Hz intramuscular stimulation of a cat soleus muscle. (Adapted from Crago et al., © 1980 IEEE.)

diameter. These terms occur in Eq. (4.31) only as ratios. As pointed out by McNeal (1976), the only effect of changing fiber diameter is therefore in the calculation of the extracellular potential terms. Because internodal length is proportional to fiber diameter, the extracellular potentials at the nodes will change with fiber diameter. In one idealized situation where an extracellular potential distribution rises linearly along a fiber, peaks over a node of Ranvier, and then declines linearly, Eq. (4.31) predicts that fiber excitation threshold will be exactly inversely proportional to fiber diameter. This theoretical dependence of excitability given a linear field along a nerve matches the experimental results of Tasaki (1953) for studies on single myelinated fibers from frog sciatic nerve. In that the focalized electric fields generated by actual stimulation systems rarely approximate such an ideal distribution, more complex dependencies of stimulus threshold upon fiber diameter are generally found. McNeal's original model predicted that, for regulated-current stimuli of 100-μs duration from a monopolar point cathode located 1 mm above a node of Ranvier, large fiber diameter thresholds were inversely proportional to the square root of fiber diameter. In the small diameter range, threshold approached an inverse square relationship with fiber diameter.

Recall from Sect. 8.2 that, in most normal physiological use, the "size principle" predicts an orderly recruitment of motor units from slow (small diameter axons from small motor neurons) to fast (large diameter axons from large motor neurons). In most electrical stimulation of myelinated fibers, however, there will be a tendency to recruit large axons at small stimulus magnitudes and then smaller axons with increased stimulus levels. Such "reversed recruitment" of myelinated nerve fibers with extracellular electrical stimulation is a well-known phenomenon (e.g., Blair and Erlanger, 1933; Petrofsky, 1978; Fang and Mortimer, 1987). A major problem in FNS for motor control restoration is that, even if sufficient spatial selectivity is achieved (primarily through electrode selection and placement), reversed recruitment of motor units will inappropriately utilize fast, more readily fatigued muscle fibers for low-force tasks. Slower, fatigue-resistant muscle fibers will only be recruited at higher stimulus levels. This also results in an undesirable, steep relation between force output and stimulus magnitude. Exacerbating this problem is the fact that, with the muscle disuse that tends to follow spinal cord injuries, many fatigue resistant muscle fibers tend to shift their metabolism towards less oxidative, and more anaerobic, more readily fatigued mechanisms. To circumvent these problems in clinical use of FNS systems, electrical stimulation "exercise" procedures are sometimes employed to alter the metabolic and physiologic properties of type IIB (FF) fibers and type IIAB (FI) towards the less readily fatigued type IIA (FR) and type I (S) profiles.

Exceptions to the reversed recruitment principle of electrical stimulation probably occur at very small electrode-nerve separation distances. Consider the idealized situation where an electric field generated by a point source very close to a node of Ranvier is so highly focused that, at adjacent nodes of Ranvier *regardless of fiber diameter* (and therefore internodal spacing), no appreciable extracellular potentials occur. In this case, the extracellular potential second-difference function of Eq. 4.31 simply reduces to the term $V_{e,n}$ for all size fibers. We would therefore expect that all size fibers would be recruited at about the same current threshold. Modeling studies of Veltink et al. (1988, 1989a), Altman and Plonsey (1989), and Sweeney et al. (1989) predict that as electrode-nerve separations decrease below the order of internodal lengths, strict reversed recruitment order disappears. This effect also has been studied experimentally by Veltink et al. (1989b) using intrafascicular electrodes.

Rate modulation in FNS differs from that in normal physiological use in one important respect. FNS systems will, by their nature, activate all motor units synchronously in phase with each stimulus at a predetermined frequency. In normal physiological use, the nervous system asynchronously activates different motor units. The result is that, at moderate average firing rates, smoothly controlled force production is possible. In synchronized FNS activation of motor units, however, smoothly fused contractions are possible only at higher average firing rates. As we have seen in Sect. 8.2, however, higher firing rates more rapidly fatigue the muscle fibers.

A variety of control strategies (including feedforward, feedback, and adaptive) which incorporate different types of machine intelligence (e.g., optimization, neural networks, fuzzy logic) have been implemented, tested, and/or proposed for FNS systems (see Chizeck et al., 1988; Crago et al., 1996). Recruitment modulation (varying pulse width or amplitude at a fixed stimulation frequency) is generally used to achieve low to moderate force output control. Rate modulation can then be used to produce high force levels after all motor units have been recruited. Simultaneous modulation of both recruitment levels and rate modulation is much more complex but offers the possibility of finer, more rapid control of force and of changes in force. In most current FNS systems, control and feedback signals are generated by externalized artificial sensors which can be difficult to build and maintain. However, it may eventually be possible to record "natural" sensor signals from implanted peripheral nerve or CNS electrodes for use in FNS feedback and control (for a review, see Hoffer et al., 1996).

Electropathology of Functional Neuromuscular Stimulation Systems

The mechanisms by which electrical stimulation might damage tissues fall into two general categories: those deriving from electrochemical processes that enable charge transfer at electrode-tissue interfaces and those that occur directly or indirectly as a result of current flow within the tissues [for a review, see McCreery and Agnew (1990)]. An implanted neuromuscular stimulation system must meet general biocompatability requirements (beyond those related to electrical stimulation) as well (Yuen et al., 1990).

Current flows in metals by electron movement, while current in the body consists of the flow of charged ions. This flow of charge can be induced by capacitive or conductive electrode mechanisms [see Robblee and Rose (1990) for a review of stimulating electrode electrochemistry]. Capacitive mechanisms (which involve charging and discharging of the electrode "double layer") are generally not sufficient to deliver the current densities needed for nerve or muscle simulation. Conductive mechanisms (which are also known as *Faradaic* and involve electron transfer across the electrode-tissue interface via electrochemical redox reactions) are therefore almost always necessary. "Reversible" conductive reactions involve chemical species that, ideally, are bound at the electrode surface and can be quantitatively reversed by alternately passing equal and opposite amounts of charge. Examples of such reactions are oxide layer formation/depletion and hydrogen atom plating. Balanced-charge biphasic (BCB) regulated-current stimuli (see Sect. 8.3) can therefore use reversible conductive reactions for electrical stimulation without introducing new electrochemical species into the tissue. "Irreversible" conductive reactions necessitate formation of new species and in general may elicit variable amounts of tissue damage depending on the body's ability to buffer or suppress deleterious effects. Examples of such processes are electrode corrosion, oxidation of chloride ions, and the electrolysis of water. Such reactions can bring about the formation of biologically toxic products and may generate large swings of pH near the electrode-tissue interface.

Although electrical stimulation of axons within their normal physiological range does not in general elicit any damage, Agnew et al. (1989) have reported that prolonged BCB electrical stimulation (in which only reversible conductive reactions were presumably utilized) of cat peripheral nerve could elicit damage in a fraction of large myelinated nerve fibers. This result depended on the stimulus parameters (current, pulsewidth, and frequency) and the total duration of stimulation. For the peroneal nerve trunk, supramaximal stimulation of $A\alpha$ nerves for 8h or more at 50 Hz consistently produced this *early axonal degeneration* (EAD) in some fibers. Depending on the stimulation parameters, damage could sometimes be found with frequencies as low as 20 Hz. In a recent series of experiments, Agnew and colleagues (see McCreery and Agnew, 1990) found that local anesthetic applied to peroneal nerves near the stimulation site prevented EAD. This complete abolition of damage by local anesthesia (which will prevent electrical excitation of the nerve fibers) strongly suggests that hyperactivity of axons may be at least partially responsible for EAD generation.

8.5 Skeletal Muscle Stimulation in Electrical Accidents[2]

The neuromuscular effects of accidental electric shocks have been a subject of study since the first development of commercial electric current sources. In particular, two possible neuromuscular responses to accidental electrical stimulation have received a great deal of attention because they can potentially contribute to injurious or fatal outcomes: the generation of involuntary movements by skeletal muscle activation and arrest of respiration. In this section, we will review existing experimental data on skeletal muscle

---

[2] Much of the material in this section is extended from an excellent review of 60-Hz contact current safety levels by Banks and Vinh (1984).

stimulation in electrical accidents. Section 8.6 analyzes contractile movements of the hand contacting a relatively low-voltage electrical source and the tetanizing condition of the so-called "let-go" threshold. In Sect. 8.7, we will review the possible effects of electric stimulation on respiration.

History and Overview

Virtually since the introduction of commercial electricity, it has been recognized that alternating current could produce tingling and pain (see Chapter 7), sensations of tissue heating, stimulation of muscle contractions, respiratory difficulty or arrest, and even cardiac fibrillation (see Chapter 6). At current levels where voluntary skeletal muscle control is lost and involuntary movements occur, "freezing" to a live conductor held in the hand is possible. As C.F. Dalziel, who performed many landmark investigations in this area, noted in 1956, electric stimulation of the forearm and hand flexors can generate very powerful grasping contractions that are not easily overcome. Dalziel reported that "in 1942, a shipyard workman in the San Francisco Bay area attempted to obtain a better view of an in-plant Christmas Eve party by climbing around an overhead crane. He slipped and, to keep from falling, grasped the 440-volt crane trolly wires. He immediately froze to the trolley wires, and his body swung back and forth until someone opened the switch. He then fell to the ground, fractured his skull, and died. A similar accident occurred in 1947 at a brass works in the Los Angeles area. A laborer lost his footing when climbing down from the roof of the factory. As he fell, he grasped the 220-volt trolley wires of an overhead crane and froze to the trolley wires. His body dangled from the wires until one employee opened the switch while another employee caught him as he fell. He received severe burns but recovered" (Dalziel and Massoglia, 1956).

Early investigators of this effect defined the current level where an individual could, through remaining voluntary control, just barely free himself from a hand-held current source as the *let-go value*. Experimental study of this value and its dependence on a number of variables [subject sex and age, contact(s) size and location, skin conductivity conditions, stimulus waveform and parameters, etc.] has dominated investigations on accidental neuromuscular stimulation effects.

Let-Go Thresholds for Hand Contact with Electrical Stimuli

In the 1920s and the 1930s, a number of studies on let-go thresholds (or similar tests) for 50- or 60-Hz stimulation were carried out (Grayson, 1931; Thompson, 1933; Dalziel, 1938; Gilbert, 1939; Whitaker, 1939). A summary of the results of these studies is presented in Table 8.2 (after Banks and Vinh, 1984). Despite the widely varying methods and end points used, the

TABLE 8.2. Summary of data from early let-go current investigations.

Report	Subject	Current path	Active electrode	Endpoint	Let-go current (mA rms)			Frequency (Hz)
					Minimum	Average	Maximum	
Gilbert (1939)	25 Both sexes	Hand-hand	20-mm brass rod	Release grip	NR	15	NR	50
Whitaker (1939)	13	Hand-hand	Pliers	Release grip	6.0	7.8	10	NR
Thompson (1933)	42 men 27 women	Hand-hand	Rotatable metal handle	Rotate handle	NR NR	8.35 5.15	20.0 8.8	60 60
Grayson (1931)	42 men	NR	NR	Not "serious discomfort"	NR	8	NR	NR
Dalziel (1938)[a]	56 men	Hand-hand	Flat metal disk	Tolerate sensation	NR	13.9	22	60
	18 men	Hand-foot	Flat metal disk	Tolerate sensation	NR	13.5	NR	60
Dalziel (1938)[b]	42 men	Hand-hand	No. 10 AWG copper wire	Release grip	NR	12.6	18	60

NR = not reported.
[a] First series.
[b] Second series.
Source: Adapted from Banks and Vinh, © 1984, IEEE.

average values for humans fell within a fairly reasonable range of about 8 to 15 mA rms.

From 1941 to 1956, Dalziel and his colleagues reported in a series of papers on human let-go currents and voltages. Despite some experimental and statistical methods that would not be acceptable by today's standards (Banks and Vinh, 1984), these reports still form the foundation of our knowledge about the let-go phenomenon and have had great impact in setting safety standards for line-frequency contact currents (see Chapter 11). The adult experimental subjects in these studies were predominantly healthy, young, male volunteers aged 21 to 25 years (with an upper age tested of 46 years). A smaller number of female subjects in their "late teens to the early 20's" were also tested (Dalziel and Massoglia, 1956). Dalziel described his female subjects as probably representing a fairly sedentary population. A small number of children were also tested. In all experiments, "an individual's let-go current [was] defined as the maximum current he [could] tolerate when holding a copper conductor by using muscles directly affected by that current" (Dalziel et al., 1943). A No. 6 or No. 7 AWG polished copper wire was grasped as the active electrode in most studies. Dalziel and colleagues felt that conductor size and shape did not have a large effect on threshold measurements. The hands of the subjects were kept moist with a "weak salt" solution, and any of three indifferent electrode sites/types were used (usually a brass plate under the opposite hand, but sometimes a plate under one or both feet, or a conducting band around the upper arm), apparently without strongly affecting the results. After allowing each subject to become accustomed to the sensations that would be encountered, the current was raised slowly to a certain value, at which point, the subject was commanded to drop the electrode. If he or she succeeded, the current values were increased by steps until the subject was no longer able to let go of the wire; if the subject failed, the test was repeated at a lower value. It is important to note that, at the let-go threshold current value, "failure to interrupt promptly the current is accompanied by a rapid decrease in muscular strength caused by pain and the fatigue associated with the severer involuntary muscular contractions" (Dalziel and Massoglia, 1956). The end point of experiments was checked by several trials thought to be insufficient in number to fatigue the individual but enough to determine the let-go value accurately (Dalziel et al., 1941). This process was undoubtedly quite unpleasant (see Fig. 8.8). Dalziel and colleagues recognized that psychological state and motivation levels certainly affected results. "Psychological factors, especially fear and competitive spirit, were the most important causes for the variations" (Dalziel et al., 1943). Table 8.3 and Figure 8.9 summarize results obtained for the 60-Hz let-go currents of 134 men and 28 women. The 99.5 percentile rank levels (9 mA rms for men, 6 mA rms for women) were considered to represent a reasonable criterion for safety (Dalziel and Massoglia, 1956).

FIGURE 8.8. An experiment subject undergoes let-go current threshold testing. (From Dalziel, (©) 1972, IEEE.)

TABLE 8.3. Summary of Dalziel's 60-Hz let-go current threshold data.

Subjects	0.5 Percentile rank	Minimum	Average	Maximum
134 men	9.0	9.7	15.9	22.0
28 women	6.0	7.4	10.5	14.0

Note: Values of let-go current in mA rms.
Source: Adapted from Banks and Vinh, © 1984, IEEE.

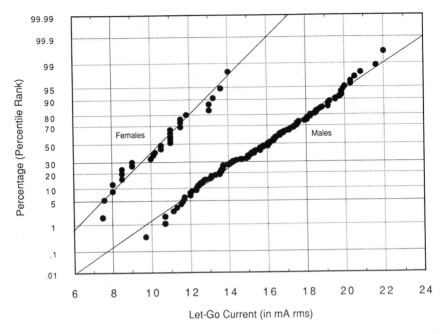

FIGURE 8.9. Let-go current percentile rank distribution curves for adult males and females given hand contact with a 60-Hz AC source. Average value for 134 males = 15.9 mA rms; extrapolated 0.5 percentile value = 9 mA rms. Average value for 28 females = 10.5 mA rms; extrapolated 0.5 percentile value = 6 mA rms. (Adapted from Dalziel, © 1972 IEEE.)

Great interest existed at this time in also establishing 60-Hz safety levels for small children. In general, Dalziel and colleagues found that obtaining let-go currents for young children was extremely difficult. In addition to the reluctance of parents to permit experimentation on their children, Dalziel's limited experiences with children indicated that children were likely to "just cry at the higher values" (Dalziel, 1972). Despite these problems, results were reported for three small boys (Table 8.4), and Dalziel recommended a "reasonably safe" current level for children as 50% of the 99.5 percentile value for normal adult males—namely, 4.5 mA rms.

TABLE 8.4. Dalziel's 60-Hz let-go threshold data for three boys.

Subject	Age	Let-go current (mA rms)
1	5 years	7
2	9 years, 3 months	7.6
3	10 years, 11 months	9
Mean		7.9

Source: Adapted from Banks and Vinh, © 1984, IEEE.

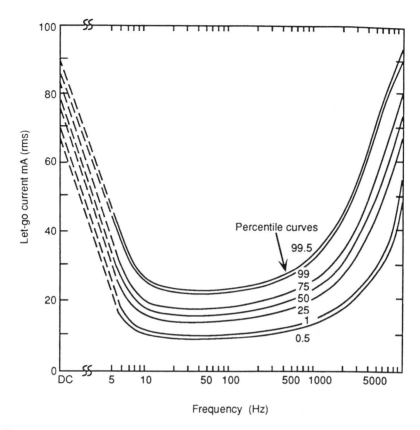

FIGURE 8.10. The effect of stimulus frequency on the let-go current threshold for adult males. Values on the left-hand axis are for DC. (Adapted from Dalziel, © 1972 IEEE, using some additional data from Dalziel et al., 1943.)

The frequency dependence of tetanizing sinusoidal current thresholds was reported in 1943 (Dalziel et al.) for groups of adult male subjects ranging in number from 26 to 30 at frequencies from 5 to 10,000 Hz as well as at DC. These results were compared with the 60-Hz study on 134 men. Figure 8.10 depicts the percentile curves for frequencies ranging from 5 to 10,000 Hz as well as at DC.[3] Tests on steady or gradually increasing DC currents "produced sensations of internal heating rather than muscular

---

[3] A somewhat unusual statistical method of "correcting" these curves, and DC data, for variations in the 60-Hz let-go threshold mean value was used. For each group of subjects tested at a given frequency, a 60-Hz let-go threshold mean was also found. A "corrected mean" for each frequency was calculated as (the actual mean of the sample group at the given frequency) × (the 60-Hz mean for the group of 134 men)/ (the 60-Hz mean for the current sample group). These corrected means were also used in calculation of percentile ranks at each frequency. The correction factors ranged from 1.00 (at 60 Hz) to 1.09 (at 5 Hz).

contractions. Sudden changes in current magnitudes produced muscular contractions, and interruption of the current produced a very severe shock. . . . The maximum [level of discomfort] a subject could take and release was termed the release current. It represents the limit of voluntary endurance rather than the let-go limit." The DC release current mean of 28 men was 73.7 mA. After "correction" for variation in the 60-Hz let-go threshold for this group, the 99.5 percentile rank "reasonably safe" DC current value for men was 62 mA. Because at 60 Hz the mean value for let-go threshold values of women was about 66% of that for men (Table 8.3), it was felt that the 99.5 percentile male values at 5 to 10,000 Hz, and at DC, could be scaled by 0.66 to obtain estimates of reasonably safe values for women.

8.6   Analysis of the Let-Go Phenomenon[4]

Analysis of factors that may be involved in the let-go phenomenon first requires consideration of forearm/hand functional anatomy (especially as regards power grip generation) and the current paths and impedances of hand-contact accidents. The fundamental principles of neuromuscular electrical stimulation presented previously in this chapter can then be utilized.

Functional Anatomy of the Hand and Wrist

A brief introduction to the anatomy and kinesiology of the hand and wrist is essential to an understanding of how powerful gripping actions occur. More detailed descriptions can be found in *Gray's Anatomy* (Gray, 1989), Kaplan's excellent functional anatomy text (1984), and Rasch (1989). The forearm (antebrachial) muscles are divided functionally into flexors and extensors (anterior and posterior), although often complex interactions are possible. Foremost among the flexors of this group that contribute to powerful gripping are the flexor digitorum profundus muscle (the medial part of which is innervated by the ulnar nerve and the lateral by the anterior interosseous branch of the median nerve) and the flexor digitorum superficialis muscle (innervated by the median nerve) (see Figs. 8.11 and 8.12). In a transverse section through the middle of the forearm (Fig. 8.12), the relatively large fraction of cross-sectional area occupied by these two muscles (52% of the area taken up by muscle, nerve, and blood vessels) attests to their potential power. Maximal flexion of the hand thus can overcome maximal extension. The intrinsic muscles of the hand are categorized into three groups—those of the thumb, those of the little finger, and those in the middle part of the hand. In gripping an object, the four small

---

[4]Portions of this section have been adapted from Sweeney (1993).

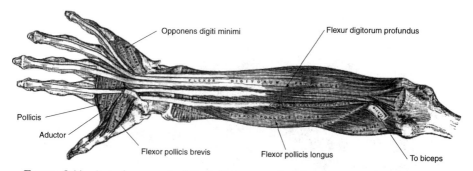

Opponens digiti minimi

Flexur digitorum profundus

Pollicis

Aductor

Flexor pollicis brevis

Flexor pollicis longus

To biceps

FIGURE 8.11. Anterior aspect of the left forearm showing deep muscles. (Adapted from H. Gray, Gray's Anatomy, 30th ed., C.D. Clemente (ed.), Lea & Febiger, 1985, p. 533.)

lumbrical muscles (innervated by the median and ulnar nerves) in particular provide flexion of the proximal segment of each finger (and under some conditions, extension of the distal finger segments) (see Gray, 1989). We can therefore hypothesize that, until the large flexor muscles of the forearm are activated, electrically activated gripping actions using only the intrinsic muscles of the hand can probably be overcome by use of the forearm extensors. The let-go threshold should occur at about the point where significant involuntary excitation of the forearm flexors has been initiated.

Body Impedances and Current Paths in Hand-Contact Accidents

We have seen in Chapter 2 that the impedance properties of dry skin are highly nonlinear and time varying. It is likely that the major source of skin nonlinearity is the corneum. At the level of electrical line voltages, a skin impedance breakdown occurs fairly rapidly. As depicted in Fig. 2.12, for example, 50-Hz AC voltage stimulation results in an initial rapid drop in impedance within a fraction of a minute, followed by a slower drop in the first minute, and then more gradual decreases in the next 7 to 8 minutes until a plateau value is achieved [see Sect. 2.2 and Freiberger (1934)]. This phenomenon may partially explain Dalziel's observation that, near the let-go threshold, failure to promptly interrupt current flow results in a rapid inability to do so.

Although the total internal body impedance, pathways, and current densities that occur in sinusoidal voltage electrical stimulation are highly dependent upon the voltage magnitudes and frequencies as well as electrode(s) size and placement, some useful generalities can be drawn. Hand-to-hand and hand-to-foot internal impedances for line voltages and frequencies are in the range of 500 to 750 ohms. As can be seen from Figs. 2.20 and 2.21, about 50% of the total internal body impedance in this situation can be attributed to the wrists and/or ankles. Cross sections

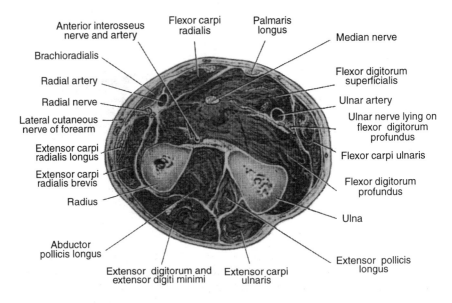

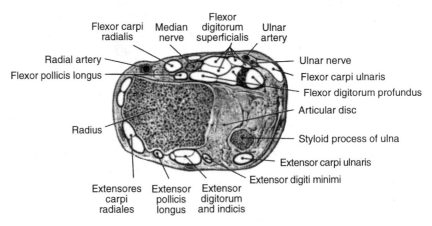

FIGURE 8.12. Transverse cross section through middle of left forearm (upper) and wrist (lower.) [Adapted from Gray, 1989, p. 620 (upper) and p. 625 (lower.)]

through these areas (e.g., Fig. 8.12) reveal that relatively poorly conducting bone, tendon, fat, and ligament dominate. Current densities in these regions should be highest in the existing higher conductivity soft tissues (nerve trunks) and blood vessels. We would therefore expect that in hand-contact electrical accidents, current density will be very focused directly beneath the electrode in the muscle and nerve of the palm and within the median and ulnar nerve trunks in the wrist.

To a first approximation, current densities beneath a cylindrical object held in the hand can be estimated by considering the surface area of the electrode (although if the rod is not completely enclosed by tissue, current densities will be highest at the edges of the conductor). The No. 6 AWG wires used by Dalziel and colleagues would have an approximate surface area of $11.1 \, cm^2$ when held by an average adult male, assuming a typical hand breadth of 8.6 cm (McConville et al., 1980) and No. 6 AWG diameter of 0.41 cm. Given a total current I (in mA), a current density of 0.09 I (mA/ cm^2) would therefore occur close to the electrode surface. For adult females, average hand breadth is approximately 7.8 cm (Young et. al., 1983), so current density would be 0.10 I (mA/cm^2). Typical hand width of a 5-year-old boy is about 5.7 cm (Synder, 1975), so current density would be 0.14 I (mA/cm^2). An estimate of current densities within the median and ulnar nerve trunks near the wrist can be obtained from Fig. 8.12. Tissues other than bone, cartilage, fat and tendon in Fig. 8.12 comprise less than 5% of the wrist cross-sectional area at this level. From anthropometric estimates of wrist circumferences (17 cm for adult males (McConville et al., 1980); 15.7 cm for adult females (Young et al., 1983); and about 11 cm for small children), we can roughly calculate wrist cross-sectional areas making the simple assumption that all wrists have the elliptical shape of Fig. 8.12. Performing these calculations yields estimates of 21.3, 18.2, and $8.9 \, cm^2$ for adult males and females and for small children, respectively. For total current I, current densities of 0.94 I, 1.10 I, and 2.25 I (mA/cm^2) result for men, women, and children, making the simple approximations that most current will flow in the nerve trunks and blood vessels at this level, that these tissues have roughly the same conductivity, and that such tissues comprise approximately 5% of the total cross-sectional wrist area regardless of sex or age. This is in contrast to densities for the more proximal forearm.

From Fig. 8.12, we see that skeletal muscle dominates the cross-section at the midforearm, and highly conductive tissues (muscle, nerve, and blood vessels) comprise about 72% of the cross-sectional area in adult males. Cross-sections of the male forearm from other anatomy sources give similar estimates of combined muscle, nerve, and blood area percentages (e.g., 73% in Sect. 57 of Morton, 1944; 73% in Sect. 56 of Carter et al., 1977). Adult females on average have a higher percentage of subcutaneous fat in the extremities as compared to adult males (25% as opposed to 12% as reported by Malina, 1975). Analysis of cross-sections in the midforearm of adult females (level Fl in Kieffler and Heitzman, 1979; Sect. 11–21 in Bo et al., 1990) as well as anthropometric estimates based on 50th percentile data for 17-year-old females (McCammon, 1970) both yield skeletal muscle cross-sectional area estimates of about 61%. Using this information from Fig. 8.12, and anthropometric data on typical midforearm circumferences [23.4 cm for adult males (McConville et al., 1980); 21.2 cm for adult females (Young et al., 1983); 16 cm for 5-year-old children (Synder, 1975)], we can estimate current densities within the relatively conductive tissues at this level. Assuming the elliptical shape of Fig. 8.12 is typical regardless of sex or age, current densities of 0.032 I, 0.047 1, and 0.069 1 (mA/cm^2) result for men, women, and children, respectively (using an estimate of 72% for combined muscle, nerve, and blood cross-sectional area fraction in children and adult males and 61% for adult females).

Excitation Threshold Estimates

Recall from Sects. 4.5 and 8.4 that, to a first approximation, changes in the extracellular electric field (i.e., the change in voltage gradients or second difference of voltage values) along an excitable fiber provide the driving force for eliciting transmembrane potential shifts. Given current density estimates (J), we can calculate voltage gradient values $(\nabla\phi))$ using the relation $\nabla\phi = J\varrho$, where ϱ is an estimate of bulk tissue resistivity.

Making use of the current density estimates derived in the previous section, we can therefore predict analogous voltage gradients if we assume a certain tissue resistivity value. Although the conductivities of both skeletal muscle and peripheral nerve trunks are anisotropic, with both a spatial and temporal dependence (see Chapter 2), we will use $200\,\Omega$ cm as an approximate value for the lumped resistivity for muscle, nerve, and blood vessels with longitudinally oriented current at power line frequencies (see Table 2.1 and Sances et al., 1983). Voltage gradients immediately adjacent to a No. 6 AWG wire (passing a current "I" in mA) held in the hand of adult males and females or small children (5-year-old boys) become 18 I, 20 I, and 28 I (mV/cm), respectively. In the nerve trunks of the wrist, we would expect voltage gradients of 188 I, 220 I, and 45 I (mV/cm); and at the mid-forearm, 6.4 I, 9.4 I, and 13.8 I (mV/cm) (for men, women, and small children, respectively).

As we have seen in Chapter 4, not only must a longitudinally oriented voltage gradient exist along myelinated nerve fibers for electrical stimulation to occur, but a *change* in voltage gradient is necessary. The terms $(V_{e,n-i} - 2V_{e,n} + V_{e,n+i})$ in Eq. (4.31) imply that a second-difference sampling of nodal extracellular voltages provides the driving force for excitation. That is, a change (difference) in two first-difference terms $(\{V_{e,n-i} - V_{e,n}\}$ and $\{V_{e,n+i} - V_{e,n}\})$ is needed. Each first-difference term is essentially a linear sampling of the extracellular voltage-gradient between two nodes of Ranvier. If the electric field along a nerve fiber is truly uniform, then in theory, excitation cannot occur. Realistically, in this situation, spatial orientation changes (e.g., bends) in nerve fibers tend to introduce apparent second-difference effects. Likewise, terminations of motor myelinated nerves at end-plates, or sensory axons at receptors, allow excitation within a uniform field (see Sect. 4.5).

The peripheral myelinated nerve fibers of α-motor neurons range in diameter from approximately 9 to 20μm (see Sect. 8.2). If we assume a factor of 100 difference between fiber diameter and internodal length (Table 4.1), large (20-μm) fibers will have internodal lengths equal to about 2 mm, while small (10-μm) fibers will have internodal lengths of about 1 mm. As we have seen in Sect. 8.4, large fibers tend to be the most easily stimulated by fairly diffuse extracellular electric field distributions, leading to the phenomenon of "reversed recruitment." With either the myelinated nerve model described in Chapter 4 or compartmental cable models of mammalian myelinated nerve (Sweeney et al., 1987), the author has found that second-differences of approximately 50 mV are needed to bring lengths of nerve fibers to threshold with long pulsewidth (rheobase) stimuli. A 20-μm diameter nerve fiber lying within a uniform voltage gradient of 25 mV/mm would experience a second-difference of 50 mV if the nerve abruptly turned by $90°$ with respect to the field (similarly, 50 mV/mm will excite a 10-μm diameter fiber). In Sect. 4.5, it is reported that, in a terminated nerve model (e.g., roughly representative of axons ending at a motor end-plate), a voltage gradient of 6.2 mV/mm suffices to excite a 20-μm diameter myelinated nerve (12.3 mV/mm for a 10-μm fiber). These lower values presumably result from the abrupt impedance transition at the cable end. All of the above voltage gradients apply for squarewave stimuli having durations several times greater than the chronaxie. For 60-Hz sinusoidal stimuli, the peak threshold current amplitudes are about 17% greater than those of square-wave stimuli (Fig. 4.15).

Directly under a hand-held electrode, or within the midforearm, we will therefore hypothesize that a voltage-gradient of approximately 7.3 mV/mm (117% of 6.2 mV/mm) or 5.2 mV-rms/mm is necessary to achieve threshold stimulation of the largest (20-μm diameter) myelinated axons terminating at motor end-plates of the lumbrical muscles in the hand or the large flexor muscles of the forearm (10.2 mV-rms/mm for 10-μm fibers). For initial excitation of the median and ulnar nerve trunks within the wrist, we will hypothesize that a rheobase voltage gradient of 29.2 mV/mm (117% of 25 mV/mm) or 20.6 mV-rms/mm is needed assuming that a "bending" of the nerve trunks with respect to the electric field occurs as the nerve trunks spread into branches, and the current distributes within the hand (41.2 mV rms/mm for 10-μm fibers).

Table 8.5 lists the current density and voltage gradient estimates derived above for an arbitrary current I (in mA) and the resultant estimates of absolute current levels needed to achieve threshold excitation in 20-μm and 10-μm diameter nerve fibers. The results in Table 8.5 predict that, at current values on the order of 1 mA (1.1 mA rms for males, 0.94 mA rms for females), typical adult males and females should experience initial, relatively small contractions of intrinsic muscles in the hand (such as the lumbricales) due to excitation of 20-μm diameter fibers in the nerve trunks of the wrist.

TABLE 8.5. Analysis of Hand-Contact skeletal muscle stimulation.

	Current density (mA/cm^2)	Voltage gradient (mV/cm)	20-μm Fiber threshold (mA rms)	10-μm Fiber threshold (mA rms)	60-Hz let-go current (mA rms)[a]
Directly beneath No. 6 AWG electrode in lumbrical muscles of the hand					
Adult males	0.09I	18I	2.9	5.7	—
Adult females	0.10I	20I	2.6	5.1	—
Small children	0.14I	28I	1.9	3.6	—
At median and ulnar nerve trunks within the wrist					
Adult males	0.94I	188I	1.1	2.2	—
Adult females	0.10I	220I	0.94	1.9	—
Small children	2.25I	450I	0.46	0.92	—
At the large flexor muscles of the midforearm					
Adult males	0.032I	6.4I	8.1	15.9	15.9
Adult females	0.047I	9.4I	5.5	10.9	10.5
Small children	0.069I	13.8I	3.8	7.4	7

[a] Estimates of Dalziel for adult males and females (Dalziel and Massoglia, 1956); single study on a 5-year-old boy as reported by Dalziel (1943).
Note: I is an arbitrary hand-contact current in milliamperes.

These estimates can be compared with Dalziel's studies on perception current thresholds for hand-contact electrical stimulation. [Recall from Chapter 7 and Sect. 8.2, that large diameter Aα sensory nerve fibers will also be present in the nerve trunks of the wrist. For example, muscle spindle primary endings are innervated by type IA fibers which average about 17μm in diameter and can be 20μm (see Fig. 3.12)]. Our predictions correspond quite well with the actual 60-Hz perception threshold mean values (1.1 mA rms for adult males and 0.7 mA rms for females) (Dalziel, 1972). Somewhat surprisingly, current densities directly beneath a No. 6 AWG electrode should result in initial muscle activation only at somewhat higher thresholds (e.g., 2.9 mA for adult males). These values probably reflect overestimations of the actual values because, in reality, the edges and proximal surface of the electrode will carry the majority of the current.

Dalziel and colleague's mean values for the 60-Hz let-go threshold currents of adult males and females (i.e., 15.9 and 10.5 mA rms, respectively) and even Dalziel's single value for the let-go threshold of a 5-year-old boy (i.e., 7 mA rms) lie within the bounds of our Table 8.5 estimates for threshold stimulation of large and small α-motor neuron fibers innervating the forearm flexors (compare the last three columns of the table for the last three rows). This is in keeping with our hypothesis that the let-go threshold should reflect the point at which forearm flexor activation reaches a level where it cannot be voluntarily overcome. Because of a combination of

anthropometric (i.e., reduced mean forearm circumference) and body composition factors (i.e., higher mean percentage of subcutaneous fat), our
estimates predict that adult females should have let-go thresholds about
69% of those of adult males. Recall that Dalziel and colleagues estimated
that female thresholds would be about 66% of those of males. Our estimate
for the let-go threshold range in a 5-year-old child (3.8 to 7.4 mA rms) is
reasonably close to the value of 7 mA rms for a 5-year-old boy as reported
by Dalziel (1943). This current range estimate is 47% of our predicted adult
male range, a factor in good agreement with the 50% factor originally
proposed by Dalziel (1943).

We can in addition predict the effects of physique on let-go threshold
values for adult males and females by constructing theoretical percentile
rank distributions for let-go currents. These theoretical distributions
can then be compared to Dalziel's data (i.e., Fig. 8.8). We first adopt the
mean let-go current values of Dalziel (15.9 mA rms for males; 10.5 mA rms
for females) as our 50th percentile estimates. For adult females, a mean
let-go current value limit of 10.5 mA rms (analogous to the 10-μm fiber
threshold value in Table 8.5 for the large flexor muscles of the forearm)
rather than 10.9 mA rms would have been obtained above if we had used
an estimate of 58.5% rather than 61% for the fraction of highly conductive tissues in the forearm. Next, we can make use of forearm circumference distribution mean and standard deviation data (adult male data
from MConville et al., 1980 and adult female data from Young et al.,
1983) to construct theoretical let-go percentile rank distributions for
the same population sample size used by Dalziel when he performed the
studies of Fig. 8.9 (i.e., $N = 134$ for men and $N = 28$ for women). These
theoretical curves as seen in Fig. 8.13 appear to incorporate much of
the behavior of Dalziel's data set. More accurate modeling of let-go
thresholds for children could also make use of known relations between
anthropometric measures and the child's sex and age. For example, it is
known that during development, the relative composition of bone, fat, and
muscle in the arm tend to change (see Johnston and Malina, 1966). As
Banks and Vinh (1984) have emphasized, a general safety standard for
children would have to take into account that children range in age and size
from infants to adults.

Thus, while Dalziel and colleagues (1956) found that their attempts to
correlate let-go currents with physical measurements of the forearm, wrist,
strength of grip, and so on, were inconclusive, our theoretical estimates
of let-go thresholds based on information from functional anatomy,
anthropometry, bioelectricity, and electrical stimulation theory appear to
indicate that such correlations should be possible. Measurements of let-go
thresholds, or at least single sinusoidal pulse electrical stimulation thresholds, in individuals where radiographs or magnetic resonance imaging
(MRI) scans of the forearm were also available, would enable more detailed future investigation of such correlations.

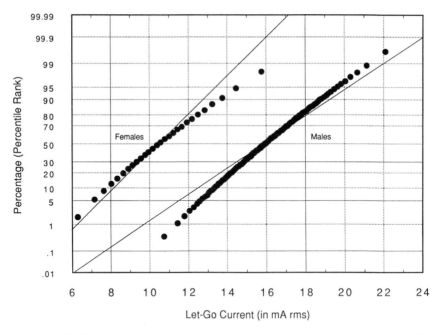

FIGURE 8.13. Theoretical let-go current percentile rank distribution curves for 134 adult males and 28 adult females. Linear regression lines are those of Dalziel and are the same as those in Fig. 8.9. (Adapted from Sweeney, © 1993 IEEE.)

8.7   Effects of Electrical Stimulation on Respiration

It has been long recognized that accidental electric shocks could inhibit or prevent the ability to breathe. In 1913, Jex-Blake (as cited by Lee and Zoledziowski, 1964) listed three processes that could lead to death from electric shock: ventricular fibrillation, prolonged respiratory arrest that outlasts the shock itself, and asphyxia resulting from tetanic contraction of the respiratory muscles during the shock. Electrical stimulation can also be used beneficially in high level quadriplegics with respiratory paralysis as well as in patients with central alveolar hypoventilation to restore respiration (see Nochomovitz, 1983; Creasy et al., 1996). In fact, electrical stimulation of the phrenic nerves for artificial respiration was postulated over two centuries ago by Hufeland (1783).

Anatomy and Physiology of Respiration

Normal low-intensity breathing is performed almost entirely by movement of the diaphragm for inspiration and by passive elastic recoil for expiration. The diaphragm is composed of skeletal muscle and dense collagenous connective tissue. It is innervated by the phrenic nerve which derives from C3 to C5 cervical plexes. Inspiration can also be achieved by contraction of skeletal muscles that raise the rib cage (sternocleidomastoids, anterior serrati, scaleni, external intercostals). The abdominal recti and internal intercostal skeletal muscles may be used during forced expiration to depress the chest cage. The neural "respiratory centers" of the brain stem provide the control and drive for respiration.

Accidental Electric Shocks and Respiration

In several of his review papers on the dangers of electrical shock, Dalziel reported anecdotally that in his group's experiments on the "let-go" phenomenon, they had observed that 60-Hz electrical stimulation with hand-held electrodes could inhibit or stop respiration. Dalziel and Lee (1968) reported that "the muscular reactions caused by commercial frequency alternating currents in the upper ranges of let-go currents, typically 18 to 22 mA or more, flowing across the chest stopped breathing during the period the current flowed, and in several instances caused temporary paralysis of the middle finger. However, normal respiration resumed upon interruption of the current, and no adverse after effects were produced as a result of not breathing for short periods." This possible danger of accidental electrical shock, as well as that of cardiac fibrillation (Chapter 6), is often cited as justifying the importance of let-go thresholds and the establishment of safety standards at currents well below the let-go range. An individual who is "frozen" to a conductor by his inability to release it cannot easily save himself. "Prolonged exposure to currents only slightly in excess of a person's let-go limit may produce exhaustion, asphyxia, collapse and unconsciousness followed by death" (Dalziel, 1959).

In the case of small children, as we have seen in Sects. 8.5 and 8.6, let-go thresholds can be on the order of 7 to 9 mA or less (see Table 8.4). In 1940, it was reported that a 4-year-old boy had died after contacting an electric fence wire (IAEI News Bulletin, 1940). It was assumed that his hand "froze" to the conductor, although Dalziel recognized that small children exposed to painful electrical stimulation might sometimes not even attempt to extricate themselves from an electrical source (Dalziel, 1972). Tests performed on the electric fence power supply indicated that a current of 9 mA was delivered when a load of 100 Ω was placed across the erminal (8.4 mA for 1,000 Ω; 8 mA for 2,000 Ω; 6.8 mA for 5,000 Ω).

In 1961, Lee reported a clinical study of 104 electrical accidents. In 30 cases, individuals were "frozen" to the conductor. "The longer the victim was held on to the circuit the greater appeared to be his chances of developing heart and chest symptoms suggestive of impending asphyxia, and of losing consciousness. . . . Artificial respiration was administered in two cases, one of whom was 'held on' and was being asphyxiated." Greenberg (1940) found that, in forelimb to hindlimb electrical stimulation of dogs, currents of up to 50 mA caused arrest of respiration during the period of current flow; however, respiration restarted as soon as the current was removed. Lee and Zoledziowski (1964) studied the effects of electric shock on the respiration of rabbits. With currents up to about 200 mA, respiratory arrest appeared to be due to muscular contraction in the chest. Only with exceptionally high limb-to-limb current levels, or with current paths that include the head, does it appear possible to electrically bring about an arrest of respiration that persists after the current is switched off (see review of Lee, 1964). It is presumed that this effect is caused by electrically induced damage to the brain stem respiratory centers.

Artificial Respiration by Electrical Stimulation

Dr. William Glenn and colleagues at Yale University, starting in 1959, developed the first practical long-term phrenic nerve stimulation system for phrenic "pacing" (e.g., Glenn and Phelps, 1985; Glenn et al., 1986 and 1988) of quadriplegics. Several commercially available phrenic nerve stimulation systems now exist which utilize implanted electrodes on or adjacent to the phrenic nerves—with over 1,000 people worldwide having received such implants (Creasy et al., 1996). An approach wherein phrenic nerve stimulation electrodes might be placed within the diaphragm (Peterson et al., 1986 and 1994a,b) has also been proposed and studied. Epidural spinal cord stimulation can be used to activate the intercostal muscles (e.g., DiMarco et al., 1987 and 1994). Geddes and colleagues (Geddes et al., 1985, 1988 and 1990; Riscili et al., 1988 and 1989a,b) have also developed an "electroventilation" method that uses surface electrodes located in the axillary region. It is thought that this technique also primarily stimulates the phrenic nerves. The same research group has also proposed the use of electromagnetic stimulation of the cervical phrenic nerves as a possible means of inducing artificial respiration (e.g., Geddes et al., 1991).

9
Stimulation via Electric and Magnetic Fields

9.1 Introduction

Electromagnetic (E-M) fields are ubiquitous in modern society. Radio frequency fields, for example, are produced by a variety of communications devices. Extremely low-frequency fields are produced by power transmission and distribution lines, as well as by home wiring and appliances. Some environmental exposures can result in measurable short-term reactions. The electric fields from high-voltage transmission lines, for example, can provide the opportunity for electric shock under some circumstances. The new generation of magnetic resonance imagers press the limits where perceptible effects might be caused by their time-varying gradient fields. Numerous research efforts have focused on biological effects associated with both short-term and chronic exposure to low-level E-M fields. Although chronic exposure issues are beyond the intended scope of this book, biophysical mechanisms proposed to account for chronic effects are treated briefly in Chapter 11.

This chapter concentrates on short-term human reactions to electric and magnetic fields whose wavelengths are long in comparison with the dimensions of the human body. By concentrating on long-wavelength phenomena, it is possible to treat electric and magnetic field effects separately. This approach differs from the study of high-frequency electromagnetic radiation, where the electric and magnetic components are intimately connected. In the long-wavelength regime, the term "electromagnetic field" is often replaced by "electric and magnetic fields."

9.2 Electric Field Induction Principles

A steady electric field produces surface charges on conducting objects, including the human body. Positive charges are accumulated on the side of a conducting body nearest to the negative source of the field, and negative charges on the side nearest the positive source. If the field is alternating, the positive and negative charges alternate in position, resulting in an alternating current within the biological medium.

It is important to distinguish between *direct* and *indirect* E-field induction effects. The former refers to charges and currents induced directly in the body. The latter refers to current or charge transfer between a person and a conducting object within the field. Before proceeding further, consider physical induction mechanisms.

Induction Mechanisms

Consider a conducting object situated above a ground plane in a vertically oriented electric field. If the object is electrically connected to ground, it will acquire a net charge, which, for simple geometric shapes (flat plate, horizontal cylinder, sphere), can be expressed by (Deno, 1975a, 1975b).

$$Q = V_s C_0 \tag{9.1}$$

where C_0 is the capacitance between the object and ground, and V_S is the space potential that would be present at the centroid location of the object if it were not present. In a low-frequency alternating field where the electrical wavelength is much greater than the dimensions of the object, one can ignore propagation effects and resort to the so-called quasistatic solution (Kaune, 1981). In that case, the induced alternating current is given by the time derivative dQ/dt, which, for a sinusoidal field, is

$$I_s = j2\pi f V_s C_0 \tag{9.2}$$

where f is the frequency of the field, and j is the complex phaser operator (indicating 90° phase shift with respect to the E-field). In a uniform field, an equivalent expression for Eq. (9.2) is obtained by noting that $V_S = Eh$:

$$I_s = j2\pi f E h C_0 \tag{9.3}$$

where E is the electric field strength, and h is the height above the ground plane at the electrical centroid of the object. The terms in Eq. (9.3) can be conveniently regrouped as

$$I_s = \left(2\pi f \varepsilon_0 E\right)\left(\frac{h C_0}{\varepsilon_0}\right) \tag{9.4}$$

where ε_0 is the permittivity of free space (8.85×10^{-12} F/m). The phasor operator j has been dropped, because we will be dealing with magnitudes. The second term in Eq. (9.4) expresses the area of a parallel-plate capacitor with small separation and having capacitance C_0 (see Eq. 2.4). This leads to the *equivalent-area* concept for describing induced current (Deno, 1975b):

$$I_s = 2\pi f \varepsilon_0 E A_e \tag{9.5}$$

For irregularly shaped objects, A_e is the area of a flat plate that would result in the equivalent induced current. Methods for evaluating Eq. (9.5) have been determined for a variety of objects and shapes (Deno, 1975a, 1975b; Deno and Zafanella, 1982; Reilly, 1979a, 1979b, 1982; Reilly and Cwicklewski, 1981).

Another parameter of interest is the *open-circuit voltage* V_0, i.e., the voltage that would be measured between the object and the ground plane in the absence of current conduction to ground. V_0 can be determined from the short-circuit current by

$$V_0 = I_s Z_0 \tag{9.6}$$

where Z_0 is the impedance measured between the induction object and ground. For an object that is perfectly insulated from ground, Z_0 is determined by its coupling capacitance to ground. In this case, V_0 achieves its maximum possible value, given by

$$V_0(\text{max}) = \frac{I_s}{2\pi f C_0} \tag{9.7}$$

where V_0 and I_s apply to magnitudes, without regard to phase [the phasor operator j has been dropped from Eq. (9.7)].

Table 9.1 provides examples of the capacitance, normalized induced current (I_s/E), and normalized open-circuit voltage (V_0/E) for several objects in a 60-Hz field of strength E. References cited in the table discuss measurements involving these objects in E-fields.

The value of I_s given by Eqs. (9.2) to (9.5) expresses the so-called *short-circuit current* (i.e., the current that would flow in a grounded conductor that contacts an otherwise insulated object). When a grounded person provides the ground path, the current that flows at the points of contact may result in a perceptible electric shock. An insulated person touching a grounded connection would also result in current flow, but here the induction object would be the person.

TABLE 9.1. Example induction parameters of various objects in 60-Hz electric fields (object on insulated surface).

Object	I_s/E (A V^{-1}m)	V_0/E (m)	C_0 (pF)
Person (1.8 m tall)	1.8×10^{-8}	0.48	100–150
Auto (midsize)	9.1×10^{-8}	0.21	200
Auto (large)	1.2×10^{-7}	0.16	2,000
School bus	3.7×10^{-7}	0.26	3,700
Commercial bus	4.9×10^{-7}	0.45	2,900
Fence wire (ht = 1 m, length = 100 m)	2.7×10^{-7}	1.0	720

Source: Data from Bracken (1976), Reilly (1982), Reilly (1979a).

We can gain some insight into the induction on a person by using the theoretical expression for short-circuit current in an upright cylinder (Reilly, 1978a):

$$I_s = \frac{2\pi^2 \varepsilon_0 f h E}{\ln\left\{(h/r)\left[(4d + h)/(4d + 3h)\right]^{1/2}\right\}} \tag{9.8}$$

where h is the length of the cylinder, r is its radius, d is the distance from the bottom of the cylinder to the ground plane, f is the frequency of the field, and E is the strength of a vertically oriented E-field. [Units in Eq. (9.8) are amperes, Hertz, meters, and volts/meter]. We have verified Eq. (9.8) with measurements on cylinders placed on a ground plane within a 60-Hz electric field (Reilly, 1978a).

If we model a person as an upright cylinder with a height-to-radius ratio of 12 and standing on a ground plane ($d = 0$), then Eq. (9.8) predicts a value of short-circuit current given by:

$$I_s = 9.0 \times 10^{-11} h^2 f E \tag{9.9}$$

When evaluated at 60 Hz, Eq. (9.9) is identical to the empirical formula determined by Bracken (1976) for people within transmission-line electric fields.

Whereas Eq. (9.9) expresses the short-circuit current at the point of ground contact, an investigator may wish to know the current density in a person who is immersed in an alternating E-field. Experimentors, using conducting models of the human body, have mapped the distribution of internally generated currents for both grounded and ungrounded humans standing within vertically oriented 60-Hz E-fields (Deno, 1977; Kaune and Forsythe, 1985). Calculations over a wider frequency range and for more complex field environments have also been carried out (Takemoto-Hambleton et al., 1988; Hart, 1992; Gandhi and Chen, 1992). Consider, for example, a person standing within a vertically oriented alternating E-field and grounded through the feet. In this case, the amount of current passing through any given horizontal cross section of the body increases as the cross-section is moved from the head toward the feet, with the greatest current passing through the feet as illustrated in Fig. 9.1 (adapted from Deno, 1977). The horizontal axis expresses the fraction of maximum current, given by Eq. (9.9), passing through a cross-section at the indicated height. The outermost curve applies to a person grounded through the feet; the other curve applies to an ungrounded person standing 13 cm above a ground plane. The maximum current in the grounded person flows through the feet. For an ungrounded person, maximum induced current flows through the midsection of the body.

The current density in a homogeneous model is the induced current divided by the cross-sectional area. One can infer from Eq. (9.9) that the

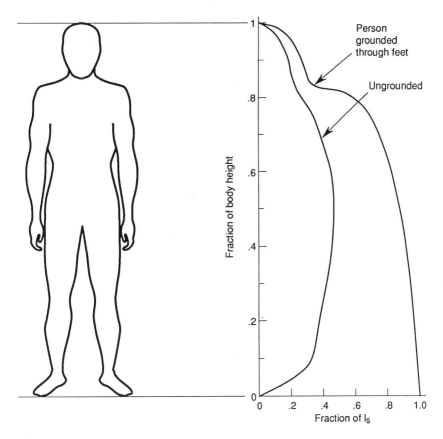

FIGURE 9.1. Distribution of current flowing in a person standing in a vertically oriented alternating E-field. Horizontal axis describes fraction of maximum short-circuit current passing through cross section at indicated height. (Adapted from Deno, 1977.)

current density induced in the upright body would not depend on a person's height, as long as the ratio h/r is held constant. This inference can be drawn by noting that the induced current and cross-sectional area of the model cylinder both grow in proportion to the square of height.

Body Surface Effects

Environmental E-field strengths are typically reported as *undisturbed* field levels (i.e., the intensity that would be measured in the absence of a biological subject). The presence of a conducting object will distort the undisturbed field, such that the field incident on the surface of the object will be significantly modified. The effect is illustrated in Fig. 9.2 for a person in a vertically oriented uniform field. The orientation of the E-field is indicated

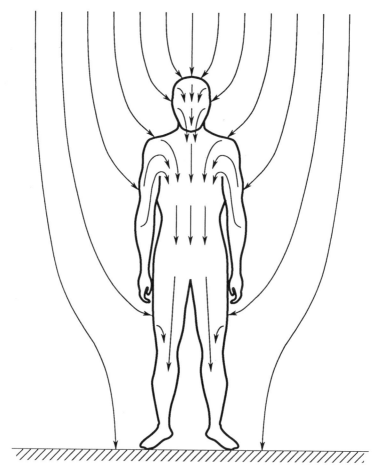

FIGURE 9.2. Perturbation of environmental electric field by a person. External electric field indicated by "flux lines" outside body. Arrows inside body roughly indicate direction and intensity of induced internal electric field and body current. Direction of arrows reverses every one-half cycle. Field lines are illustrative and not drawn to scale.

by the direction of the *flux lines*, and its intensity by their density (not drawn to scale in the figure). The flux lines are normal to the surface of the body. The lines drawn within the body roughly represent the internally induced current.

As suggested by the figure, the E-field on the surface of the body will be enhanced in certain regions, particularly on the upper body, and reduced in other regions. The average field, E_a, over the entire body is related to the short-circuit current by (Kaune, 1981)

$$E_a = \frac{I_s}{2\pi f \varepsilon_0 A_b} \qquad (9.10)$$

where A_b is the body surface area. Notice the similarity of Eqs. (9.5) and (9.10).

E_a will be exceeded at particular regions of the body, particularly at elongated surfaces on the uppermost body. Table 9.2 expresses field-enhancement factors applying to the average, or to the topmost region of humans and animals. The table also includes normalized values of short-circuit current, applicable to 60-Hz fields. The enhancement factor indicates the value of the incident field as a ratio of the undisturbed field. For example, the field at the top of a person's head will be 18 times the undisturbed field measured in the absence of the person. If the person were to raise one hand above the head, the electric field on the raised hand would be enhanced even further.

Table 9.2 applies to the situation where the source of the E-field is several body dimensions distant from the subjects. If, instead, the source were close to the subject, there would exist a further enhancement of the incident field because of the interaction of the charges on the body surface, and the charges on the conducting object.

Induced Electric Shock Principles[1]

Figure 9.3 illustrates a conceptual model of ELF electric field induction. An AC high-voltage source, such as an electric power transmission line, is capacitively coupled to a charging object through C_s. The object is also capacitively coupled to ground through C_0, and may include leakage resistance, indicated by R_0. The maximum sinusoidal current that would flow in a low-impedance connection to ground is the short-circuit current I_s discussed above.

TABLE 9.2. Electric field enhancement and induced current for several species.

| Species | Field enhancement | | I_s/E |
	Avg	Top	[A/(V/m)]
Human (standing)	2.7	18.0	1.6×10^{-9}
Swine	1.4	6.7	7.0×10^{-10}
Rat (resting)	0.7	3.7	1.2×10^{-11}
Rat (rearing)	1.5	*	2.4×10^{-11}
Horse	1.5	*	2.7×10^{-9}
Cow	1.5	*	2.4×10^{-9}

* Data not available.
Note: "Avg" refers to field enhancement averaged over body. "Top" refers to field enhancement at top of body. I_s/E is normalized short-circuit current at 60 Hz.
Source: Data from Kaune (1981).

---

[1] Portions of this section have been adapted from Reilly and Larkin (1987).

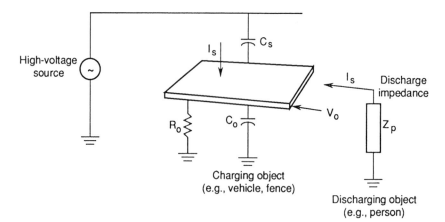

FIGURE 9.3. Equivalent circuit model of electric field induction.

The voltage and charge developed across C_0 can be discharged through a conductive path to ground, such as through a grounded person, represented by the impedance Z_p. The capacitor discharge may be conducted through direct contact with the skin, or, with sufficient voltage, through a spark. The critical field strength required to support an electrical discharge in air depends on the material of the two conductors, and, in accordance with Paschen's law, on the product of electrode separation and atmospheric pressure (Howatson, 1965). The voltage needed to conduct a spark discharge to dry, intact human skin is in the vicinity of 500 V (see Sect. 2.5). After direct contact is made with a charged object, a sinusoidal current I_p will flow into the body. For many cases of practical interests, $I_p \approx I_s$ (Reilly, 1979b), where I_s is the short-circuit current.

Figure 9.4 shows examples of voltage waveforms measured on an object energized by a 60-Hz source and contacted by a subject's finger. These examples were produced using equipment that simulates waveforms observed in field conditions (Reilly and Larkin, 1987). Each record shows a combination of capacitor spark discharges, indicated by abrupt discontinuities in the voltage trace and a period of continuous sinusoidal current while the finger rests on the electrode. Current spikes can be seen in coincidence with the abrupt changes in voltage, but the magnitude of these spikes is not represented accurately, because the sampling rate in these records ($100 \mu s$) was relatively slow. In Fig. 9.4a, V_0 is 1,000 V rms. A capacitor discharge occurs upon contact with the electrode at $t \approx 20$ ms. In Fig. 9.4b, V_0 is 2,000 V rms. Again, there is a single prominent contact discharge, but there are also several discharges just as the finger is withdrawn ($t \approx 57$ ms). Expanded waveforms of capacitive discharges are shown in Figs. 2.29 to 2.33. If the voltage at the moment of contact is high enough (about 500 V), the discharge occurs through a spark; below the critical voltage, the discharge occurs only at the moment of physical contact with the electrode.

Both the capacitor discharge voltage and the sinusoidal contact current are important components of the AC field-induced stimulus. These quantities are related in a well-defined way:

$$\frac{V_0}{I_s} = Z_0 \tag{9.11}$$

where Z_0 is the impedance of the charging object. For many cases of interest, $C_s \ll C_0$ and Eq. (9.11) can be written:

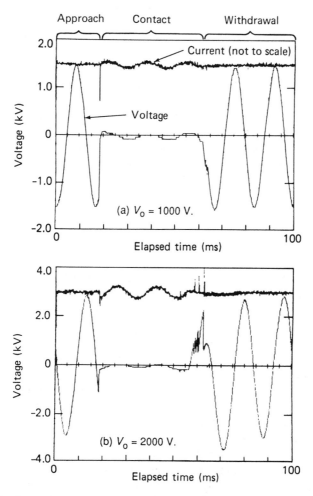

FIGURE 9.4. Example waveforms for AC field-induced stimuli, with tapped finger contact of energized electrode $C_0 = 200\,\mathrm{pF}$.

$$\left[\frac{V_0}{I_s}\right] = \left[\frac{1}{1\big/R_0^2 + \left(2\pi f C_0\right)^2}\right]^{1/2} \tag{9.12}$$

where f is the sinusoidal oscillation frequency of the electric field. Absolute values are indicated because Eq. (9.12) applies to the magnitude of V_0 and I_s without regard to their relative phases.

The condition of greatest induced voltage on the charging object occurs when leakage resistance is large, that is, $2\pi f R_0 C_0 \gg 1$. In this case, Eq. (9.12) reduces to

$$\left|\frac{V_0}{I_s}\right| \cong \frac{1}{2\pi f C_0} \tag{9.13}$$

In many practical cases, however, R_0 may be small enough that the induced voltage is substantially below its theoretical maximum (Reilly, 1979b). In such cases, the ratio V_0/I_s, given in Eq. (9.12), becomes smaller than its theoretical maximum (i.e., the open-circuit voltage becomes less prominent). From a sensory point of view, the sinusoidal current in the contact phase of the stimulus may take on greater significance than the capacitive discharge component.

In this chapter, the effects of leakage resistance are expressed in terms of the factor $k = |V_0/I_s|(2\pi f C_0)$. A functional expression for k can be obtained by multiplying Eq. (9.12) by $2\pi f C_0$:

$$k = \left[1 + \frac{1}{\left(2\pi f R_0 C_0\right)^2}\right]^{-1/2} \tag{9.14}$$

For a well-insulated object, $k = 1$, which is the value attained as $R_0 \to \infty$. The value $k = 1/2$, for example, implies that leakage resistance reduces the induced voltage to one-half that of a well-insulated object.

A very slow approach to an AC-charged source provides conditions in which a sequence of spark discharges can occur. Both the average number of discharges and the average voltage at which they occur will increase as the approach velocity is reduced and as the peak sinusoidal voltage is increased. Figure 9.5 illustrates typical discharge patterns associated with very slow approaches using a device that could move an electrode toward a stationary fingertip with a fixed, controllable velocity. In each example, the charging voltage is 1,900 V rms, and the capacitance is 100 pF. Each discontinuity in the waveform represents a single spark discharge. Notice that the discharges tend to occur at lower voltages in successive cycles of the sinusoidal waveform, but that nearly all of the discharges extinguish at a minimum voltage of about 600 V associated with the spark plateau described in Sect. 2.5. The number of discharges is greatest at the slowest approach. In each

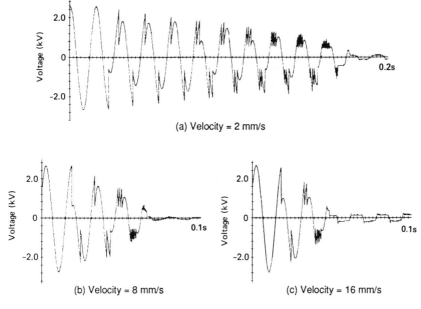

FIGURE 9.5. Examples of stimulus patterns for various electrode approach veloci-ties. Energizing voltage is 1.9 kV rms. Horizontal axis is elapsed time; $C_0 = 100$ pF. (*a*) Velocity = 2 mm/s. (*b*) Velocity = 8 mm/s. (*c*) Velocity = 16 mm/s. (From Reilly and Larkin, 1987.)

trace, discharges begin to occur singly on the individual half-cycles of the 60-Hz waveform. Then, as electrode distance is shortened, discharges occur in clusters of increasing number and decreasing peak amplitude.

To gain some perspective about the factors that influence multiple discharges, we recorded a large number of wave-forms while subjects tapped electrodes of different sizes and energized at different voltages. The number of discharges occurring at finger approach and withdrawal were counted, and each discharge was measured. These counts and measure-ments were also made with an electrode moving at a controlled low velocity toward the skin. The results of these measurements were compared with a computational model, in which a grounded electrode was assumed to ap-proach an energized electrode in random phase with respect to the 60-Hz charging voltage (Reilly et al., 1983).

Table 9.3 illustrates averages of withdrawal and approach data for two subjects, each experiencing 1,200 tapped electrode trials. The important findings for a tapped electrode can be summarized as follows: (1) Below 900 V, multiple discharges are rare. Typically, only one discharge is ob-served, in the approach stage. (2) As the AC voltage is raised above 900 V, the number of discharges increases. The increase is greater in finger withdrawal that in approach. (3) There are generally more withdrawal

TABLE 9.3. Statistical summary of discharge number for AC field stimuli.

| Voltage (V rms) | Mean number of discharges per tap | | | |
| | Ball electrode | | 1-mm electrode | |
	Approach	Withdraw	Approach	Withdraw
600	0.95	0.23	0.95	0.69
900	0.97	0.37	0.94	0.93
1,200	1.05	1.04	1.01	1.38
1,500	1.14	1.39	1.12	1.54
1,800	1.23	1.75	1.15	1.90
2,400	1.57	2.71	1.36	2.98

Note: Electrode tapped with the fingertip, using light tap force.
Source: Reilly and Larkin (1987).

discharges for a small electrode than for a large one. There is little dependence on electrode size in the approach stage. (4) When multiple discharges occur, they cluster within individual quarter-cycles of the sinusoidal waveform.

For very slow approach velocities (less than 20 mm/s), and adequate voltage (>500 V peak), a series of spark discharges may be obtained on individual half-cycles of the alternating induced voltage function. Table 9.4 lists statistical data on the number of discharges for voltages from 1.09 to

TABLE 9.4. Stimulus characteristics for very slow approaches to electrode.

Voltage (kV rms)	Velocity (mm/s)	n_d	n_c
1.09	2	20.8	7.6
	4	11.7	3.9
	8	4.4	2.0
	16	1.7	1.3
1.31	2	27.2	8.8
	4	18.0	5.2
	8	10.1	3.1
	16	4.0	1.9
1.57	2	49.5	16.7
	4	24.8	7.6
	8	16.4	4.7
	16	6.2	2.2
1.89	2	60.6	18.2
	4	33.8	9.8
	8	20.9	5.5
	16	8.3	3.4

Note: n_d = average number of discharges during stimulus pattern; n_c = average number of waveform half-cycles during which discharges were present.
Source: Reilly and Larkin (1987).

1.89 kV rms. In Chapter 7 we show how the sensory system integrates these repeated stimuli.

9.3   Direct Perception of ELF Electric Fields

Mechanisms for Human Detection

Although power frequency E-fields induce internal body currents, such currents can he easily dismissed as having any ability to stimulate excitable tissue. This statement can be easily demonstrated by considering the E-field necessary to excite peripheral nerves. Consider, for example, a person of 1.8 m height standing within a 60-Hz field, and grounded through both feet. If we consider the combined cross-section of the two lower legs (less bone) as 35 cm^2, and the minimum current density threshold as 0.1 mA/cm^2 (see Sect. 7.6), then we calculate from Eq. (9.9) that an undisturbed field strength of 200 kV/m would be required to stimulate peripheral nerves in the ankles. Considering that the undisturbed field is enhanced at body surfaces (18 times, for example, at the head, as noted in Table 9.2), the body would be in a state of severe corona at the stated field level.

Nevertheless, low-frequency E-fields can be perceived directly by humans and animals through other means. The mechanism for human detection is thought to be primarily the result of the vibration of body hair or clothing. Hair has been observed to vibrate at both 60- and 120-Hz rates in a 60-Hz E-field; these rates suggest two different mechanisms of hair vibration (Deno and Zafanella, 1982; Cabanes and Gary, 1981). When the hair is hydrated, it appears to be sufficiently conductive that charges induced on the surface of the skin are free to move along the hair shaft. This phenomenon would result in a mutual repulsion between individual hair follicles, with a maximum force occurring on each one-half cycle of the electric field. The result would be a 120-Hz vibration rate in a 60-Hz field. Dry hair, in contrast, is a very poor conductor, on which environmental ions may be held relatively immobile. The force on dry hair would consequently respond directly to the E-field, resulting in a 60-Hz vibration within a 60-Hz field. Low temperature and humidity decrease human sensitivity to ELF field detection, whereas moisture caused by humidity or perspiration increases sensitivity. At low stimulation levels, subjects typically perceive a 60-Hz E-field as a gentle breeze on exposed body surfaces (Reilly, 1978; Kato et al., 1989). A tingling sensation between body and clothes is also observed. At higher levels, subjects describe a distributed tingling or crawling sensation on the skin.

Tests on human perception of 60-Hz E-fields were carried out at the high-voltage research facility known as "Project UHV." During two months of testing, approximately 122 men and 8 women walked within an electric field produced by an overhead transmission line, stopping at selected points corresponding to a gradually increasing electric field. At each stopping point, participants considered the sensation on their raised hands, head, head hair, or tingling between body and clothes (hands at sides). The participants recorded one of three responses: no feeling, perception, or annoyance. Figure 9.6 (from Reilly, 1978b) summarizes the responses of the men. A few participants reported perception at remarkably low field levels ($\leq 2 \, kV/m$). It would require more carefully controlled tests to be sure that participants actually respond to low field levels rather than confuse other stimuli such as a breeze. For the group of 8 women, the median threshold of stimulation on the raised hands was $17.5 \, kV/m$, versus $6.7 \, kV/m$ for a group of 60 men tested on the same day. The higher threshold for women is thought to be because their body hair is typically shorter than men's. The fact that E-field detection thresholds are indistinguishable in male and female mice (Stern and Laties, 1985) tends to make other gender-related explanations less plausible.

Other tests conducted with 50-Hz fields at an indoor test facility (Cabanes and Gary, 1981) have found higher thresholds for field perception than shown in Fig. 9.6. In tests with 75 subjects (65 male, 10 female), median

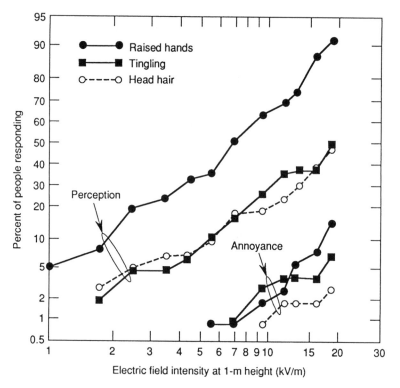

FIGURE 9.6. Response thresholds of men to the direct effects of 60-Hz electric field. Temperature = 50–65 °F, both dry and wet conditions included. (From Reilly, 1978.)

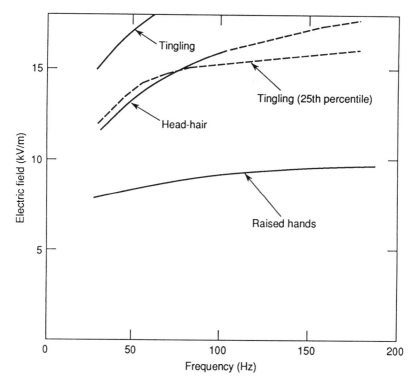

FIGURE 9.7. Median values of threshold of perception of hair sensations of people as a function of the frequency of the electric field. For "tingling" the 25th percentile also is given. (From Deno and Zafanella, 1982. Copyright 1987. Electric Power Research Institute. EL-2500. *Transmission Line Reference Book: 345 kV and Above,* 2nd ed. Reprinted with permission.)

thresholds were found to be 25 kV/m with the hands at the sides, and 12.5 kV/m with one raised hand. The fact that these thresholds are higher than those for outdoor tests (Fig. 9.6) might be explained by a less humid climate in the indoor facility, although relevant climatological data were not published for the indoor tests.

Figure 9.7 illustrates the connection between the electric field frequency and perception levels (from Deno and Zaffanella, 1982). Perception thresholds were least at the lowest tested frequency (30 Hz), with a gradual loss of sensitivity as the frequency was increased. For the data represented in Fig. 9.7, the rate of hair vibration was not reported; as noted earlier, it could be either at the frequency of the applied field, or twice the frequency, depending on the degree of hair hydration.

Cutaneous vibration sensitivity has a frequency response counter to that indicated for E-field sensitivity. Hair receptors are examples of *rapidly adapting* mechanoreceptors that are responsible for detection of vibration on hairy skin. These receptors have a U-shaped threshold curve when plotted against the frequency of excitation, as shown in Fig. 3.18. A complete model for electric field detection may have to include the mechanical compliance and electric impedance properties of the hair follicles.

Human perception is greatly enhanced when a strong DC field is combined with an AC field, as shown in tests by Clairmont and colleagues (1989). Test subjects recorded reactions within electric field environments produced by conductors energized at various AC or DC potentials. Individuals recorded sensation levels at measurement locations where the AC field, DC field, and ion current densities were monitored. Sensation levels were reported on a six-point scale defined as follows: (0) not perceptible; (1) just perceptible; (2) definitely perceptible, but not annoying; (3) slightly annoying; (4) very annoying; (5) intolerable.

Figure 9.8, illustrating head hair perception, shows that alternating and static fields have a synergistic effect—the combined fields can produce a much stronger reaction than either field taken separately. Such synergism might be explained by a mechanism where ions are produced by the DC conductors and collected on body surfaces; the alternating component of the fields would provide an oscillatory force for hair vibration. Although

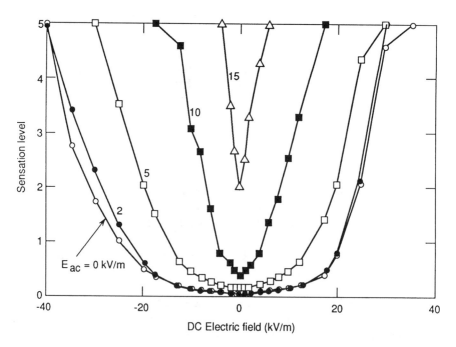

FIGURE 9.8. Reaction levels for exposure to mixed DC and 60-Hz AC electric fields; stimulation of head hair while standing in field. Indicated field levels apply to undisturbed field. Horizontal axis applies to DC field strength; curve parameters apply to AC field strength. (From Clairmont et al., ©1989, IEEE.)

such an explanation appears plausible, note that even a pure static field is perceptible. Detection of a static field may also be the result of hair vibration, which has been reportedly observed beneath a DC high-voltage transmission line. Such hair movement is not clearly understood but may be associated with corona on the tips of the hair shafts. Laboratory studies by Kato and colleagues (1986) failed to detect hair movement within a static field. Differences between laboratory and environmental observations might be associated with the density of ambient air ions, and relative humidity.

In other laboratory experiments, Kato and colleagues (1989) exposed the hairy arm and the palm of the human hand to 50-Hz electric fields. The median perception threshold was 50 kV/m on days that the relative humidity ranged from 65% to 78%. On dry days with humidity ranging from 22% to 33%, nearly all subjects failed to perceive fields up to 115 kV/m, although one exceptional subject detected a field at 115 kV/m. In all cases, perception was associated with hair movement. When the hairless hand was exposed, the field could not be detected up to the maximum level used in the experiments (115 kV/m).

The field levels cited in Kato's experiments may seem large in comparison with those indicated in Figs. 9.6 to 9.8, but may be accounted for by differences in the definition of exposure levels. In Kato's laboratory experiments, the reported fields are those on the surface of the skin; in the environmental measurements (Figs. 9.6 to 9.8), the reported levels apply to the *undisturbed* field (i.e., the rms value of the field in the absence of a person). The difference between these two definitions is significant, as suggested by Table 9.2.

Various studies have examined detection of ELF fields by laboratory mammals (Stern et al., 1983; Stern and Laties, 1985; Sagan et al., 1987; Weigel et al., 1987). Weigel's experiments with cats provide significant insight into mechanisms of transduction in E-field perception. In these experiments, both normal and hairless cats' paws were exposed to strong 60-Hz E-fields. Thresholds as low as 160 kV/m were noted, where the field value applies to a measurement directly on the surface of the exposed body part. That threshold value is similar to the perception threshold noted in Fig. 9.5 for human detection, if one accounts for the field enhancement on local body parts.

Weigel and colleagues isolated receptor types and their receptive fields in the cat's hind limb by monitoring action potential activity from the spinal dorsal roots that supply those regions. They concluded that the most sensitive mode of transduction was via rapidly adapting receptors. It was thought that hair vibration could activate hair follicle receptors directly, or could vibrate the skin and activate other types of nearby receptors, such as pacinian corpuscles. In some cases, the later mode appeared to be the more sensitive mechanism for transduction. But even in the hairless paw, electric field stimulation was still noted, although at somewhat increased threshold levels. Stimulation in the hairless paw was thought to be from vibrational displacement of the skin due to dielectric forces across the corneum, similar to the dielectric mechanism discussed in Sect. 7.2.

One should exercise great caution in comparing electric field detection thresholds of humans and laboratory mammals because of differences in body hair, as well as body shape, size, and orientation. Comparisons with nonmammalian experimental species require even more caution, because of the possibility of differences in detection mechanisms. Certain highly sensitive fish, for example, are able to detect E-fields within the marine environment as low as 5 nV/cm (Kalmijn, 1990). The existence of a unique receptor and signal-processing mechanisms are said to be responsible for this remarkable capability.

9.4   Human Reactions to AC Electric Field-Induced Shock[2]

Stimuli produced through alternating E-field induction are more complex than the single capacitor discharges discussed in Sect. 7.3. We have seen in Sect. 9.2 that AC stimuli may contain one or more capacitor discharges when physical contact with the induction object is made and broken, as well as sinusoidal current during the period of maintained physical contact.

The individual discharges of the AC field-induced stimulus are identical to the capacitor discharges discussed in Chapter 7. Thus, the factors that affect sensitivity to individual discharges are also important with AC stimuli. With AC field-induced stimuli, however, additional factors are important, including multiple discharges and sinusoidal contact current. The strength of the individual discharge, relative to the contact current, will depend on both the capacitance and leakage resistance of the AC field-charged object, as suggested by Eq. (9.12). It is therefore important to evaluate AC field stimulus sensitivity throughout a range of capacitance and leakage-resistance values.

Perception Thresholds

The sinusoidal contact component of the AC stimulus has a peak amplitude that is considerably below the peak of the capacitive discharge current. Nevertheless, the contact component can be perceptually potent—experimentally, we find that 60-Hz sinusoidal currents can be detected with large-area fingertip contact at an average current of about 0.25 mA (Reilly et al., 1983). This value is consistent with 60-Hz touched-contact thresholds noted in other studies (see Chapter 7).

---

[2] Parts of this section have been adapted from Reilly and Larkin (1987).

To further compare sensitivity to stimuli induced by alternating E-fields with single-discharge stimuli, we tested the two kinds of stimuli randomly interleaved during the same testing session. Capacitance was also varied randomly, among values of 100, 800, and 6,400 pF. At each capacitance value, we determined the individual's threshold ratio, defined as the voltage threshold for a single capacitor discharge divided by threshold rms voltage for the AC field stimulus. For the six subjects tested, there were wide differences (about 2:1) in absolute levels of sensitivity. Nevertheless, the threshold ratios were highly similar at each capacitance. The implication of this result is that, by knowing an individual's sensitivity to one type of waveform, it is possible to determine his or her sensitivity to another waveform using a simple multiplicative constant. The constant will, of course, depend on the details of the two waveforms being compared.

Figure 9.9 plots the perception threshold voltage for contact with an object charged by a 60-Hz E-field; thresholds decrease with increasing

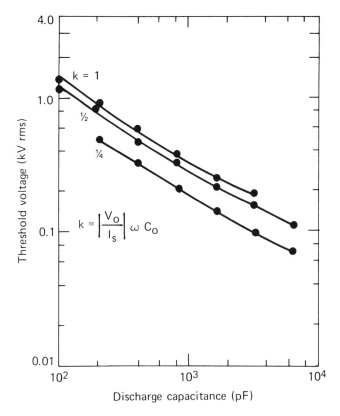

FIGURE 9.9. Perception threshold for 60-Hz AC field-induced stimuli. Tapped electrode. Curves for $k = \frac{1}{2}$ and $\frac{1}{4}$ indicate effect of leakage resistance. (From Reilly and Larkin, 1987.)

capacitance, and with decreasing leakage resistance. [See Eq. (9.14) for a definition of the leakage-resistance parameter k.] The resistance effect stems from the greater sensory contribution of the sinusoidal contact current component as leakage resistance is reduced. The fact that thresholds are lower with reduced leakage resistance does not mean that reduced resistance makes the shock exposure worse. If the induction E-field is held constant, decreased leakage resistance reduces the induced voltage even more than it does the threshold. For example, comparing $k = \frac{1}{4}$ with $k = 1$, the induced voltage is reduced by the multiple 0.25, but the perception threshold at $C = 3{,}200\,\text{pF}$ is reduced by the multiple 0.52 (see Fig. 9.9).

Superthreshold Effects

Suprathreshold reactions were tested for alternating E-field induction stimuli in a manner similar to that described in Chapter 7 for single capacitive discharges. In the magnitude-estimation procedure, perceived magnitude for 60-Hz E-field stimuli on the fingertip was found to grow at the 2.7 power of induced stimulus voltage at 6,400 pF, and at the 2.2 and 2.8 powers at 800 and 200 pF, respectively. These exponents are not very different from those observed for single-discharge stimuli (Chapter 7).

Table 9.5 lists, for six subjects, the mean affective response category thresholds as multipliers of individual perception thresholds. The multiples are for 60-Hz field-induced stimulation of the fingertip in active electrode touching. Compared with the corresponding multiples for single discharges (Table 7.2), 60-Hz stimuli reach the "unpleasant" category at approximately the same multiple; however, "pain" and "tolerance" categories are reached with generally lower multiples.

Multiple Discharges

When a person grazes a metallic object that is energized by an alternating E-field, there is a possibility for multiple discrete discharges. As demon-

TABLE 9.5. Multiples for superthreshold categories with 60-Hz E-field-induced stimuli.

Response category	Capacitance		
	200 pF	800 pF	6,400 pF
Unpleasant	2.1	2.2	1.8
Pain	2.9	2.9	2.4
Tolerance	4.3	6.2	3.5

Note: Tabulated values indicate multiples relative to perception thresholds for given response categories. Data apply to fingertip stimulation.
Source: Reilly and Larkin, 1987.

strated in Sect. 9.2, even a seemingly slow approach produces stimulus patterns not unlike a deliberate direct touch with the fingertip. It is very difficult to generate more than two to four discharges with a deliberate grazing motion of the limb. However, a continuous stream of discharges can be easily produced by pressing a shirt sleeve against an energized electrode. In our tests, continuous sparking through a 0.18-mm-thick fabric was easily produced at 1.3 kV and above. Multiple discharges occur because the fabric prevents electrical contact with the skin, which would preclude further capacitive discharges. The duration of such discharges depends largely on the reaction delay for the person to pull the arm away. A variety of studies on human reaction time show that about 0.25 s is needed to move the hand in response to a stimulus (Woodworth and Schlosberg, 1954). This period corresponds to 30 half-cycles of a 60-Hz stimulus.

To produce multiple discharges reliably in the laboratory, a device was used to move an electrode toward the skin with controlled motion. Figure 9.5 gives examples of the discharge patterns. Sensory magnitude estimates for repeated discharges were obtained in a manner similar to that described for single-discharge stimuli (Chapter 7). On each test trial, electrode approach velocity was randomly set at either 2, 4, 8, or 16 mm/s.

Figure 7.11 illustrates perceived sensation magnitude (geometric means for three subjects), plotted against n_C, the number of 60-Hz half-cycles during which capacitive discharges were produced. The magnitude estimates were related to the effective categories "neutral" and "pain" in a separate procedure. Figure 7.11 suggests that a perceptible but neutral stimulus can become highly unpleasant if presented as a repetitive train. A rationale for representing n_C as the independent variable is that there is a tradeoff between the amplitude and number of pulses in each half-cycle packet (Fig. 9.5), and that it approximately characterizes the overall stimulus. The values of n_c are the means for each velocity that was used in the procedure. The straight lines in this format indicate power functions with an exponent of 0.85, indicating nearly a one-to-one correspondence between n_C and the mean perceived magnitude.

The dynamic range from the threshold of perception to pain is rather narrow (Table 9.5). The limited dynamic range is especially striking when one considers that a stimulus of low sensory potency can become quite unpleasant when presented as a repetitive train. Consequently, conditions giving rise to perceptible alternating field-induced shocks may, under some circumstances, also present conditions for creating unpleasant effects.

Extrapolation to Other Frequencies

If field strength is held constant (resulting in constant V_0), the induced current will be proportional to frequency [refer to Eq. (9.13)]. At the 50-Hz power frequency common in Europe, the contact current would be reduced by a factor of 50/60 (8.3%) relative to that at 60 Hz. This would result in an upward shift of 50-Hz thresholds, but the shift should not exceed 8.3% of the values plotted in Fig. 9.9.

If the induction frequency is increased beyond 60 Hz, the sinusoidal contact current takes on increasingly greater sensory importance in comparison with the capacitive discharge component, and, for sufficiently high frequencies, will become the dominant sensory factor. Above 1 kHz, perception thresholds for sinusoidal currents rise approximately in proportion to frequency (Sects. 4.6 and 7.5), reaching a plateau at about 100 kHz where heating effects predominate (Fig. 7.12). Since the current induced by the alternating electric field also rises in proportion to frequency, the E-field level for perception of induced contact current should be nearly constant between about 1 and 100 kHz.

The opportunity for multiple discharges can increase with the induction frequency. When individual capacitive discharges can be perceived, multiple discharges can greatly increase sensory potency, as noted in Fig. 7.11. Sensory effects with multiple discharges may be intensified at frequencies somewhat above 60 Hz because, sensory magnitude is encoded in part through the action potential repetition frequency (see Chapter 3), but this effect would be expected to saturate at an AP rate on the order of 200/s.

Skin Erosion from Low-Energy Discharges

The energy in a spark discharge to dry skin can be dissipated in a very small volume because of the small effective diameter of the spark contact and the fact that the resistivity of the dry corneum is high. The volume of dissipation is especially small where the corneum is thin. Because of the high concentration of dissipated energy, relatively small discharge energies are capable of causing skin erosion. The effects of single or low frequency discharges to the corneum are discussed in Sect. 2.6, where it is shown that discharge energy typical of carpet sparks is capable of minor skin erosion when repeatedly presented to the same spot on the dry skin of the leg. Radio frequency discharges are discussed in Sect. 11.4, where it is shown that relatively low voltage contacts ($\approx$200 V peak) can produce minor erosion of corneal tissue in a small area. These contacts may be associated with a stinging sensation. Such sensations may be due to the stimulation of thermal receptors due to a localized heat rise. Another possible explanation involves nonlinear impedance characteristics of the skin, which produces low frequency modulation currents capable of exciting sensory neurons electrically (see Sect. 4.6).

9.5 Time-Varying Magnetic Field Induction

Magnetic fields, produced by all current-carrying devices, induce electric currents in conductive material, including the human body. Under most conditions of environmental exposure, the induced currents are far too small to stimulate excitable tissue. Nevertheless, fields capable of stimulation may be associated with certain industrial or research facilities, or with medical diagnostic devices.

In this chapter, magnetic fields will be characterized in terms of *magnetic flux density*, B, having units of tesla (T). That quantity is related to *magnetic field strength*, H, by $B = \mu H$, where μ is the permeability constant of the medium, and H is the field strength in units of amperes per meter (A/m). The permeability of air (μ_0) in this system of units is $4\pi \times 10^{-7}$. The permeability of biological tissue differs little from that of air (see Chapter 2). Consequently, the value of B within the biological medium may ordinarily be taken as that measured in air in the absence of the biological specimen. The system of units used above is the *International System* (SI). Flux density, B, is sometimes reported in the cgs system in units of *gauss*. To convert between the two units, note that 1 gauss $= 10^{-4}$ tesla.

To calculate the induction in a volume conductor, such as the human body, a simple approach treats the volume as if it were made up of concentric rings normal to the direction of the field, as illustrated in Fig. 9.10a. In this simple picture, the field is assumed to be oriented along the long axis of the body. The magnetic field induces an E-field within the body, having roughly circular paths, as indicated in the figure. The induced E-field, in turn, produces circulating *eddy currents*, which follow the path of the electric field in a medium of homogeneous conductivity.

In accordance with Faraday's law, the internally induced electric field **E** is related to the time rate of change of flux density **B** by

$$\oint \mathbf{E} \cdot \mathbf{dl} = -\frac{\partial}{\partial t} \iint \mathbf{B} \cdot \mathbf{ds} \tag{9.15}$$

The first integral is taken over a closed path, and **ds** is the element of area normal to the direction of **B**. Each bold-face term is a vector quantity; elsewhere in this chapter the standard fonts are used to represent scalar quantities. If B is uniform over the region inside a closed path of radius r, the induced electric field strength calculated from Eq. (9.15) is

$$E = -\frac{r}{2}\frac{dB}{dt} \tag{9.16}$$

where the direction of the induced E-field is along the circumference of the circle. In some applications the magnetic field varies sinusoidally as $B = B_0 \sin 2\pi ft$, and Eq. (9.16) becomes

$$E = \left(r\pi fB_0\right)\cos 2\pi ft \tag{9.17}$$

where the term in parentheses is the peak induced E-field during the sinusoidal cycle, and B_0 is the peak magnetic field. Frequently, calculations for sinusoidal magnetic fields express only the magnitude term in Eq. (9.17), leaving off the cosine term. For induction in a concentric ring model, such as that shown in Fig. 9.10a, Eq. (9.16) suggests that the outermost rings would have the greatest E-field strengths. According to this simple model,

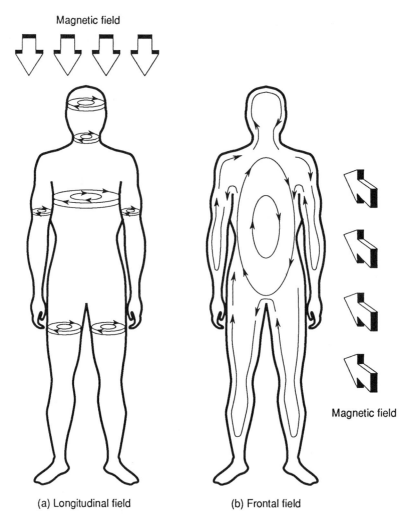

(a) Longitudinal field   (b) Frontal field

FIGURE 9.10. Distribution of internally induced electric fields from whole-body exposure to time-varying magnetic field. Magnetic field direction is parallel to long axis of body in part (a), and perpendicular to front of body in (b). Direction of internal E-field and induced current reverses every half-cycle of alternating magnetic field.

the maximum E-field for whole-body exposure would be computed from Eq. (9.16) with r being the maximum circle radius that can be drawn on the body in a plane perpendicular to B.

When the magnetic field is perpendicular to the long axis of the body, the pattern of induced fields and eddy currents is somewhat more complex, as suggested in Fig. 9.10b. Treating the human body as a prolate spheroid as in Fig. 9.11, a long-wavelength solution was published by Durney and colleagues (1975), and expressed in applied form by Spiegel (1976). The long-wavelength assumption is that the field wavelength is much greater than the dimensions of the biological subject. The complete solution for the induced E-field because of electromagnetic exposure involves one component as a result of the incident electric field, and another as a result of the incident magnetic field. The internally induced electric field for three orientations of a sinusoidal magnetic field is given by

$$E_1 = \frac{-j2\pi f B_x \left(a^2 y \mathbf{a}_z - b^2 z \mathbf{a}_y \right)}{\left(a^2 + b^2 \right)} \tag{9.18}$$

$$E_2 = \frac{j2\pi f B_y \left(a^2 x \mathbf{a}_z - b^2 z \mathbf{a}_x \right)}{\left(a^2 + b^2 \right)} \tag{9.19}$$

$$E_3 = j\pi f B_z \left(y \mathbf{a}_x - x \mathbf{a}_y \right) \tag{9.20}$$

where E_1, E_2, and E_3 are the magnitudes of the sinusoidal electric fields that are induced by magnetic field exposure along the x, y, and z axes, respectively; B_x, B_y, and B_z are orthogonal magnetic field magnitudes; a and b are semimajor and semiminor axes of the spheroid; $\mathbf{a}_x$, $\mathbf{a}_y$, and $\mathbf{a}_z$ are unit vectors along the x, y, and z axes; f is the frequency of the field; and j is the phasor

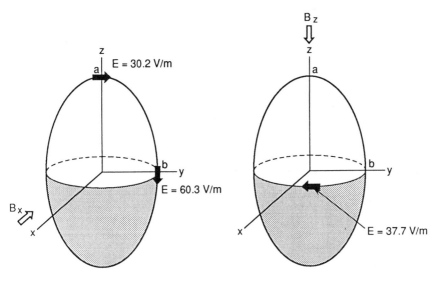

(a) field perpendicular to long axis   (b) Field parallel to long axis

FIGURE 9.11. Prolate spheroid model of uniform magnetic field induction. Bold arrows indicate direction of induced electric field at outer boundaries axis $a = 0.4\,\text{m}$ and semiminor axis $b = 0.2\,\text{m}$.

operator (indicating $90°$ phase shift with respect to the sinusoidal phase of the B field). Equations (9.18) to (9.20) define spatial vector components of the induced electric field in terms of the magnitude of the sinusoidal variation [the term in parentheses in Eq. (9.17)]. The vector magnitude in Eq. (9.20) is identical to the amplitude term of Eq. (9.17) because $r = (x^2 + y^2)^{1/2}$ along a circular path.

A general expression for pulsed fields of arbitrary waveshape can be derived from Eqs. (9.18) to (9.20) using arguments based on Fourier synthesis. Designating A_n as the magnitude of the nth Fourier component of flux density, its derivative is $j2\pi f_n A_n$. It follows that the electric field induced in an elliptical cross-sectional area is

$$ E = -\frac{dB_W}{dt}\left[\frac{a^2 u \mathbf{a}_v - b^2 v \mathbf{a}_u}{a^2 + b^2}\right] \tag{9.21} $$

where $\mathbf{a}_u$ and $\mathbf{a}_v$ are unit vectors along the minor and major axes respectively, (u, v) is the location within the area, and B_w is the magnetic flux density in a direction perpendicular to the elliptical cross section. For a circular area, $a = b = r$, and Eq. (9.21) is equivalent to Eq. (9.16) on the perimeter of the circle.

Consider, for example, a spheroidal representation of the torso of a large person with $a = 0.4$ m and $b = 0.2$ m, and a uniform 60-Hz magnetic field of strength 1 T. If the field is oriented along the z axis, then Eq. (9.20) specifies that the induced electric field at the equator of the ellipsoid is 37.7 V/m. For orientation along the x axis, Eq. (9.18) specifies that the field is 30.2 V/m at the upper extreme of the spheroid, and 60.3 V/m at the lateral extreme. The induced E-field components for this example are illustrated by bold arrows in Fig. 9.11.

The internally induced electric fields result in circulating eddy currents. The current density, J, is related to the induced E-field by the conductivity of the medium, σ, in accordance with $J = E\sigma$. For induction in living subjects, calculation of the current distribution is complicated by the widely differing conductivities of various body components (muscle, bone, blood vessels, fat, etc.). The model used in this chapter treats the body as if it were of homogeneous conductivity. The force responsible for stimulation is taken to be the induced E-field within the biological medium (Chapter 4), and the actual current density is of secondary interest.

In some applications, a knowledge of the current density is essential. One such application concerns the distribution of internal heating caused by E-M field exposure. Here, the distribution of current density must be known in order to determine the patterns of power deposition within the body. Gandhi and colleagues (1984) evaluated the distribution of power deposition resulting from magnetic field exposure by modeling the body as a series of rectangular current loops, as illustrated in Fig. 9.12. Each loop incorporates the complex impedance appropriate to the tissue in a particular loca-

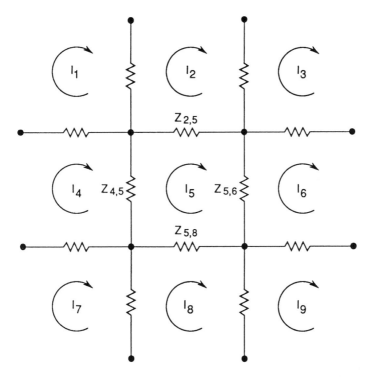

FIGURE 9.12. Impedance mesh model for calculating magnetically induced currents in nonhomogeneous medium. For simplicity of notation, only nine magnetic loops are indicated. Induced electromotive force in each loop is calculated according to Eq. (9.15). (Adapted from Gandhi et al., 1984, © 1984 IEEE.)

tion of the body. The induced electric field in each loop can be calculated in accordance with Eq. (9.15); the current value is thereby inferred using the complex conductivity (including anisotropic properties) of the elements in the loop. By simultaneously solving the loop equations for the entire matrix, one can determine the overall current distribution.

Figure 9.13 illustrates the power deposition, calculated with the above method, through a cross section containing the liver cut in a plane perpendicular to the long axis of the body. The magnetic field is assumed to be perpendicular to the cross section (as in Fig. 9.10a), and to have a sinusoidal variation at either 27.1 or 13.6 MHz. The figure indicates that power deposition generally rises when the measurement location moves from the center of the body to its outer perimeter. This overall behavior is a consequence of the radial dependency of the induced field, indicated in Eq. (9.16). For a perfectly homogeneous medium, the power absorption would vary as r^2 (since power is proportional to the square of current). In Fig. 9.13, deviations from an r^2 dependency arise from variations in conductivity. The

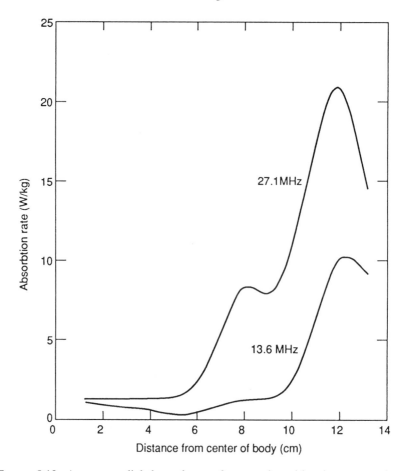

FIGURE 9.13. Average radial dependence of power deposition in cross-section of human liver; cross-sectional plane perpendicular to long axis of body; distance measured from center of body. Magnetic field perpendicular to cross section; magnetic flux density is 56μT; frequency of field indicated on curves. (From Gandhi et al., 1984, © 1984 IEEE.)

sudden drop near the edge of the body $(r = 12\,\mathrm{cm})$ results from a sudden decrease in conductivity at the outer layers comprising fat and skin.

9.6   Principles of Excitation by Time-Varying Magnetic Fields

This section emphasizes excitatory effects of magnetic fields in which the exposure is over a large portion of the body, such as the entire torso or whole head. Magnetic stimulation may also be provided over localized areas of the body, using pulsed fields from small magnetic coils. Applications of local magnetic stimulation are treated in Sect. 9.9.

Pulsed Fields: Large Area Exposure

The subject of magnetic field excitation has been extensively studied with application to patient exposure in magnetic resonance imaging (MRI). In these imaging systems, three types of fields are applied at levels of intensity much greater than what typically is found in most other environments. MRI fields consist of an intense static field, a radio frequency field, and a switched gradient field. The biological effects and mechanisms of interaction associated with these fields are very different from one another. Biological mechanisms associated with MRI static field include magnetohydrodynamic effects, which are treated in Sect. 9.10. Radio frequency exposure concerns in MRI have largely centered around thermal effects, which are discussed in Sects. 9.5 and 11.5. Switched gradient field effects are mainly associated with excitation of peripheral nerves, which is the main focus of this Section. A treatment of various hazards in MRI is given by Shellock and Kanal (1994). We shall also treat other applications of large area exposure to magnetic fields, not necessarily confined to MRI.

Pulsed magnetic waveforms induce *in-situ* E-fields that are proportional to the time rate of change of flux density, dB/dt. Figure 9.14 shows examples of pulsed gradient field magnetic waveforms used in magnetic resonance imaging (MRI) consisting of gated trapezoidal and sinusoidal functions and associated dB/dt waveforms. Typical MRI gradient field pulse duration's may be tenths of milliseconds to several milliseconds long, and may include sequences of pulses separated by intervals from tenths of milliseconds to several milliseconds long.

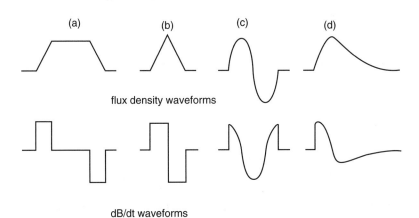

FIGURE 9.14. Example of flux density, and dB/dt waveforms.

To determine magnetic thresholds for nerve and heart excitation, we must consider several factors, including the rheobase thresholds, the appropriate strength–duration (S–D) time constants, and the anatomical arrangement of the excited tissue.

Minimum Excitation Thresholds

As seen in Chapter 4, the E-field aligned with the long axis of a nerve or muscle fiber governs excitation. Since the induced E-field is proportional to dB/dt, the spatial and temporal properties of dB/dt will determine the excitation potential. With magnetic induction, dB/dt is necessarily biphasic and charge-balanced (i.e., the area under the positive and negative-going phases are equal). However, the dB/dt waveform of a pulsed B-field can be treated as monophasic for the purposes of nerve excitation if the biphasic reversal of the induced field is gradual or delayed, as seen in Fig. 9.14 (a) and (d). Recall from Chapters 4 and 6 that nerve and heart S–D relationships dictate that E-field thresholds are lowest for stimuli that are long in comparison to the S–D time constant, and that thresholds for monophasic pulses are typically lower than for biphasic waveforms of equal duration. Consequently, minimum magnetic excitation thresholds can be analyzed using an assumption of a monophasic square-wave dB/dt pulse.

As noted in Sect. 4.5, a nerve fiber may be excited at one of three loci: a nerve terminus, a sharp bend, or a location where the spatial gradient of the E-field is sufficient (Fig. 4.15). With magnetic exposure over a large portion of the anatomy, the spatial derivative of the induced E-field is typically small, and stimulation at a nerve terminus will determine the lowest threshold. Based on a myelinated nerve fiber situated in a uniform field oriented along the long axis of the fiber, E-field thresholds are calculated as 6.2, 12.3, and 24.6 V/m for fiber diameters of 20, 10, and 5 μm respectively (Table 4.2). The threshold applicable to a 20-μm fiber will be used here for a conservative calculation.

Recall from Chapter 6 that the cardiac excitation threshold is a minimum during the unexcited (diastolic) interval of the cardiac cycle; the fibrillation threshold is a minimum during the vulnerable (partially refractory) period. The relatively lower excitation threshold will be used here to evaluate magnetic stimulation of the heart. For prolonged exposure, the cardiac excitation threshold at the one- and 50-percentile ranks are estimated at 5 and 10 V/m respectively with large-area electrical stimulation (Sect. 6.10). Note that the presumed one-percentile cardiac excitation threshold is very close to the median nerve threshold. To facilitate the calculations in this section, a reference threshold of 6.2 V/m will be used for both the nerve and the heart.

Strength-Duration Time Constants

The S–D time constants of cardiac and nerve tissue differ widely. As noted in Chapters 4 and 7, a theoretical model for myelinated nerve with uniform field excitation indicates $\tau_e = 0.12\,\mathrm{ms}$, a value consistent with in-vivo nerve stimulation in animal experiments. As we shall see presently, that theoretical value is close to experimental time constants derived with large area magnetic stimulation. In contrast, the experimental average of τ_e for cardiac excitation with large-area exposure is about $3\,\mathrm{ms}$—a factor more that 10 greater than that for nerve stimulation. Large area exposure is assumed for most cases of magnetic induction because of the nonfocal nature of the induced electric field.

The S–D time constant determines the thresholds for stimuli that are brief in comparison to τ_e. In the short pulse regime, minimum thresholds are defined by the peak value of B rather than dB/dt. As a result of S–D principles defined in Sect. 4.2, it follows that $B_o = \dot{B}_o\,\tau_e$, where $\dot{B}_o$ is the minimum dB/dt threshold for $t \ll \tau_e$, and B_o is the rheobase B threshold for $t \gg \tau_e$.

Anatomical Considerations

Body geometry is illustrated in Fig. 9.15 for a large man. Body dimensions in applied situations may differ from this example, and calculations would vary accordingly. Furthermore, the location of the heart within the torso will vary as the body position is changed from prone to erect, and also will move during the cardiac pumping cycle. Calculated E-fields have been carried out by representing body cross sections as ellipses, as indicated in Fig. 9.15. Numbered points indicate positions where the E-field (shown in V/m) has been calculated according to Eq. (9.21) with $dB/dt = 100\,\mathrm{T/s}$.

Calculated E-fields generally increase as the location is moved toward the perimeter of the body, and are especially great along the minor axis of the ellipse. Points of maximum E-field on the perimeter of the body are points 14, 9, and 3 in the three cross sections; the maximum E-field on the heart occurs at points 10, 6, and 1.

An underlying assumption in the above is that the incident magnetic field is constant in both magnitude and phase over the torso cross-section. A somewhat more conservative calculation can be made by assuming the entire body is uniformly exposed to the B field by using an equivalent body ellipse equal to the height of the person. A calculation for sagittal exposure was made, using semimajor and semiminor axes of 0.9 and 0.17 m respectively. In this case, the calculated E-field was somewhat elevated relative to the case where only the torso was exposed. The increase was 11% at point (10), and 14% at point (14).

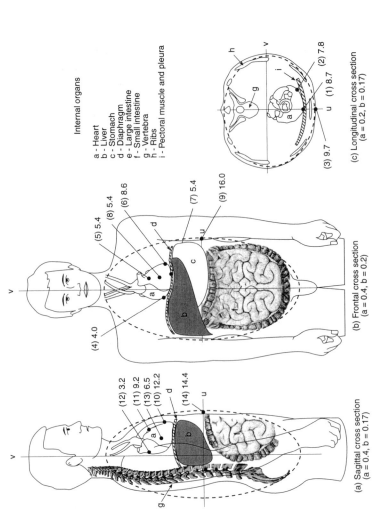

Internal organs

a - Heart
b - Liver
c - Stomach
d - Diaphragm
e - Large intestine
f - Small intestine
g - Vertebra
h - Ribs
i - Pectoral muscle and pleura

(3) 9.7   (1) 8.7   (2) 7.8

(c) Longitudinal cross section
(a = 0.2, b = 0.17)

(5) 5.4   (8) 5.4   (6) 8.6   (7) 5.4   (9) 16.0

(4) 4.0

(b) Frontal cross section
(a = 0.4, b = 0.2)

(12) 3.2   (11) 9.2   (13) 6.5   (10) 12.2   (14) 14.4

(a) Sagittal cross section
(a = 0.4, b = 0.17)

FIGURE 9.15. Example of position of the heart within three cross sections of the body; a and b are semimajor and minor axes of equivalent cross-section ellipsoids. Numbered locations are identified with calculated electric field intensity (V/m) for magnetic exposure within the ellipse of $dB/dt = 100\,\mathrm{T/s}$. Direction of field is perpendicular to the cross-section. View of internal organs shows conductive paths through the heart.

Another implicit assumption is that the internal conductivity of the body is homogeneous. In reality, the body is composed of various organs having diverse conductivities, as indicated in Fig. 9.15, and these variations will distort the internal E-field with respect to a homogeneous structure. For instance, in the sagittal cross section, the E-field will be enhanced between the skin and the spinal column as induced eddy currents are crowded around the low-conductance vertebrae and the curvature of the torso in the lower lumbar region. Indeed, Schaefer and colleagues (1994) report that sensory perception from magnetic stimulation is enhanced in regions where boney projections are just below the surface of the body. It should be clear from Fig. 9.15 that the heart is part of a large electrically conductive circuit that includes the blood vessels, diaphragm, liver, and intestines, and should not be treated as an isolated conducting organ suspended in non-conductive air (lungs) as some have opined. With longitudinal magnetic exposure (Fig. 9.15c), the heart is part of a circumferential electrical circuit that includes the pericardium and muscle lining the thorax.

Calculated Nerve and Cardiac Excitation Thresholds

In accordance with the preceding, a reference value of 6.2 V/m will be assumed as the median rheobase threshold for nerve excitation, and also as the one-percentile rheobase threshold for cardiac excitation. Table 9.6 lists dB/dt needed to induce 6.2 V/m at selected points on the body model of Fig. 9.15 according to Eq. (9.21). The table also lists B_o, the minimum threshold

TABLE 9.6. Magnetic field parameters necessary to induce threshold E-field at selected body loci (*calculated by elliptical cross-section model*).

B-field orientation	Ellipse b,a (cm, cm)	Body locus	Minimum threshold dB/dt (T/s)	B (mT)
Longitudinal	17, 20	(1) heart	71.3	213.9
		(3) nerve	62.6	7.51
Frontal (torso)	20, 40	(6) heart	68.1	204.3
		(9) nerve	38.8	4.66
Sagittal (torso)	17, 40	(10) heart	48.8	146.4
		(14) nerve	43.1	5.17
Sagittal (whole body)	17, 90	(10) heart	44.0	132.0
		(14) nerve	37.8	4.54

Calculations are for human geometry of Fig. 9.19. Minimum dB/dt thresholds based on $E = 6.2$ V/m for long-duration pulses. Minimum B thresholds determined by $B_o = \dot{B}_o \, \tau_e$ for short-duration pulses, with $\tau_e = 0.12$ ms (nerve), and 3 ms (heart); $\dot{B}_o$ is the minimum dB/dt threshold.

for short pulses, using $B_o = \dot{B}_o \tau_e$, where $\dot{B}_o$ is the rheobase dB/dt threshold; $\tau_e = 0.12$ and $3.0\,\mathrm{ms}$ for nerve and heart respectively. In the case of sagittal exposure, an additional conservative calculation is shown for exposure to an elliptical representation of the entire body.

Figure 9.16 illustrates dB/dt and B thresholds as strength-duration curves for a magnetic field normal to the sagittal cross section of the human body; the S–D time constant τ_e is shown as a parametric variable. The curves apply

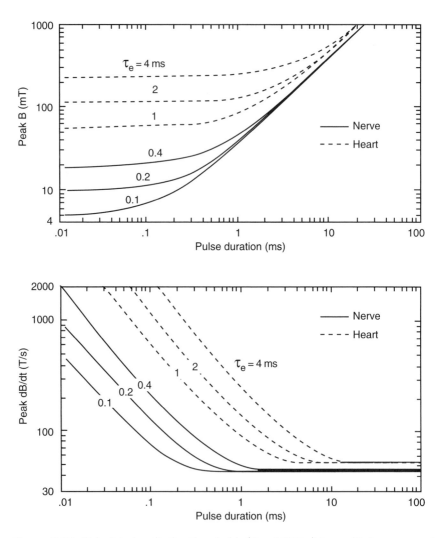

FIGURE 9.16. Calculated excitation thresholds ($E = 6.2\,\mathrm{V/m}$) for sagittal exposure of torso to pulsed magnetic field. Curve parameter τ_e applies to strength–duration time constant of excitable tissue. Upper panel: peak flux density; lower panel: peak dB/dt. (From Reilly, 1992.)

to the most sensitive loci on the heart or peripheral nerve, namely points 10 and 14 on Fig. 9.15. At long durations, dB/dt thresholds (lower panel) converge to a minimum value (rheobase) independent of τ_e. At short durations, B thresholds (upper panel) converge to a minimum value that is directly proportional to τ_e. Consequently, choosing a small value of τ_e provides a conservatively low estimate of the B-threshold.

As a general rule, the more central position of the heart within the torso in comparison with the more eccentric position of peripheral nerves requires a greater magnetic field to achieve a given induced E-field. The relative geometric advantage of the heart depends on the direction of the incident field with respect to the body. Nerve and cardiac thresholds are predicted to be most disparate when the incident field is perpendicular to the frontal cross-section of the body, and least disparate when the field is perpendicular to the sagittal cross-section. The lowest absolute thresholds are predicted for peripheral nerves with frontal exposure, and for the heart with sagittal exposure.

A further separation of cardiac and nerve excitation thresholds can be realized by taking advantage of the relatively longer excitation time constants of cardiac tissue. Cardiac and nerve tissue thresholds become more disparate as the stimulus phase duration is made shorter than the S–D time constant of the heart (about 3 ms).

Sinusoidal Fields: Large Area Exposure

Excitation thresholds, when plotted versus sinusoidal frequency, describe a *strength–frequency* (S–F) curve. Excitation thresholds for sinusoidal and rectangular stimuli have been compared in Chapter 4 using a theoretical model for myelinated nerve fibers. At low frequencies ($\approx 100\,\text{Hz}$), sinusoidal thresholds are minimum, and exceed the pulsed monophasic stimulus thresholds by 17%, which would place the minimum sinusoidal E-field threshold at 7.25 V/m for a 20-μm fiber. Thresholds rise at both lower and higher frequencies according to the S–F relationships described by Eq. 4.35, in which the parameters f_o and f_e describe the low- and high-frequency corners, respectively. At high frequencies, E-field thresholds eventually rise in proportion to frequency. The low-frequency corner was found to be $f_o \approx 10\,\text{Hz}$. As for the high frequency corner, available data suggest a wide spread of values. The myelinated nerve model indicates $f_e = 5{,}400\,\text{Hz}$ for axonal stimulation (Chapter 4). Electrocutaneous sensory stimulation indicates a range of values, having a median of about 500 Hz (Chapter 7). With cardiac excitation, the median experimental value of f_e is 120 Hz (Chapter 6).

The upper transition frequency f_e is significantly lower for electrocutaneous stimulation as compared with the myelinated nerve model. That observation is consistent with the finding that experimental values of τ_e are higher with electrocutaneous data. In developing criteria for magnetic stimulation, the relevance of electrocutaneous data is not clear— possibly a model based on axonal stimulation may be preferable. This speculation is suggested by the good agreement of model and experiment when an axon is excited subcutaneously (Reilly et al., 1985). Unfortunately, available frequency sensitivity data, reviewed in Sect. 9.7, are inadequate to provide clear guidance on the most appropriate value of f_e for magnetic stimulation.

Figure 9.17 illustrates S–F curves for magnetic excitation of nerve and cardiac tissue for sagittal exposure, with the parameter f_e varied. As with Fig. 9.16, the curves apply to the most sensitive loci on the heart or periphery of the body, namely points 10 and 14 on Fig. 9.15. At low frequencies, dB/dt thresholds converge to a minimum value independent of f_e (lower panel); at high frequencies, B thresholds converge to minimum value that is directly proportional to f_e (upper panel).

Criteria for Nerve and Heart Excitation: Large Area Exposure

The S–D curves of Fig. 9.18 have been used in to develop standards for patient exposure to pulsed and sinusoidal fields in magnetic resonance imaging (see Sect. 11.3). For pulsed stimuli, the horizontal axis is interpreted as the duration of a single monophasic phase of dB/dt; for sinusoidal stimuli, it is the duration of a half-cycle of the sinusoidal wave. The figure indicates dB/dt values required to induce $6.2 \, \text{V/m}$ at points on the body model (Fig. 9.15) as indicated in Table 9.6. Recall from Sect. 4.6 with sinusoidal stimulation of multiple cycles, thresholds eventually converge to that of a single monophasic pulse of duration equal to the half-cycle time of the sinusoid. Consequently, one can conservatively use Fig. 9.18 as threshold criteria for both pulsed and sinusoidal fields, regardless of the number of cycles of sinusoidal variation. As discussed in Sect. 11.3, the criteria associated with these figures formed the basis for MRI exposure standards of the FDA in the United States, the NRPB in the United Kingdom, and the IEC in continental Europe.

Standards pertain to exposure of patients without implanted metallic devices. With such implants, an enhancement of the induced E-field can occur in regions near the implant (Reilly and Diamant, 1997).

Body Size Scaling

One can apply the methods presented above to subjects of various body sizes using an elliptical representation of the torso dimension of the subject, and solving Eq. 9.21 for the value of dB/dt needed to induce the threshold E-field. One can simplify this procedure with an approximate method based

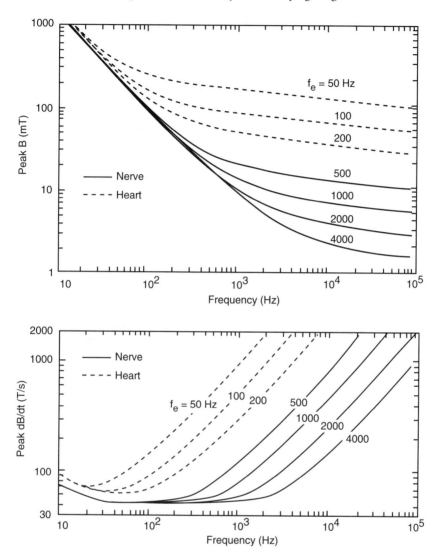

FIGURE 9.17. Calculated excitation thresholds for sagittal exposure of torso to sinusoidal magnetic field. Curve parameter f_e applies to upper frequency constant of excitable tissue; lower frequency constant $f_o = 10\,Hz$. Upper panel: peak flux density; lower panel: peak dB/dt. (From Reilly, 1992.)

on body weight. Note that the induced E-field is directly proportional to the dimensions of the ellipse (i.e., if we halve the ellipse dimensions, the maximum induced E-field will be halved as well). Further, note that body weight varies as the cube of the linear dimensions of the subject. Therefore, assum-

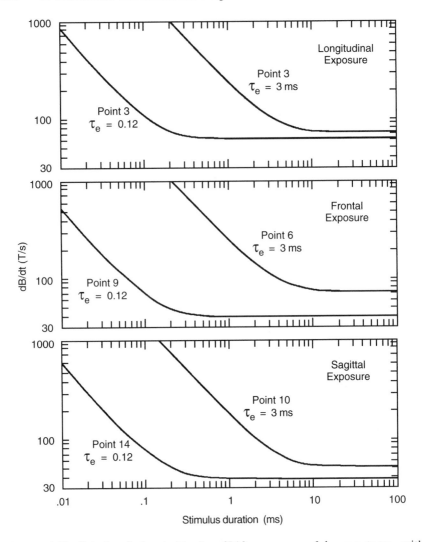

FIGURE 9.18. Calculated thresholds for dB/dt exposure of human torso, with induced E-field of 6.2 V/m as a criterion; reference points identified in Fig. 9.15.

ing that subjects have a similar body shape, we can scale B-field thresholds from one subject size to another by using a body weight scaling factor

$$K_B = \left(\frac{W_o}{W}\right)^{1/3} \qquad (9.22)$$

where W_o is a reference body weight for which the threshold is known, and W is the weight of another subject. For the subject depicted in Fig. 9.15, one may ascribe a body weight of about 90 kg. If we apply the body weight scaling formula to a small child of body weight 10 kg, we obtain $K_B = 2.08$. Consequently, thresholds appearing in Table 9.6 would be multiplied by that factor for application to the small child, provided that the excitation threshold of 6.2 V/m applies equally to both the large and small subject. One might question that assumption, since cutaneous electrical thresholds are subject to a body size scaling law (Eq. 7.8) in which small individuals have lower thresholds for contact currents than large individuals. The role of body size in magnetic thresholds has not been systematically studied.

9.7   Experimental Investigations of Magnetic Excitation: Large Area Exposure

When I first published theoretical models for magnetic excitation of peripheral nerve, most available experimental data were derived from small coil systems in which exposure was limited to a small area of the anatomy (McRobbie and Foster, 1984; Polson et al., 1982a; Polson et al., 1982b; Ueno et al., 1984; Irwin, 1970). My review of these early studies (Reilly, 1989; 1992) suggested that if one properly accounts for factors associated with the temporal and spatial distribution of the magnetic stimulus, the experimental findings are seen to be reasonably consistent with theoretical models. Since publication of my early predictive models, numerous experiments have been conducted using large coil systems; most of those studies were motivated by the desire to verify acceptability limits for patient exposure in MRI examinations. As will be demonstrated below, experimental findings for both nerve and cardiac excitation are surprisingly close to the predictive models—an outcome that I ascribe to equal measures of inspiration and good luck, considering the many simplifying assumptions used in the models.

Thresholds for Nerve Excitation

Table 9.7 summarizes experimental data on nerve excitation by large area magnetic exposure. The first column roughly describes the excitation waveform. The phase duration (t_p) is that for a single phase of the dB/dt waveform—for sinusoidal waveforms, $f = 1/(2t_p)$. For experiments in which t_p was varied, the largest value used in the experiments is listed. The magnetic field direction is that of the principal spatial component with the convention that x, y, and z indicate magnetic field directions perpendicular to sagittal, frontal (coronal), and longitudinal (axial) cross sections of the subject, respectively. The threshold dB/dt applies to the listed value of t_p; the first listed value indicates the mean or median, ±indicates a standard deviation, and ( ) indicates a minimum observed threshold.

The last two rows of data in Table 9.7 apply to excitation of the phrenic nerve, leading to respiratory reactions. The magnetic thresholds for

TABLE 9.7. Nerve excitation thresholds for large area magnetic exposure of human and animal subjects.

Waveform of dB/dt	phase duration (μs)	n	subject	mag. field direction (x,y,z)	reaction locus	reaction	threshold dB/dt (T/s)	Reference
Damped sine, 1 ~	1,190	20	Human	z	Torso, extremities	Sensation	75 (45)	Bourland et al., 1990
Pulse	200	13	Human	x y z	Torso	Sensation	99.7 ± 13.5 86.7 ± 15.9 101.8 ± 12.3	Schaefer et al., 1994
Pulse	368	11	Human	y z	Torso	Sensation	63.7 ± 13.5 94.9 ± 22.4	Schaefer et al., 1995
Sine, 4.7 ms	394	10	Human	z	Abdomen	Sensation	60 ± 6.0	Budinger et al., 1991
Sine, 46 ms	238	2	Human	z	Torso	Sensation, twitch	61.0	Cohen et al., 1990
Sine, 32 ~	400	—	Human	y	Head	Sensation	71.0	Yamagata et al., 1991
Pulse	1,000	52	Human	y	Torso	Sensation	14.4*	Bourland et al., 1997
Damped sine, 1 ~	572	—	Dog	x y z	Torso	Twitch	92 114 179	Nyenhuis et al., 1990
Pulse	560	12	Dog, 17–32 kg	z	Torso	Twitch	370 ± 357	Bourland et al., 1991a
Pulse	530	—	Dog	z	Torso	Inspiration 50% vol.	1,000 (450)	Bourland et al., 1991b
Pulse	450	1	Dog, 21 kg	z	Phrenic nerve	Inspiration	450	Mouchawar et al., 1991

Notes: First dB/dt value indicates mean or median; ± indicates standard deviation; ( ) indicates minimum; — indicates information not supplied. Phase duration is maximum used in experiment. Direction: x = sagittal; y = frontal; z = longitudinal. * Revised metric (see text).

respiratory effects are, as expected, greater than those for peripheral nerve excitation because of the more central anatomical location of the stimulated organs.

To compare the experimental data with theoretical predictions, we must account for several factors, including the waveshape and the size of the subject or of the exposure area. Consider, for instance, the second row of data from the experiments of Schaefer and colleagues (1994). If we consider the S–D time constant $\tau_e = 120\mu s$ as suggested by the theoretical model, then by applying the S–D law (Eq. 4.6), we expect that rheobase dB/dt would be a factor of 1.23 below the listed threshold values, namely 80.8, 70.3, and 82.5 T/s for x, y, and z exposure directions respectively. If we assume that $\tau_e = 150\mu s$ as indicated by experimental data (described below), the multiple would be 1.36, and we would estimate rheobase thresholds as 73.4, 63.9, and 75.0 T/s for x, y, and z directions respectively. Compare those values with the theoretical rheobase predictions in Table 9.6 of 48.8, 38.8, and 62.6 T/s for a large man and for a monophasic square-wave dB/dt pulse. Further adjustments for body size and waveform factors would suggest even closer correspondence of theory and experiment.

As a second example, consider the row of data applying to Nyenhuis and colleagues (1990). Using an estimated body weight of 20 kg for the canine subject, the scaling factor to a 90-kg man (Eq. 9.22) would be 1.65. Application of that scaling factor to the canine dB/dt thresholds results in thresholds of 55.7, 69.1, and 108 T/s for application to a human.

The experimental conditions applying to Table 9.7 do not exactly correspond to the assumptions in the theoretical models. For instance, the theoretical model assumes a monophasic square wave dB/dt pulse, and constant value of the peak flux density and its phase over the exposure area. Experimental deviations from those assumptions would tend to produce higher experimental thresholds. Furthermore, the theoretical model assumes a large person of homogeneous conductivity, whereas actual subjects vary in size, and internal conductivities are not homogeneous. Experimental deviations from those assumptions would produce lower or higher experimental thresholds. Nevertheless, even a rough interpretation of the experimental data shows that the predicted thresholds do not differ greatly from experimental findings.

The most recent data from Bourland and colleagues (1997) indicate a much lower dB/dt rheobase than previous data from the same laboratory. This results from a change in the exposure metric used by the investigators (Nyenhuis et al., 1997). In earlier studies, the investigators reported the peak field, whereas in recent studies they reported a spatial average.

Of course, the relevant exposure metric is the E-field induced within the subject. Under conditions where the induced E-field could be effectively determined, the rheobase E-field for stimulation of the forearm was found to be 5.9 V/m (Havel et al., 1997)—a value that is quite close to the theoretical minimum value of 6.2 V/m.

Thresholds for Cardiac Excitation

The experimental data discussed here apply to acute effects with large-area, short-term exposure of humans and animals. In addition to these effects, there is evidence of more subtle cardiac reactions with much lower intensities and longer durations of exposure. A statistically significant slowing of the human heart by a few percent has been reported with simultaneous exposure to a 60-Hz environmental electric field of 9 kV/m and a magnetic field of 20 μT, but without effect at lower (6 kV/m; 10 μT) or higher (12 kV/m; 30 μT) fields (Graham et al., 1994). At present, these findings cannot be explained on the basis of understood biophysical models.

Experimental data on magnetic excitation of the heart have been available since about 1991, with the introduction of experimental devices for advanced MRI echo-planar devices. As might be expected, human data on magnetic excitation of the heart does not exist; experiments to date have been performed with dogs as summarized in Table 9.8. As with stimulation of the heart by conducted current (Chapter 6), magnetic stimulation is most effective when it is delivered during the T-wave of the cardiac cycle (Yamaguchi et al., 1994).

Thresholds for two of the experiments in Table 9.8 (Mouchawar, 1992; Yamaguchi, 1992) are given in terms of the induced E-field on the heart, rather than the incident magnetic flux density. The authors each calculated those E-fields for their coil systems situated over an infinite, homogeneous medium. The data of Mouchawar show an average excitation threshold of 124 V/m at a stimulus phase duration of 530 μs. If we considered the stimulus as a monophasic dB/dt pulse, and use a value of $\tau_e = 3$ ms for cardiac tissue, we would infer an average rheobase of 21.5 V/m, which is a factor of 1.7 greater than the rheobase threshold of 12.4 V/m used as the median cardiac excitation threshold in the model described in Sect. 9.2. For reasons that have been analyzed previously (Reilly, 1993), even the inferred minimum value of 21.5 V/m in Mouchawar's experiments, as well as Yamaguchi's threshold of 30 V/m are thought to exceed attainable rheobase. One contributing factor is the biphasic nature of the waveforms used by both experimenters. The phase durations in these experiments (0.571 and 2.13 ms) are below the expected S–D time constant of 3 ms. Under such conditions, a biphasic reversal, such as evident in the waveforms used by the experimenters, would increase thresholds relative to a monophasic pulse, or a biphasic pulse with a delayed or more gradual phase reversal.

Excitation thresholds for the heart should be substantially greater than that for nerve as long as the pulse duration is sufficiently less than the S–D time constant of the heart (3 ms). Consequently, the avoidance of peripheral sensations in a patient should provide a conservative safety margin with respect to cardiac excitation. The margin between nerve and heart excitation thresholds was examined in canine subjects by Bourland and associates

TABLE 9.8. Cardiac excitation parameters for magnetic stimulation of animal subjects.

Waveform of dB/dt	phase duration (μs)	subject	n	mag. field direction (x,y,z)	threshold dB/dt (T/s)	threshold E-field (V/m)	nerve ex. multiple	Reference
Pulse	530	Dog, 17–32 kg	12	z	4,242 ± 678	—	14.1 ± 6.7	Bourland et al., 1991a
Pulse	530	Dog, 21–32 kg	10	z	3,900	—	—	Bourland et al., 1991c
Pulse	530	Dog	9	z	2,135 ± 457	—	—	Bourland et al., 1992
Pulse	571	Dog	11	x	—	124	—	Mouchawar et al., 1992
Damped sine*, 1~	540	Dog	12	z	2,155 ± 286	—	—	Nyenhuis et al., 1992
Damped sine $\tau = 4.2$ ms	2,130	Dog, 10, 15 kg	2	x	—	30	—	Yamaguchi et al., 1992
Damped sine	490	Dog, 9 kg	1	x	—	159	—	Hosono et at., 1992

Notes: First dB/dt value indicates mean or median; ± indicates std. dev.; — indicates info. not supplied.
Phase duration is maximum used in experiment.
* Static field of 1.5 T added.
Direction: x = sagittal; y = frontal; z = longitudinal.

(1991a), who reported a threshold ratio of cardiac to neuromuscular excitation of 14.1 ± 6.7. We can compare that ratio with one calculated for human subjects using as rheobase a threshold of 62.6 T/s for nerve, and 142.6 T/s for the heart as suggested in Table 9.6 for longitudinal exposure. (As suggested by Table 6.12, the listed 1% cardiac threshold has been multiplied by 2 to give the 50% cardiac threshold). From the S–D formula, with $\tau_e = 0.12$ and 3.0 ms for nerve and heart respectively, we derive a theoretical heart/nerve excitation ratio of 14.0, which is quite close to the experimental ratio.

Stimulus Duration and Frequency Effects

The strength duration time constant, τ_e, allows us to scale from the rheobase threshold at long stimulus durations to the threshold at shorter durations through the S–D law (Eq. 4.6). As seen in Chapter 4, τ_e differs substantially not only with respect to tissue type, but also with the spatial distribution of stimulus field. It is therefore important to assess τ_e under appropriate conditions of stimulation.

Table 9.9 summarizes τ_e data from experiments that used magnetic stimulation of nerve and cardiac tissue in humans and animals. Some of these data were obtained with much smaller coil systems than used in the experiments listed in Tables 9.7 and 9.8. But even for these smaller coils, the spatial derivative of the induced E-field is sufficiently small that the experimental τ_e values can be attributed to large coil systems. In Table 9.9, the

TABLE 9.9. Strength-duration time constants determined with magnetic stimulation.

Subject	n	Exposed area	Magnetic field device	Reaction	τ_e (μs)	Reference
Human	10	Motor cortex	8.5 cm dia. coil	EMG-arm	152 ± 26	Barker et al., 1991
Human	12	Wrist	8.5 cm dia. coil	EMG-arm	150 ± 55	Barker et al., 1991
Human	4	Wrist	4.8 cm dia. coil	Perception	146 ± 7.5	Mansfield & Harvey, 1993
Human	14	Forearm	11 cm dia. coil	Perception	395 ± 79	Havel et al., 1997
Dog	12	Thorax	z-coil	Muscle twitch	148 ± 49	Bourland et al., 1991a
Dog	—	Thorax	x-coil y-coil z-coil	Twitch	291 213 240	Nyenhuis et al., 1990
Dog	2	Heart	22-cm dia. coil	Ec. beat	≈3,000	Yamaguchi et al., 1992

Direction: x = sagittal; y = frontal; z = longitudinal.

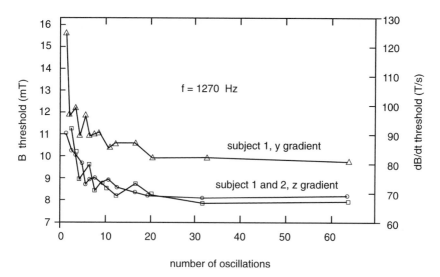

FIGURE 9.19. Variation of perception threshold with number of cycles of magnetic stimulation at 1,270 Hz. Z-gradient exposure was constant over longitudinal cross section of body; y-gradient varied over cross-section. (From Budinger et al., 1991.)

majority of τ_e values do not differ greatly from the value of 120 μs derived with the SENN model. However, the data of Nyenhuis and of Havel (both with the same laboratory) indicate larger τ_e values for reasons that remain unexplained. τ_e determined by Yamaguchi and colleagues for cardiac excitation is close to the value used in the theoretical models.

Another aspect of duration sensitivity was studied by Budinger and colleagues (1991). In their experiments, 10 healthy men were positioned with the pelvic region in the maximum field of a z-gradient coil. A sinusoidal magnetic waveform could be varied in frequency as 600, 960, 1,270, and 1,950 Hz; also varied were the number of cycles of duration. The variation of perception thresholds with the number of stimulus cycles, illustrated in Fig. 9.19, shows characteristics of the theoretical SENN model as in Fig. 4.19—the threshold is maximum with one cycle of stimulation and gradually diminishes to a minimum plateau as the number of cycles is increased. Similar dependency on the number of cycles of sinusoidal stimulation has been reported by Schmitt and colleagues (1994), who demonstrated a minimum plateau at 25 cycles of stimulation at a frequency of 1.0–1.3 kHz, and by Yamagata and associates (1991), who reported a minimum plateau at 30 cycles of stimulation at 1,250 Hz.

Budinger and colleagues also evaluated the variation in perception threshold with the frequency of stimulation. For a field in the y direction, the threshold expressed in B units diminished with increasing frequency, reaching a minimum plateau at high frequencies ($\approx$1,950 Hz) as expected by the theoretical models discussed above. However, for fields in the x

or z direction, a plot of thresholds vs. frequency showed sharply declining thresholds at the higher frequencies, in contrast to theoretical expectations. Others who have tested perception sensitivity thresholds have found frequency dependency in accord with theoretical models, in which a minimum B-field plateau is found at frequencies above about 3 kHz (Mansfield and Harvey, 1993).

Suprathreshold Nerve Excitation Reactions

Suprathreshold reactions can be reported as multiples of perception thresholds. For cutaneously applied currents, ratings of unpleasantness or pain are typically elicited at a perception threshold multiple of about 2 for sinusoidal or repeated stimuli (Table 7.3). Suprathreshold reactions were investigated by Budinger and associates (1991) using z-coil exposure to the torso of two human subjects. The ratio of pain to perception was 1.3 when averaged over stimulus frequencies of 960, 1,270, and 1,950 Hz. Similar findings were reported by Bourland and associates (1997) using a much larger sample size ($n = 52$). For instance, with y-axis exposure, Bourland's subjects reported definite discomfort at threshold multiple of 1.44, and intolerable pain at a multiple of 1.89. These reported multiples for discomfort and pain are considerably smaller than that in electrocutaneous experiments. They might be explained if subjects were to experience more aversive reactions associated large area magnetic stimulation as compared with more focal electrocutaneous stimulation, although this conjecture remains unproven.

9.8 Visual and Auditory Reactions to Electromagnetic Exposure

Direct Perception of Magnetic Fields

In certain animal species, specialized mechanisms for magnetic field detection can be demonstrated (Tenforde, 1989). If the dominant mechanism of human perception is due to peripheral nerve stimulation, then magnetic thresholds can be determined with the models of Sect. 9.6. At 60 Hz, for example, magnetic thresholds for the most sensitive neural structures are estimated to be 120 mT (peak) for frontal exposure of the torso and 190 mT for longitudinal exposure.

Magnetic perception thresholds were tested in now-classic experiments by Tucker and Schmitt (1978). These experiments established that humans cannot detect 60 Hz fields as high as 0.75 mT when applied longitudinally to the whole body, or as high as 1.5 mT when applied locally to the head. An interesting finding was the extreme care that must be exercised to avoid artifactual perception due to extraneous cues in the experimental procedure, such as vibration, acoustic cues, or dimming of lights. The exposure in these experiments was below theoretical nerve excitation thresholds by a factor of about 200. Apparently, if some detection mechanism exists in humans other than peripheral nerve stimulation, it requires whole-body exposure above 0.75 mT to be effective.

Visual Sensations

A particularly sensitive mode of detecting magnetic fields occurs through the perception of *phosphenes*, which are perceived light patterns induced by nonphotic stimuli, such as pressure on the eyeballs, or electric energy. They are often referred to as electro- and magnetophosphenes when induced by electric currents or by magnetic stimulation, respectively. In an interesting historical account, Marg (1991) reports the initial discovery of electro-phosphenes in 1755 through the discharge of static electricity from a Leyden jar, but the first report of magnetophosphenes was not made until the late 1900s by d'Arsonval.

A unique feature of electric phosphenes is their low excitation threshold and sharply defined frequency sensitivity as compared with other forms of neural stimulation. Figure 9.20 compares thresholds for magneto- and electrophosphenes from the experiments of Lövsund et al. (1980a, 1980b). Magnetophosphenes were elicited in individuals with the head placed between the poles of a large electromagnet (temple to temple), and electrophosphenes from current through electrodes placed on the temples. Magnetophosphenes are shown in Fig. 9.20 on an absolute scale (Lövsund et al., 1980b); electrophosphenes are shown on a relative scale (Lövsund et al., 1980a). Somewhat lower magnetophosphene thresholds have been reported by Silny (1986). The threshold of phosphenes depends on the level of background illumination (Barlow et al., 1947b), and this parameter must be controlled in experimental procedures.

A third curve in Fig. 9.20 shows electrical thresholds divided by the frequency; this display facilitates a comparison of the curve shapes of magneto- and electrophosphenes. A rationale for this representation is that the induced E-field (and current density) is directly proportional to the frequency of the magnetic field, as suggested by Eq. (9.17). Electrical thresholds, modified in this manner, ought to conform to the curve shape for magnetic thresholds. The two curves do, in fact, appear quite similar, suggesting that the magnetic effect is the result of induced current, rather than a direct action of the magnetic field.

As shown in Fig. 9.21 electrophosphene thresholds tested by Adrian (1977) increase as f^n, with $n = 3.5$, in contrast to $n \approx 1$ for peripheral nerve excitation (Chapter 4); the figure also shows thresholds for auditory effects, which will be discussed presently. Adrian also used stimuli consisting of two frequencies. Phosphenes were produced as long as the difference frequency was near the most sensitive single frequency (≈ 20 Hz), even though the individual frequencies would have been ineffective if presented singly. This

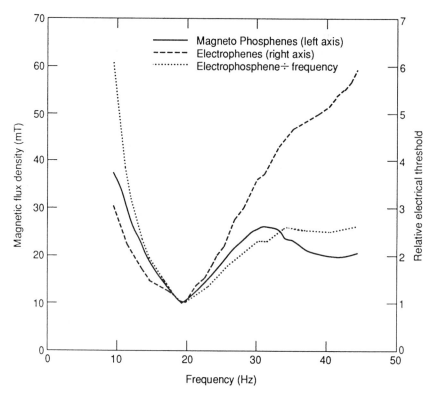

FIGURE 9.20. Comparison of phosphene thresholds for magnetic and electric excitation. Magnetic threshold on absolute scale (left vertical axis); electric threshold on relative scale (right vertical axis). Background illumination at 3cd/m^2. (Adapted from Lövsund et al., 1980b.)

finding points to a nonlinear mechanism for the activation of action potentials. Although the neural membrane is highly nonlinear when depolarized near the action potential threshold (Chapters 3 & 4), it is only slightly nonlinear when substantially below threshold. Whether the nonlinear aspects of the neural membrane can explain the findings of Adrian has not been explored.

Using Lövsund's observed magnetophosphene thresholds, we can estimate corresponding induced E-fields using an ellipsoidal model of the head as in Eq. (9.21). Such a treatment is only approximate because the experimental field in Lövsund's experiments was not uniform over the entire cross-section of the head, as is assumed in the theoretical treatment. Consider sagittal exposure of a head model with semimajor and semiminor axes of 0.13 and 0.1 m respectively, and a minimum threshold of 10 mT at 20-Hz frequency. With these parameters, the maximum E-field induced within the head is calculated to be 0.079 V/m. At the location of the retina, the calculated field is 0.053 V/m, which is consistent with the current density threshold of $0.008 \, A/m^2$ at the retina determined for electrophosphenes (Lövsund et al., 1980b), assuming that the conductivity of the brain is 0.15 S/m (Table 2.1).

A somewhat greater E-field threshold was determined in experiments by Carstensen and associates (1985), in which sinusoidal current was introduced into the eye through a saline-filled cup electrode held against the eyeball. Carstensen calculated the E-fields within the head using a 3-D finite element model of the head and diverse tissues. At the most sensitive frequency tested (25 Hz), the current threshold was 0.04 mA, which corresponded to an E-field at the retina of 0.2 V/m.

Phosphene thresholds are quite unlike those for peripheral nerve excitation. For one thing, the internal E-field corresponding to phosphene perception at the optimum frequency is a factor of 100 or so below minimum

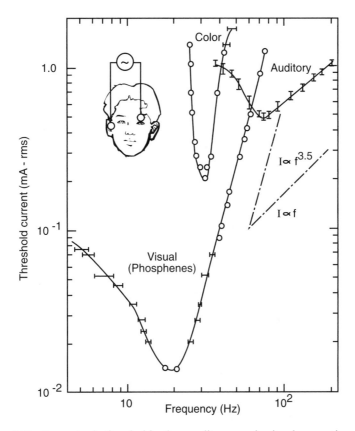

FIGURE 9.21. Perceptual thresholds for auditory and visual sensations from transcranial stimulation by electric current. Electrode placement (1.75 cm dia.) as indicated. Insert shows f^n slopes with $n = 1$ and $n = 3.5$. (From Adrian, 1977.)

thresholds for neural stimulation (Table 4.2). Furthermore, the frequency sensitivity differs substantially from that for neural stimulation (Fig. 9.17). These observations suggest that some mechanism other than afferent neural stimulation is responsible for the production of phosphenes.

Mechanisms for Phosphenes

Although visual sensations from stimulation of the visual cortex has been demonstrated (Brindley and Lewis, 1968; Brindley and Rushton, 1977; Ronner, 1990), experimental evidence demonstrates that phosphenes with much lower electrical thresholds can be traced to excitation of structures in the retina, rather than the optic nerve or the visual cortex of the brain. The visual field for phosphenes occurs in the opposite quadrant relative to the area of stimulation on the retina, whether through magnetic (Barlow et al., 1947) or contact current stimulation (Brindley et al, 1955), as would be expected if the site of excitation were in the retina. A subject who was blind because of the removal of both eyeballs could not experience phosphenes (Lövsund et al., 1980a). However, another subject who was blind as a result of *retinitus pigmentosa*, a disease of the photoreceptors and epithelium, did experience electrical phosphenes. This subject was thought to have had functional bipolar and ganglion cells in the retina.

Brindley and colleagues (1955) performed a remarkable series of experiments to explore the mechanism of electrical phosphenes. One has to marvel at his dedication in accepting the passage of electric current through electrodes positioned on his retina—all in the pursuit of scientific inquiry. Brindley found that phosphenes were most sensitive to current in a direction radial to the retina, but were difficult to produce when the current traveled in a tangential direction. This would imply that the affected structures are most likely cells with a radial orientation, possibly photoreceptor cells, although the aforementioned experiments with a blind subject would suggest bipolar or ganglion cells might be likely (see Fig. 3.24). From Brindley's data with capacitor discharges, we can infer a threshold charge density of about $0.09 \mu C/cm^2$, which is about a factor of 100 below the nerve excitation threshold for cutaneously applied capacitor discharges (Fig. 7.5).

A likely locus of stimulation is the presynaptic junction of cells within the retina (Lövsund et al., 1980a; Knighton, 1975a; 1975b). This view seems reasonable in light of the fact that small changes in the potential of presynaptic cells can be greatly magnified in the potential and excitability of post synaptic cells (see Sect. 3.7). That could account for the low excitation threshold of phosphenes in comparison to nerve excitation. Knighton estimated that depolarization of the presynaptic membrane by only $60 \mu V$ is sufficient to evoke a visual response. In contrast, in order to excite a peripheral nerve fiber, membrane depolarization by about $15 mV$ is necessary (Chapter 4)—a factor of 250 times greater.

Duration and Frequency Relationships for Visual Effects

The sensitivity of phosphenes to temporal aspects of a retinal stimulus differs markedly from that for the nerve and heart. Experimental data show that the chronaxy or S–D time constant for phosphenes using electrodes on the temples is about 14 ms (Baumgardt, 1951; Bergeron et al., 1995). An interesting aspect of phosphene S–D curves is that the threshold does not monotonically decrease with increasing stimulation duration. Rather, the S–D curve has a long-duration plateau as with neural S–D curves, but exhibits a dip in the threshold by as much as 30% just before the rising phase at short durations. The slope of the S–D curve from Bergeron's experiments was approximately $t^{-0.5}$, in contrast to t^{-1} for nerve and muscle S–D characteristics, where t is the duration of a monophasic stimulus. This finding is complicated by the fact that Bergeron used a regulated voltage rather than a regulated current device for stimulation, which can affect the S–D slope because of the non-linear impedance properties of the skin (see Sect. 2.2).

In the experiments described previously, Knighton developed S–D curves for electrically evoked responses, and found S–D time constants, τ_e, in the range 14 to 36 ms (Fig. 9.22). These values are consistent with the phosphene data described above, but are about 100 times greater than corresponding values for excitation of peripheral nerve (Chapters 4 and 7),

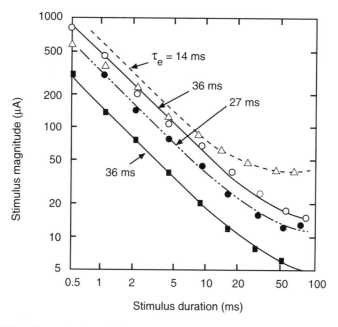

FIGURE 9.22. Strength–duration curves for electrically evoked potentials in the retina of the frog's eye. Curves represent various experimental procedures. (Adapted from Knighton, 1975.)

and are about 10 times greater than those applying to excitation of cardiac and skeletal muscle (Chapters 6 and 8). The relatively large values of τ_e for visual responses are consistent with strength-frequency data for phosphenes, which show an upper transition frequency ($f_e \approx 20\,\text{Hz}$) that is much lower than that for nerve ($\approx 1,000\,\text{Hz}$) or cardiac ($\approx 100\,\text{Hz}$) excitation.

Phosphenes were tested in five human subjects by Budinger and colleagues (1984) using a pulsed magnetic field having a biphasic dB/dt waveform with a phase duration of 10 ms. The threshold dB/dt was found to be 1.3 T/s for younger subjects, and 1.9 T/s for older men; subjects were most sensitive to a repetition rate of 15 per second. These dB/dt thresholds are roughly a factor of two higher than those obtained by Lövsund with sinusoidal magnetic fields having a similar phase duration. If the S–D time constant is assumed to be 14 ms (the lower value obtained by Knighton), one would conclude that Budinger's thresholds are a factor of 2.0 above a presumed rheobase for the same subjects.

One should distinguish between the stimulation of the retina and of the visual cortex, both of which can elicit phosphenes. Phosphenes induced through stimulation of the visual cortex have τ_e values typical of nerve excitation (i.e., on the order of 0.2 ms) (Ronner, 1990). In addition, rheobase thresholds for visual cortex stimulation are typical of direct neural stimulation, in contrast to the much lower thresholds with retinal stimulation.

Electrical phosphenes persist beyond the duration of the applied stimulus, in contrast to neural or cardiac action potentials, whose duration are on the order of 1 and 200 ms respectively. The duration of phosphene persistence increases with the level of the both electric and magnetic stimuli (Barlow, 1947). At threshold levels, phosphenes persist less than 1 s; if the stimulus is increased by a factor of 4, durations may exceed 16 s. Following the cessation of phosphenes, there is a refractory period in which thresholds for subsequent stimulation is increased. With the application of strong stimuli, refractory periods up to 60 s have been observed (Barlow, 1947).

Implications of Phosphenes for CNS Synaptic Interactions

Although photoreceptors are not found in the central nervous system (CNS), the brain and spinal column are rich in neuro synaptic junctions. It is logical to inquire whether CNS interactions are possible at the low phosphene thresholds pertaining to retinal stimulation. While we lack a clear answer to this question, experimental evidence shows that CNS interactions are indeed possible with magnetic stimulation of the brain at intensities well below levels necessary to excite neurons. For instance, Silny exposed the human head to sinusoidal magnetic fields, and recorded visual evoked potentials (VEP) on the surface of the scalp in response to a visual stimulus. With a 50 Hz stimulus at a flux density of 60 mT, Silny observed significantly altered VEP patterns in 12 of 15 test subjects. It is remarkable that these alterations persisted for as long as 70 minutes after cessation of the stimulus. In addition, his subjects reported headaches and "indisposition" above 60 mT exposure. We may compare these values with excitation thresholds of cortical neuron through exposure of the head by a 50-Hz magnetic field. For example, the excitation threshold of cortical neurons would be approximately 1.4 T-peak, assuming a 11-cm spherical diameter for the brain, and a threshold E-field of 12.3 V/m for a 10-μm neuron (Table 4.2).

The ability of subexcitation fields to alter neuronal response has also been reported by Bawin and associates (1984; 1986) who exposed hippocampal slices from the rat brain to magnetic fields. In these experiments, neuronal excitability was inferred with measurements of the compound action potential evoked by a monophasic stimulus. It was found that neuronal excitability could be significantly altered by applying sinusoidal fields at 5 and 60 Hz with in situ intensities as low as 0.5 V/m-rms. Both increases and decreases in excitability could be effected by the sinusoidal fields. The authors noted that long-term increases in excitability were of comparable magnitude for 5 and 60 Hz fields, and generally lasted from 10 minutes to hours after the cessation of the sinusoidal field.

Cook and colleagues (1992) reported changes in auditory evoked brain potentials in human subjects exposed simultaneously to a 9 kV/m electric field and a 20 μT magnetic field. The authors concluded that *changes* in exposure level may be more important than the duration of exposure. It should be noted that these effects were elicited at exposure levels well below the synaptic effects thresholds discussed above.

Auditory Sensations

The ability to electrically stimulate auditory nerves within the cochlea has lead to a well-developed technology of cochlear implants for hearing impaired individuals (Leake et al., 1990). It is also possible to stimulate auditory sensations with less invasive electrical means, which will be reviewed here.

Stimulation at Audio Frequencies

Auditory perception from stimulation at audio frequencies has been called *electrophonics* (Flottorp, 1953; Adrian, 1977). Thresholds for electrophonics determined by Adrian are shown in Fig. 9.21, along with electrophosphene thresholds for the same electrode arrangement. It can be seen that the slope of phosphene thresholds versus frequency is similar to that for neural stimulation and considerably distinct from electrophosphenes. A further finding of Adrian was that threshold sensitivity responded to the difference frequency when two frequencies were mixed in the stimulus, much like similar properties for phosphenes as noted above.

One cannot rule out the possibility that Adrian's electrophonics may result from neuromuscular excitation that couples mechanically into normal auditory channels since the electrophonic thresholds shown in Fig. 9.21 are close to neurosensory and neuromuscular stimulation thresholds with the indicated electrode size (Chapters 7 and 8). In an investigation of a variety of methods for stimulating electrophonics, Flottorp (1953) concluded that the mechanism of transduction was of extra choclear origin. One method requires a dry electrode to be rubbed against dry skin adjacent to the ear. The underlying mechanism in this case is likely due to electrostatic vibration of the skin, as described by Eq. (7.1). Other mechanisms involving electrodes external to the ear were thought to be the result of neuromuscular stimulation. The fact that the perceived electrophonic frequency was double the stimulating frequency would be consistent with this explanation.

Stimulation at Microwave Frequencies

So-called *microwave hearing* was first reported in World War II by radar operators who heard clicks or buzzing sounds when they were within the beam of pulsed radar energy. Since that time, much research has been conducted on the properties and mechanisms of microwave hearing. A review of the subject and original research have been presented by Lin (1990). This research persuasively establishes that microwave hearing is the result of a small transient temperature rise in brain tissue, which, through thermoelastic expansion, launches an acoustic wave that reaches the inner ear through normal auditory processes. The temperature rise is extremely small ($\approx 10^{-6}\,°C$), as is its duration ($\approx 10\,\mu s$). Nevertheless, the result is sufficient to be detected by the inner ear.

Auditory thresholds depend on pulse width, as seen in Fig. 9.23, which shows perceived loudness as a function of pulse width for radiation at 800 MHz. The auditory threshold is inversely proportional to perceived loudness. The figure shows resonances which arise from the roughly spherical resonant acoustic cavity formed by the head. In this example, the first resonance occurs at about $50\,\mu s$. For short pulses, thermoelastic displacement or pressure in the head is approximately proportional to the product of power density and pulse width (i.e., the energy density per pulse). A "short" pulse is defined such that $2\pi f t_d \ll 1$, where f is the fundamental resonant frequency, and t_d is pulse duration.

The frequency of vibration is independent of the microwave absorption pattern within the head and depends only on the equivalent radius of the brain and its acoustic properties. Theoretical models with an ideal spherical skull establish the fundamental frequency of vibration as

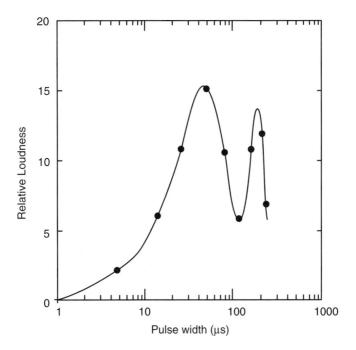

FIGURE 9.23. Perceived auditory loudness from microwave exposure at 800 MHz as a function of pulse width, constant incident power density. Data from 18 subjects. (From Lin, 1990.)

$f = kn/r$, where n is the velocity of propagation ($\approx 1{,}460$ m/s) r is the radius of the sphere, and $k = 0.72$ and 0.5 for a constrained and unconstrained brain mass, respectively. Experimental values closely follow this relationship. For instance, with a brain radius of 8 cm, the vibration frequency is calculated to be 9.1 kHz for a constrained mass. It is significant that microwave hearing occurs only in individuals with the ability to hear sound through bone conduction at frequencies above 8 kHz (Lin, 1978).

Table 9.10 lists microwave hearing thresholds as a function of pulse width for various measures of exposure. It can be seen that auditory thresholds occur at a constant value of incident energy density [i.e. product of peak power and duration (third column) or absorbed energy per gram of tissue (last column)]. The lowest threshold of peak power density reported by Lin at an optimum pulse width was 250 mW/cm^2, which would correspond to a specific absorption rate (SAR) of about 0.1 W/g. These exposures are quite large when compared with standards for electromagnetic exposure (reviewed in Chapter 11), which limit rms SAR typically to 0.4 W/kg, and rms power density in certain frequency regimes to 10 mW/cm^2.

TABLE 9.10. Threshold for microwave-induced auditory effect in human subjects, 45 dB background noise, 2,450 MHz carrier frequency.

Pulse width (μs)	Peak power (W/cm^2)	Energy density (μJ/cm^2)	Peak SAR (W/g)	Absorbed energy (mJ/g)
1	40.00	40	16.00	16
2	20.00	40	8.00	16
4	10.00	40	4.00	16
5	8.00	40	3.20	16
10	4.00	40	1.60	16
15	2.33	35	0.93	14
20	2.15	43	0.86	17
32	1.25	40	0.50	16

Notes: SAR calculated for equivalent spherical model of the head. Quantities are per pulse.
Source: Lin (1990).

9.9   Local Magnetic Stimulation[3]

In recent years, considerable attention has been devoted to using localized time-varying magnetic fields for stimulating excitable tissue without the use of electrodes. The main advantage of this technique is that it provides a noninvasive means of stimulation without causing pain (Barker et al., 1985). One potential application is magnetic stimulation of the brain. Recent results from magnetic brain stimulation include measurement of central motor conduction time (Barber et al., 1987; Cracco, 1987; Hess et al., 1987; Mills and Murray, 1985), monitoring of spinal cord function during spinal surgery (Levy, 1987; Shields et al., 1988; Krauss et al., 1994), and the ability to produce motor twitches for a single digit (Amassian et al., 1988). Nonmotor brain areas have been stimulated as well. Some examples of stimulating cognitive areas are the suppression of visual perception by stimulating the visual cortex (Amassian et al., 1989a) and eliciting a sense of movement in a paralyzed limb (Amassian et al., 1989b). Stimulation of peripheral nerve is also possible, but has not received the same attention as brain stimulation, except in a few special cases such as facial nerves (Evans et al., 1988; Maccabee et al., 1988b; Schriefer et al., 1988). Magnetic stimulation of lumbosacral roots has also shown utility (Tsuji et al., 1988; Chokroverty and DiLullo, 1989). Applications have also included magnetic stimulation of the heart (Nyenhuis et al., 1992, 1994) and of the inspiratory nerves (Voorhees et al., 1990).

---

[3] This section was written by H. A. C. Eaton and J. P. Reilly.

Small Coil Stimulators

The circuitry used to produce a magnetic stimulus is straightforward, but there are many variations (Davey et al., 1988; Hallgren, 1973; Merton and Morton, 1986; Polson et al., 1982a; Barker, 1994) Typical devices rapidly discharge a capacitor into a small coil placed near the excitable tissue as in Fig. 9.24. The figure omits the charging circuit necessary to produce the initial charge on the capacitor and the inherent series resistance of the coil, thyristor, capacitor, and cabling. These resistances are of considerable practical concern and cannot be neglected. Resistor R_s is used only to limit the current flowing into the gate of the thyristor during firing. Figure 9.24 shows the simplest possible trigger control; many other trigger methods are possible.

After the switch is pressed in Fig. 9.24, the thyristor is triggered, discharging the capacitor through the coil. Current that flows through the capacitor until the voltage at the capacitor terminals reaches zero is described by a series RLC equation:

$$i(t) = \frac{V_c \sin\left(t\sqrt{1/LC - (r_1/2L)^2}\right)}{\sqrt{L/C - r_1^2/4}} \exp\left(-tr_1/2L\right) \qquad (9.23)$$

Where r_1 is the sum of the series resistances of the coil, capacitor, thyristor, and cabling. This equation is valid for the time interval from $t = 0$ (when the thyristor is triggered) until $t = t_s$ when the diode begins to conduct, where t_s is given by

$$t_s = \frac{\tan^{-1}\left\{\frac{\left[2L\sqrt{1/LC - (r_1/2L)^2}\right]}{(r_1 - 2r_2)}\right\}}{\sqrt{1/LC - (r_1/2L)^2}} \qquad (9.24)$$

where r_2 is the sum of the parasitic resistances to the right of the diode. A four-quadrant arctangent function must be used for Eq. (9.24). The time t_s is slightly later than the time at which the current is maximal. After t_s, the current circulates through the diode, thyristor, and coil and is governed by a series RL circuit equation:

$$i(t > t_s) = I_s \exp\left[-\frac{(t - t_s)r_2}{L}\right] \qquad (9.25)$$

Where I_s is the coil current at the time that the diode begins to conduct and is given by

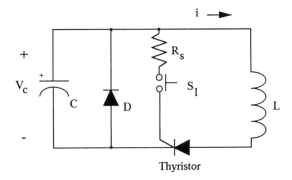

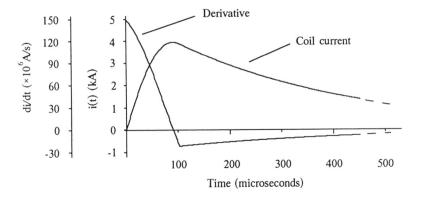

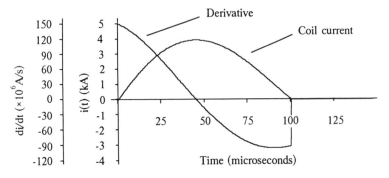

FIGURE 9.24. Upper: a simple magnetic stimulator circuit employing a thyristor switch. The coil, L, is the stimulus coil located near the excitable tissue. Middle: The coil current waveform from the circuit, and its derivative. Lower: The coil current and derivative waveforms that result when the diode is removed from the circuit.

$$I_s = \frac{V_c \exp\left(-t_s r_1/2L\right)}{\sqrt{L/C + r_2^2} - r_1 r_2} \tag{9.26}$$

Figure 9.24 shows the shape of the coil current as well as its derivative. The scales are for a hypothetical circuit having $C = 200\mu F$, $V_c = 750\,V$, $L = 5\mu H$, and $r_1 = 50\,m\Omega$. These are realistic circuit parameters, although far different values are possible. The coil current rises rapidly, followed by a long decay period. The induced electric field in the tissue, proportional to dI/dt, is biphasic; such is always the case with magnetic stimulation because the coil current always returns to zero. With the circuit of Fig. 9.24, the stimulus is a short-duration, high-amplitude pulse. The energy stored in the capacitor is dissipated primarily as heat in the coil resistance. The diode prevents reverse charging of the capacitor.

If the diode in Fig. 9.24 is removed, the coil current will be governed by Eq. (9.23) alone, which is valid since the thyristor will allow current to flow in only one direction. If the diode is removed from the circuit in Fig. 9.24, the thyristor switch discharges the capacitor through the coil, which then recharges the capacitor in the opposite polarity. Using this technique, much of the energy originally stored in the capacitor is returned to it, and less heat is dissipated in the coil. Figure 9.24 also shows the waveform of coil current and its derivative when the diode is deleted from the circuit. The stimulus and repolarization pulses are of roughly equal amplitude and duration.

For a practical circuit, the capacitor may be several hundred μF, and may be charged to over $1\,kV$. The inductor may range from a few μH to hundreds of μH, depending on the coil arrangement. Peak current of several kA is typical. Special care must be taken to keep the inherent series resistances of all components to a minimum (usually under $0.1\ \Omega$), so that the LC circuit is as underdamped as possible. Stray inductance in the cabling connecting the circuit must be kept as small as possible, especially in the (typically long) cable between the stimulator electronics and the coil. The ringing frequency of the RLC circuit is usually in the 1- to 10-kHz range. Considerations of the S–D curve of the target tissue aid in choosing the optimal ringing frequency (see Sect. 9.6). Because this frequency is governed only by the values of r, L, and C, it is difficult to adjust the stimulus duration while maintaining a constant stimulus amplitude. Usually, a single fixed duration is used, in which case the amplitude of the stimulus is easily adjusted by changing the initial voltage on the capacitor.

A large amount of energy is stored in the capacitor before each discharge. Most circuits used to recharge the capacitor build up this charge slowly, resulting in a long interpulse period (typically a few seconds). Considerable heating of the stimulating coil results after delivery of many pulses, and coil temperature is usually monitored on commercial magnetic stimulators. Multiple action potentials are difficult to produce. One or more seconds are typically required to recharge the energy storage capacitor, which prevents stimulation at higher rates. Allowing an LC circuit to "ring" continuously is not an effective means for producing multiple action potentials. If a low ringing frequency is used, the magnitude of dI/dt will be low unless very high peak currents are used. Such currents are difficult to handle, and resistive losses will damped the oscillations quickly, making this scheme impractical. If a high ringing frequency is chosen, the excitation threshold of the nerve is increased. More sophisticated circuits for charging and discharging of the capacitor coupled with special coil designs can produce magnetic pulse trains capable of producing multiple action potentials (Davey et al., 1988).

Induced Electric Field

It is not possible to design a coil that would focus stimulation to any desired location, depth, and orientation of the induced current. The inherent shape of various body structures places limitations on the possible orientations of the induced currents. For example, in the near-spherical regions of the head, it is impossible to induce significant radially oriented (perpendicular to the skull) currents with a magnetic stimulator (Eaton, 1992). Despite this limitation, it is possible to excite fibers of diverse orientations within the brain's cortex because of the positions of neurons within the convolutions of the cortex, as suggested by Fig. 9.25. The figure shows that an induced tangential E-field may be oriented for preferential stimulation at the ends and bends of the cortical neurons, in accord with mechanisms described

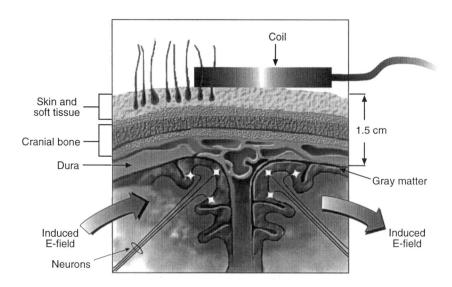

FIGURE 9.25. Anatomical structures and induced E-Field with magnetic stimulation of the brain. Neurons have diverse orientations with respect to tangential E-field because of convolutions of the cortex.

in connection with Fig. 4.15. Although one cannot focus excitation at will within the brain, it is still possible to target excitation to small regions of the cortex using dual coil systems (Ueno, 1994), as will be described presently.

Coil current, geometry, and the tissue boundary conditions determine the pattern of the induced currents. For the case of a magnetic stimulator, the coil size and its distance from the point where the field is being computed are much smaller than one wavelength. This allows the approximation that the phase shifts due to propagation delays can be ignored. Tissue conductivities are small enough that the skin effect can also be ignored. With these approximations, the induced electric field is computed as (Eaton, 1992)

$$\mathbf{E} = -\mu_0 \frac{di}{dt} \int_{coil} \frac{d\mathbf{l}}{4\pi R} - \nabla V \qquad (9.27)$$

where R is the distance between the current filament and the point in the medium where E is being computed, i is the instantaneous current, and $d\mathbf{l}$ is an element of the current-carrying conductor. The current is assumed to be confined to a thin filament centered on the coil wire. The term ∇V arises from the charge density appearing on the boundaries between different tissues, or other inhomogeneities. In this analysis, induced currents in the tissue are assumed to have a negligible contribution to the total magnetic field, a quite realistic assumption for biological materials. In a homogeneous, isotropic medium, either of infinite extent or having appropriately symmetric boundaries for the coil under consideration, the charge density will be zero and, consequently, only the integral term above contributes to the electric field. Several investigators have examined the induced electric fields for the special case of zero charge density for different coil arrangements. Much work in the area concerns producing a focal region of high electric field (Maccabee et al., 1988a; Rösler et al., 1989).

The boundary conditions are found from the following expression:

$$\sigma \nabla \cdot \mathbf{E} + \mathbf{E} \cdot \nabla \sigma = -\varepsilon \nabla \cdot \frac{\partial \mathbf{E}}{\partial t} - \frac{\partial \mathbf{E}}{\partial t} \cdot \nabla \varepsilon \qquad (9.28)$$

In a region where ε and σ do not vary spatially, this leads to an exponential decay of ∇ E and charge density. Because the integral term in the expression for E already has zero divergence, $\nabla^2 V$ must equal zero in such a region. Thus Laplace's equation is solved to find V in a homogeneous region. On the boundaries of the homogeneous region, E must satisfy the above relation. For most purposes, the boundary conditions used to determine V can be computed for the stationary case of $di/dt = 1$, and the resultant V-field is then multiplied by di/dt. This neglects the $d\mathbf{E}/dt$ terms above. Omitting these terms neglects the time lag between the application of the field and the resultant rearrangement of charge; however, this approximation is good, even for highly capacitive tissues (Eaton, 1992). The result of this simplification is that the spatial distribution of the field does not vary with time.

Equation (9.27) can be used to predict the induced electric field inside the body for nearly any practical coil arrangement. The closed line integral is performed along the path of the coil winding. For the case of a circular coil, there is no closed-form solution to the integral, but it can be represented in terms of the complete elliptic integrals (Jackson, 1962). If simple body models are used, the determination of the V-field is not difficult. The finite-element approach has been used for simple-body models (Ueno et al., 1988), and can be applied to more complicated problems as well.

Local Stimulation of the Brain Cortex

A simple model for the brain cavity is that of a uniform, spherically shaped conductor bounded by an insulator, representing the skull. Various theoretical models have been applied to this problem (Roth et al., 1991; Grandori and Ravizzani, 1991; Eaton, 1992). Using the model of Eaton, Fig. 9.26 illustrates the magnitude of the electric field along a great circle that encloses the cortex at a distance 2 cm below the center of the coil (1.5 cm below the surface of the scalp). The cortex is assumed to have a diameter of 11 cm. An insert in the figure shows the top view of a spherical skull on which is placed one 6-cm coil (upper panel) or two 5-cm coils (lower panel). The vertical axis gives the induced E-field on an arbitrary scale that depends on the rate of change of current in the coil. The field magnitude is represented without regard to its direction. The E-field just beneath the coil mainly follows circular patterns approximating the shape of the coil when shown in the top view, and is primarily circumferential when shown in the side view. At greater depths the E-field would exhibit reduced peak E-fields, and would be less focal (since the depth of affected tissue is at a greater proportion of the coil dimension).

To illustrate the vertical scaling in the upper panel, a peak current of 3,700 A with a rise time of 0.2 ms into a single winding ($dI/dt = 1.85 \times 10^7$ A/s) would induce a peak E-field of 18.75 V/m, which is the theoretical E-field threshold (according to the SENN model) for a 10-μm fiber excited by a 0.2 ms pulse. The lower panel shows E-field patterns induced by a so-called figure-of-eight (F8) coil, in which the current is in a reinforcing direction at the interface of the two coils. To scale this figure, note that a peak coil current of 2,600 A with a rise time of 0.2 ms (1.3×10^7 A/s) in each coil loop would produce a peak E-field of 18.75 V/m. Not only is the F8 coil more efficient, but it is more focal, i.e., the E-field pattern is concentrated in a smaller region of the cortex. In this example, we have cited thresholds applicable to a terminated fiber having a diameter of 10 -μm. In the human pyramidal tract, 1.7% of myelinated fibers have diameters of 10μm or greater—some as large as 20μm (Lassek, 1942). Consequently, the cited thresholds would apply to the minority of fibers. For smaller fibers, thresholds are expected to rise in inverse proportion to fiber diameter (see Chapter 4).

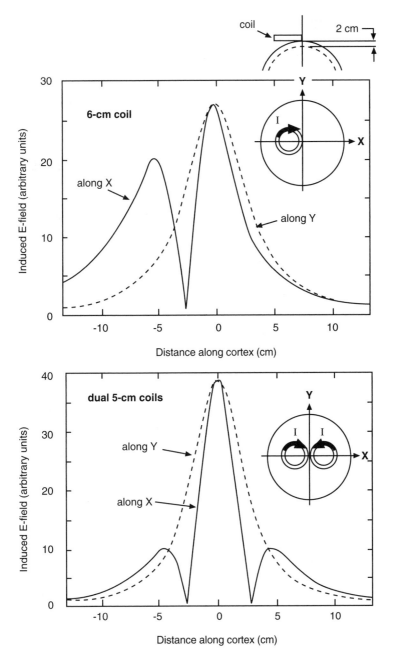

FIGURE 9.26. Magnitude of peak E-field along cortex of brain 2 cm beneath stimulating coils. Upper: single 6-cm coil; lower: dual 5-cm coils. Vertical axis is on relative scale that depends on coil dI/dt.

For many diagnostic applications of brain function, it is often important to confine the area of stimulation to as small a region as possible (Krauss et al., 1994). In attempting to increase the focality of magnetic stimulation, one confronts limitations dictated by the physics of magnetic induction. We cannot necessarily achieve greater focality by simply making the coil smaller, because the induced E-field fields at the cortex may actually become less focal as a result of the fact that the target tissue is at a greater depth in proportion to the coil size. Furthermore, there would be a less efficient coupling area of a small coil, thereby increasing the requirements on peak coil current. While the F8 coil does achieve greater focally and is more efficient than a single coil, it may be desirable to increase the focality even more for clinical applications.

One can achieve greater focality with a unique F8 coil design that makes use of the nonlinear electrodynamics of the neural membrane (Reilly et al., 1993). A subthreshold rectangular conditioning pulse (CP) is provided to one loop of an asymmetric F8 coil either before or during a sinusoidal stimulus, which is applied to the second loop. The interaction effects of a conditioning pulse with a sinusoidal stimulus are shown in Fig. 4.24, which

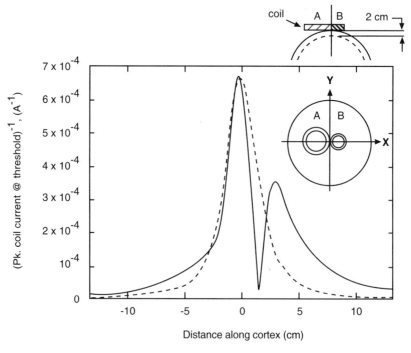

FIGURE 9.27. Excitation threshold requirements with dual-function stimulation. 5-cm coil has 0.2 ms current ramp @ 75% threshold. Vertical axis shows inverse of current in 2.5-cm coil needed to excite neurons at various locations along brain cortex as indicated on horizontal axis. Small coil has sinusoidal current at 80 kHz frequency.

Table 9.11. Threshold response with various coil configurations.

Coil configuration

coil size				
coil 1 (cm)	6	5	–	6
coil 2 (cm)	–	5	2.5	2.5
waveform				
coil 1	0.2 ms ramp	0.2 ms ramp	–	0.2 ms ramp
coil 2	–	0.2 ms ramp	80kHz, t=1 ms	80kHz, t=1 ms
Min. threshold current				
coil 1 (A - pk)	3703	2597	–	2800
coil 2 (A - pk)	–	2597	2632	1538
Spatial res. @ 130% T				
Δx (cm)	2.52	2.12	4.65*	1.22
Δy (cm)	3.85	3.19	2.32	2.12
$\Delta x \Delta y$ (cm^2)	9.70	6.76	10.79	2.59

* width encompassess two peaks beneath coil

is derived from the SENN model. Results from a particular implementation of this technique is shown in Fig. 9.27, in which a 0.2 ms CP is applied to a 5-cm diameter coil at 75% of the excitation threshold (2,800 A peak current, 0.2 ms duration) of a 10-μm nerve located in the highest induced field (see Fig. 9.25), and a sinusoidal wave at 80 kHz and 1 ms duration is applied to an adjacent 2.5-cm coil. The vertical axis in Fig. 9.27 indicates the inverse of peak current in the small coil needed to bring a 10-μm fiber to threshold if it is located at the position indicated on the abscissa. It is assumed that the fiber is a terminated one, with an orientation optimized for stimulation. For instance, the current threshold for a fiber located at the peak in the figure would require a peak sinusoidal coil current of 1,500 A in the 2.5-cm coil to achieve threshold.

As a measure of focality, we extracted the width of the excitation region when the stimulus is raised to 130% of the threshold of a 10-μm nerve in the optimum location. A justification for this arbitrary metric is that one cannot precisely target the threshold of a most sensitive neuron at the peak of the induced E-field, and that a group of neurons with diverse thresholds and orientations would likely be brought to excitation over a region of the cortex. Results using this metric are shown in Table 9.11 for several stimulus arrangements, including a 6-cm coil, a 5-cm F8 coil, a 2.5-cm coil, and a dual function stimulator with 6- and 2.5-cm coils. The table indicates the coil sizes, the current waveforms in each coil, the peak threshold current for a 10-μm neuron in the most sensitive location, the spatial width of the excitation region when the stimulating current is raised to 130% of its minimum threshold value, and the approximate excited area. This last parameter indicates the focality of the various schemes. The spatial resolution of excitation with the dual function stimulator is significantly smaller than with the other three configurations. For instance, the excited area with the dual function apparatus is 0.38 times that of the 5-cm F8 coil.

9.10   Scales of Reaction: Power Frequency Magnetic Field Exposure of the Head

Figure 9.28 illustrates a scale of reactions to power frequency (50, 60 Hz) magnetic field exposure. It is assumed that the adult human head is uniformly exposed to the field. The right-hand column indicates reactions that may occur with stimulation of neurons within the retina or the brain; the magnitude of the stimulus is placed at a vertical position such that the relevant threshold can be read on four scales that indicate various metrics of the exposure. The right-most numerical scale indicates the magnetic field flux density of a 60 Hz field expressed in milligauss (mG) and Tesla (T) units; the next scale to the left shows the temporal derivative of the field (units: T/s); the next scale expresses the maximum induced E-field (units: V/m) at the site of stimulation; the left-most scale expresses the current density (units: A/m^2) at the site of stimulation, assuming a brain conductivity of 0.15 S/m.

The thresholds indicated in Fig. 9.28 are approximate rms values at the median of a subject distribution under the stated conditions. The statistical distribution of thresholds among a large sample of subjects is lacking, and one can only estimate the median value and its variation. However, if the variation of thresholds for synaptic effects parallels that for peripheral nerve excitation (Table 7.10), one would anticipate that sensitivity at the one percentile rank among a large sample of subjects would be about a factor of 2 or 3 below the median. Indeed, Silny (1986) reported the lowest phosphene thresholds at 5 mT, applying to a frequency of 20 Hz, in contrast to the presumed median value of 10 mT at the same frequency from the data of Lövsund.

The four scales of Fig. 9.28 can be tied together only for magnetic exposure with a specified frequency and spatial distribution—in this case, uniform exposure of the head by a 50 to 60 Hz field. If we were to change the frequency, or the spatial distribution of the applied field, the thresholds and the interrelationships among the four scales would change, in accord with the principles expounded in Sect. 9.5 and 9.6.

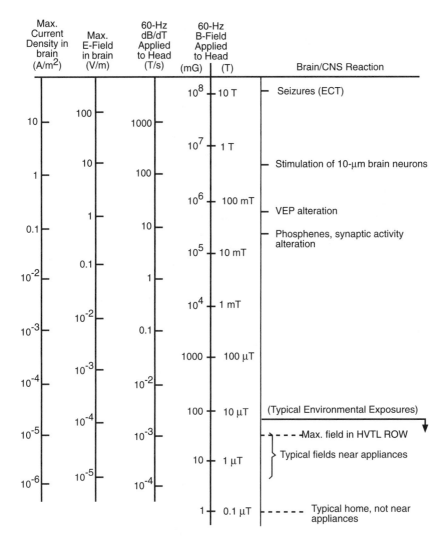

FIGURE 9.28. Human responses to power frequency magnetic exposure of the adult human head; spacially uniform magnetic flux density. Reaction thresholds are approximate rms values for median response.

The lowest reaction indicated is that of visual phosphenes. The phosphene threshold for 50 to 60 Hz stimulation is shown at a flux density of 20 mT-rms. Experimental evidence discussed in Sect. 9.8 strongly suggest that the responsible mechanism is the alteration of synaptic activity within the retina. A similar mechanism might be active at neural synapses within the brain, and therefore the phosphene threshold might apply to other central nervous system interactions. Evidence for this conjecture was provided by Silny (1985), who reported that 50 Hz magnetic exposure to the head can alter visual evoked potentials (VEP) at flux densities of 60 mT—somewhat above the phosphene threshold, but well below neural excitation thresholds. Silny also reported that his experimental subjects experienced headaches and "indisposition" with 60 mT exposure at a frequency of 50 Hz. Other evidence for CNS effects with sub-threshold electrical stimulation has been developed with experiments using in-vitro animal brain preparations (Wachtel, 1979; Bawin et al., 1984, 1986). Those studies demonstrated that an in-situ E-field below the threshold of neural excitation is capable of altering the excitability of brain neurons.

The next threshold shown in Fig. 9.28 applies to stimulation of neurons having an axon diameter of $10\mu m$, for which the rheobase rms E-field threshold is 8.8 V/m (12.4 V/m peak) according to the myelinated nerve model described in Chapter 4. The excited fiber is assumed to reside in the cortex, where the induced E-field would be greatest within the brain. The assumed fiber diameter is a relatively large one among the distribution of fibers to be found in the brain, although a small minority of fibers as large as $20\mu m$ have been reportedly found in the human pyramidal tract (Sect. 3.4). Excitation thresholds are inversely proportional to fiber diameter in accord with nerve excitation models (Sect. 4.5).

The highest threshold shown in Fig. 9.27 applies to the induction of seizures. Although I am not aware of any reports of seizures from magnetic stimulation, I have estimated the seizure threshold based on experience with surface electrodes used in electro convulsive therapy (ECT). For this purpose, I estimated the E-field threshold within the brain using the current density distribution data derived from models of surface electrodes on the head (Rush and Driscoll, 1968; Weaver and Rush, 1976), along with the applied current thresholds reported for ECT (Sackeim et al., 1987). The resulting magnetic threshold is a conservative estimate. Indeed, magnetic stimulation of the brain without untoward results has been demonstrated by many researchers (Ueno, 1994a), as described in Sect. 9.9.

While ECT has been useful in treating depression, the treatment has some significant side effects. As an alternative, repetitive transcranial magnetic stimulation (rTMS) has been explored for treatment of depression without the inducement of seizures (George et al., 1995). The authors reported significant improvements in some drug-resistant patients when rTMS was applied to the left prefrontal cortex of the brain using an F8 coil (see Sect. 9.9). The level of stimulation was 80% of the threshold needed to evoke a motor twitch of the abductor pollicis brevis muscle when the stimulator was placed over the motor cortex. Twenty-minute treatments were performed daily over several weeks. The main side effect, observed in two patients ($n = 6$), was the development of mild headaches that responded to aspirin or acetaminophen.

In principle, it is possible to create measurable short-term thermal effects within the brain with magnetic field exposure. For instance, consider the current density of $45\,A/m^2$ at the seizure threshold shown in Fig. 9.28. The corresponding specific absorption rate (SAR) is 12 W/kg in accord with Eq.

(11.8). As indicated by Eq. (11.11), a temperature rise of $1\,°C$ would require a duration of exposure of 333s at the indicated current density. In this example, a $1\,°C$ temperature rise requires a very high level of magnetic exposure for a long duration. However, if the frequency of the magnetic field is increased, the same temperature rise would be associated with a smaller magnetic field, since the induced E-field increases with frequency. Thermal considerations with high-frequency magnetic field exposure are treated in Sect. 11.5.

At the lower portion of Fig. 9.28 are listed typical environmental exposures in nonoccupational settings. The examples show field levels that may be encountered within the rights of way of high voltage transmission lines, near household appliances (Gauger, 1985), or within a typical home. These environmental exposures are several orders of magnitude below the lowest reaction threshold shown in the figure. In certain occupational settings, however, much higher magnetic field exposures may be encountered. For instance, workers in the electrosteel industry or welding industries can encounter magnetic fields at power frequencies that approach the indicated phosphene thresholds (Lövsund et al., 1982).

9.11   Magnetic Forces on Moving Charges

Magnetohydrodynamic Effects

When a charged particle moves within a magnetic field, it experiences a force

$$F = Q\mathbf{v} \times \mathbf{B} \qquad (9.29)$$

where $\mathbf{F}$ is the force, Q is the charge on the particle, $\mathbf{v}$ is its velocity, and $\mathbf{B}$ is the magnetic flux density; bold type indicates vector quantities. According to the cross-product expressed in (9.29), the vector quantities are in a mutually perpendicular orientation (i.e., velocity, field, and force oriented along x, y, and z directions, respectively). The magnitude of the force is given by

$$F = QvB \sin\theta \qquad (9.30)$$

where θ is the angle between $\mathbf{v}$ and $\mathbf{F}$. This force, known as the *Lorentz force*, causes moving charges to drift within a magnetic field in a direction perpendicular to their flow. Because of this drift, there results a separation of charges, which can be detected as a voltage known as the *Hall effect*. The magnitude of the Hall effect voltage is

$$V = v\,B\,d\,\sin\theta \qquad (9.31)$$

where d is the distance between two plates used to sense the voltage. This principle has lead to many useful devices, including a blood flow meter (Kolin, 1952). In a biological medium, blood flow is an example of moving charges, which, in a magnetic field, would be expected to experience a Lorentz force, and a Hall voltage. To illustrate the application of Eq. (9.31), consider a flow rate of 0.6 m/s in the human aorta within a field of 1 T. Assuming a vessel diameter of 2.5 cm, the maximum induced potential across the aortic cross section is 15 mV according to Eq. (9.31).

The Hall effect voltage can be observed in an ECG as an enhancement of the T-wave amplitude in rats exposed to strong static magnetic field (Gaffey and Tenforde, 1981). The average minimum field strength at which a measurable enhancement occurred was reported as 0.3 T; at 2 T, the average enhancement was 408%. The authors determined that this enhancement was caused by the superposition of a Hall effect potential on the natural ECG signal during the repolarization (T-wave) phase of the heart. During this phase, ejection of blood into the ascending aorta would give rise to a flow potential according to Eq. (9.31), thereby augmenting the T-wave. It was observed that the T-wave returned to normal immediately after cessation of the magnetic field, even with exposures up to 5 h. The authors observed no arrythmias, changes in heart rate, or changes in respiration rate for exposure up to 2 T. Similar studies with primates demonstrated no measurable changes in blood pressure for exposure up to 1.5 T (Tenforde et al., 1983). Others have demonstrated no significant difference in cardiac excitation thresholds with pulsed magnetic stimulation with or without a concurrent 1.5 T static field (Nyenhuis et al., 1992). A transient increase in blood flow was observed in mice exposed to an 8 T field (Ichioka et al., 1998).

Other investigators (Jeneson et al., 1988) were able to detect a statistically significant increase in the cardiac cycle length of 17% in resting human subjects within a longitudinal 2-T static field. The authors suggested that these effects might have been seen in human subjects but not animal subjects as a result of the relatively larger human dimensions. The authors opined that the observed effect is probably harmless in healthy subjects, but that its safety in dysrhythmic patients was not certain.

Magnetohydrodynamic forces can also exert a drag on blood flow in very intense static magnetic fields. Theoretical models predict drag forces will affect vascular pressure by less that 0.2% in a field of 10 T (Keltner et al., 1990); others have predicted a change in axial velocity of blood flow by a few percent in a 1 T field (Dorfman, 1971).

Additional effects attributed to magnetohydrodynamic forces include vertigo, taste sensations and nausea in human subjects exposed to static fields of 4.0 T (Schenk et al., 1992). Sensations of vertigo and taste were also reported by a few subjects in a 1.5-T field. In all cases, the reported sensations were associated with rapid head movement within the field. Some subjects also reported phosphenes during rapid eye movement in the 4-T field. For subjects lying stationary within the field, the effects were absent. The cause for the reported effects was not considered in this study, but one may hypothesize mechanisms involving mechanical forces on the inner ear, or induced electrical potentials.

Ion Resonance

An unrestricted ion traveling within a magnetic field will travel in a circular path because of the Lorentz force. If a static field is combined with an alternating field, a resonant condition will exist when the alternating field has a frequency

$$f_c = \left(\frac{1}{2\pi}\right)\left(\frac{q}{m}\right)B_0 \qquad (9.32)$$

where f_c is the frequency of the alternating field, (q/m) is the charge-to-mass ratio of the ion, and B_o is the magnitude of the static field. This condition is known as *ion cyclotron resonance* (ICR).

Certain biological reactions have been observed to be sensitive to extremely low frequency alternating magnetic fields at frequencies that follow ICR conditions within a geomagnetic field (Postow and Swicord, 1996). For instance, the $^{40}Ca^{2+}$ ion, with a q/m ratio of $4.82 \times 10^6 C/kg$, has a resonance frequency of 34.8 Hz in a static field of $50\mu T$ (Liboff and Parkinson, 1991).

The existence of a pure ICR mechanism in biological systems is not a realistic expectation because of the requirements of a large, unrestricted path for the ion. Proponents of ion resonance mechanisms do not suggest that cyclotron resonance is actually occurring in the biological medium, but rather that some mechanism is present that follows the ICR relationship. One suggested mechanism involves *ion parametric resonance* (IPR), as proposed by Lednev (1991), and later clarified by Blanchard and Blackman (1994). The IPR model predicts that a resonant condition is created by the combination of a static and an alternating magnetic field, which influences transitions of energy states of ions, and that these transitions can affect biological activity within cells. The critical IPR frequencies coincidentally occur at multiples of the cyclotron resonance frequency. The IPR theory predicts that resonance conditions depend on the ratio of flux densities of the alternating and static magnetic fields, whereas the magnitude of the AC field does not appear in the ICR formulation (Eq. 9.32). Empirical tests with cellular preparations show the response of neurite outgrowth can be affected by application of static and alternating magnetic fields that follow the IPR formulation (Blackman et al., 1994).

10
High-Voltage and High-Current Injuries[1]

MICHAEL A. CHILBERT

10.1 Introduction

Injuries resulting from electrical accidents can include tissue destruction, cellular excitation, and trauma secondary to the passage of current. Thermal injuries in the extremities can lead to amputation because of the deep nature of the burn. Cell lysis can also destroy tissue if there is a sufficient electric potential across the cell membrane. The effects of lysis are sometimes delayed. Cellular excitation of muscle and nerve can lead to cardiac fibrillation or transient neural dysfunction. Secondary injuries result from flash burns, falling, or gross contraction of muscles.

Electrical accidents result in nearly 1,100 deaths per year in the United States, 10% of which are caused by lightning. One-third of these deaths are caused by voltages below 1,000 V occurring in the home and workplace. Low voltages account for more than half of the industrial deaths (Dalziel, 1978), High-voltage deaths are typically from industrial accidents. About 2% of low-voltage accidents and 10% of high-voltage accidents are fatal.

The number of accidents resulting in survivable injury are not consistently documented, because victims are not admitted to the same critical-care areas within a given facility, and those not surviving at the scene are not admitted. Victims of severe burns are, of course, admitted to burn-care units, whereas those afflicted only with cardiac maladies are not. Typical injury reports include only the burn victims. The evaluation of the burn's extent is very difficult and may require multiple procedures. Electric burn injuries account for 4% to 6% of all admissions to burn-care facilities (Hammond and Ward, 1988; Hunt et al., 1980; Rosenberg and Nelson, 1988). These burns cover an average of 12% of the body surface area (BSA), but result in limb amputation for 50% to 70% of the cases. The

---

[1]This research was supported in part by NIH research grant GM34856 and the Department of Veterans Affairs Medical Center Research Center Research Funds, Milwaukee, Wisconsin.

average BSA resulting in subsequent death is only 34%, compared to a much higher BSA percentage for thermal burns.

This chapter emphasizes electrical burn injury and its theoretical basis. The physical parameters affecting the character of injury and including impedance considerations and current distribution throughout the body are also included. Discussions of thermal and nonthermal trauma follow. Lightning injuries often have unique characteristics and traumatic sequelae; these are to be discussed as well.

10.2 Modes of Injury

The following sections provide background for the forms of trauma seen clinically. Thermal trauma has the greatest consequence to survivors of electric accidents; loss of limb is often the result. Nonthermal trauma has the form of lesions in neural and connective tissues. Lightning injuries often involve lesions exclusive of burns. Other trauma affects the heart rhythm and can result in fibrillation.

Thermal Injury

Thermal injury is caused by heating of tissue from the passage of current. The amount of heat generated in the tissue depends on spatial and temporal patterns of current density and tissue resistivity. The current density and resistivity are, in turn, affected by the heat generated in the tissue. Consequently, thermal injury involves a feedback process.

Heat generated in the tissue is given in terms of energy density:

$$Q_j = J^2 \varrho t \tag{10.1}$$

where Q_j is the thermal energy density (J/cm³), J is current density (A/cm²), ϱ is the resistivity (Ωcm), and t is the duration of current. In its simplest form the thermal energy density can be related to the change in temperature in the following manner:

$$Q_j = pc\Delta T \tag{10.2}$$

where p is the tissue density (g/cm³), c is the specific heat of tissue (J/g°C), and ΔT is the temperature change (°C). Equation (10.2) assumes that there are no thermal losses by conduction to blood and adjacent media, or by convection or radiation into air. In other words, all boundaries of the specific volume are adiabatic, and there are no internal heat sinks or sources in the volume. By solving for the change in temperature using Eqs. (10.1) and (10.2),

$$\Delta T = \frac{Qj}{pc} = \frac{J^2 \varrho t}{pc} \tag{10.3}$$

Thermal injury depends on the duration of exposure and the temperature in accordance with Figure 11.12 (Henriques, 1947; Henriques and Moritz, 1947). Cutaneous burns occur when the temperature is elevated for a sufficient length of time: 45 °C requires more than 3 h, 51 °C requires less than 4 min, and 70 °C requires less than 1 s for injury. Temperature levels that cause injury in other tissues are similar, as is the injury rate for a given temperature. Electrically induced thermal injury of muscle begins at 43 °C with a 1-A current in the limb for 15 min, and at 46 °C the tissue damage is much greater (Chilbert et al., 1985b). Electrically induced thermal damage to nerve has been noted to occur at 48 °C after several seconds. Necrosis limits have been measured for various tissue types in humans and animals (Héroux, 1992). For example, with 2.75 min. of exposure, the temperature for necrosis was found to be 48.1 °C in human skin, 50.9 °C in porcine skin, and 49.3 °C in porcine muscle.

The deposition of heat depends on the current density and resistivity. Resistivity decreases with increasing temperature, causing further increases in current density and temperature. This aspect of resistivity change is rarely considered in burn models because of the resulting nonlinear equations. Heat is removed from the tissue by conduction into adjacent tissue, by convection and radiation into the air at the surface, and by blood flow. To account for conduction and blood flow effects, one can expand Eq. (10.2) into:

$$pc\frac{\partial^2 T}{\partial t^2} = k\frac{\partial^2 T}{\partial x^2} + k\frac{\partial^2 T}{\partial y^2} + k\frac{\partial^2 T}{\partial z^2} + wp_b c_b\left(T_b - T\right) + q_m + q_j \quad (10.4)$$

where T is the tissue temperature, k is the thermal conductivity of tissue, w is the blood perfusion rate, p_b is the blood density, c_b is the specific heat of blood, T_b is the initial temperature of blood, q_m is the volumetric rate of metabolic heat generation, and q_j is the volumetric rate of electrical heating. Note that the first three terms of the equation describe the tissue conduction, and the fourth term describes the effects of vascular convection. Equation (10.4) is termed the *bioheat equation*, which was expressed in its initial form by Pennes (1948) and has been modified extensively for various applications (Song et al., 1988). Table 10.1 lists measured values of thermal

TABLE 10.1. Thermal conductivity values (k) for skin, fat, and muscle.

Medium	Conductivity, $k \times 10^3$ (W/cm °C)
Amorphous	2.09
Skin	3.40
Fat	3.23
Muscle	5.00
Water (40 °C)	6.32

Source: Data from Pennes (1948), Olsen et al. (1985), Song et al. (1988).

conductivity for skin, fat, and muscle of human and hog tissues. These values have been used in thermal models. Values for the thermal conductivity of water are included for comparison. The thermal conductivity of blood used by Song et al. (1988) was 5.0×10^3 W/cm °C. Other parameter values used were $p_b = 1.05$ g/cm^3 and $c_b = 3.8$ J/g °C. The value for w is 2 to 12(ml blood)/(ml tissue)/s for dermal perfusion.

Lee and Kolodney (1987b) developed a unidimensional, axisymmetric model for the heating resulting from a high-voltage electrical contact in the upper limb. The limb was modeled as a coaxial cylinder having a 10-cm diameter and included layers to represent bone, muscle, fat, and skin of 1.0-, 3.5-, 0.3-, and 0.2-cm thicknesses, respectively. The bioheat equation used for this model included the term for electrical heating while neglecting metabolic heat. Lee and Kolodney show that tissue perfusion is critical for the removal of heat in the tissue, but does not significantly modify the heating process during the application of current.

Thermal injury depends on the current path, which is determined by the contact points on the body (sometimes termed entry and exit sites). Inside the body, current distribution is determined by tissue resistivity. Early perceptions of current in the body stated that it would flow equally through all tissues or that it would flow along the path of least resistance (in the vessels and nerves). Both are wrong, but are still referred to in clinical reports. Experimental evaluation of the current path through the body has shown that the resistivity of each tissue in a given cross-section determines the current distribution through the cross section (Chilbert et al., 1983, 1985b, 1988, 1989; Sances et al., 1981a, 1983).

Tissue resistivity may be anisotropic; that is, the resistivity can vary with current direction (Sect. 2.1). For example, muscle has a lower resistivity along its fibers than across them, and skin has a high resistivity across its layers and a lower resistivity along its layers. Some tissues can be considered isotropic. The implications of anisotropic resistivity in thermal burns is that the direction of current through a given tissue may explain selective tissue trauma seen away from the contact site. This selective tissue destruction is well documented and is seen in muscle tissue near the bone (Artz, 1974, 1979; Hammond and Ward, 1988; Luce et al., 1978; Moncrief and Pruittion, 1971; Pontionen et al., 1970; Rosenberg and Nelson, 1988; Sances et al., 1979; Wang et al., 1984, 1985, 1987; Zelt et al., 1988).

Selective thermal trauma can also be related to tissue resistivity changes with temperature and events at the contact site. As the tissue temperature increases, the resistivity decreases, altering the current distribution and subsequent temperature changes in the tissue. Contact site events that change the current distribution are desiccation and arcing. Desiccation of skin at the contact will cause the total current to decrease rapidly. This limits the contact time at voltages below 1,000 V and the trauma is usually limited to the contact region. With higher voltages, arcing can occur over the desiccated tissue and can alter the current path.

Electroporation

When the electric field strength across the cell membrane is sufficiently large, the cell membrane will rupture and cause cell lysis (Lee and Kolodney, 1987a; Lee, 1992; Weaver, 1992). The process is called *electroporation*, which is an increase in membrane permeability as a result of development of aqueous pores in the membrane (Lee et al., 1988). Electroporation of the cell membrane leads to rupture when the pore size becomes large enough or when fusion of several pores occurs. Calculation of the membrane potential uses cable theory, which is discussed in Chapter 4, and depends on the cell length, cell diameter, the internal fluid conductivity, the membrane conductivity, and the thickness of the cell membrane, and has been derived for electrical injury situations by Gaylor et al. (1988).

The membrane potential causes the pore size to increase proportionately until an irreversible state is reached. This occurs when the pore size reaches a diameter of one-half the thickness of the membrane. The membrane potential causing rupture is in the range of 800 to 1,000 mV. Reversible electroporation has been noted to occur below a 200-mV membrane potential. The incidence of pores has been evidenced by experimentally observed increases in membrane permeability, but the exact mechanism of electrical membrane failure has not been determined, although it is thought to be initiated as molecular defects in the bilayer component of the cellular membrane (Lee and Kolodney, 1987a; Lee et al., 1988).

Electroporation can be detected in culture as a sudden reduction of impedance across the cellular membrane. The membrane voltage required for electroporation depends on the duration of the stimulus, as illustrated in Fig. 10.1. In this example, a minimum breakdown potential signaling reversible electroporation is achieved in a giant algal cell for stimulus durations in excess of about 100 μs; breakdown can occur with shorter stimulus durations in accordance with a strength–duration (S–D) law not unlike that observed for neural excitation effects (Chapter 4). Reversible electroporation has applications in cellular biology research (Neuman et al., 1989; Teissie and Rols, 1994), as it can provide a means for inserting chemicals into the interior of a cell, while otherwise leaving the cell intact.

The role of electroporation in electrical injury is related to the electric field strength in the biological medium. Studies of muscle cells in culture by Lee et al. (1988) have shown that an electric field strength between 50 and 300 V/cm can disrupt cells of 1 mm length. This is consistent with values of the membrane potential of his earlier study (Lee and Kolodney, 1987a). For smaller cells, such as fibroblasts of 10-μm diameter, the electric field strength that causes rupture exceeds 1,000 V/cm. The relation of this data to accident victims is given in Sect. 10.5. For small cells, the membrane potential depends on the cell length, cell diameter, and membrane thickness (Gaylor et al., 1988); for long cells, the membrane potential depends on the length constant and the electrical field strength (see Chapter 4). Muscle

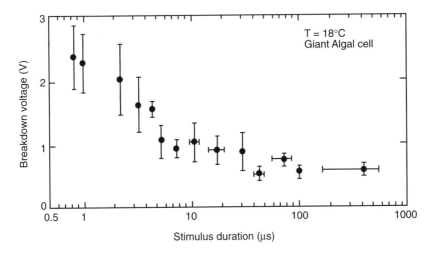

FIGURE 10.1. Strength–duration relationship for reversible cellular membrane breakdown. Vertical axis indicates potential across cell membrane. (Reprint from Benz and Zimmerman, © 1980 with permission from Elsevier Science, The Boulevard, Langford Lane, Kidlington OX5 1GB, UK.)

cells greater than 1 cm in length can be considered electrically long. The electric field needed for electroporation decreases as temperature increases. Consequently, the tissue trauma caused in a region of elevated temperature is related to both electroporation and temperature.

Fibrillation

The effects of electricity on the heart are treated in Chapter 6. This chapter will focus on short-duration, high-voltage, and high-current applications. Power-line accidents typically result in severe burns, whereas household voltages result in fibrillation (Sances et al., 1979). Most deaths at power-line voltages are attributed to burns or to secondary trauma such as falling. Severe burns or complications arising from them are the causes of death in the hospital. Rarely is instantaneous death attributed to fibrillation at high voltages, yet several case reports of latent cardiac anomalies have been published (Ahrenholz et al., 1988; Dixon, 1983; Guinard et al., 1987; Hammond and Ward, 1988; Rouse and Dimick, 1978; Wilson et al., 1988). However, fibrillation can occur at high current levels with very short application times, suggesting that these levels are not constant and may vary in a similar fashion to the minimum fibrillation level.

Ferris et al. (1936) published a comprehensive study on fibrillation from currents up to 17 A for a duration of 0.3 s. Until recently, most studies of fibrillation have used current levels below 10 A to delineate minimum electrical thresholds for the cardiovascular system. Kouwenhoven (1964) reported that fibrillation does not occur above 7.5 A when the contact lasts longer than 1 s. Since 1980, this author and coworkers have been involved with the study of electrical injuries and current pathways associated with high voltages and currents (Chilbert et al., 1983, 1985a, 1985b, 1988; Prieto et al., 1985; Sances et al., 1979, 1981a, 1981b, 1981c, 1983). It was noted that fibrillation often occurred at current levels above 10 A; one study investigated high-current fibrillation in hogs (Chilbert et al., 1989).

In the high-current fibrillation studies of the author and coworkers, the current delivery system supplied 1.5 to 200 cycles of 60-Hz currents between 1 and 51 A at voltages between 500 and 6,000 V. The current was applied during the preselected cardiac cycle through a triggering circuit synchronized with the cardiac waveform (Chilbert et al., 1988) as indicted in Fig. 10.2. Nineteen fibrillations occurred in 67 runs performed on 20 animals. Results show that 18 out of 32 short-duration runs involving the T wave caused fibrillation; 9 fibrillations out of 14 runs occurred below 10 A and 9 out of 18 occurred above 10 A. As the current increases, the likelihood of defibrillation increases (see Chapter 6). This also indicates that the likelihood of fibrillation decreases with current amplitude and has been noted by others (Daiziel, 1968, 1972; Geddes et al., 1986). The short-duration applications used here were always less than one complete cardiac cycle (excluding the run of 200 cycles).

The work of Ferris et al. (1936) has shown that, for current durations of 30 ms (about 2 cycles at 60 Hz) during the sensitive intervals, fibrillation can be induced by 15-A, 60-Hz currents in sheep. They also noted that the high-voltage, short-duration shocks do not show a cumulative effect (i.e., increase fibrillation's rate of occurrence), and that normal cardiac rhythm returned within 5 min in the nonfibrillating runs. Of 913 runs in 132 sheep, only 100 resulted in fibrillation. They recorded 11 fibrillations out of 16 runs at 4 A, 6 fibrillations out of 17 runs at 12 A, and one fibrillation out of 51 runs at 24 A. This shows a decrease in the likelihood of fibrillation as the current level increases; the occurrence at 25 A is about 10 times less than that at 5 A. This is consistent with the hog data above. For currents of 4 to 14 A applied for 30 ms during the other intervals of the cardiac cycle, no fibrillations were recorded; however, 4-A currents applied for 150 ms to 260 ms did cause occasional fibrillations.

Ferris tabulated body weight and heart size of sheep, hogs, dogs, and calves. One important aspect of body weight and heart size is the ratio between the two (Table 10.2). For similar body weights, the hog heart is closest in size to the human heart, because the body weight and the ratio are very similar. The thoracic circumference of hogs and humans is also comparable for specimens of the same weight, although the cross-sectional arrangement of organs and other tissues is different. The similarities of ratio and circumference suggest that the hog is an excellent model for fibrillation in humans. Also, the results of Ferris et al. (1936) show that hogs tend to fibrillate at lower-than-average current levels, thus making the hog a conservative model for ventricular fibrillation.

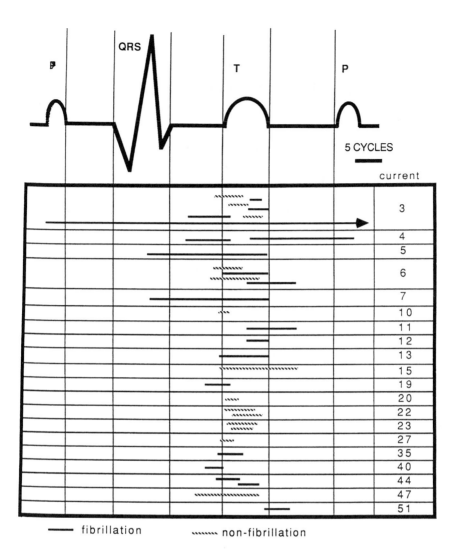

— fibrillation   ﹏﹏ non-fibrillation

FIGURE 10.2. Occurrence of fibrillation during specific periods of the cardiac cycle. Current applications causing fibrillation all are associated with the T-wave period. Fibrillation occurs less frequently at higher current levels. All currents involve the cardiac cycle shown if normal cardiac rhythm is extrapolated past the onset of current. (From Chilbert et al., © 1989 IEEE.)

TABLE 10.2. Heart/body ratio of animals and humans.

Species	N	Average weight Heart (g)	Body (kg)	Ratio (%) heart/body
Human	*	280	70	0.40
Hog	9	300	79	0.38
Dog	10	170	22	0.77
Sheep	25	270	56	0.48
Calf	10	420	70	0.60

Source: Data from Ferris et al. (1936); Chilbert et al. (1989).

419

10.3   Impedance Considerations and Current Distribution in the Body

Total Body Impedance

The current that flows through the body in high-voltage and high-current injuries is determined by the total body impedance and its change with time. The voltage is typically constant in electrical injuries. The minimum total impedance typically assumed for low-voltage electrical accident analysis is $500\,\Omega$ (Dalziel, 1978; Taylor 1985), and variations to this value are discussed in Chapter 2. At high voltages the total impedance of the body is less in short-duration contacts, while for longer durations the values increase with time.

Figure 10.3 depicts the current-time relationship for electrical contact at three low-voltage levels. The current amplitude can be separated into three phases determined by its rate of change with time. The initial phase occurs

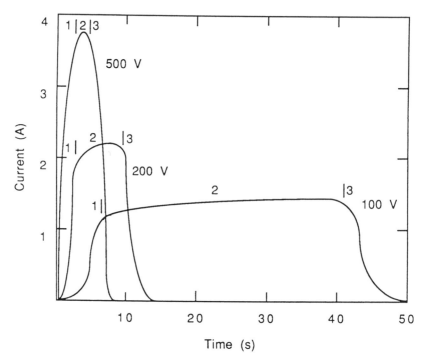

FIGURE 10.3. Relationship of current to time for a 5-cm disk in contact with the skin of a hog at different voltages. Note that an increasing voltage results in a decrease in the effective application time. Each trace shows the triphasic characteristics for current, which are the initial phase (1), middle phase (2), and final phase (3), as described in the text.

TABLE 10.3. Measured changes with voltage in the hog.

Volts	n	Effective application time (s)	Maximum applied current (mA)	Minimum total resistance (Ω)	Total energy (J)
75	13	205	721	104	8,773
100	14	92	806	124	6,008
150	15	21	1,087	138	2,789
200	13	11	1,227	163	2,208
250	11	5.6	1,471	170	1,734

Note: Values for a 2.5-cm-diameter disk applied to the hindlimb with a large plate electrode on the other hindlimb.

as the contact is established (the contact impedance is rapidly reduced); this is sometimes termed the breakdown of the electrode-skin interface. The middle phase is the phase of maximal current flow. Increases during the middle phase can be attributed to heating of the tissue near the electrode contact and in cross-sections where large current densities occur. The final phase occurs when the tissue under the electrode becomes highly resistive because of desiccation or charring of the skin. The rate at which these three phases occur decreases as the voltage increases (Fig. 10.3). The three phases are present over a range of application times (Carter and Morely, 1969a, 1969b; Prieto et al., 1985; Sances et al., 1981a, 1981b). Application times for five voltage levels are listed in Table 10.3 and show the decrease in time with increasing voltage. With increasing voltage, the minimum total impedance increases and the total energy decreases. The changes in impedance and energy are caused by the rate of charring at the electrode site. The contact phenomenon of Fig. 10.3 is applicable for low voltages that do not support arcing.

High voltage and current are usually associated with arcing at the electrode site (Sances et al., 1979, 1981c). The initial and middle phases of the current-time plot can occur within milliseconds; the onset of arcing is the final phase. Voltages above 1,000 V are sufficient to arc through air and desiccated skin. Once arcing is established, the arc length and internal impedance limit the current. Figure 10.4 shows that current falls with time when an arc is established, indicating that the arc length increases as tissue is burned away. The arcing appeared to occur over regions of uncharred tissue, taking the path of minimum resistance (Sances et al., 1981c). The tissue removal by the arc continues until the circuit is interrupted or the limb is transected. At 7,200 V (Fig. 10.4), soft tissue was removed from the limb within 5 s; the remaining 11 s were required to transect the bone. At 14,400 V, limb transection time is about 3 s. Circuit interruption can occur when the nonconduction path length exceeds the voltage's ability to maintain an arc or by a ground-fault interrupter, circuit breaker, or fuse (Sect. 11.7).

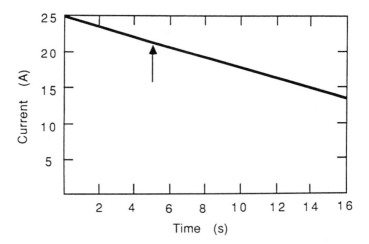

FIGURE 10.4. Current versus duration for limb transection at 7.2 kV. Arrow indicates point of soft tissue removal; remaining time is needed to transect the bone. The contact was between a wire contact on the hindlimb and a large plate under the other hindlimb. [From Sances et al. (1981b), © 1981 IEEE.]

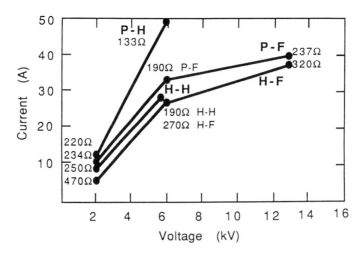

FIGURE 10.5. Current versus voltage for first contact in the hog. P-H is large plate electrode-to-hindlimb wire configuration, P-F is large plate electrode-to-forelimb wire, H-H is hindlimb-to-hindlimb wire contacts, and H-F is hindlimb-to-forelimb wire contacts. The plate was located under the hindlimb opposite the wire contacts. Total body impedances are shown for each measurement point and were determined from the average current for each run. [From Sances et al. (1981b), © 1981 IEEE.]

TABLE 10.4. Total body resistance (Ω) for band electrodes located on the hindlimb and analogously on the forelimb (8-cm disk electrode values d included for comparison).

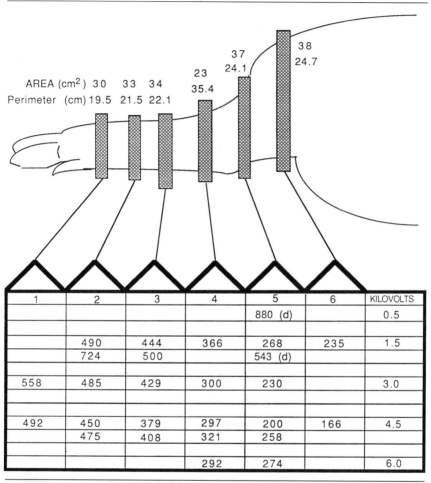

Source: Adapted from Chilbert et al. (1989).

Experimental studies of electrical burn injury have been performed in the hog, since this animal approximates the human in skin, anatomy, weight, and cross-sectional dimensions better than other species. The total impedance of the hog at high voltages for different electrode types and locations is given in Fig. 10.5. This figure shows that a large contact (plate electrode) on the hindlimb decreases the total impedance and that forelimb-to-hindlimb impedance is greater that hindlimb-to-hindlimb impedance. This is further illustrated in Table 10.4, which shows a decreasing impedance with increasing contact area and perimeter. As the location of the band electrodes is moved up the limbs, the impedance decreases, which is consistent with observations made previously (Prieto et al., 1985; Sances et al., 1981b). The impedance also changes with the applied voltage level.

Impedance depends on both applied voltage and its duration (see Fig. 10.10). At low voltages applied for long time periods (those that exceed the second phase), the impedance increases with voltage while the total energy decreases (Fig. 10.6a). The decrease in total energy with increasing voltage

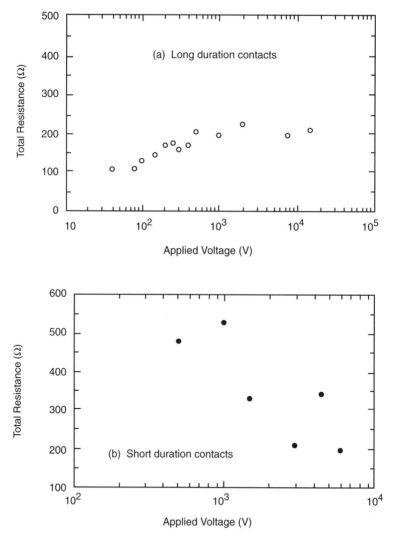

FIGURE 10.6. Minimum total body resistances for voltages applied between the hindlimb and a large plate on the opposite hindquarter. (a) Voltages applied for long time periods (through the third phase of the current-time waveform). (b) Voltages applied for short time periods (the first phase of the current-time waveform).

indicates less tissue heating at higher voltages (Table 10.3). For short periods of time ($t < 1$s), the impedance depends on the initial phase of the current waveform (Fig. 10.3). The short-duration impedance at low voltages will therefore decrease as the voltage increases, which is evident in Fig. 10.6b. High-voltage impedance changes are similar to those of low-voltage changes. Short-duration contacts occurring in the first phase are typically less than 100 ms, and long-duration contacts would be denoted by the onset of arcing.

Body impedance varies with voltage, duration, and contact parameters. Increasing the voltage increases the current and decreases the total impedance for short application times. The longer application times increased the peak current at lower voltages, because the tissue resistivity decreases with increasing temperature near the electrode sites. The electrode contact site on the limbs determined the contact area and the perimeter for the band electrodes (Table 10.4), and this affects the total current flow as well. In general, increasing the area and perimeter increases the current for a given voltage; consequently, the largest parts of the limbs had a greater contact area and perimeter and allowed greater current passage with the band electrodes. Also, the location of the contact on the body will affect the total body impedance, since different current paths have different impedances.

Contact Impedance and Segmental Body Impedance

Contact impedance will limit the level of current in the body, its duration, its path, and the probable extent of trauma. The remaining portion of the total impedance can be divided into individual impedances for the various segments of the body. These segmental body impedances can be modified to account for tissue heating away from the contact points. Tissue heating decreases impedance, thereby increasing current. By combining information from the contact impedance and the segmental body impedances, one can better model the effects of electrical trauma.

Experimental evaluation shows that, for circular and elliptical contacts, the contact impedance is related to contact area and perimeter (Prieto et al., 1985), where the contact impedance can be expressed as

$$R_c = KA^{-1/4}P^{-1/2} \tag{10.5}$$

where R_c is the contact resistance, K is a constant of resistivity, A is the electrode contact area, and P is the electrode contact perimeter. For a circular electrode $R_c = 0.3 K/r$, where r is the electrode radius. The constant K depends on the applied voltage, temperature, and skin condition (wet, dry, abraded, etc.). For dry electrode contact at 100 V, Prieto et al. (1985) derived a value for K of 264 for areas between 0.3 cm^2 and 20.3 cm^2 and perimeters between 2 and 28 cm. The result of a least-squares linear regression fit of this equation to the experimental data gives a correlation coefficient of 0.94. As the electrode size increases in this model, the contact

resistance approaches zero, leaving only the internal impedance to limit current.
Further analysis of the contact has been performed by using finite-element analysis on an axisymmetric model in two dimensions. Figure 10.7 shows the boundary and arrangement for the analysis of a circular disk containing layers for skin, fat, and muscle. The model is based on a 5-cm-diameter disk with a potential of 100 V contacting the tissue, which is a nonhomogeneous, anisotropic, semi-infinite medium. Model parameters are listed in Table 10.5. The model includes surface and volume heat transfer effects and temperature-dependent resistivities. The skin and fat layers have isotropic resistivities; the muscle resistivity is anisotropic with its transverse resistivity defined in the z direction and its longitudinal resistivity defined in the r direction. Thermal properties are isotropic in all tissues.
The model determines the current density and temperature distribution in the tissue. Plots of the thermal isoclines resulting after 5 s and 9 s of

TABLE 10.5. Finite-element analysis parameters for a disk electrode in contact with a nonhomogeneous, anisotropic, semi-infinite medium used to model electric burn injury at the contact site.

Dimensions (cm)	
Disk radius	2.5
Skin thickness	0.15
Fat thickness	0.35
Dirichlet boundary conditions	
Ambient air temperature (°C)	25
Initial tissue temperature (°C)	37
Disk potential (V)	100
Reference potential (V)	0
Material properties	
Electrical resistivity (Ωcm)	
Skin	280
Fat	375
Transverse muscle	650
Longitudinal muscle	290
Thermal conductivity (W cm^{-1}°C^{-1})	
Skin	0.037
Fat	0.020
Muscle	0.042
Density (g cm^{-3})	
Skin	1.00
Fat	0.85
Muscle	1.05
Specific heat (J g^{-1}°C^{-1})	
Skin	3.2
Fat	2.3
Muscle	3.8
Heat transfer coefficient: 13.45×10^{-6} W cm^{-2}°C^{-1}	

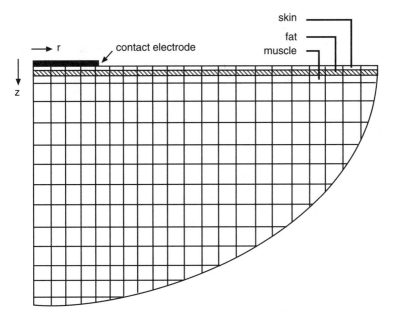

FIGURE 10.7. Geometric arrangement for an axisymmetric finite-element analysis of a disk on a three-layered, semiinfinite medium.

contact are shown in Fig. 10.8a and 10.8b. The thermal gradient emanates from the edge of the disk and proceeds somewhat concentrically into deeper tissue. Surface temperatures indicate a peak value at the edge of the disk (Fig. 10.9), showing the effect at the edge. The electrode edge effect of increased temperature has been shown experimentally by Sances et al. (1981) in saline and in the hog using a thermographic camera. The total current entering the tissue through the disk electrode is shown in Fig. 10.10. The model compensates for changes in skin that exceed 100°C by rapidly increasing the resistivity, resulting in a drop in the current after 9s. This simulates desiccation of the skin comparable to experimental observations of Prieto et al. (1985). The analysis shows that the greatest current density and temperature increase occurs under the edge of the disk. More than 50% of the current entering the tissue enters in the outer 15% of the disk radius or in the outer 28% of the total area. These results indicate the importance of the circumference to the level of current that will flow into the body. Disk electrodes at high voltages of similar size to those used previously at household voltage levels show similar changes due to contact area and edge length, indicating that the results of the finite-element analysis are valid at high voltages as well (Chilbert et al., 1989; Prieto et al., 1985). Theoretical studies by Caruso and colleagues (1979) showed that current is concentrated at the edge of an electrode to a degree that depends on the conductivity layers beneath the electrode (see Fig. 2.9).

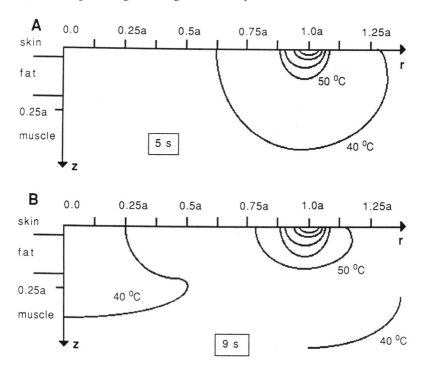

FIGURE 10.8. Results of the finite-element analysis showing thermal isoclines at (a) 5 s and (b) 9 s. Isocline increment is 10 °C per contour. The disk potential was 100 V, and the model continued until tissue temperatures exceeded 100 °C. Current density levels under the disk are very similar to the thermal isoclines.

Internal impedance can be represented by segmental resistances of the body. Each body segment can be modeled by either a constant resistance or a temperature-varying resistance. Figure 10.11 shows segmental resistance percentages for the human body based on the internal hand-to-hand values being 100%. The calculation of segmental impedance is

$$R_s = R_{hh}\left(\frac{S}{100}\right) \tag{10.6}$$

where R_s is the segmental resistance, R_{hh} is the internal hand-to-hand resistance, and S is the segmental percentage in Fig. 10.11 for the desired body segment. Chapter 2 gives some values for internal resistance in humans, but the evaluation of the contact impedance is usually quire different in those studies. The hand-to-hand resistance is used as a reference because most accidents involve a hand-to-hand contact. Experimental studies in the hog have shown that the total internal resistance is 370 Ω, the forelimb is 150 Ω, the hindlimb is 170 Ω, and the body impedance is 45 Ω. These measurements were made at voltages around 10 V and at a constant 60-Hz current of 7 mA.

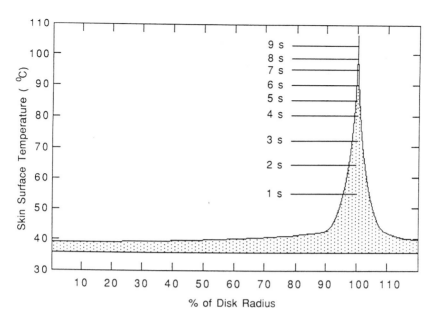

FIGURE 10.9. Temperature along the surface of the skin during the first 10s of current flow. The graph approximates the transient temperature distribution for an isopotential disk of 100 V contacting a nonhomogeneous, semiinfinite medium. The peak temperature at the disk's edge is given for each second of contact.

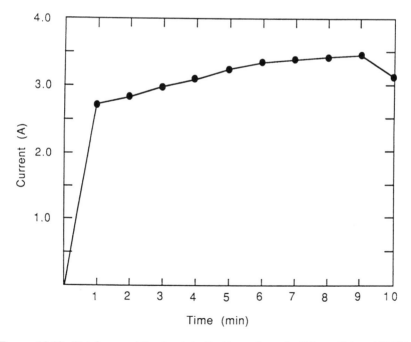

FIGURE 10.10. Total current flowing into the tissue from the 2.5-cm disk at 100 V for a 10-s contact. The current falls off after 9s because the skin temperature would exceed 100 °C, when tissue desiccation would occur.

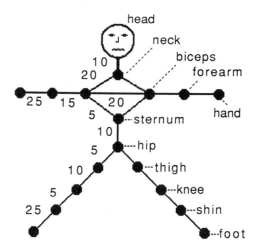

FIGURE 10.11. Segmental body resistances for the human body. Numbers indicate the percentage of the total internal hand-to-hand resistance between each of the designated parts of the body. Figure derived from Figs. 2.39 and 2.40 and from Biegelmeier (1985).

Tissue resistivity is inversely related to temperature. The slow rise in current during phase 2 (Fig. 10.3) is likely the result of decreasing resistivity with increasing temperature. Body segments in contact with the electrodes, or otherwise subject to temperature change, can be modeled with a variable resistance value that is temperature dependent. Each segmental body resistance can be modified to change with temperature using the following equation:

$$R_{ts} = R_s\left(1 - 0.025\Delta T\right) \qquad (10.7)$$

where R_{ts} is the resistance modified by the temperature change ΔT. Experiment shows that increasing temperature decreases resistivity at a rate of 2.5% per °C, which is about $5\,\Omega\text{cm/}°\text{C}$ for muscle tissue (Chilbert et al., 1983, 1985b). At low voltages, heating occurs in the vicinity of the contacts, if at all. High-voltage tissue heating occurs around limb joints, at the contact sites, and in regions of small cross-sectional area. Changes in the overall resistance will increase the total current flowing through the body and likewise increase the tissue current density. Because tissues in a given cross-section have different resistivities, there will be different current densities in the different tissues; thus, specific heating in the tissues will also be different. This leads to nonuniform heating and tissue trauma.

Tissue Current Densities and Current Distribution

The extent of thermal trauma and the likelihood of fibrillation depend on the current density in tissue and current distribution in the body. Current densities have been measured in the tissues of the limbs in hogs using a constant-current source (Chilbert et al., 1985a, 1985b, 1989; Sances et al., 1981a, 1983). Figure 10.12 gives the limb current densities for given applied currents between to the hindlimbs measured in the region of the hindlimb shown in Fig. 10.13. Table 10.6 lists values of current density and resistivity at specific applied current levels between the hindlimbs. The listed voltage values are initial values, because the voltage drops with time when the current is held constant. Table 10.7 gives tissue resistance, resistivity, current, current density, and energy density for one current application of 0.9 A with an initial voltage of 400 V. This table shows how the current is distributed through the cross-section of Fig. 10.13 before tissue trauma and heating alters the current path. Current density in the limbs is more likely to change at high current levels than in the body because significant heating occurs almost exclusively in the limbs and at the contacts.

Current density measured in the body structures is given in Table 10.8, along with the corresponding resistivities. Linear extrapolation of current densities (as in Fig. 10.12) in the body for higher levels of current is accurate, because the tissue temperature in the body changes minimally. The largest current densities are in the back region. The back muscle at the level of the upper lumbar region has a current density of 0.223 mA/cm^2, and the spinal cord current density is approximately 0.280 mA/cm^2 for the same

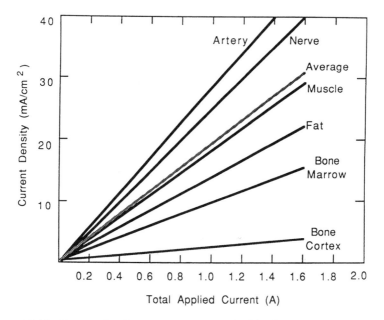

FIGURE 10.12. Current density versus current in the hindlimb of the hog measured approximately 7-cm proximal to the ankle joint; electrodes applied across the hindlimbs. The average line represents the average current density in all the tissues at that level.

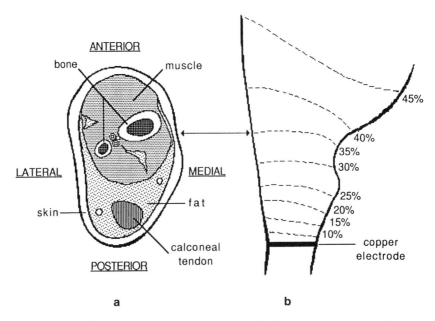

FIGURE 10.13. (a) Cross section of the hog hindlimb where current densities were measured. (b) Hindlimb location of measurement region shown with isopotential percentage lines of the applied voltage for electrodes placed on the hindlimbs.

TABLE 10.6. Tissue current density and resistivity in the limb versus applied current from hind limb to hindlimb in the hog.

Constant applied current (mA)	Initial applied voltage (V)	Artery		Nerve		Muscle		
		J	ϱ	J	ϱ	J_L	ϱ_L	ϱ_T
10	6	0.32	147	0.26	201	0.18	296	512
30	15	1.18	145	0.86	209	0.50	282	525
100	45	3.4	152	3.0	191	2.0	295	483
300	135	9.8	140	8.7	197	7.0	287	501
600	260	21.0	150	15.2	196	12.0	292	492
1,000	415	35.9	155	27.1	200	19.5	290	650

(mA)	(V)	Fat		Bone marrow		Bone cortex	
		J	ϱ	J	ϱ	J	ϱ
10	6	0.12	375	0.10	550	0.03	1828
30	15	0.43	360	0.27	531	0.09	1880
100	45	1.5	352	1.0	535	0.30	1832
300	135	4.9	377	3.1	547	0.85	1859
600	260	9.3	366	5.3	542	2.0	1876
1,000	415	13.4	386	8.1	525	2.8	1836

Note: J in mA/cm^2, ϱ in Ωcm, J_L = longitudinal current density, ϱ_L = longitudinal resistivity, ϱ_T = transverse resistivity of muscle.
Source: From Sances et al. (1983).

TABLE 10.7. Typical values measured in the hindlimb cross-section of the hog.

Tissue	Tissue resistivity (Ω cm)	Tissue resistance (Ω)	Current density (mA/cm^2)	Tissue current (mA)	Tissue area (cm^2)	Energy density (J/cm^3)
Vessel	155	911.76	51.61	8.77	0.17	412.90
Nerve	200	1,666.67	40.00	4.80	0.12	320.00
Muscle (L)	290	19.24	27.59	415.72	15.07	220.69
(T)	650	—				
Fat	380	31.77	21.05	251.79	11.96	168.42
Bone marrow	550	495.50	14.55	16.15	1.11	116.36
Bone cortex	1,850	898.06	4.32	8.91	2.06	34.59
Tendon	398	196.19	20.09	40.78	2.03	160.70
Dermal layers	432	50.94	18.52	157.04	8.48	148.15
Average/total	363	8.85	22.05	903.95	41.00	176.3

Note: Applied voltage = 400 V between the hindlimbs, current = 0.9 A. Electric field strength = 8 V/cm. Resistivity and current density were measured experimentally. Cross-sectional tissue resistance = (resistivity) (tissue area)/(length = 1 cm), tissue current = (current density) (tissue area), tissue energy = (current density)^2 (resistivity) (time = 1 s). Cross-sectional areas were determined by digitization of a limb section taken through the measurement region. Averaged values are resistivity, current density, and energy density. Total values are resistance, current, and area. Total resistance determined by adding conductances of individual tissue elements.

TABLE 10.8. Current density and resistivity in the hog for 100 mA applied from left forelimb to right hindlimb.

	n	J (mA/cm^2)	S.D.	ϱ (Ωcm)	S.D.
Ventral intestine	4	0.071	0.077	511	43.1
Dorsal intestine	4	0.077	0.049	610	96.9
Back muscle:					
Neck	1	0.021	—	—	—
Upper thoracic	1	0.061	—	—	—
Middle thoracic	1	0.141	—	270(L)	—
Upper lumbar	4	0.223	0.079	1,006(T)	58.6
Lumbosacral	1	0.055	—		
Kidney	4	0.097	—	447	61.2
Liver	4	0.065	—	570	37.5
Lung	4	0.058	0.033	1,605	42.8
Heart (transverse to axis)	4	0.073	0.044	800	—
Abdomen, midline (skin, fat, muscle layers)	4	0.065	0.032	547	28.6
Abdomen, side (skin, fat, muscle layers)	4	0.076	0.028	560	42.0

Note: Current density in back muscle as measured in the longitudinal direction. L = longitudinal; T = transverse resistivity.
Source: From Sances et al. (1983).

region when a current of 100mA is applied from the forelimb too the opposite hindlimb. Various current density levels in the spinal cord are reported in Table 10.9 for three different contact orientations. High current levels in the spinal cord can lead to latent neural dysfunction (Sances et al., 1979). At a level of 10 A applied for 30s, the spinal-cord temperature was elevated only 1.5 °C, showing that temperature in the cord is not an important factor in its trauma (Sances et al., 1983). The cross-sectional area of the spinal cord is 0.07% of the body cross section, but the cord carries 0.12 to 0.15% of the total current.

Current distribution in the limb changes with time and temperature and is shown by the change in tissue current densities given in Table 10.10. Initial values of resistivity and current density are also shown in Table 10.7 for all tissues. Current shifts from artery, nerve, tendon, and dermal layers to muscle and fat because of increased temperature and tissue degradation. The decrease of current in the other tissues is the result of an increase in resistivity or a slower decrease in resistivity. Nerve and vessel resistivities increase with temperature once tissue damage occurs, whereas muscle resistivity decreases with tissue damage. The ultimate distribution of tissue trauma is related to the peak temperature in the tissue, which is determined from the energy density as affected by the resistivity and current distribution. However, the resistivity changes with temperature, thus altering the rate of temperature change. For muscle, where the resistivity decreases with temperature, the tissue damage occurs more selectively, as different parts of the muscle become damaged at different rates and enhance the current flow through the damaged regions. This phenomenon is the likely cause for selective muscle groups that are completely burned being next to groups having only minor trauma.

TABLE 10.9. Spinal cord current density for 100mA applied across the limbs.

	Transverse J (mA/cm^2)	Longitudinal J (mA/cm^2)
Forelimb–forelimb		
Cervical	0.0155	0.0247
Thoracic	0.0022	0.0058
Lumbar	—	0.0011
Forelimb–hindlimb		
Cervical	0.0075	0.0708
Thoracic	0.0100	0.299
Lumbar	0.0080	0.257
Hindlimb–hindlimb		
Cervical	—	—
Thoracic	0.0008	0.0020
Lumbar	0.0057	0.0040

Note: Transverse spinal cord resistivity = 1,970 Ωcm, longitudinal resistivity = 214 Ωcm.
Source: From Sances et al. (1983).

TABLE 10.10. Simultaneous current density and temperatures 7 cm proximal to distal tibia versus time with 1-A current applied from hindlimb to hindlimb.

Time (min)	Applied voltage (V)	Artery		Nerve		Muscle	
		J (mA/cm^2)	T (°C)	J (mA/cm^2)	T (°C)	J (mA/cm^2)	T (°C)
0	355	30.0	39.0	24.3	38.5	13.3	39.0
2	293	28.7	41.0	22.8	41.5	12.5	40.5
4	265	27.1	43.0	21.9	44.5	12.5	42.5
6	253	26.5	46.0	21.0	47.5	12.2	44.5
8	235	25.5	47.0	20.5	49.5	12.2	46.5
10	220	25.0	48.5	20.0	52.0	12.2	47.5

(min)	(V)	Fat		Bone marrow		Bone cortex	
		J (mA/cm^2)	T (°C)	J (mA/cm^2)	T (°C)	J (mA/cm^2)	T (°C)
0	355	9.5	38.5	3.0	39.0	1.8	39.0
2	293	10.5	42.0	3.1	40.0	1.9	40.0
4	265	12.8	45.5	3.1	42.5	2.0	42.5
6	253	14.4	49.0	3.2	44.0	2.0	44.0
8	235	14.8	52.0	3.1	46.0	1.9	46.0
10	220	15.0	54.0	3.1	47.0	1.9	47.0

Source: From Sances et al. (1983).

10.4   Thermal Trauma

Thermal burn injury is the most common trauma that occurs at high current and high voltage levels, although electroporation may also cause tissue damage. In this section the resultant electrical trauma will be referred to in terms of thermal burns. This section will discuss the anatomical and physiological aspects of electrical burns, which occur at the contact site, in the deeper tissues, around the joints, and in the spinal cord. Thermal trauma depends on the current density in the tissue more than the resistivity, as indicated by Eq. (10.1).

Heating at the Contact Site

Most burns are at or near the point of contact. When the injury current is high, then trauma can also be seen farther away from the contact site. The discussion here will relate the transient temperature to the severity of burn and the effects of blood flow on the extent of injury. The superficial burn is characterized by three regions of tissue destruction and charring at the electrode contact edge (Fig. 10.14). The first region of tissue destruction, which is adjacent to the electrode contact, contains desiccated and denatured tissue, indicated by erupted blisters, shrinkage of the tissue, and charring at the edge of the contact. The second region is the ischemic region, where the tissue is devoid of blood flow but retains its fluid content. The outermost region, represented by a darkened ring caused by hemor-

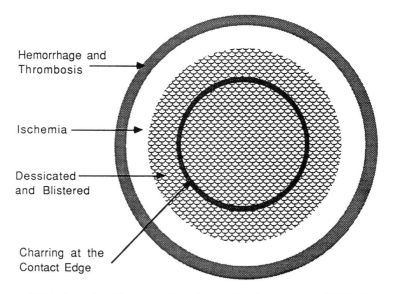

Hemorrhage and Thrombosis

Ischemia

Dessicated and Blistered

Charring at the Contact Edge

FIGURE 10.14. Superficial burn resulting from a circular contact at 100 V. There are three distinct regions of trauma: a desiccated region adjacent to the electrode having erupted blisters, an ischemic region that maintains its fluid content, and a hemorrhagic region darkened by thrombosis. Also shown is a ring of charring that occurs at the electrode edge.

rhage and thrombosis of the microvasculature, appears several minutes after current is stopped. These burn regions extend into deeper layers of the dermis as well (Fig. 10.15) and extend into underlying muscle tissue when the current is sufficiently high. Desiccation is deeper at the electrode edge than in the center. Histological evaluation of the dermis shows that the trauma from electrical burns is very similar in character to nonelectrical thermal damage. The extent of trauma can be determined by the transient temperature response of the tissue.

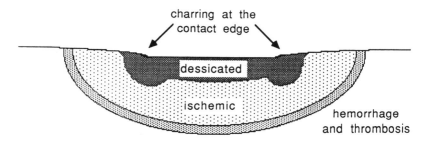

charring at the contact edge

dessicated

ischemic

hemorrhage and thrombosis

FIGURE 10.15. Dermal cross section beneath a circular contact at 100 V. The resultant injury is a full-thickness dermal burn. The three regions of Fig. 10.14 are present. Desiccated tissue extends deeper into the tissue at the electrode edge than at the center.

Henriques (1947) and Henriques and Moritz (1947) have investigated the thermal tissue response in terms of the time-temperature relationship. As seen in Fig. 11.12, the time necessary for trauma decreases exponentially with peak temperature. Skin will burn at 45 °C after an hour but at 60 °C only a few seconds are required. Experimentally measured temperatures are given in Fig. 10.16 for a circular electrode of 5-cm diameter having a 100-V potential and applied through the third phase of the current-time waveform (Fig. 10.3). The measured temperatures corresponds well with the finite-element model results of Fig. 10.8. In general, the peak temperature delay from the time of peak current increased primarily with distance from the edge of the electrode (Fig. 10.17). A secondary effect increasing the time delay of peak temperature is the depth from the skin surface, and regions without blood flow (ischemic regions of Figs. 10.14 and 10.15) also had longer delays. The secondary effects are the result of heat radiation and convection at the surface and to slower heat conduction in ischemic tissue.

Tissue perfusion by blood is an important factor in the generation of electrical burns. Comparing the above in-vitro studies to studies in situ, the results show significant variations in the total applied energy, contact time, and rate of tissue cooling. The parameters of voltage, electrode size, average power, and minimum total resistance were the same. The total energy applied was 24% less in situ than in vivo, the contact time was 21% less in situ that in vivo, and the rate of tissue cooling was 50 to 75% less in situ than in vivo. This factor is important to the evolution of electrical burn trauma, since the data indicate that blood flow helps lessen the severity of the trauma in short-duration contacts.

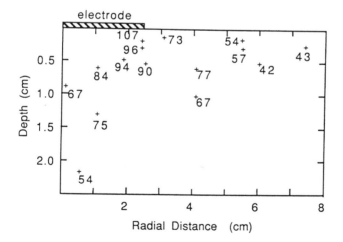

FIGURE 10.16. Peak tissue temperatures measured experimentally beneath a 5-cm-diameter circular contact at 100 V in five hogs. The voltage was applied until the current ceased from desiccation of the contact site (phase 3 of the current-time waveform). (Adapted from data of Prieto et al., 1985.)

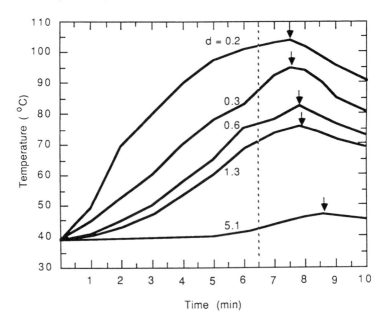

FIGURE 10.17. Plots of tissue temperatures beneath the electrode at different depths (cm) with time, 100 V contact. Peak temperatures do not occur simultaneously, but at times related to depth below the electrode edge following the current falloff of phase 3, as indicated by arrows. Broken line indicates current cessation.

Heating of the Tissues

When currents are high and/or application times are long, trauma extends beyond the contact site, requiring surgical intervention and possibly amputation. The trauma affects primarily the limbs and is typically located in muscle tissue. Peripheral nerves and blood vessels are also affected, because they are usually adjacent to affected muscle. Bone is usually the least affected, but adjacent muscle is affected more than distant muscle. Deep tissue injury depends critically on the current pathway. As noted by Eq. (10.3), the distribution of current is based on the electric field strength in a given cross-section and the tissue resistivity.

Heating of tissue and subsequent resistivity changes have been investigated in hogs and dogs (Sances et al., 1983; Chilbert et al., 1985b). Table 10.10 shows that current density in fat increases with temperature but decreases in artery, muscle, and nerve. Tendon and dermal current density decreases with temperature. The current path will shift due to changes in the resistivity. This change is the result of a direct temperature dependence of resistivity and to tissue degradation. Tissue degradation causes increased resistivity in nerves and vessels while decreasing muscle resistivity. Neural tissue increases 80 to 100% in resistivity when thermally injured, while muscle decreases as much as 80%. The changes in muscle resistivity become more important when one considers that a large change in impedance could be indicative of burn severity.

Studies performed by the author (Chilbert et al., 1985a 1985b) indicate that tissue temperature, resistivity, and the severity of trauma can be correlated in muscle tissue. Figure 10.18 shows the measurement sites in the dog gracilis muscle for temperature and resistivity. Electric burn trauma was induced by passing 1-A currents at 60 Hz until the distal measurement site reached a temperature of 60 °C. Peak temperatures along the muscle decreased proximally as the limb cross-section increased (Fig. 10.19). Regions 1 and 2 exceeded 50 °C, while the remaining regions remained below 47 °C. Severe trauma was noted in regions 1 and 2, while regions 6 and 7 had edema with minimal structural changes. Regions 3, 4, and 5 were in a transition zone between viable and nonviable tissue. The resistivity increased proximally, but all values were less than control values (Fig. 10.20). The initial resistivities at the onset of the I-A current increased 35% because of muscular contraction; they then declined with increasing temperature. The resistivities continued to decline even as the tissue cooled, indicating further progression of tissue injury.

Thermal trauma to peripheral nerves is also related to the maximum temperature reached. The temperature in a nerve depends on the temperature of the surrounding tissue. The peak temperature of nerve in Table 10.10 shows that it is only 2 °C less than the tissue in which it was measured (i.e., fat). Neural activity (indicated by averaged evoked response)

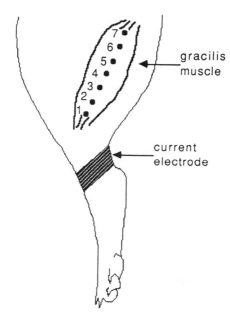

FIGURE 10.18. Measurement sites in the gracilis muscle of the hindlimb of a dog. Temperature and resistivity were measured at each site. Site 1 is the most distal and site 7 is the most proximal.

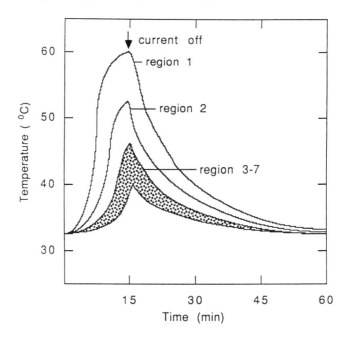

FIGURE 10.19. Temperature increase caused by the flow of a 1-A current with time. Current was applied for 15 min. Temperatures at regions 1 and 2 were elevated above 50 °C, while the other regions were between 40 and 46 °C and are grouped together.

decreases with temperature as neurons are damaged. Degradation of the peripheral nerve response with a thermal injury (not electrically induced), caused by slowly increasing the temperature, shows the development of trauma (Fig. 10.21). Important features of the evoked response indicate that certain neurons are more sensitive to increased temperature, noted by a larger decrease at 45 °C and at 50 °C. Which neurons are affected at these temperature levels in not presently known. For rapid heating of the nerve the temperature of trauma is above 50 °C, where recovery of the evoked response is usually rapid once the temperature reaches that of the body. However, heating of the nerve with electric current causes permanent damage at lower temperatures. The evoked response is always immediately affected by current, at levels above 3.5 mA/cm^2 along the nerve, causing blocking of the response from 5 to 20 min. Once temperatures exceed 45 °C, permanent reductions in the evoked response may occur, and above 50 °C the response never returns. This suggests that the current may affect neurons directly, exclusive of temperature, possibly by electroporation. Limited studies in the spinal cord are consistent with peripheral nerve information, but the temperature threshold for electrically induced trauma is lower.

Thermal trauma to bone is seldom seen except at very high voltages (Sances et al., 1981b). The high resistivity of cortical bone prevents significant direct heating and protects the marrow as well. Table 10.10 shows that bone has the lowest temperature in cross-section. Muscle tissue near the bone is often burned, while muscle tissue away from the bone is not affected. Some clinicians have erroneously attributed this phenomenon to heating of the bone, but it is the result of higher current densities in the muscle near the bone (Sances et al., 1983) because heating is more dependent on current than resistance.

In cross-section, the joints have a large quantity of bone compared to other more conductive tissues. This creates very high current densities in the more conductive tissue, consequently causing severe trauma at the joint while leaving tissues by the long bones viable (Sances et al., 1983; Zelt et al., 1988). Also, the joint capsules have been known to disrupt explosively from the buildup of steam (Sances et al., 1981b, 1983). However, if the victim were to flex the joint during electrical contact, the current will have an alternate path through the skin surfaces above and below the joint, thus preserving the joint. This has been noted clinically around the elbow, where burn marks appear on the skin proximally and distally.

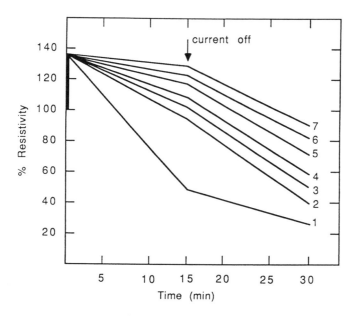

FIGURE 10.20. Changes in resistivity measured during and following the application of a 1-A current. Resistivities of muscle were measured in the transverse direction and perpendicular to the direction of current flow and are given in percent of control values. Initial increase of resistivity is due to the contraction of the muscle by stimulation from the current. Initial resistivity is for contracted muscle at time = 0 min. Muscle resistivity continues to decrease with decreasing temperature after current is turned off at $t = 15$ min in region 1.

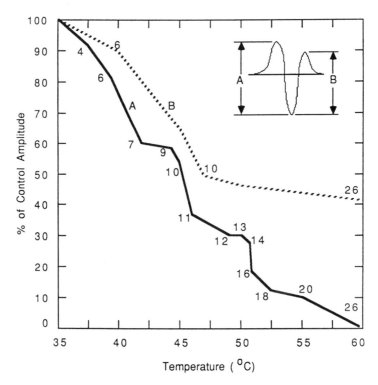

FIGURE 10.21. Changes in the evoked response of the tibial nerve in the hog with temperature. The reductions in the peak amplitudes were permanent. The first component of the evoked response is denoted by A, and the second is denoted by B. Numbers on curves indicate time in minutes after the temperature increase was started. Amplitude is given in percent of normal evoked response amplitude.

10.5 Nonthermal Trauma

Lesions in tissue not caused thermally are likely caused by electroporation (Lee et al., 1988; Weaver, 1992; Gaylor et al., 1992). Typically, the electric field strength needed to rupture muscle cells or cause nerve damage must exceed 100 V/cm. Also, it has been noted that elevated temperature increases the likelihood of electroporation. Studies in the peripheral nerve of hogs show that no significant alterations in the evoked response occur at current densities up to 167 mA/cm². This current density level results in an electric field strength of 33.2 V/cm, which is below the minimum level of 100 V/cm cited by Gaylor et al. (1989). To attain such field strengths in the limb, the applied voltage would have to exceed 5,000 V. For spinal cord injury to occur, the minimum applied voltage would need to be between 35 kV and 50 kV for a hand-to-foot path. These values were predicted from measurements in the hog given in Tables 10.4 and 10.9 and in Fig. 10.3.

From Table 10.9 one notes that a maximum current density of $0.3\,mA/cm^2$ results when a current of $100\,mA$ is applied from the forelimb to the hindlimb and the spinal cord has a resistivity of $214\,\Omega cm$. Since the electric field strength should be above $100\,V/cm$, the current density in the cord needed to cause irreversible electroporation would have to be $500\,mA/cm^2$. To obtain this current density in the spinal cord, an applied current of $167\,A$ between forelimb and hindlimb is needed. To estimate the voltage level for $167\,A$, refer to Table 10.4: At electrode location 4, nearly $20\,A$ will flow at $6\,kV$. A simple extrapolation of the voltage-current ratio ($6\,kV$-$20\,A$) shows that $50\,kV$ is needed to generate $167\,A$ in the body. However, resistance decreases with increasing voltage, therefore using the values of $7.2\,kV$ and $25\,A$ from Fig. 10.3 and extrapolating to the 167-A level, the applied voltage needed is $35\,kV$. The value for humans may be lower, since reports in the clinical literature have cited spinal cord dysfunction at $25\,kV$ (Kanitkar and Roberts, 1988).

10.6   Lightning Injuries

Lightning accidents occur infrequently relative to other electrical incidents, but often enough to be well documented. According to Biegelmeier (1986), one-third of all lightning injuries are fatal. The immense power of lightning is shown by the effects of its millions of volts and 5,000 to 200,000 A (Sances et al., 1979; Strasser et al., 1977). Reviews of lightning injury have been given by Taussig (1968) and Silversides (1964). Persons struck by lightning who were clinically dead have been revived, and maladies seen afterward eventually disappear in most cases (Taussig, 1968).

The afflictions commonly encountered with lightning are mostly neurologic. Burns can occur at the entry points and may be accompanied by internal lesions. The burns often have a splashlike or arborlike appearance, assumed to be caused by current tracking over the body (Artz, 1979). Skeletal fractures occur when the person falls or is thrown by muscular contractions. Effects on the nervous system can occur immediately, can be delayed for several hours, or can occur days after.

Immediate effects of lightning strike depend on the severity of the incident. The victims usually undergo a total arrest of all body functions, which slowly return over a given length of time (Taussig, 1968). Fatalities can occur when the victim has a long respiratory and cardiac recovery period and no assistance is administered such as CPR, (Strasser et al., 1977; Taussig, 1968). Abdominal lesions are also notably fatal, especially when internal hemorrhage is present (Artz, 1979). For the less serious accidents, the heartbeat is present with respiration intact or soon returning, leaving the victim in an unconscious state. Upon arousal, the victim may act disoriented and may have paresis of some or all the limbs (Critchley, 1934; Silversides, 1964). Other, less common deficits are visual, auditory, and speech (Critchley, 1934). Often amnesia occurs with lightning insults (Strasser et al., 1977; Taussig, 1968). Hypertension is normally observed, with anxiety and, at times, neurotic behavior. These symptoms usually vanish within a week (Strasser et al., 1977).

Secondary effects appear within the first few days after the accident. Except for an occasional change in the electrocardiogram or trauma to internal organs, all secondary effects are neurologically oriented, consisting of paralysis (usually of the legs), muscle pain throughout the body, photophobia from the intense light, and various autonomic disturbances. These effects normally vanish within a week. Latent effects (seen after 5 days) usually are resolved within a month after the accident. Neurologic symptoms are thought to be associated mainly with the progression of vascular disease or dysfunction. Neurologic symptoms in this stage are also common to high-voltage power-line injuries, which will be discussed later (Silversides, 1964).

Soft tissue lesions are found more frequently with lightning accidents than in high-voltage accidents (Massello, 1988; Sharma and Smith, 1978). Neurologic lesions often occur where there is an interface between regions of different resistivities. In lightning accidents one commonly sees splitting of the cortical layers of the brain and petechial and subarachnoid hemorrhage (Critchley, 1934; Silversides, 1964). Several explanations of the lesions have been suggested, but all lack supportive evidence. One hypothesis suggests that the lesions are caused by the heating effects, which produce pockets of gas or steam. This is not generally accepted as a good explanation, because the heat generated is probably insufficient to cause boiling of the tissue. Fluid electrolysis has also been suggested; however, the time involved and current necessary may exclude this as a possibility. The most acceptable theory is electroporation causing cellular lysis at or near the tissue interfaces. The lesions have been seen more often at tissue interfaces, where charge accumulation can occur (Critchley, 1934; Strasser et al., 1977).

10.7   Clinical Observations

Of all electrical injuries, most occur to those working for electric utility companies. In one report, 95% worked for utility companies, of which 50% or more were linemen (Butler and Gant, 1978). The voltage levels for most of these injuries exceed 1,000 V and are at the standard commercial frequency. Reviews of the general aspects of electrical injury have been presented by many (Dixon, 1983; Sances, 1979; Skoog, 1970; Wu, 1979; Lee et al., 1992, 1994). Clinical reviews have shown common observations among patients (Haberal, 1986; Hammond and Ward, 1988; Luce et al., 1978; Sharma and Smith, 1978; Solem et al., 1977). Some authors have focused reviews on the neurologic sequelae noted in electric accidents (Critchley, 1934; Silversides, 1964; Strasser et al., 1977). Others have reviewed the aspects of electric burns (Artz, 1974; Butler and Gant, 1978; Rosenberg and Nelson, 1988; Rouse and Dimick, 1978).

The various forms of trauma seen with electrical injury are the same as those seen in thermal burns with the addition of others unique to electrical injury. Some aspects of electrical injuries are similar to crush injuries. Clinical reports and reviews present observations in the general areas of burns, lesions, neurologic effects, and cardiovascular effects. The reports and reviews usually focus on one of the areas, while the patients may present symptoms from all areas.

Electrical Burns

The pattern of electrical burns can vary greatly from those seen in thermal injuries (Artz, 1979). Electrical burns follow the current path, whereas thermal burns start at the surface from one location and radiate into the surrounding tissue (Janzekovic, 1975; Luce et al., 1978; Ponten et al., 1970; Sances et al., 1979). Deep tissue necrosis is often evident. Although tissue may appear viable, it may be damaged and require secondary procedures. The transformation of visibly viable tissue to necrotic tissue has been termed progressive necrosis, but experimental studies indicate that the visible viable tissue is actually necrotic from onset of the trauma (Zelt et al., 1988). Artz (1979) has reviewed and presented the various aspects of electrical injury, especially those associated with electric burns. Also, reference is made to internal lesions and renal sequelae. The following is a summary of burn wound observations:

Contact-site wounds usually signify deep tissue destruction locally. The contact sites can be defined in terms of the primary site, which is the contact with the energized source, and the secondary site, which is the contact with the ground or neutral source. The primary wound is usually charred and depressed. The secondary wound is dry, depressed, and has the appearance of the current exploding outward. Massive swelling is evidence of extensive tissue damage caused by heat and electroporation.

Skin injury varies from small circular spots to large areas of charring. Adjacent to the charred tissue is a whitish-yellow ischemic region. Surrounding the ischemic skin is an area darkened by vascular hemorrhage and thrombosis. All three regions are relatively cold and without sensory perception.

Vessel damage may extend beyond the general area of injury. Thrombosis has been noted away from the burn injury. Typically, extreme vascular spasms, thrombosis, and necrosis of vessel walls are observed. In non-thrombotic, damaged vascularization, hemorrhaging may result and lead to serious complications.

Muscular trauma is caused by the direct heating of current or by occlusion of the arterioles supplying the muscle. Damage to the muscle is usually uneven, affecting groups of fibers while not affecting adjacent areas.

Uneven damage is characteristic away from the contact sites and may not be noticed initially (Zelt et al., 1988).

Burns caused by arcing imply high temperatures at the contact site and extensive deep tissue destruction. Traumatic limb amputation has been documented (Sances et al., 1979). Cursory burns are caused when arcing ignites the victim's clothing.

Renal failure is more common in electrical injuries than in thermal burns. Renal damage is caused by direct electric involvement of the kidneys and/or renal vessels, or more often by the abnormal breakdown of protein from other injured tissues. Devitalized muscle in electrical injury causes renal complications similar to those seen in severe crush injury (Artz, 1974). There is also a greater occurrence of hemoglobinurea and hematinuria in the electrically injured patient. One critical complication is acute tubular necrosis, probably caused by myoglobin breakdown.

Oral burns are most often seen in children and cause severe burns to the lips, tongue, and dentition. This form of electrical burn has been reported extensively particularly in recent years, because of the special treatment needed (Barker and Chiaviello, 1989; Donly and Nowak, 1988; Palin et al., 1987; Sandove et al., 1988; Silverglade, 1983). Of particular concern is tissue contraction of the lips and cheeks, which is prevented by special splints (Sandove et al., 1988).

Most pediatric electrical injuries admitted to the hospital occur to the mouth (35–60%) and most victims of oral burns are under the age of 4 (Baker and Chiaviello, 1989; Palin et al., 1987). Oral burns account for almost all severe burns. Electrical trauma in children usually results from defective equipment. Nearly all injuries occurring to children are preventable, especially with routine equipment maintenance.

Tissue Lesions

Tissue lesions are evident at the tissue interfaces and are commonly associated with neural dysfunction and abdominal complications. Autopsy upon electrical accident victims often reveals submucosal hemorrhages dispersed throughout the gastrointestinal tract. Abdominal lesions are associated with a high mortality rate. They are difficult to detect and treat. Delayed fatality is usually associated with abdominal lesions of the gastrointestinal tract (Artz, 1979). Typically the patient is comatose, and vital signs degrade over several days until death occurs.

Lesions have the most significance in neural tissues. Once a nerve has been severed, its function seldom returns. Lesions may also be responsible for latent neural dysfunctions that appear up to 3 years after the injury (Sances et al., 1979; Silversides, 1964). When lesions are noted in the spinal cord, they are seldom complete transections. Clinical signs of these lesions are spastic paresis with little or no sensory deficit. The lesions usually are not suspected until the onset of patient ambulation (Baxter, 1970). These afflictions, like many associated with the nervous system, are not well understood.

Neurologic Sequelae

Neurologic sequelae are manifest at all levels of the nervous system and are either permanent or transient. Permanent effects are the result of thermal trauma or lesions, while transient disorders can disappear within days or last for months (Hooshmand et al., 1989).

Damage to peripheral nerve can be caused by electroporation (lesions) or excessive heating of the tissue. The ulnar, radial, and femoral nerves have the highest incidence of injury due to heating. Occasionally lesions are formed in the peripheral nerves, causing sensory or motor deficits. Some cases have been reported as progressive with time, the victim slowly losing sensation in the limbs and other clinical abnormalities developing (Kinnunen et al., 1988). Injury to the peripheral nerves, as stated above, is permanent, and recovery is minimal. Peripheral nerve damage is usually associated with adjacent tissue injury.

Transient disorders often seen are neurovascular, neuromuscular, and sensory. The neurovascular disorders include vascular constriction and spasmodic reactions, both of which reduce blood flow and vascular dilation (Christiansen et al., 1980; Hooshmand et al., 1989). Vascular spasms cause a reduction of blood flow in an area, usually associated with tingling, numbness, and, at times, paralysis. Vasodilation is believed to be the cause of fainting spells sometimes encountered after electrical injuries. Neuromuscular disorders are paresis, paralysis, hypertension, and muscular pain.

Current having a magnitude great enough to affect the spinal cord or nerve roots has a transthoracic path, most commonly hand to hand, or hand to foot. Head-to-extremity paths, although not very common, will also affect the spinal cord. With hand-to-hand contact, current is greatest in the cervical region (Table 10.9). Hand-to-opposite-foot current paths mostly affect the heart and thoracic spinal cord. Head-to-extremity current paths affect the upper brain centers and cervical spinal cord (Butler and Gant, 1978; Sances et al., 1979; Solem et al., 1977).

Permanent damage to the spinal cord may be caused by vertebral displacement secondary to falling, by muscular contractions, or by lesions in the cord or nerve roots. Skeletal fractures in electrical injury are normally attributed to falling or being thrown by muscular contraction. Lesions in the spinal cord account for many of the permanent disabilities of motor function. Lesions are seen mainly in cases of power-line contact above 11,000 V (Butler and Gant, 1978). The magnitude of the lesions in the spinal cord is not readily apparent until the onset of ambulation (Baxter, 1970; Solem et al., 1977).

Transient effects of current on the spinal cord appear at all stages of recovery: immediate, secondary, or latent. The immediate effects are loss of consciousness, respiratory arrest, and neural effects on the heart rate. Unconsciousness may cease immediately after current flow has stopped or linger for a day or more (Silversides, 1964). Respiratory arrest occurs from the sustained contraction of intercostal muscles and interruption of the neural respiratory centers. Sustained contractions of the intercostal muscles last only while the current is applied. Spontaneous breathing has been delayed for several hours in some victims, but usually returns within 20 min. Secondary effects of electric injury involving the spinal cord are temporary paralysis of the limbs, vascular spasms, and muscle pains (Silversides, 1964).

Immediate effects of electrical accidents involving the head may include convulsions, coma, cerebral edema, hysteria, amnesia, tinnitus, deafness, and visual disorders. Convulsions are induced in electrical injury from the passage of current through the cerebral cortex, analogous to those induced in electroconvulsion therapy. Coma is observed in patients who have been in cardiopulmonary arrest for relatively long periods of time. Clinical signs of electric coma are dilated, unresponsive pupils, along with no reflex response. Cerebral edema is seen only in the severest cases (very high voltages and lightning), typified by a softening of the tissue (Critchley, 1934). Hysteria is mentioned throughout the literature; its severity is minimal. Specific hysteric symptoms are agitation, confusion, amnesia, and transitory auditory and visual dysfunction (Critchley, 1934; Silversides, 1964). Amnesia is the most common symptom. Often the cranial nerves are directly affected; tinnitus and deafness are observed in victims, particularly when the victim does not lose consciousness. Permanent deficits in cranial nerves are associated with lesions along the cranial nerves.

Visual disorders include blindness, blurred vision, photophobia, and cataracts (Silversides, 1964). Blindness is caused as a result of current effects on the optic nerve or interference with the photoreceptors in the retina (Al-Rabiaed et al., 1987). Permanent blindness is caused by lesions along the optic tract or separation of the retina from the choroid. Blurred vision results when the innervation of the lens musculature and iris have been affected. This condition is typified by dilated pupils. Photophobia occurs when the retina is exposed to intense light from an arc or lightning flash. Cataracts have been known to develop after lightning strikes and very high-voltage contacts with the head (Moriarty and Char, 1987).

Secondary conditions associated with the cerebrum involve the prolonging and delayed development of the immediate symptoms discussed above. Secondary effects are mainly hysterical complications. In addition to those stated above are speech loss, disorientation, and narcolepsy (Critchley, 1934; Silversides, 1964).

Latent effects includes psychosis, hemiplegia, aphasia, epilepsy, choreoathetosis, and other hysterical conditions (Critchley, 1934; Silversides, 1964). Psychosis is rare and may be related to preaccident manifestations. Hemiplegia and aphasia (not as common) are seen in conjunction with cerebrovascular disturbances. Prior existence of vascular disease correlates well with hemiplegia and aphasia resulting from electrical accidents. Posthemiplegic Parkinson's disease has been known to develop, either the unilateral or the asymmetric bilateral form (Silversides, 1964). The initial convulsion in a developed epileptic condition may have started during the current flowing in the body. If the current path includes the head, the formation of an epileptic focus is likely. Actual lesions are rare, leaving a diffuse focus as the cause. Victims who developed epilepsy respond well to drug therapy, which usually is withdrawn over a period of time.

Cardiovascular Effects

Electrical accidents have also produced latent cardiovascular disorders that are usually transient. Some of these affects may be associated with fibrillation and defibrillation of the victim (Wilson et al., 1988) or with severe burns. Because these effects are usually less life-threatening than the complications of burn, they are seldom reported. However, the occurrence of cardiovascular complications is frequent (Guinard et al., 1987). High current levels flowing through the chest can cause cardiac arrest, which is then followed by a somewhat normal cardiac rhythm. Most cardiovascular effects follow high-voltage electrical injuries. Latent effects include atrial fibrillation, premature ventricular contractions, bradycardia, tachicardia, ventrical and atrial ectopic foci, conduction branch block, and nonspecific S-T interval and T-wave changes (Butler and Gant, 1977; Jones et al., 1983; Skoog, 1970; Solem et al., 1977).

The causes of latent cardiac effects and delayed arrhythmias is unknown. Some suggested causes include the residual effects of fibrillation, enzymal reactions with the cardiac tissue, neural dysfunction, or coronary artery spasms. The fibrillation of the heart or a sustained myocardial contraction by the current may alter the metabolic activity of the heart and result in altered cardiac rhythm (Solem et al., 1977). Certain enzymes are released from necrotic muscle tissue, such as various forms of creatine kinase, which can alter the function of the cardiac cells or mitochondria and again alter the cardiac rhythm (Ahrenedolz et al., 1988). Because various forms of transient neural anomalies have been noted in other areas of the nervous system, neural alterations in the inervation of the heart may also occur (Skoog, 1970). Similarly, the vascular spasms noted in the periphery may also occur in the coronary arteries and lead to altered cardiac rhythm (Luce and Gottlieb, 1984; Skoog, 1970). Fortunately, these disturbances are usually transient and disappear within a week (Jones et al., 1983; Luce and Gottlieb, 1984).

Latent vascular effects, such as spasms or peripheral coldness, have been attributed to neural disturbances (Kinnunen et al., 1988). Similarly, transient neural disturbances have been attributed to vascular spasms (Silversides, 1974; Skoog, 1970). Some latent effects are attributed to loss of endothelial cells and elasticity in the vessel wall, which leads to vascular impairment (Wang and Zoh, 1983). Also, the vasculature may be impaired by arachidonic acid metabolites, such as thromboxane, that cause vasoconstriction. The latent effects are often a result of the initial trauma, but appear as healing occurs.

10.8 Clinical Treatment

Treatment of electrical injury begins at the scene of the accident with removal of the victim from the electrical source and the observations of vital signs. A lack of pulse or respiration requires immediate cardiopulmonary resuscitation. One should note specific details of the accident, such as the contact voltage, whether the victim fell, if there are fractures, and if there are burns evident at the contact sites. Rapid transportation of the victim to an adequate trauma center is essential (Dixon, 1983).

The potential for considerable fluid loss coincident with deep tissue damage is great, therefore fluid replacement must be started as soon as possible. Intravenous isotonic fluids are administered to maintain the urine output at approximately 100 ml/h. Traditional burn formulas cannot be used to calculate fluid replacement requirements. These formulas underestimate the required fluid volume because they are based only on the skin surface area involved; fluid replacement principles for a crush injury are more appropriate (Artz, 1979; Dixon, 1983). If myoglobinuria is present, sodium bicarbonate is administered to alkalinize the urine, which inhibits pigment precipitation. Also, an osmotic diuretic, usually mannitol, can be used to flush the renal tubules (Baxter, 1970; Dixon, 1983, Hunt et al., 1980; Moncrief and Pruitt, 1971).

The electrocardiogram should be monitored, and echocardiography may be useful in assessing cardiac damage. Monitoring the hemoglobin and hematocrit will indicate the degree of hemolysis, and arterial blood-gas measurements are used to determine the acid-base state of the victim (Artz, 1974; Dixon, 1983).

The external appearance of the electrical burn wound is not indicative of the extent of underlying tissue damage. The electrical burn may be masked by flame or flash burns (Moncrief and Pruitt, 1971). Initially, the wounds are thoroughly cleansed. If the initial examination reveals a charred limb, an absent or diminished pulse in a limb, a loss of sensory or motor nerve function in a limb, or evidence of considerable limb swelling, then an escharotomy and fasciotomy should be performed. Fasciotomies have both a therapeutic and diagnostic function, as they may restore circulation and permit the surgeon to inspect the underlying musculature visually (Baxter, 1970; Wang 1985). Amputation of at least one extremity is frequently required following high-voltage injuries. If a limb is clearly necrotic, then amputation should be performed immediately (Parshley et al., 1985). Amputations are generally done in the first week postinjury, after the patient's condition has stabilized. Arteriograms may be helpful in determining the initial level of amputation (Hunt, 1979).

Although there is general agreement that obviously necrotic tissue should be debrided as soon as possible, there is still controversy over the treatment of tissue that is questionably viable (Barnard and Bostwick, 1976; Baxter, 1970; Rouse and Dimick, 1978). One approach involves early exploration and debridement, with repeated surgical procedures after 24 to 72 h; until all nonviable tissue has been identified and removed (Artz, 1974; Hunt et al., 1980; Luce and Gottlieb, 1984; Parshley et al., 1985). A more conservative approach may be used, where the debridement procedure is delayed until a definite delineation between viable and nonviable tissue becomes evident, usually after 8 to 10 days. Proponents of the aggressive approach claim earlier wound closure and less incidence of infection, while proponents of the conservative approach feel that unnecessary multiple surgical procedures are avoided, the local blood supply is not altered by surgical trauma, and viable tissue is less likely to be removed. Zelt et al. (1988) showed experimentally that all tissue injury is histologically evident at the onset of injury; progressive necrosis does not occur.

Several methods have been advanced to determine the viability of questionable tissue so that excision of the necrotic tissue can be conducted as early as possible. Arteriography has been used to assess vascular damage, but this method does not consistently visualize the smaller arterial and arteriolar vessels in the muscles (Hunt et al., 1980). Thus, only occlusion of the larger vessels can be reliably determined with arteriography. The technetium-99 m pyrophosphate scan has been used to identify the extent of cardiac and skeletal muscle damage, including deep areas of muscle necrosis that may be missed by visual inspection during a fasciotomy (Hunt, 1979). Other studies have suggested that electrical impedance or nuclear magnetic resonance spectroscopy can be useful in determining tissue necrosis (Chilbert et al., 1985a, 1985b).

Tissue resistivities were compared with histologic changes as well as results of ^{31}PNMR spectroscopy. Abnormalities in anatomic as well as metabolic measures were shown to correspond well to resistivity changes. Measurements of impedance at several frequencies were made to compute the phase-plane plot of muscle tissue (Fig. 10.22). Impedance is unique for each tissue as well as its condition, reflecting traumatic injury, viability, edema, and so on (Ackmann and Seitz, 1984). Likewise, changes in the phase angle are indicative of alterations in tissue, but are more proportionately related to the number of cells present. The relaxation frequency occurs at the point of maximum reactance. These boundary changes (caused by cell lysis) alter the relaxation frequency; loss of cells increases the relaxation frequency. With electrical burn injury, the amount of cell lysis is coincident with the temperature rise in the tissue. Figure 10.20 shows how muscle resistivity changes with temperature and has been correlated to trauma (Chilbert et al., 1985b). The level of trauma, as indicated by

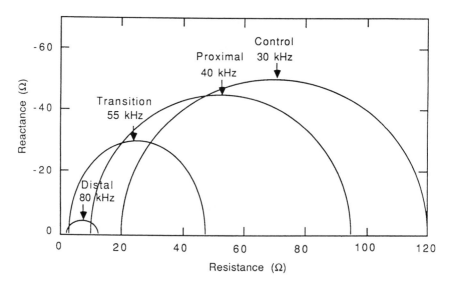

FIGURE 10.22. Impedance and relaxation frequency changes seen in muscle tissue with electric burn. The proximal tissue impedance curve corresponds to tissue that is only mildly edematous, the transition tissue impedance curve corresponds to tissue showing greater edema and some loss of cell structure, and the distal tissue impedance curve corresponds to tissue that shows gross destruction of cell structure and extensive damage.

^{31}PNMR spectroscopy and histology, shows a 75% decrease in resistivity with severely burned tissue, whereas tissue only slightly affected by the electricity showed a decrease of 25% or less. Using impedance techniques to determine tissue viability allows separation of compartmental changes (edema) and boundary changes (cell lysis) so that tissue vitality is better determined. Figure 10.22 indicates both the impedance change and relaxation frequency change seen in electric burns. Once the probability of tissue survival is determined for a given change in impedance and relaxation frequency, clinical measurements can be performed and appropriate excision of nonviable tissue can be made. This will eventually lead to reduced morbidity in the electric burn patient by minimizing surgical procedures and chances for infection.

The bacterial population of the wound must be controlled to prevent infection, especially clostridial myositis, which may develop as a result of inadequate excision of necrotic tissue (Artz, 1979; Hunt et al., 1980). Topical antibacterial agents are used rather than systemic agents, because the blood flow is often compromised in the region of the wound. Temporary wound closure can be achieved with porcine xenografts, autografts, or homografts, whereas permanent wound closure is usually achieved with an autograft. The use of a free muscle flap, utilizing the latissimus dorsi, for immediate coverage of deep electrical injury has been advocated. However, if the debridement prior to grafting is inadequate, then muscle necrosis or infection may be masked by the vascularized muscle flap. A viable cutaneous homograft may be a useful tool to test the readiness of a wound for surgical closure or definitive autografting. Vein grafts, using autotransplantation of the saphenous vein to replace the radial and ulnar arteries, have been used to restore circulation to the hand and prevent limb necrosis (Wang et al., 1984, 1985).

Electrical injuries involving only the scalp may be managed with the same procedures used for soft-tissue electrical burns of other parts of the body (Worthen, 1982). Contrary to earlier procedures for treating an electrical burn of the scalp and skull, a full-thickness scalp flap should be used to cover the exposed skull as soon as possible, regardless of the depth of damage to the skull. The vascularized cover promotes skull regeneration, as ingrowth and revascularization occur on both the pericranial and meaningeal surfaces (Worthen, 1982). Alternatively, split-thickness skin grafts can be applied to the skull after the cortical bone is removed to expose cancellous bone, although later resurfacing with rotational scalp flaps is required to restore the hair. A graft consisting of autogenous greater omentum covered with autogenous split skin has been used to cover a large irregular area of exposed skull following a high-voltage electrical injury. Extensive destruction of the face requires a multistage reconstructive procedure, which may extend over a period of several years.

Advances in the understanding of cell damage because of electroporation have suggested possible therapies for reversal of cell damage. It has been observed that following electroporation of cells in vitro, membrane repair can be promoted by the application of surfactant compounds within a couple of hours of the injury (Lee et al., 1994). Whether this would be a practical therapy in vivo remains to be demonstrated.

11
Standards and Rationale

11.1 Introduction

The previous chapters provide an account of electrical forces that can exert measurable influences on biological systems. Although the bioelectric mechanisms treated in this book can be used in a controlled fashion to exert a beneficial effect, the same forces, when presented in an uncontrolled fashion, can sometimes be detrimental. For that reason, various agencies have developed protective standards and guidelines.

There are many such agencies whose purposes are often directed to specialized environments, applications, or user groups. This chapter reviews standards and guidelines concerning electromagnetic radiation exposure to the general populace, to individuals in occupational settings, and to patients undergoing magnetic resonance imaging. We also review standards and protective measures developed to protect individuals from electric shock in consumer products. The rationale for these standards and related bioelectric mechanisms will be discussed.

11.2 Electromagnetic Field Exposure Standards

Standards of the IEEE and ANSI

In the United States, the most widely applied standards for exposure to electromagnetic fields above 3 kHz are those of the Institute of Electrical and Electronic Engineers (IEEE), which were adopted by the American National Standards Institute (ANSI) in 1992. ANSI is a private, nonprofit organization that does not write specifications. Rather, this organization provides an organizational structure for the development, review, and publication of standards, such as the standards on electromagnetic radiation known as IEEE/ANSI Standard C95.1 (IEEE, 1992). ANSI standards are considered advisory, as the agency has itself no enforcement or regulatory powers. Nevertheless, many of the ANSI standards, and the IEEE C95.1

electromagnetic standard in particular, are adopted as official regulations by many authorities. The technical work and development of the IEEE electromagnetic standards are carried out by the IEEE Standards Coordinating Committee-28 on Non-Ionizing Radiation (SCC-28). The current standard was issued by the IEEE in 1991 and adopted by ANSI in 1992 (IEEE, 1992) as a revision of a previous standard that existed in 1982 (ANSI, 1982). The 1982 standard covered the frequency range from 300 kHz to 100 GHz; the 1991 standard covered the frequency coverage from 3 kHz to 300 GHz. At the time of its publication, the membership of the committee responsible for the 1991 standard consisted of more than 125 members from research (72%), industry (10%), consultants for industry (3%), government administration (4%), and independent consultants and the general public (11%). The disciplines of the members included the physical and biophysical sciences (33%), life sciences (43%), medicine (13%), and others, such as law, medical history, safety, and so on (11%) (see Petersen, 1995).

Rationale for IEEE/ANSI C95.1 Standards

The standards before 1982 limited incident power density to 10 mW/cm^2 across the applicable frequency band based on the assumption that biological effects were related to a thermal mechanism in the exposed organism (Petersen, 1991). Better dosimetric information was available for the 1982 standards, and it was determined that the data on biological effects correlated better with the specific absorption rate (SAR), than with the incident power density. The SAR metric is stated in watts per kilogram of tissue (see Sect. 11.5).

The IEEE committee considered the SAR metric superior because the absorbed power averaged over the whole body for a given incident power density was found to vary with frequency, as seen in Fig. 11.1 (Durney et al., 1985). This figure shows the incident power density consistent with a SAR value of 0.4 W/kg averaged over the whole body. A vertically polarized (electric field vector orientation) propagating field is assumed, which is the worst case polarization with respect to absorbed energy. The figure shows that the incident power density at a constant value of SAR displays a resonance at frequencies above 3 MHz, and the frequency at which the resonance effect is greatest depends on the size of the subject. The figure shows resonance curves for human body weights ranging from 10 to 70 kg— a range that would include small children and adults.

After applying prescreening criteria, the database used in the 1982 ANSI standard consisted of 321 papers; an additional 60 papers were included in the 1991 standard (IEEE, 1992). These papers could be classified into the following categories: environmental factors, behavior and physiology, immunology, teratology, central nervous system, cataracts, genetics, human studies, thermoregulation, biorhythms, endocrinology, development,

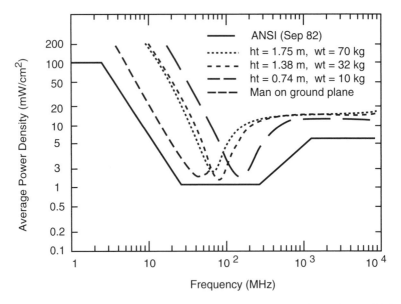

FIGURE 11.1. Power densities that limit human whole-body SAR to 0.4 W/kg compared to ANSI standard of 1982. (From Durney et al. 1985.)

evoked auditory response, hematology, and cardiovascular effects. The category associated with the lowest thresholds of a possible detrimental effect was that of behavioral disruption of complex tasks (in rats) corresponding to an SAR value of 4 W/kg. Whether observed laboratory effects were the result of a thermal mechanism or some other mechanism was not determined in the C95.1 rationale. In terms of the human metabolism, 4 W/kg is expended by moderate activity such as housecleaning or driving a truck, and falls well within the normal range of thermal regulation. The committee applied a "safety factor" of ten, which brought the whole-body SAR limit to 0.4 W/kg.

The solid boundary in Fig. 11.1 is drawn to encompass the family of curves for various body sizes. By so limiting the allowable incident power density, one ensures that a SAR value averaged over the body does not exceed 0.4 W/kg in any individual. This boundary was chosen as the basis for exposure limits in the 1982 IEEE/ANSI radiation standards. The exposure at frequencies below 3 MHz is limited to an incident power density of 100 mW/cm^2, whereas the absorption curves in Fig. 11.1 indicate that the corresponding field rises without limit as the frequency falls below 3 MHz. A cap was placed on the incident power density in order to limit the possibility of induced shock, or spark discharges in high intensity fields (see Sect. 11.4).

We can relate electric and magnetic field intensities to the incident power density using the relationships for a propagating electromagnetic wave:

$$Z = \frac{E}{H} = 377\Omega \tag{11.1}$$

$$S = \frac{E^2}{Z} = H^2 Z \tag{11.2}$$

where E is the electric field (V/m), H is the magnetic field (A/m), S is the power density (W/m^2), and Z is the impedance of the propagation medium (Ω). In "free space" (a vacuum), $Z = 377\,\Omega$ which is the assumption used in the radiation standards; the impedance of air is essentially the same as that of free space. The relationships expressed in Eqs. (11.1) and (11.2) apply to a propagating wave in the "far field" many wavelengths distant from the source of energy.

In many applications, standards are quoted with respect to the magnetic flux density, B (units: Tesla, T), rather than magnetic field intensity, H (units: Amperes per meter, A/m). These two quantities are related by

$$B = \mu H \tag{11.3}$$

where μ is the permeability of the medium. In free space or air, $\mu = 4\pi \times 10^{-7}$ H/m, and we have the relationship that 1 A/m corresponds to 1.257 μT.

Electric and Magnetic Field Criteria for IEEE/ANSI C95.1

Figure 11.2 illustrates the limits in C95.1 over the frequency range 3 kHz to 10 GHz. Separate limits are shown for *controlled* and *uncontrolled* environments which comprise a so-called "two-tier" standard. Although not shown in the figure, the standards extend the high-frequency cap of 10 mW/cm^2 to 300 GHz for both controlled and uncontrolled environments. Controlled environments are locations where people are aware of the potential for exposure as a result of employment, by other cognizant persons, or where exposure is incidental as a result of transient passage through areas in which the exposure may be between the controlled and uncontrolled environment limits. Uncontrolled environments are locations where individuals have no knowledge or control of their exposure, such as in living quarters or work places where there is no expectation of exposure.

Before 1983, exposure standards in the United States did not distinguish between exposure of the general public and those in occupational or controlled environments (Osepchuk, 1994). In 1983, the Massachusetts Public Health Department adopted a two-tier policy in which exposures to the public were one-fifth of the levels stated in the then-existing ANSI C95.1 of 1982. The averaging time was 30 minutes for the lower tier and 6 minutes for the upper tier. Consequently, the two tiers are equivalent for exposure

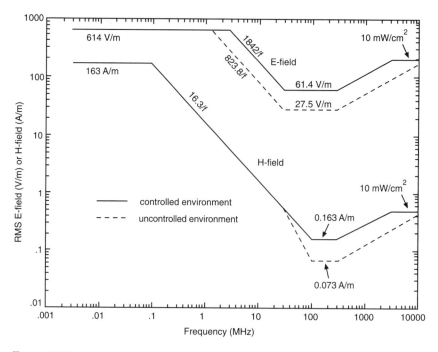

FIGURE 11.2. Maximum permissible exposure to electric and magnetic fields according to IEEE C95.1 (1991).

durations of 6 minutes or less. The lower limit was intended to protect the general public who might have chronic exposure when living near EMF sources. The factor one-fifth was derived from the presumed ratio of maximum occupational weekly exposure (40 hr/wk) to that for the chronically exposed public (168 hr/wk). The two-tier Massachusetts limits were essentially adopted by the National Council on Radiation Protection and Measurements (NCRP, 1986).

One criterion applied to the 1991 standard was the resonance curve for SAR = 0.4 (Fig. 11.1). In conformance with existing NCRP practice, the IEEE committee included an additional safety factor in the minimum resonance region for uncontrolled environments, resulting in a minimum SAR of 0.08 W/kg. Based on studies of absorption of electromagnetic energy in humans, it was concluded that the electric field component coupled energy into the biological medium more efficiently than did the magnetic field (Gandhi, 1988). Consequently, the electric field limits were selected according to Eq. (11.2), with incident power density limited by the solid line of Fig. 11.1; the magnetic field limits were selected to permit greater exposure at frequencies below 100 MHz, in recognition of the less efficient coupling of the magnetic field in this frequency regime. For instance, in a controlled environment, the E- and H-fields in the minimum plateau region are consistent with an incident power density of $1\,mW/cm^2$ ($=10\,W/m^2$), namely $E = 61.4\,V/m$ and $H = 0.163\,A/m$, as calculated by Eq. (11.2). In an uncontrolled environment, the allowed power density in the resonance region is $0.2\,mW/cm^2$. At frequencies above $15\,GHz$, the standards have a cap of $10\,mW/cm^2$ for both controlled and uncontrolled environments. By observing both the electric and magnetic field limits, one ensures that an average SAR value of $0.4\,W/kg$ will not be exceeded at any frequency, whether the exposure is in the near or far field of the source. Note that the SAR limits contained in C95.1 apply only at frequencies between $0.1\,MHz$ and $6\,GHz$.

Although average whole-body SAR limits may be so limited, individual regions within the body will exhibit absorption rates above and below the average rates. Implicit in these limits is the assumption that peak SAR within limited regions of the body may exceed the average. Accordingly, IEEE C95.1 provides that SAR limits may be relaxed in any $1\,g$ of tissue to $8\,W/kg$, except for the hands, wrists, and feet, where the SAR limit is limited to $20\,W/kg$ as averaged over any $10\,g$ of tissue in the controlled environment and one-fifth of that value in uncontrolled environments.

The averaging time applicable to IEEE C95.1 is 6 minutes for frequencies below $15\,GHz$; at higher frequencies, the averaging time goes down to a low of 10 seconds in controlled environments. For pulsed fields, IEEE C95.1 specifies additional criteria which would limit the peak exposure (see Sect. 11.3). Averaging time rules are specified in uncontrolled environments for frequencies in the resonance region (30–$100\,MHz$) in which the averaging time is $6\,min.$ at and below $30\,MHz$, and $30\,min.$ at $100\,MHz$ and above. Considerations involved in averaging time are addressed by Osepchuk (1989).

Induced Current and Electric Shock Considerations

Based solely on absorbed power, one would conclude that permissible fields can increase without limit in inverse proportion to frequency below the resonance region. However, in order to avoid problems of induced electric shock at low frequencies, IEEE C95.1 imposes additional constraints on the maximum permissible fields, and also on induced current density within the body as discussed below. Thermal perception of electric currents is discussed in Sect. 7.5.

As seen in Fig. 11.2, the permissible E-field is capped at $614\,V/m$ below $3\,MHz$. That cap was imposed in recognition of the possibility of spark discharges to exposed individuals. Note that C95.1 standards do not preclude the possibility of spark discharges under some conditions (see Sect. 11.4). The magnetic field below $100\,kHz$ is capped at $H = 163\,A/m$, which is equivalent to $B = 0.205\,mT$. Compare that value with Fig. 9.17, which shows magnetic thresholds calculated for whole body exposure of a large person.

TABLE 11.1. Maximum induced current limits from
IEEE/ANSI C95.1.

	0.003–0.1 MHz	0.1–100 MHz
Controlled environment		
Both feet	$2,000f$	200
Each foot	$1,000f$	100
Contact	$1,000f$	100
Uncontrolled environment		
Both feet	$900f$	90
Each foot	$450f$	45
Contact	$450f$	45

Limits are stated in mA-rms, with f in MHz. Averaging time =
1 s. Foot limits assume a free-standing individual (no metallic
contacts). Contact limits are for touching a metallic object.

Even if we use the worst case assumption of the parameter f_e in Fig. 9.17, we
see that the C95.1 magnetic field limit is well below the threshold for
peripheral nerve excitation.

The C95.1 standard also imposes limitations on induced currents as noted
in Table 11.1, which were intended to limit the severity of induced electric
shock or burns. The intent of the Standards Committee was to limit the
possibility of perceptible contact current with a grasping contact. However,
these limits would allow perceptible current with a touch contact under
some conditions. Consider, for instance, the limits on contact current.
At a frequency of 10 kHz, the standard would allow a contact current of
10 mA in the controlled environment, and 4.5 mA in the uncontrolled
environment. At a frequency of 100 kHz, the standards would allow 100 mA
in the controlled environment and 45 mA in the uncontrolled environment.
Comparing these values with the reaction thresholds in Fig. 7.12, we conclude that the standards below 100 kHz would allow perceptible shock
in the uncontrolled environment and painful shock in the controlled environment, particularly for small individuals, if subjects were to access the
current through a touch contact with an energized conductor. Above
100 kHz, the standards would permit sensations of heating that could be
perceptible with a touch contact in the uncontrolled environment and painful in the controlled environment. For small subjects or children, even
stronger sensations might occur across the range of frequencies shown in
Table 11.1 (see Sect. 7.10). The perceived reactions would be reduced
relative to a touch contact if the subject were to contact the energized
conductor with a large area, such with by grasping with the whole hand
(refer to Fig. 7.15, and Chatterjee et al., 1986), as was assumed in the C95.1
standard.

Other Standards on Electromagnetic Exposure

Another group with national stature in the United States is the National Council on Radiation Protection (NCRP). This group was chartered by Congress in 1928 to develop information, and issue exposure guidelines on all forms of radiation. NCRP issued guidelines on radio frequency electromagnetic fields in 1986 (NCRP, 1986). The standing committee members and consultants on the radio frequency standards consisted of 16 experts at the time. NCRP adopted the then-existing ANSI standards of 1982 for occupational settings and applied an additional acceptability factor of 5 for the general public. Subsequently, IEEE/ANSI C95.1 issued in 1991 echoed the NCRP two-tier philosophy with "controlled" and "uncontrolled" environment limits and applied the additional safety factor of 5 to the uncontrolled environment. The NCRP reasoned that the lower limits should be applied to the general public who may consist of particularly vulnerable individuals and others who may be unaware of, or have no option about their exposure. The additional safety factor of 5 was arrived at by rounding the ratio of a 40-hour work week to the total number of hours in a week $(40/168 \approx 0.2)$. The guidelines of the NCRP are more conservative than those of the IEEE/ANSI at the higher frequencies. Above 1,500 MHz, NCRP limits top off at a power density of 0.2 and 5 mW/cm^2 for general and occupational populations respectively. The IEEE/ANSI limits reach a ceiling of 10 mW/cm^2 for both populations.

Electromagnetic exposure standards and guidelines have been published by other world-wide agencies covering the general public, occupational, and military populations. Of particular interest are the standards of the National Radiation Protection Board (NRPB) in the United Kingdom (NRPB, 1993; McKinlay et al., 1993). These have been recently published, their rationale is well documented, and the standards cover a wide frequency range—from zero to 300 GHz. With respect to static fields, the NRPB recommendations are designed to prevent acute direct effects, such as vertigo and nausea. Restrictions on ELF electric and magnetic fields are intended to avoid acute effects on the central nervous system. Restrictions on radio-frequency and microwave fields were intended to avoid adverse responses as a result of increased heat load and elevated tissue temperature.

The NRPB specifies "basic restrictions" as listed in Table 11.2—these are the fundamental dosimetric quantities of limitation in the standards. For frequencies from 1 to 10 GHz, the basic restrictions specify current density and SAR limits within the human body. From 10 to 300 GHz, the restrictions specify incident power density, becuase the depth of penetration of energy at these high frequencies was assumed to be small. SAR limits are specified with respect to the volume of tissue for averaging.

From the basic restrictions, the NRPB derives "investigation levels," which are environmental measureables for investigating whether compli-

TABLE 11.2. Basic restrictions on exposure to electromagnetic fields as specified by the NRPB.

Frequency range (Hz)	Basic restriction	Averaging area	Comments
0–1	200 mT	—	24-h average
	2 T	—	maximum, body
	5 T	—	maximum, limbs
	100 mA/m^2	—	Current density
1–10	100/f mA/m^2	—	Current density
10–10^3	10 mA/m^2	—	Current density
10^3–10^5	f/100 mA/m^2	—	Current density
10^5–10^7	0.4 W/kg, 10 g	body	SAR, 15 min avg.
	10 W/kg	10 g	SAR, 6 min. avg., head & fetus
	10 W/kg	100 g	SAR, 6 min. avg., neck & trunk
	20 W/kg	100 g	SAR, 6 min. avg., limbs
	f/100 mA/m^2	—	Current density, head, neck, trunk
10^7–10^{10}	0.4 W/kg	body	SAR, 15 min. avg.
	10 W/kg	10 g	SAR, 6 min. avg., head & fetus
	10 W/kg	100 g	SAR, 6 min. avg., neck & trunk
	20 W/kg	100 g	SAR, limbs
10^{10}–3 × 10^{11}	100 W/m^2	—	Power density, max. on body

Note: f specified in Hz.
Source: McKinlay et al. (1993).

ance with the basic restrictions is achieved. If the measured quantity is below the investigation level, one assumes that compliance with the basic restrictions is achieved. If the measured quantity is above the investigation level, it does not necessarily imply that the basic restrictions are exceeded. According to the NRPB, factors that might be considered in such an assessment include the efficiency of the coupling of the person to the field, the spatial distribution of the field across the volume of space occupied by the person, and the duration of exposure (NRPB, 1993). The NRPB investigation levels are similar to the IEEE/ANSI C95.1 standards (Fig. 11.2) in that they recognize the body resonance phenomenon, and the associated SAR and current density values are in a similar range.

An international agency also specifying EMF guidelines is the International Commission on Non-Ionizing Radiation Protection (ICNIRP), which was estalished in 1992 as a separate entity under the International Radiation Protection Association (IRPA). This commission, which has as many as 18 representatives from 11 countries (depending on year of commission), publishes EMF guidelines in association with the World Health Association (WHO). Separate publications cover static fields (ICNIRP, 1994), 100 kHz to 300 GHz (IRPA, 1988), and 50/60 Hz (IRPA, 1990). The most recent ICNIRP publication (1998), which was published as this book was in press, covers the frequency range from 0 Hz to 300 GHz.

General reviews of international standards have been provided by Gandhi (1990), Polk and Postow (1996), and Klauenberg and colleagues (1995). More recent standards have been published by the Netherlands (1997). In general these standards recognize the resonance phenomenon discussed above and provide for minimum exposure at frequencies within the resonance region. The minimum fields allowed within the resonance region typically are within a factor of two of the IEEE/ANSI standard. The standards differ much more, however, at frequencies below 1 MHz. For instance, in the frequency range 10 to 100 kHz, the E-field limits (in V/m) for continuous exposure by the general public, in the order greatest to least are: 1,500 (Germany), 1,000 (NRPB, United Kingdom); 614 (IEEE, United States); 614 (IRPA); 280 (Canada); 87 (Netherlands); 25 (Soviet Union). The maximum magnetic field limits (in A/m) for the same frequency range are: 350 (Germany); 163 (IEEE, United States); 64 (NRPB, United Kingdom); 64 (IRPA); 4 (Netherlands); 1.8 (Canada). An equally wide disparity exists among other standards at frequencies below 10 kHz, as noted in the next section.

Electromagnetic Exposure Limits at Low Frequencies

Much public discussion has occurred in past years concerning electromagnetic exposure at very low frequencies and particularly at power frequencies (60 Hz in the North America, and typically 50 Hz elsewhere). This attention has been largely driven by concerns about public health issues associated with power transmission and distribution lines—a subject that remains controversial. There are presently no IEEE standards for frequencies below 3 kHz, although the IEEE C95.1 committee is presently considering standards in that frequency regime. However, other standards do specify exposure limits at very low frequencies, and these will be reviewed here. The reader is also directed to Sect. 11.7 for an analysis of established bioelectric mechanisms as a basis for low frequency standards.

Figures 11.3 and 11.4 illustrate very low frequency magnetic and electric field limits that are specified by national and international agencies. One agency is the American Conference of Governmental Industrial Hygienists in the United States (ACGIH, 1996), which is a private professional group. The limits specified by ACGIH are intended to apply to professionals in industrial and occupational settings, and not necessarily the general public. The limits are intended to protect against hazards in the workplace. The ACGIH states that biological effects for E-fields below the specified limits have been demonstrated in the laboratory, but these reactions were not thought to be hazardous. In the case of the B-field limits, an exposure to the arms and legs could be increased by a factor of 5 in the frequency regime 1 to 300 Hz; for the hands, a factor of 10 could be applied. A second agency is the European Agency for Electrotechnical Standards (CENELEC, 1995). The CENELEC standard shown in Figs. 11.3 and 11.4 is considered a

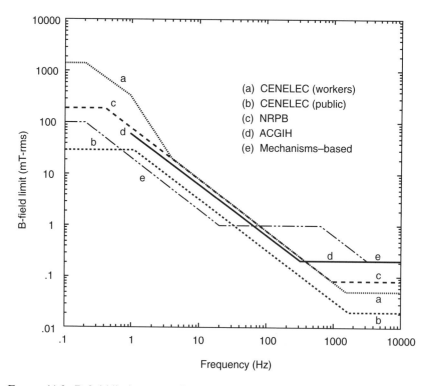

FIGURE 11.3. B-field limits at very low frequencies: whole body exposure, extended durations (e.g., whole work day). Curves (a)–(d): existing standards. Curve (e): hypothetical mechanisms—based standard for short-term exposure in uncontrolled environments.

"prestandard," subject to further review. These limits are intended to prevent adverse short-term effects in humans caused by current densities that exceed $10 \, mA/m^2$—a criterion derived from laboratory demonstration of magnetophosphenes, possible nervous system effects, and bone healing effects. A third agency represented in the figures is the National Radiological Protection Board in the United Kingdom (NRPB, 1993; McKinlay et al., 1993). The stated NRPB rationale of the low-frequency standards is to limit induced current density in the head, neck, and trunk as indicated in Table 11.2 so as to avoid effects on the central nervous system functions, such as control of movement, posture, memory, reasoning, and visual processes. The most recent guidelines of ICNIRP (1998) cover the frequency range 0 Hz to 300 GHz.

The magnetic field limits shown in Fig. 11.3 are all below the threshold for phosphenes, which at the most sensitive frequency (20 Hz) is 10 mT (Fig. 9.20); the limits are also well below thresholds for excitable tissue stimulation (Fig. 9.18). Thus, one would not expect acute reactions for exposures constrained by the indicated B-field standards.

Figure 11.3 includes curve (*e*), which is a hypothetical standard for short-term exposure in uncontrolled environments. Curve (*e*) is based on established mechanisms for reactions to short-term magnetic field exposure as developed in Sect. 11.7.

The E-field limits shown in Fig. 11.4, however, are sufficiently great that they would not preclude significant reactions from induced shock, spark discharges, and corona. For instance, with fields in the range 20 to 30 kV/m, corona discharges could easily be produced from body surfaces, and severe spark discharges could easily occur with ordinary activity in which a person touches grounded objects (see Sects. 9.2–9.4). It is difficult to see how one could function in such a high field environment without taking special precautions to avoid these effects. In recognition of the difficulties in high E-field environments, ACGIH recommends that workers avoid situations where spark discharges may be produced in fields above 5 kV/m and recommends the use of protective clothing in fields above 15 kV/m.

Considering the significant public debate about health effects from power frequency fields, it is useful to focus on EMF limits at 50/60 Hz, which are listed in Table 11.3. Table 11.3 also includes limits of the International Commission on Non-ionizing Radiation Protection (ICNIRP, 1998). In the

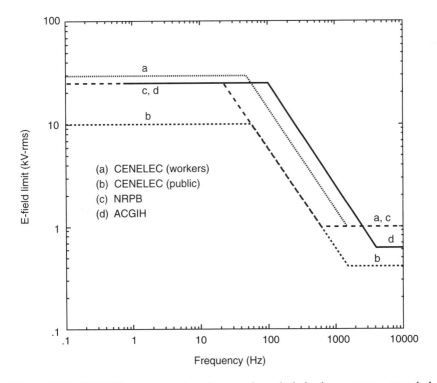

FIGURE 11.4. E-field limits at very low frequencies: whole body exposure, extended durations (e.g., whole work day).

TABLE 11.3. Other standards evaluated at power frequency.

Agency	E-field (kV/m)		B-field (mT)	
	50 Hz	60 Hz	50 Hz	60 Hz
ACGIH (1996)				
whole body	25	25	1.2	1
arms & legs	25	25	6	5
hands & feet	25	25	12	10
ICNIRP[c] (1998)				
workers	10	8.33	0.5	0.42
general public	5	4.17	0.1	0.083
NRPB (1993)	12	10	1.6	1.33
CENELEC (1995)				
workers, whole body[b]	30	25	1.6	1.33
workers, limbs	30	25	25	20.8
gen. public, whole body	10	10	0.64	0.53
gen. public, limbs	10	10	10	8.33

Notes: (a) Limits are stated as rms quantities; (b) exposure duration (hr) $\leq$ 80/E. (c) Exceedance of ICNIRP limits requires analysis to test for compliance with current density restrictions.

frequency range below 1 kHz, the ICNIRP magnetic field limits are intended to provide an adequate safety margin below an induced current density of $100 \, mA/m^2$, which was taken as a threshold for acute changes in central nervous system excitability or visual evoked potentials. Safety factors of 10 and 50 were provided for occupational groups and the general public, respectively, thus providing basic restrictions of 10 and $2 \, mA/m^2$.

Additional power frequency guidelines have been developed for application to electric power transmission and distribution lines, as shown in Table 11.4 (after Goellner et al., 1993). Note that the B-field units in Table 11.4 are in μT, rather than in mT, as in Table 11.3. While there are no applicable federal standards in the United States, a number of states and local regulatory bodies have specified limits of exposure to the general public from power lines. Table 11.4 also lists power line environmental limits adopted in other countries. The power frequency magnetic field limits of state and local agencies in the United States are well below corresponding limits in other countries, and well below the international limits listed in Table 11.3. One should distinguish between an environmental limit and a standard. An exposure standard generally specifies the area of the body being exposed, the allowed level, the duration, as well as other exposure conditions. An environmental limit or a product performance standard would generally not specify such conditions, although they might include considerations other than human exposure (e.g., product interference).

Additional product performance guidelines on low frequency EMF emissions have been specified by the Swedish federal government for application to video display devices, such as computer monitors, and television sets (Mild and Sandström, 1994; MWN, 1995). Existing Swedish guidelines in the frequency range 5 Hz to 2 kHz specify an E-field of 25 V/m measured at a distance of 50 cm in front of the display, and a B-field of 250 nT measured 50 cm all around the display. In the frequency range 2 to 400 kHz, the corresponding values are 2.5 V/m and 25 nT. Proposed guidelines specify field values in categories, and the manufacturer would be required to label the equipment as to its category of compliance (MWN, 1995). According to Mild and Sandström, the guidelines were developed without clear proof of health risks, but as a "prudent avoidance" strategy that was felt to be readily achieved by manufacturers.

The exposure limit discussed above encompass an enormous range. For instance, at 60 Hz, the various guidelines and standards specify E-field limits ranging from 10 V/m to 30 kV/m, and B-field limits ranging from $0.2 \mu T$ to 21 mT. Although part of this range can be attributed to distinctions between standards, performance limits, and environmental limits, the wide range

TABLE 11.4. Power frequency exposure standards for transmission and distribution lines.

Agency	E-field (kV/m)		B-field (μT)	
	In ROW	Edge ROW	In ROW	Edge ROW
US State/local				
California (Irvine)	—	—	—	0.4
Florida[a]				
≤230 kV	8	3	—	15
500 kV, single	10	2	—	20
500 kV, double	10	2	—	25
Minnesota	8	—	—	—
Montana[b]	—	1	—	—
New Jersey	—	3	—	—
New York PSC	—	1.6	—	20
North Dakota	9	—	—	—
Oregon	9	—	—	0.4
Tennessee	—	—	—	0.4
Outside US[d]				
Australia[e]				
occupational	—	2	—	500
general public	10	2	—	100
Germany	—	—	—	5000
Italy				
extended exposure	—	5	—	100
limited exposure	—	10	—	1000

Notes: (a) More stringent standards apply to Lake Tarpon line; (b) ≥69 kV; (c) limits are stated as rms quantities; (d) 50 Hz; and (e) whole day exposure;
Source: Goellner et al. (1993).

also reflects the controversy and lack of agreement concerning chronic-exposure EMF health effects, especially at frequencies below a few kHz. Most EMF standards recognize constraints imposed by acute human reactions to EMF exposure, which have been treated in the previous chapters of this book. The mechanisms responsible for these acute effects are widely recognized, reasonably well understood, and experimentally verified. On the other hand, most standards do not acknowledge hazards for exposures below the limits where acute reactions are known to exist; exceptions to this statement apparently apply to some transmission line standards, and the Swedish VDT emission standards, which limit emissions to very low values. The mechanisms that might be responsible for these presumed weak field effects are not well-established. Although a thorough treatment of weak-field effects is beyond the intended scope of this book, we will treat this subject in brief in Sect. 11.6.

Static Electric and Magnetic Fields

Strong static magnetic fields can cause biological reactions through their influence on moving charges, as described in Sect. 9.11. Examples would include blood flow, and the movement of the biological system itself, such as with eye or body movement. Table 11.5 summarizes limits on static fields as recommended by several agencies. The limits in the upper part of the table apply to extended exposure in occupational groups, or by the general public. ICNIRP guidelines for static magnetic field exposure state that detrimental effects do not occur below 2 T according to current knowledge. The ICNIRP exposure guidelines in Table 11.5 reflect a safety factor of 10 for occupational exposure, and 50 for the general public. For the CENELEC occupational limits, the duration of exposure is specified as $t \leq 112/E$, where t is the duration (hours) in any 8-hour period, and E is the E-field in kV/m. For instance, at the upper limit of 42 kV/m, an exposure duration of 2.67 hour per 8-hour period would be allowed. The NRPB reduces limits by a factor of 10 for chronic (24 hr) exposure because of the lack of chronic human exposure data, rather than specific adverse findings related to chronic exposure. The lower part of Table 11.5 applies to patients undergoing magnetic resonance imaging. It is assumed that for those individuals, the duration of exposure would be relatively brief, and not often repeated.

11.3   Pulsed Electromagnetic Fields

It is possible to stimulate peripheral nerves and even the heart with a sufficiently strong time-varying magnetic field. For instance, we have seen that peripheral nerves can be excited with a magnetic stimulus of about 40 T/s oriented normal to the sagittal or frontal cross section of a large

TABLE 11.5. Static field criteria.

Agency	Max. E-field (kV/m)	Max B-field (T)	Ref.
General Exposure Criteria			
NRPB			
max. body	25	2.0	NRPB, 1993;
limbs only	—	5.0	McKinlay, 1993
24-h avg.	—	0.2	
ICNIRP[(b)]			ICNIRP, 1994
workers-avg. in work day	—	0.2	
workers-maximum	—	2.0	
workers-limbs	—	5.0	
gen.public-continuous	—	0.04	
CENELEC			CENELEC, 1995
workers-whole body	42	2	
gen. public	14	0.04	
workers-limbs	—	5	
gen. public-limbs	—	0.1	
MRI Exposure Criteria			
IEC		2.0[a]	IEC, 1995
NRPB			NRPB, 1991
trunk/head	—	2.5[a]	
limbs	—	4.0	
IRPA			IRPA, 1991
trunk/head	—	2.0	
limbs	—	5.0	
FDA	—	2.0	FDA, 1992

Notes: (a) Limits may be extended to 4 T under specially controlled conditions; (b) Warnings expressed below 0.5 mT for individuals with implanted pacemakers. See also ICNIRP (1998).

person, provided the duration of dB/dt is sufficiently long (Table 9.6). Although such fields would be considered extremely large as compared with environmental exposures, even in occupational settings, such fields are attainable in magnetic resonance imaging machines. When magnetic resonance imaging technology was first introduced, such fields were not considered attainable. However, subsequent advances in MRI echo planar technology have drastically increased the limits of attainable fields to the point that future development will be limited not by attainable technology, but by the need to avoid nerve and heart excitation.

Pulsed Field Limits in Magnetic Resonance Imaging

The new generation of fast imaging echo planar MRI devices involve significantly greater switched gradient fields than used previously. To understand the conditions required for peripheral nerve stimulation with this advanced technology, the U.S. Food and Drug Administration (FDA) commissioned two studies that established theoretical predictions of nerve and heart excitation by MRI switched gradient fields (Reilly, 1989, 1991). These studies formed the basis of advisory guidelines on MRI switched gradient fields adopted by the FDA in the United States (FDA, 1992) and subsequently by the International Electrotechnical Commission (IEC) in Europe (IEC, 1995). Subsequent revisions to the FDA guidelines (FDA, 1995) identifies previous FDA limits as "levels of concern," above which the manufacturer must establish through human volunteer studies that painful stimulation will not occur. A decade ago the limits expressed in the FDA and IEC guidelines would have imposed no real limit on MRI technology. Today, these limits impose significant constraints on operating parameters for the new generation of MRI machines.

The studies cited above were theoretical ones that combined the nonlinear models of myelinated nerve, described in Chapter 4, with magnetic induction models, as described in Chapter 9. At the time the first FDA-sponsored study was completed, there was little experimental verification for these predictions, although focal nerve excitation by small coil systems had already been demonstrated. Subsequently, various experimental studies with large coil systems (reviewed in Chapter 9) demonstrated that thresholds for both nerve and the heart are consistent with theoretical predictions.

Subsequent to their adoption by the FDA, these guidelines were adopted by the National Radiological Protection Board in the United Kingdom (NRPB, 1991; Saunders, 1991), and later by the International Electrotechnical Commission (IEC, 1993, 1995). Figure 11.5 illustrates the IEC Guidelines, on which are superimposed several reaction curves. The IEC curves (a) and (d) separate three operating regions. The lowest region is considered an *uncontrolled exposure zone*, in which only routine patient monitoring is required. In the middle region, designated the *first controlled operating zone*, deliberate action and medical supervision are required. In the uppermost region, the *second controlled operating zone*, specific security measures are required to prevent unauthorized operation in this zone, and patient exposure is permitted only under a human studies protocol approved according to local requirements.

Curve (b) represents expected thresholds for peripheral nerve excitation based on the studies described in Chapter 9, in which the magnetic field is perpendicular to the sagittal or frontal cross-section of the patient; for a field oriented along the longitudinal dimension of the body, the nerve excitation threshold would be about 50% greater. Curve (c), representing the threshold of discomfort or pain, is elevated above the perception curve by a factor of two in accordance with similar multiples obtained from human electrocutaneous stimulation (Table 7.3). More recent experimental data with human subjects excited by experimental MRI devices suggest that painful reactions may be somewhat closer to the perception curve–elevated above it by a factor of about 1.3 (Budinger et al., 1991; Bourland et al.,

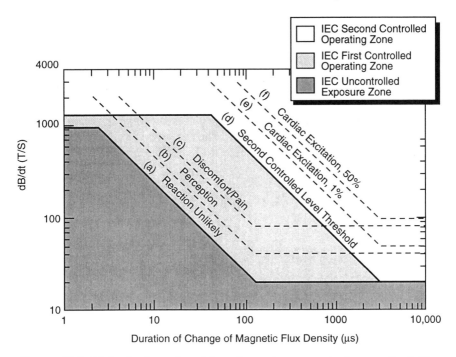

FIGURE 11.5. IEC criteria and anticipated reactions to pulsed magnetic field— whole-body exposure of large adult. (a) Upper limit of IEC uncontrolled exposure zone; (b) threshold of nerve stimulation; (c) threshold of discomfort or pain; (d) upper limit of IEC first controlled operating zone; (e) one-percentile threshold for cardiac excitation; (f) 50-percentile threshold for cardiac excitation. Limit criteria from IEC (1995); reaction threshold from Reilly (1992).

1997). Curves (e) and (f) are one- and 50-percentile thresholds for cardiac excitation (see Chapter 9). The upper limits on curves (a) and (d) were intended to limit whole-body heating effects.

The curves appearing in Fig. 11.5 are asymptotic expressions of the strength–duration (S–D) relationships for neural and cardiac excitation by rectangular, monophasic dB/dt pulses (as in Fig. 4.2). The S–D time constant, τ_e, which defines the corner of the lower asymptote, is taken as $120\,\mu s$ for nerve stimulation as determined by a myelinated nerve model for a terminated or sharply bent axon (see Sect. 4.5); a time constant of 3ms has been chosen for cardiac stimulation, which is considered applicable to stimulation over a large area of the myocardium (see Sect. 6.3). Although these curves specifically apply to rectangular monophasic dB/dt pulses, they also represent conservative lower limits to excitation thresholds for pulsed or continuous sinusoidal stimuli, provided that one interprets the horizontal axis as the duration of a half-cycle of the sinusoidal waveform (Reilly, 1989).

The lower plateaus of the curves in Fig. 11.5 represent the minimum excitation thresholds for long duration stimuli $(t_p > 10\tau_e)$. For short duration stimuli, $(t_p < 0.1\tau_e)$ theoretical thresholds are inversely proportional to duration even into the submicrosecond range. Indeed, experimental data supports this interpretation for durations down to a small fraction of a microsecond (see Chapter 7). However, curves (a) and (d) are limited to upper plateaus in order to avoid unacceptable tissue heating associated with eddy currents from high dB/dt values. The placement of the upper plateau depends on the temperature rise one is willing to accept, and the duty factor one assumes for the pulsed stimulus.

The standards discussed here do not consider the possibility of patient implants. For a treatment of that subject, see (Reilly and Diamant, 1997; Buechler et al., 1997).

Extrapolation of MRI Exposure Limits to EMF Standards

The maximum permissible exposure limits of IEEE/ANSI C95.1 (Fig. 11.2) are stated as rms values applicable to continuous, sinusoidal radio frequency fields. A difficulty arises for pulsed sinusoidal waveforms of low duty factor, for which the allowable peak field may rise above the nerve excitation threshold, yet still satisfy the rms limit of the standard. If B_c represents the allowable peak field for continuous exposure at a particular frequency, then the allowable peak field B_o consistent with a constant rms value, for an arbitrary duty factor d_f is given by:

$$B_o = \frac{B_c}{d_f^{1/2}} \tag{11.4}$$

where d_f is the fractional duration of the on time during the repetition period (assuming an integer number of half-cycles of the sinusoidal variation). Equation (11.4) is a criterion based on constant energy.

To avoid nerve stimulation from pulsed fields, it is necessary to impose a constraint on the peak value of the allowable field. Since pulsed field limits have already been specified for MRI exposure, we proposed the application of MRI criteria to IEEE/ANSI standards (Reilly, 1998).

As seen in Fig. 11.5, MRI exposure criteria are stated in terms of the duration of dB/dt. In order to make a connection with sinusoidal fields, we need to define the relationship between the peak field and its time derivative. For a pulsed field consisting of a linear increase over the pulse duration (trapezoidal pulse), the relationship is:

$$B_o = \dot{B}_o t_p \quad \text{(pulsed trapezoidal field)} \tag{11.5}$$

where $\dot{B}_o$ is the value of dB/dt during the pulse, and τ_p is the phase duration, defined as the pulse duration for a rectangular pulse, or the half-cycle time for a sine wave (see insert of Fig. 4.15). For a sinusoidal field, the relationship is

$$B_o = \frac{1}{\pi} \dot{B}_o t_p \quad \text{(sinusoidal field)} \tag{11.6}$$

In this case, $\dot{B}_o$ is the peak of dB/dt during the sinusoidal cycle.

Figure 11.6 plots the nerve excitation limit of Fig. 11.5 (curve (a) in both figures), which is read on the left vertical axis of Fig. 11.6. Curves (b) and (c) plotted on Fig. 11.6 show the corresponding peak flux density according to Eqs. (11.5) and (11.6), and are read on the right vertical axis. Curve (c) may be interpreted as a limit to the peak value of the flux density for sinusoidal fields.

Figure 11.7 shows curve (a) of Fig. 11.6 expressed on a format consistent with the magnetic field standards of C95.1; also plotted is the limit imposed by C95.1 on the peak allowable field for a 100% duty factor waveform (d_f = 1.0). Peak flux density, B_o, is expressed on the left vertical axis of Fig. 11.7 in units of mT; the right vertical axis gives the corresponding peak magnetic field strength, H_o, in A/m, in conformance with the units of C95.1. The horizontal axis has been expressed in frequency units using the relationship

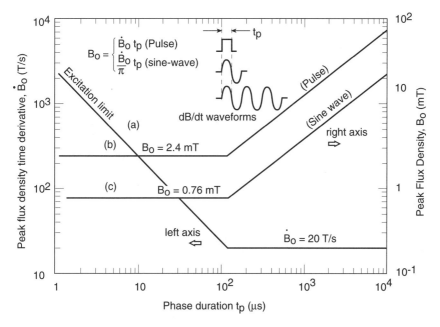

FIGURE 11.6. Magnetic field exposure criteria—adapted from MRI exposure limits for switched gradient fields. (a) = IEC criteria on B (MRI); (b) = curve (a) limits on B for pulsed dB/dt; (c) = curve (a) limits on B for continuous sine dB/dt. (From Reilly, 1998.)

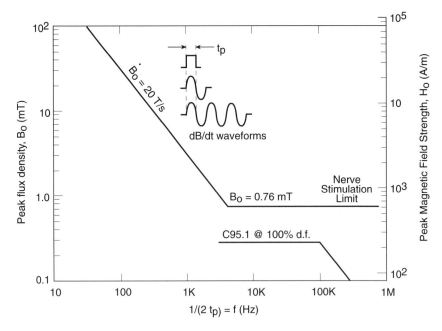

FIGURE 11.7. Magnetic field exposure criteria that avoids peripheral nerve stimulation (including safety margin) for pulsed sinusoidal fields. Nerve stimulation limit has safety factor of 2X below threshold for pulsed or CW sine wave. (From Reilly, 1998.)

$$f = \frac{1}{2t_p} \tag{11.7}$$

For frequencies above $3\,\mathrm{kHz}$, $B_o = 0.76\,\mathrm{mT}$, or equivalently, $H_o = 605\,\mathrm{A/m}$ according to the MRI-based limits. The peak value of H_o from C95.1 with $d_f = 1.0$ is $231\,\mathrm{A/m}$. It can be seen from Eq. (11.4) that the IEEE/ANSI rms limits would exceed the peak MRI-based limits for $d_f \leq 0.15$.

One might question whether the avoidance of peripheral nerve stimulation with exposure to pulsed magnetic fields is a proper criterion for standards that are intended to avoid hazardous conditions, not merely perceptual ones. Although a reaction at the threshold of perception is unlikely to be hazardous, most would agree that unpleasant or painful stimuli could be hazardous under many conditions. Since the threshold multiple from perception to painful reactions is about 1.4 (Sect. 9.7), the multiple implicit it the proposed peak limits is about 2.7 for a large subject with frontal exposure of the torso and 4.4 with longitudinal exposure (refer to Table 9.6). For comparison, an acceptability margin of 10 or more has been incorporated into the SAR limits that underlie existing C95.1 standards. Note, however, that since SAR is an energy metric, it is proportional to the square of the field (Eqs. 11.2; 11.9). Therefore, a margin of 10 in SAR corresponds to a margin of 3.16 in the associated field.

Limits of Applicability

For stimulation by sinusoidal currents, thresholds of excitation rise approximately in proportion to frequency beyond a critical frequency (a few kHz for neurosensory or neuromuscular stimulation), provided that the stimulus consists of multiple sinusoidal cycles (see Sect. 4.6). This fact is responsible for the lower plateau of nerve excitation threshold beyond a few kHz when expressed in magnetic flux density units as in Fig. 11.7—as we increase frequency, the threshold increases in proportion to frequency, as does dB/dt (and consequently, current density). The two effects compensate one another, resulting in a constant magnetic flux threshold of excitation versus frequency.

As we increase frequency, the current density for nerve excitation increases proportionately and can become so great that electrical excitation thresholds will exceed requirements for perception due to heating. For sinusoidal currents that are at least a significant fraction of a second in duration, the cross-over between electrical and thermal perception thresholds occurs at about 100 kHz (Sect. 7.5). The thermal cross-over accounts for the diminishing limits beyond 100 kHz in the C95.1 standard, as seen by the lower curve in Fig. 11.7—the indicated limit ensures that the current density within the medium is maintained to a constant value that does not exceed thermal perception thresholds. However, for pulsed sinusoidal waveforms, it is possible to electrically excite sensory neurons without thermally exciting thermal receptors. This statement can be understood in terms of excitation models presented in previous chapters. For instance, we have seen that in Fig. 4.19 that the perception threshold to a high-frequency sinusoidal stimulus reaches a plateau at durations beyond a millisecond or so. Consequently, by reducing the duration of a continuous sinusoidal stimulus to a millisecond, for example, one significantly reduces its rms value (i.e., heat producing capacity) without significantly reducing the sensory perception threshold associated with that stimulus.

The question arises: how high in frequency can we extend the nerve stimulation plateau shown in Fig. 11.7? As noted in Sect. 7.5, human perception experiments with relatively long duration sinusoidal currents demonstrate the proportionality between electrical thresholds and frequency to 100 kHz, at which point thermal perception thresholds dominate, suggesting a frequency limit to at least 100 kHz. Experiments with anesthetized rats, which allowed significant tissue heating, demonstrated neuromuscular stimulation with approximate frequency-proportional thresholds to 1 MHz—the highest frequency tested (Fig. 4.22). From this observation, it would appear reasonable to extend the plateau of Fig. 11.7 to at least 1 MHz. And from human sensory experiments with brief conducted current pulses, one could infer experimental correspondence to perhaps 10 MHz. This statement is based on the fact that human electrocutaneous perception thresholds are inversely proportional to the duration of brief monophasic currents to about $0.1\,\mu s$—the shortest tested duration (see Sect. 7.3).

11.4   Consideration of Spark Discharges in EMF Standards

We have seen in Chapters 2 and 9 that spark discharges may occur when a person contacts metallic objects in high-intensity electric fields. Spark discharges induced by alternating fields may be felt and can be unpleasant or even painful if strong enough. Besides causing sensory effects, spark discharge can cause local skin erosion or burns because of the very small effective area of contact, particularly when conveyed to dry skin in an area where the corneal layer is thin. In those areas the energy of the spark is primarily dissipated in a very small volume of tissue. Section 2.6 discusses corneal degradation as a result of spark discharges.

To support a spark discharge to dry, intact skin, an object charged by a static or ELF field must attain a critical voltage of about 500 V relative to the intact, dry skin, as noted in Sect. 2.6. That voltage represents the minimum potential necessary to initiate a spark to the skin and is approximately the potential below which a spark to intact skin will extinguish. A plateau voltage as low as 330 V has been measured for spark discharges to the skin of the forearm on which the corneal layer has been stripped away, suggesting that 330 V is also a minimum spark initiation potential for moist or damaged skin.

The preponderance of published research concerning effects of spark discharges on humans has been conducted with static charges and at 60 Hz. Lacking specific laboratory data applicable to frequencies above 60 Hz, it is difficult to state the amount of skin erosion or damage that would occur with discharges from higher frequency fields. Unless the speed of approach between the person and the energized conductor is high in relation to the oscillation frequency, a spark will initiate at the peak of the sinusoidal waveform, regardless of frequency. For frequencies substantially above 60 Hz, the number and severity of sparks to the skin during a discharge episode will depend on the frequency, and the discharge time constant (see Fig. 2.34). Also, the potential required for the breakdown of air decreases at frequencies substantially above 60 Hz. This occurs when the transit time of positive ions or electrons is less than the half-cycle time of the oscillating field (Craggs, 1978). At radio frequencies, the reduction of breakdown potential across small metallic electrode gaps ($<1\,mm$) in air is about 15% to 20%. At frequencies of several GHz, more substantial reductions can occur.

Table 11.6 lists the peak E-field needed to achieve a spark discharge from various objects using a spark initiation criterion of 330 and 500 V; values are calculated using the parameters listed in Table 9.1. Table 11.6 applies to situations where the discharging object has dimensions much smaller than

TABLE 11.6. Electric field parameters for spark discharge induction.

Charged object	C_o (pF)	V_o/E (m)	Required E: V_o = 500 V (V/m)	Required E: 330 V (V/m)
Person (h = 1.8 m)	150	0.48	1042	638
Auto (mid size)	200	0.21	2,380	1,571
Auto (large)	2,000	0.16	3,333	2,062
School bus	3,700	0.26	1,923	1,269
Commercial bus	2,900	0.45	1,111	733
Fence wire (h = 1 m, L = 100 m)	720	1.00	500	330

Table applies to quasi-static condition with vertically polarized field.
Discharging object is several object dimensions from the source of the field. Field wavelength is much larger than major dimension of object.

the wavelength of the field and is several object dimensions distant from the source of the electric field. The last column of Table 11.6 lists the minimum undisturbed E-field (i.e., the field in the absence of the discharging object) needed to support a spark discharge. Except for the fence wire, the minimum E-field required for spark discharges exceed the low-frequency E-field limits of IEEE/ANSI C95.1, except as time-averaging permits higher peak fields with pulsed EMF.

It is important to recognize the potential hazards from burns to personnel who contact metallic objects in fields produced by radio frequency (r-f) transmitters. An example would be on an aircraft carrier, where a communications antenna may be placed near an aircraft that is standing on a metallic deck. A voltage may be induced on the aircraft relative to the deck that is sufficient to cause a skin burn when an individual touches the aircraft. Note that there may be significantly different voltages developed at various points on the aircraft if the wavelength of the energizing field is less than the dimensions of the induction object (a typical situation). The U.S. Navy uses a voltage criterion of 140 V-rms in r-f fields to define a potentially hazardous situation that could cause a person pain, visible skin damage, or could cause an involuntary reaction (NAVSEA, 1982). The navy recognizes that because of the many variables involved, it is not uncommon to encounter significantly higher voltages that do not result in a burn problem. In tests conducted near a 1 kW transmitting antenna operating in the 5 to 10 MHz range, I experienced very small skin lesions on the fingertip and knuckle when touching a fighter aircraft having an open circuit voltage of 140 V-rms at the point of contact. The contacts were attended by a stinging sensation, and lesions appeared as localized white spots on the corneum.

The navy voltage criterion amounts to a peak voltage of 198 V, which is substantially below the spark discharge criterion of 330 to 500 V mentioned above for ELF potentials. If we reduce ELF breakdown potentials by 20% for application to r-f, one would infer a peak breakdown potential in the range 233 to 354 V based on the spark discharge data—still well above the navy burn criterion. It is possible that the r-f burns observed at 140 V-rms do not involve air breakdown, but rather corneal breakdown when direct contact is actually made. The result might be essentially the same from the point of view of skin damage—a very small area of current conduction would be involved in either case. While an open-circuit voltage criterion of 140 V-rms might be protective of r-f burns for most cases, it is likely to be overly conservative in situations where the current conducted by an individual is small. A criterion combining open-circuit voltage and conducted current is probably needed to adequately address r-f burn hazards.

11.5   Absorbed Energy and Thermal Considerations in EMF Standards

Specific Absorption Rate (SAR) Limits

Exposure standards in IEEE/ANSI C95.1 are related to the specific absorption rate (SAR), which is the energy absorbed per unit mass of biological tissue. SAR may be related to the internal current density by

$$SAR = \frac{J^2}{\sigma p} = \frac{J^2 \varrho}{p} \tag{11.8}$$

or, in terms of the internal electric field by

$$SAR = \frac{\sigma E_i^2}{p} = \frac{E_i^2}{\varrho p} \tag{11.9}$$

where E_i is the rms in-situ electric field, J is rms current density, σ is tissue conductivity, $\varrho = 1/\sigma$ is the tissue resistivity, and p is tissue density. Traditionally, SAR is specified in units of Watts per kilogram.

We can relate SAR to tissue temperature rise using the heat equation (see Eq. 10.3), with the following result

$$\frac{\Delta T}{t} = \frac{SAR}{pc} \tag{11.10}$$

where ΔT is the temperature rise (°C), t is the duration of exposure, p is tissue density ($g\,m^{-3}$), c is tissue specific heat ($J g^{-1} C^{-1}$), and SAR is in units (W/g). If we express SAR in W/kg, and use $c = 3.8$ and $p = 1.05 \times 10^6$ for muscle tissue (see Table 10.5), the relationship is

$$\frac{\Delta T}{t} = 2.5 \times 10^{-4} SAR\left(W/kg\right) \tag{11.11}$$

Equation (11.11) is a worst-case expression in which there are no thermal losses from the tissue, thereby implying that the temperature would rise indefinitely as long as the energy input were provided. In reality, thermal losses will occur locally through conduction to adjacent tissue and vascular cooling, and will occur over the body through thermal radiation from external body surfaces, respiration, and evaporative processes. As an illustration of temperature rise, Eq. (11.11) states that with $SAR = 1\,W/kg$, a temperature rise of $1\,°C$ would require 4,000 s (1.1 hr) of continuous exposure under worst-case conditions.

The value of SAR expressed by Eqs. (11.8) and (11.9) will vary with frequency for a given in-situ E-field due to the variation of tissue conductivity (σ) with frequency. For instance, at frequencies of 10^3, 10^8, and $10^9\,Hz$, σ (parallel) for muscle tissue is approximately 0.5, 0.9, and 1.4 S/m respectively (see Table 2.1). Consequently, if we calculate the value of E_i associated with $SAR = 1\,W/kg$ in muscle tissue, we obtain values of 45, 35, and 27 V/m at frequencies of 10^3, 10^8, and $10^9\,Hz$ respectively.

Endogenous Field Limits to EMF Interactions

Biological systems experience endogenous electrical forces that provide an ultimate limitation on the minimum induced electrical force that can have a measurable biological effect. One of these forces is a consequence of the thermal agitation of charges within the biological medium. Other endogenous electrical forces include "$1/f$ noise" from cell membrane activity, "shot noise" associated with the discrete nature of electronic charge, and fields from electrical activity of nerve and muscle (Barnes, 1996). The following development demonstrates some of the considerations concerning thermal limitations to bioelectric interactions (after Weaver and Astumian, 1990). Whereas thermal noise represents one basic limit on bioelectric interactions, the other sources of noise can only add further limitations. In fact, $1/f$ noise (so-called for its frequency spectrum) may be the dominant source of noise at cellular membranes for frequencies below 160 Hz (Barnes, 1996).

The mean-squared voltage generated by random thermal fluctuations (so-called Johnson-Nyquist noise) is given by

$$\overline{V_{kT}^2} = 4RkT\Delta f \qquad (11.12)$$

where R is the resistance across which the voltage is measured, k is Boltzman's constant ($1.8 \times 10^{-23}\,JK^{-1}$), T is absolute temperature, and Δf is the noise bandwidth.

In the case of electromagnetic field interactions, it is widely believed that the site of interaction is at the cellular membrane, where we will consider the application of Eq. (11.12). The noise bandwidth of the membrane is related to its time constant τ_m, by $\Delta f \approx 1/(4\tau_m)$, where $\tau_m = RC$, R is the membrane resistance, and C is its capacitance. For a circular cell of radius r and membrane thickness d, $C = 4\pi\varepsilon_o\varepsilon_r r^2/d$, where ε_o is the dielectric permittivity of free space ($8.85 \times 10^{-12}\,Fm^{-1}$), and ε_r is the relative permittivity of the membrane. We therefore can write Eq. (11.12) as

$$\overline{V_{kT}^2} = \frac{kTd}{4\pi\varepsilon_0\varepsilon_r r^2} \tag{11.13}$$

If we evaluate Eq. (11.13) for a circular cell with $T = 310\,\mathrm{K}$, $d = 5\,\mathrm{nm}$, $r = 10\,\mu\mathrm{m}$, and $\varepsilon_r = 2$, and taking the square-root, we obtain $V_{kT} = 2.8 \times 10^{-5}\,\mathrm{V}$, which is the rms noise voltage present across the cellular membrane.

If an induced field it to have a cellular effect, it must not be swamped by endogenous thermal noise. Consider therefore the intracellular E-field necessary to create a transmembrane voltage equal to the noise voltage (i.e., signal-to-noise ratio = 1). The maximum voltage V_m developed across the membrane of a spherical cell in response to an intracellular field E_i is $V_m = 1.5\,E_i r$, where E_i is the in-situ or intracellular E-Field (see Sect. 4.3). Therefore, the minimum in-situ E-field that would just equal thermal noise ($V_m = V_{kT}$) in the example spherical cell is $E_i = 1.9\,\mathrm{V/m}$. For an elongated cell of length L, the maximum induced membrane voltage is $V_m = E_i L/2$. Following a procedure similar to the above example but for an elongated cell with $L = 150\,\mu\mathrm{m}$ and $r = 25\,\mu\mathrm{m}$ (e.g., a fibroblast or muscle cell), the in-situ field that would just equal thermal noise is calculated as $E_i = 0.08\,\mathrm{V/m}$. These examples would be restricted to frequencies below a few kHz, because of the bandwidth limitation of the membrane.

Others have calculated thermal noise limitations using different assumptions about the site of the effect. For instance, a value $E_i = 4\,\mathrm{mV/m}$ for unity signal-to-noise ratio was calculated by Polk (1993) by assuming that the pertinent mechanism of action is the redistribution of charge at the counterion layer of the cell surface, rather than the transmembrane voltage.

We expect that a measurable cellular response would require a signal-to-noise ratio greater than unity—information processing systems typically require signal-to-noise voltage ratios of at least 4 for meaningful signal detection and false alarm probabilities. Consequently, it is likely that a similar multiple applies to the noise-equivalent fields cited above for a credible biological effect.

Consider the thermal noise limits in the above examples in comparison to induced in-situ fields associated with typical environmental fields at 60 Hz. For example, consider a magnetic field of $10\,\mu\mathrm{T}$, and an electric field of $5\,\mathrm{kV/m}$—values that are at the upper limits of the fields that may be encountered beneath high voltage transmission lines. Using the principles and models described in Sect. 9.5, we calculate that the magnetic field would induce a maximum E-field of $0.6\,\mathrm{mV/m}$ at the periphery of the torso of a large adult. And using the principles described in Sect. 9.3, the electric field is calculated to induce $15\,\mathrm{mV/m}$ in the ankles of that person, and much smaller fields elsewhere in the body if grounded through the feet.

It can be seen in the above examples that even relatively strong ELF environmental fields induce in-situ fields below, or at best marginally above the most conservative thermal noise limits discussed above. Many experimenters report biological effects in which ELF in-situ fields are below 10 mV/m, and in some cases are as low as $10 \mu V/m$ (i.e., less than presumed thermal noise limits) (Weaver and Astumian, 1990; Polk, 1993; Tenforde, 1993).

It has been argued that thermal noise limitations cannot be overcome by any known biophysical mechanism, and therefore that biological effects from weak environmental fields are impossible (Adair, 1991). What then are we to make of the many reports of experimentally determined biological effects that result from induced fields below calculated thermal noise limits? Should we dismiss all such reports as violating thermal limits, and ipso-facto flawed as some have suggested? Or is it possible that other biophysical mechanisms might be responsible for overcoming thermal limitations? In order to address these questions, the next section discusses established and proposed mechanisms that may be involved in biophysical reactions.

11.6   Consideration of EMF Interaction Mechanisms in Standards Setting

Categories of Mechanisms of Bioelectric Interaction

When setting limits on electromagnetic exposure, it is often necessary to make assumptions about underlying mechanisms. With a mechanism model, one can specify appropriate dosimetric measures and parametric relationships, such as frequency or temporal sensitivity factors. For instance, if we were to assume that the biological action responds to tissue heating, then we would conclude that the energy deposition is the appropriate dosimetric parameter, which can be measured by SAR units. Furthermore, parametric relationships involving the frequency of the electromagnetic energy, and its temporal and spatial distribution can be adequately modeled. On the other hand, when dealing with membrane polarization phenomena, such as nerve excitation, it is the in-situ electric field that produces the effect. Furthermore, the spatial and temporal structure of electrical forces that influence excitable membrane effects will be quite different than those that determine a thermal effect.

Numerous mechanisms have been advanced to explain a variety of bioelectric phenomena. With application to standards criteria, it is necessary to differentiate between *established* and *proposed* mechanisms. We define an *established* mechanism as one for which interaction with a person is well established, and for which thresholds of reaction are understood. On the other hand, a *proposed* mechanism is one which is not sufficiently understood to define the threshold of interaction in a living person, or where experimental results have not yet been established with confidence.

The classification of a mechanism as "proposed" does not necessarily mean that biological activity connected with that mechanism is in doubt, but, of course, in some cases it may mean just that. It is possible that a mechanism might be well established at a cellular level, for instance, but that its application to the whole intact person is not presently understood. Or a mechanism might be well established in a nonhuman species, but its application to humans may be uncertain. From the point of view of standard setting, we would classify such a mechanism as "proposed." Only established phenomena applicable to living humans should form the basis for standards. On the other hand, developing knowledge on proposed mechanisms should be carefully monitored to ascertain whether reclassification to the "established" class is warranted.

Tables 11.7 and 11.8 provide a listing of various mechanisms that have been advanced to explain electric and electromagnetic interactions with biological systems. Table 11.7 lists those mechanisms that can presently be characterized as *established*; Table 11.8 lists *proposed* mechanisms. Two columns indicate whether the effect is considered to be associated with the in-situ electric field or magnetic field. In Table 11.7, an attempt has been made to provide a numerical estimate of the lowest in-situ field that is expected to have a measurable biological effect under most favorable conditions; we will apply the term "rheobase" to these minimum thresholds. The column labeled "condition" provides a brief indication of conditions

TABLE 11.7. Established mechanisms of human interaction with E&M fields.

Mechanism	In-situ E-field (V/m)	In-situ B-field (T)	Condition	Section reference
1. Synapse activity alteration by membrane polarization (phosphenes)	0.05	NA	$f = 20\,\text{Hz}$	3.7, 9.8
2. Peripheral nerve excitation via membrane depolarization	6	NA	$t > 1\,\text{ms}$	3.3, 4.4
3. Muscle cell excitation by membrane depolarization (skeletal)	6	NA	$t > 10\,\text{ms}$	8.3
(cardiac)	12	NA	$t > 10\,\text{ms}$	6.2–6.5
4. Electroporation (reversible)	50	NA	$t > 0.1\,\text{ms}$	10.2
(irreversible)	300	NA	$t > 0.1\,\text{ms}$	
5. Resistive (joule) heating (e.g., $SAR = 1\,\text{W/kg}$ in muscle for $\Delta t \approx 1\,\text{hr}$, $\Delta T \approx 1\,°\text{C}$)		NA	$t \approx \text{continuous}$	11.5
$f = 10^3\,\text{Hz}$	45	—		
$f = 10^8\,\text{Hz}$	35	—		
$f = 10^9\,\text{Hz}$	27	—		
6. Audio effects via thermoelestic expansion (Peak $SAR \approx 100\,\text{W/kg}$ in brain)	300	NA	$f = 0.1$–$10\,\text{Ghz}$ $t \approx 1\,\text{ms}$	9.8
7. Magneto hydrodynamic effect	NA	1.5	$f \approx 0$	9.11

References are to book sections.

TABLE 11.8. Proposed mechanisms of human interaction with E&M fields

Mechanism	In-situ E-field (V/m)	In-situ B-field (T)	Condition	Reference
1. Soliton mechanisms through cell membrane proteins	√	NA		Lawrence & Adey, '82
2. Spatial/temporal cellular integration	√	NA		Litovitz et al., '94
3. Stochastic resonance	√	NA		Krugilikov & Derfinger, '94
4. Temperature mediated alteration of membrane ion transport	√	NA	mm waves	Alekseev et al., '94, '97
5. Plasmon resonance	√	NA	$f = 10^9 - 10^{11}$ Hz	Fisun, '93
6. Radon decay product attractors	√	NA	$E_a \approx 1\,\mathrm{kV/m}$	Henshaw et al., '96
7. Rectification by cellular membranes	√	NA		Astumian et al., '95
8. Ion resonance	NA	√	$f(e/m, B_{dc}/B_{ac})$ $f \approx 10-100\,\mathrm{Hz}$	Liboff, '85; Lednev, '91; Blanchard & Blackman, 94
9. Ca^{++} oscillations	NA	√		Liboff, '93
10. Nuclear magnetic resonance	NA	√		Blackman et al., '88
11. Radical Pair Mechanism	NA	√		Steiner & Ulrich, '89; Grissom, '95
12. Magnetite interactions	NA	√		Kirschvink, '89

Notes: E_a: ambient electric field; *e/m*: charge-to-mass ratio of resonant ion; B_{dc}/B_{ac}: steady/alternating magnetic field.

which are necessary to attain the rheobase threshold. The last column refers to book sections where the mechanism is treated in some detail.

Table 11.8 lists proposed mechanisms of human interaction with electric and magnetic fields. In this table, the action of the mechanism is classified as being mediated by the in-situ electric or magnetic field, as indicated by a check in the appropriate column. A numerical value for human reaction is not provided in Table 11.8 because of the lack of sufficient knowledge concerning the mechanism. However, an attempt has been made to indicate exposure conditions where the mechanism is thought to be active, where such information can be determined. The last column of Table 11.8 refers the reader to a few citations where the mechanism is defined. The listing is not meant to be a comprehensive bibliography, as that would require an inordinate amount of space.

The following paragraphs give a brief description of each mechanism. The references that are included in the descriptions are intended to provide minimal background on the particular mechanism.

Established Mechanisms for Human Bioelectric Response

The first four mechanisms listed in Table 11.7 are membrane polarization effects, which are produced by an in-situ electric field that exerts a polarizing voltage across the nerve membrane (See Sect. 4.3). For elongated cells, such as nerve cells, the most favorable orientation for membrane polarization effects is when the E-field is aligned with the long axis of the cell, in which case the maximum polarization will occur at the terminus of the cell. Items (5) and (6) are thermal mechanisms. Item (7) is a magnetic field effect. The following paragraphs provide a modicum of detail for Table 11.7

Synapse Interactions (Sects. 3.7 and 9.8)

Post-synaptic membrane potentials vary with a complex of excitatory and inhibitory cells that synapse on a given cell. Small changes in the presynaptic resting potential can be greatly multiplied in the postsynaptic membrane and make an effective synapse inoperable, or a weak synapse highly effective in producing a postsynaptic action potential (see Sect. 3.6).

An example of this effect is attributed to the phenomenon of electro- and magneto-phosphenes, which are the visual effects that are produced when electric currents or magnetic fields are applied to the head (see Sect. 9.8). Researchers have concluded that phosphenes are generated through modification of synaptic potentials in the receptors or neurons of the retina, rather than direct stimulation of post retinal pathways or the visual cortex.

Using the observed magneto-phosphene thresholds at the most sensitive frequency (20 Hz), the induced E-field at the location of the retina is calculated to be about 50 mV/m—a value that is a factor of about 100 below the rheobase nerve excitation thresholds (refer to Sect. 9.8). It does not necessarily follow that such low thresholds would apply to other neural synapses, because of the highly specialized configuration of neurons in the retina.

Peripheral Nerve Excitation Through Membrane Depolarization (Sects. 3.3 and 4.4)

Nerve excitation is initiated by depolarization of the neural membrane, which activates voltage-gated ion channels, thereby producing a propagating action potential (see Chapter 3). In contrast to the graded response of synaptic activity mentioned above, nerve excitation is a threshold response, sometimes referred to as an *all-or-nothing* phenomenon. Nerve excitation, characterized by a propagating action potential, is triggered when the membrane is depolarized by about 15–20 mV. The external field necessary for nerve excitation depends on the duration of the excitation. A time constant in the range 100 to 200 μs typically applies to S–D curves for nerve excitation. A minimum threshold of about 6 V/m is found for the largest myelinated nerves ($\approx$20 μm diameter), with a long duration ($\approx$2 ms) monophasic E-field pulse (see Chapter 4).

Muscle Cell Excitation Through Membrane Depolarization (Sects. 6.2 to 6.5 and 8.3)

Although electrical excitation of skeletal muscle typically occurs through excitation of motor neurons, it is possible to directly stimulate muscle cells (see Chapter 8). Strength–duration time constants of muscle cells are typically in the range of 1 to 10 ms, which is a factor of 10 greater than the time constants for nerve excitation. Table 11.7 lists a rheobase threshold for skeletal muscle and cardiac muscle cells based on experimental and theoretical data. The rheobase field for muscle excitation is similar to that for nerve stimulation i.e., 6 V/m. The median rheobase for cardiac excitation is about 12 V/m; at the one-percentile level, the cardiac rheobase is about 6 V/ m (see Chapter 6). In order to achieve these rheobase thresholds, it is necessary to stimulate the tissue with a monophasic pulse having a duration greater than about 10 ms.

Electroporation (Sect. 10.2)

The electric field normally developed across a cellular membrane is very large. With a membrane potential of 0.1 V and a membrane thickness of 10^{-7} m, the electric field would be 10^6 V/m across the membrane. If the cell is sufficiently hyperpolarized relative to its normal resting potential, the intense electric field developed across the membrane can produce pores (see Chapter 10). Hyperpolarization to a membrane potential around 200 mV will produce temporary pores, which revert to a normal condition with the cessation of the hyperpolarizing field. When the membrane potential is raised to about 800 mV, the individual pores become enlarged or can fuse and become irreversible. The electric field values listed in Table 11.7 apply to electric field pulses of sufficient duration ($t > 0.1$ ms) that can lead to reversible or irreversible electroporation of muscle cells.

Resistive Tissue Heating (Sect. 11.5)

Tissue heating results from the passage of electric current through resistive material. The appropriate measure of heating is energy deposition, which can be quantified by the SAR measure (see Sect. 11.5). For example, with $SAR = 1$ W/kg in muscle tissue, it would require 4,000 s (1.1 hr) of continuous exposure to achieve a temperature rise of 1 °C. The in-situ electric field associated with this value of SAR in muscle tissue ranges from 45 to 27 V/m for frequencies from 10^3 to 10^9 Hz, as noted in Table 11.7.

Audio Effects Through Thermoelastic Expansion (Sect. 9.8)

Auditory sensations can be perceived by a person whose head is exposed to microwave energy. The effect is explained by the development of a thermoelastic wave that results from a brief temperature rise because of the absorption of pulsed microwave energy within the head. The brief temperature rise causes an expansion of the absorbing tissue that launches an acoustic wave within the skull and is perceived mechanically through the normal auditory mechanism of the ear. The threshold stimulus is characterized by energy density in the vicinity of $40\mu J/cm^2$. The corresponding SAR value will depend on the pulse width. The threshold E-field listed in Table 11.7 corresponds to a value of $SAR = 100\,W/kg$ in the brain, which is the lowest threshold attributed to microwave hearing. Most data on microwave hearing have been collected at 2.45 GHz; data at other frequencies are scarce.

Magnetohydrodynamic Effect (Sect. 9.11)

When ions flow in a direction orthogonal to a magnetic field, a voltage is produced in a mutually orthogonal direction (e.g., velocity in x-direction, magnetic field in y-direction, induced E-field in z-direction). For example, assuming a flow rate of 0.6 m/s in the human aorta, the authors calculate that a potential of 15 mV would be produced in a field of 1 T, as noted in the table. Magnetohydrodynamic effects result in vertigo or taste sensations at 1.5 T (noted in Table 11.7). In addition to the magnetohydrodynamic generation of electrical potentials mentioned above, there will also exist a drag force on a conducting fluid that is flowing within a magnetic field. For a field of 5 T/s, the pressure within human vasculature will be affected by less than 1%.

Proposed Mechanisms for Human Bioelectric Response

Numerous mechanisms have been proposed to account for human reactions to low-level electromagnetic exposure. The placement of some of these mechanisms in the "proposed" category does not necessarily imply that the particular mechanism is in doubt, but rather that there presently does not exist a quantifiable and verified theory through which one may specify reaction thresholds in the human organism. It must be anticipated, however, that some of these proposals will be shelved, or simply ignored for lack of supporting development. Listed below are several mechanisms from Table 11.8 that have been proposed to account for human reactions to low level electromagnetic radiation. The frequency thought to result in measurable biological activity ranges from the ELF to the microwave regimes.

Soliton Mechanisms (Lawrence & Adey, 1982)

According to this theory, a nonlinear wave (soliton) is created in proteins associated with membrane channels, which propagates along the protein and through the membrane, thereby supplying energy for chemical events within the cell. These processes are thought to be affected by the electric field external to the cell.

Spatial/Temporal Cellular Integration (Litovitz et al., 1994a, 1994b)

Various experimenters have reported bioelectric effects from exposures that produce in-situ E-fields that are below the endogenous field within the biological medium. One possible explanation is that spatial and temporal integration processes exist that can differentially enhance spatially or temporally coherent applied fields with respect to incoherent endogenous thermal noise fields.

Stochastic Resonance (Krugilikov & Dertinger, 1994)

A weak signal may be amplified by system noise itself in a bistable or multistable system. The cellular membrane is thought to provide the non-linear characteristics necessary for such enhancement in biological systems.

Temperature Mediated Alteration of Membrane Ionic Transport (Alekseev et al., 1994, 1997)

The authors have determined that in-situ exposure of pacemaker neurons to millimeter waves (e.g., $f = 75\,GHz$, $SAR = 4.2\,mW/kg$) cause significant changes in firing rates and spike amplitude. The effect was attributed to a brief, transient temperature rise of $0.0025\,°C/s$ at the neural membrane, resulting in altered membrane ionic transport. The authors found that with millimeter radiation, local SAR hot spots on cellular membranes can exceed $1,000\,W/kg$ at $10\,mW/cm^2$ incident power density.

Plasmon Resonance Mechanisms (Fisun, 1993)

This mechanism refers to sheets of surface charge (plasmons) on cellular membranes, which display resonance properties in response to an applied field. The advocates of this mechanism speculate that the associated redistribution of surface charges can alter transmembrane potential, thereby significantly affecting membrane ion transport, conformation of intra-membrane proteins and lipid bilayers, with potentially pathological consequences. According to theoretical models, resonant plasmon activity would apply to fields in the frequency range 10^9 to $10^{11}\,Hz$.

Radon Decay Product Attractors (Henshaw et al., 1996)

Henshaw and colleagues have found that radon decay products are attracted to common sources of electric fields, such as power lines and electric appliances. They observed that the electric fields produced around power sources can enhance the deposition of radon decay products on environmental surfaces. They speculate that an external electric field as low as 1 kV/m might enhance the deposition of aerosols containing decay products in the air passages of the mouth, neck, and chest. They suggest that these observations may provide a link between ELF fields and human cancer.

Rectification by Cellular Membranes (Astumian et al., 1995)

Proteins in cellular membranes are said to rectify oscillating electric fields, thereby allowing a DC response to accumulate from an AC field. The development of a DC membrane voltage bias in excitable tissue can be demonstrated in response to a moderately strong sinusoidal membrane voltage in the vicinity of 1 mV (see Sects. 4.6 and 6.5). However, this action as an explanation for biological response to very weak fields has not been established.

Ion Resonance (Liboff, 1985; Lednev, 1991; Blanchard & Blackman, 1994)

Ion resonance models suggest that a magnetic field will resonate with ionic motion at a frequency that depends on the charge-to-mass ratio of particular resonating ions, as well as the copresence of a static magnetic field, such as the earth's field (see Sect. 9.11). So-called "paramagnetic" formulations state that the resonant frequency also depends on the ratio of magnitudes of the steady and alternating magnetic field components.

Ca^{++} Oscillations (Liboff, 1993)

ELF electric fields surrounding a cell are said to alter the frequency of Ca^{++} oscillations in the cellular cytosol (liquid medium of the cytoplasm). These oscillations are believed to convey information to the cell based on the frequency of oscillation of Ca^{++} ions.

Nuclear Magnetic Resonance (Blackman et al., 1988)

This mechanism is suggested to affect $^{45}CA^{++}$ efflux from brain tissue. The theory suggests that a nonzero nuclear spin is required to have an effect, whereas most biological isotopes have zero spin. Therefore, a NMR explanation for coupling by weak ELF may be questionable.

Radical Pair Mechanism (Steiner & Ulrich, 1989; Walleczek, 1994; Grissom, 1995)

A magnetic field can modify the electron valence spin states through quantum mechanical mechanisms during free radical formation, thereby altering radical-dependent reactions. Such states are independent of random thermal interactions. Consequently, in principle, there should not exist a thermal noise limit governing such magnetic field interactions.

Magnetite Interactions (Kirschvink, 1989, 1996)

Magnetite (ferromagnetic compounds) are found in migratory birds, and other animals, including mammals. It is found in trace amounts in the human brain, and possibly other tissue. It has been speculated that these trace compounds may provide a link for magnetic field interactions with brain tissue. At microwave frequencies in the range 0.5 to 10 GHz, the mechanism of interaction is said to be through the process of ferromagnetic resonance, which results in enhanced absorption of electromagnetic energy by cells.

11.7 ELF Magnetic Field Standards Derived from Established Mechanisms

We have seen in Sect. 11.2 that the IEEE/ANSI C95.1 standards do not currently specify limits below 3 kHz, although the responsible IEEE committee is presently at work on that problem. This section will consider exposure limits in this frequency regime based on the excitation mechanisms that are discussed above.

Threshold Criteria

Figure 11.8 illustrates asymptotic limits to S–D curves for pulsed magnetic fields derived from principles and measurements discussed in previous chapters. The rheobase dB/dt and the S–D time constant for each curve is given in Table 11.9. Curve (a) is based on excitation of a 10-μm nerve fiber in the cortex of the brain, using a rheobase E-field of 12.3 V/m (Table 4.2), along with the assumption that the brain can be represented by a sphere of 11-cm diameter. In this case, the threshold dB/dt is calculated according to Eq. (9.16). The assumed diameter of 10 μm is a rather large one within the distribution of fibers to be found in the human brain, although a small fraction of brain neurons may exceed this diameter (Sect. 3.3). Curve (b) applies to the median cardiac excitation threshold and is obtained by multiplying the one-percentile threshold in Table 9.6 by a factor of 2. Curve (c) represents the threshold of a 20-μm peripheral nerve fiber as listed in Table 9.6. Curve (d) is a median threshold limit for phosphenes, which has been derived from the sinusoidal thresholds shown in Fig. (9.20), using the principles described below.

The pulsed field thresholds of Fig. 11.8 can be converted to sinusoidal thresholds by applying principles and assumptions that have developed previously. One assumption is that at the excitation threshold, the peak of a pulsed stimulus is equivalent to the peak of a sinusoidal stimulus having a duration of many cycles (Sect. 4.6 and Fig. 4.19). Another assumption is that the S–D time constant, τ_e, and the strength-frequency constant, f_e, are

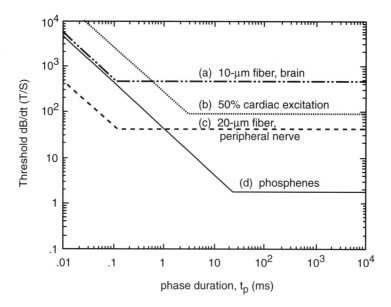

FIGURE 11.8. Strength-duration curves for stimulation by monophasic dB/dt pulses, whole body exposure, short-term reactions, large adult.

related by $f_e = 1/(2\tau_e)$. A further relationship between the peak flux density, B, and the peak dB/dt is defined for a sinusoidal field in Eq. (11.6). A final relationship is that the rms value of a sinusoid is its peak divided by $\sqrt{2}$. Figure 11.9, obtained by applying these relationships to Fig. 11.8, illustrates magnetic field thresholds for sinusoidal fields. It is apparent that phosphenes define the lowest thresholds at frequencies below 430 Hz and that peripheral nerve stimulation defines the lowest thresholds above that frequency.

Limit Criteria

The thresholds depicted in Fig. 11.9 are intended to represent approximate median values among a population of healthy individuals. To derive protec-

TABLE 11.9. Pulsed magnetic field threshold parameters.

Reaction	$\dot{B}_o$ (T/s-pk)	τ_e (ms)
10-μm brain neuron excitation	448	0.12
50% cardiac excitation	98.0	3.0
20-μm peripheral nerve excitation	37.8	0.12
phosphenes	1.78	25

$\dot{B}_o$ = rheobase dB/dt; peak values listed.

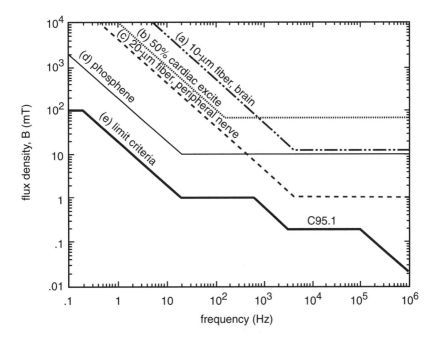

FIGURE 11.9. Thresholds for short-term reactions to sinusoidal magnetic fields—whole body exposure, large adult. Curves (a)–(d): human reaction thresholds; (e): derived limit criteria with acceptance factor.

tive standards from thresholds, it is customary to apply an acceptability factor to account for particularly sensitive individuals, those in a pathological state, and for uncertainties in the methodology of determining thresholds. Although we do not know the statistical distribution of thresholds for the reactions shown in Fig. 11.9, it would not be unreasonable to assume a log-normal distribution, similar to threshold distribution for electrocutaneous sensitivity (Table 7.10), or cardiac excitation (Table 6.12). Accordingly, a reasonable guess is that magnetic thresholds at the one-percentile level would be a factor of perhaps 2 or 3 below the median for healthy individuals. Further allowances should be made for individuals in a pathological state, which can lead to greater population variance (e.g., Fig. 6.17). Considering these factors, an acceptability factor of 10 will be applied to the thresholds of Fig. 11.9.

Note that the acceptability factor of 10 applied here is a conservative one when applied to a field magnitude, rather than an SAR metric as in IEEE C95.1. This is true because the SAR value is proportional to the square of the field (Eqs. 11.2, 11.9). Consequently, a factor of 10 applied to SAR is equivalent to a factor of $\sqrt{10}$ in the associated electric or magnetic field under free-field propagation conditions.

If we apply a factor of 10 to the lowest thresholds in Fig. 11.9, we obtain the acceptance curve (e) shown in the figure. At frequencies above 3 kHz,

TABLE 11.10. Sinusoidal magnetic field threshold parameters.

Reaction	B_o (mT-rms)	f_e (Hz)
10-μm brain neuron excitation	12.1	4,170
50% cardiac excitation	66.2	167
20-μm peripheral nerve excitation	1.02	4,170
phosphenes	10.0	20

the acceptance curve has been merged with the C95.1 curve by extrapolating the C95.1 low frequency plateau on a slope inversely proportional to frequency below 3 kHz. With this procedure, the C95.1 plateau is a factor of 5.0 below curve (c) in the region 3 to 100 kHz; below 3 kHz, it is a factor 10 below curve (d). The acceptance curve below 0.2 Hz has been capped at 100 mT-rms (141 mT-peak). This limit is a factor of 10 below a peak field of 1.41 T, in consideration of human reactions (vertigo, taste sensations, cardiac rate changes) reported at similar intensities of a static field (Sect. 9.10). Table 11.11 provides a numerical description of curve (e).

The purpose of a standard is to protect against a detrimental effect, not just a perceptible one. Consider first the peripheral nerve excitation threshold curve (c) from that perspective. Although nerve stimulation is not a disturbing experience at the threshold of perception, unpleasant or painful sensations are experienced with magnetic stimuli that exceed the peripheral nerve perception threshold by only 40% or so (Sect. 9.7). Consequently, curve (e) at frequencies above 615 Hz is approximately a factor of 7 below a detrimental effect. Curve (d) is based on phosphene perception. While this phenomenon has not been reported to be disturbing in a laboratory setting, it is not clear whether this would be the case for affected individuals in an uncontrolled environment. Consequently, induction of phosphenes is assumed to be a situation that should be avoided in an uncontrolled environment.

TABLE 11.11. Magnetic field exposure limits based on mechanisms of short-term reactions.

Frequency range (Hz)	B_o (mT-rms)	H (A/m-rms)
<0.2	100	7.95×10^4
0.2–20	$20/f$	$1.59 \times 10^4/f$
20–615	1	795
615–3×10^3	$615/f$	$4.89 \times 10^5/f$
3×10^3–10^5	0.205	163
10^5–10^6	$2.05 \times 10^4/f$	$1.63 \times 10^7/f$

Criteria above 3 kHz based on existing IEEE/ANSI C95.1. Criteria below 3 kHz based on threshold reactions, with acceptability factor of 10.

The mechanisms-based limits are compared in Fig. 11.3 with existing standards on magnetic field exposure in the frequency 0.1 Hz to 10 kHz. The mechanisms-based curve (e) provides conservative limits with respect to the other standards in the frequency range 1 to 20 Hz. At higher frequencies, existing standards are equal to, or below curve (e).

Other Considerations for ELF Standards

The acceptance curve shown in Fig. 11.9 is derived from established bioelectric mechanisms, as defined in Sect. 11.6. A mechanisms-based approach is not the only basis to be used in setting electromagnetic field acceptability standards. Other information to be considered includes epidemiological data, and laboratory studies that are deemed reliable, but for which a mechanistic explanation may not presently be established. Such information must be judged as to its reliability, application to an intact human, and evaluated hazard risk. Consequently, curve (e) should be considered as an upper limit on acceptable exposure to magnetic fields in uncontrolled environments.

For particular applications in controlled situations, it may be acceptable to allow greater exposure than indicated by curve (e). For instance, in magnetic resonance imaging procedures, higher exposure levels are permitted by existing guidelines (Sect. 11.3). Furthermore, with exposure to a limited region of the anatomy, greater exposure may be tolerated. For instance, with exposure to the head only, thresholds of peripheral nerve excitation [curve (c) in Fig. 11.9] would be elevated considerably, and acceptance levels in the frequency regime above 500 Hz could be increased accordingly.

Note that the acceptance limits shown in Fig. 11.9 are hypothetical ones. These have been included in this chapter to demonstrate a methodology for deriving limits based on established mechanisms in the ELF regime. One can also see that existing ELF standards (Fig. 11.3) are similar to the mechanisms-based standards over a broad portion of the frequency range depicted. The methodology and the derived limits for curve (e) resemble those applied in European standards (Bernhardt, 1985; 1988).

11.8   Standards in Consumer Products and Installations

Standards Setting Agencies

In addition to the electromagnetic exposure standards discussed previously, standards and performance guidelines have also been developed by various agencies to protect against electric shock hazards in electrical products and installations. For instance, the National Electric Safety Code, developed by the IEEE and published as an ANSI Standard (IEEE, 1990), contains rules and provisions for the installation of supply and communication lines, equipment, and associated work practices employed by a utility for electric supply, communication, railway, or similar activity. The IEEE also publishes as an ANSI standard a guide for the safe grounding practices in AC substation design (IEEE, 1986). The National Electrical Code, published by the National Fire Protection Association (NFPA, 1990) provides guidance for the installation of electrical wiring in homes and businesses. These codes are advisory ones from private or professional organizations without enforcement powers. Nevertheless, the codes are widely applied, and many enforcement agencies in North America require conformance to these rules.

Various agencies are involved in electrical safety standards for consumer products. For instance, trade organizations in the United States that develop advisory criteria for product safety are the National Electrical Manufacturers Association (NEMA) and Underwriters Laboratories (UL). UL is widely recognized since many consumer products carry the UL seal of approval, which means that the product has been manufactured to conform to UL safety standards. The Consumers Product Safety Commission (CSPC) was established by Congress in 1972 under the Consumer Product Safety Act. The main activity of the CPSC is to monitor safety problems in consumer products, to disseminate such information, and to work with other agencies in developing safety standards, although the agency has written a few mandatory standards itself. Outside of North America, the International Electrotechnical Commission (IEC) develops electrical safety standards for European products.

Other agencies that issue guidelines and standards for consumer products are the Center for Devices and Radiological Health (CDRH) of the Food and Drug Administration (FDA) under Public Law 90-602 of the Radiation Control for Health and Safety Act of 1968, and the IEEE Standards Coordinating Committee-34. There is a verbal understanding that FDA does not wish to issue more performance standards or regulations on consumer products like color TV or cellular phones. Instead, they support SCC-34. Current projects under SCC-34 are cellular phones, and a marine radar—the latter is just in the planning stage.[1]

Safety Criteria for Consumer Products

Table 11.12 lists criteria of UL and the IEC for electrical products. In this table, the UL data are intended as "limits" (i.e., recommended maximum currents); the IEC criteria are described as "minimum thresholds" (i.e., minimum values where a physiological effect is expected to occur). The difference between the two is that a "limit" may include an additional safety

---

[1] Information from John Osepchuk of Full Spectrum Consulting, August, 1997.

TABLE 11.12. Conducted current criteria for consumer products

Effect	IEC[a] (mA-rms)	UL[b] (mA-rms)
Startle	0.5	0.5[c] 0.75[d]
Let-go	10	5
Ventricular fibrillation	35	20

Notes: (a) IEC thresholds for 15–100 Hz, per publication IEC 479; (b) UL limits for 60 Hz; (c) applies to portable appliances; (d) applies to fixed appliances.

factor applied to a "threshold." The startle current criteria apply to leakage current from portable appliances. The intent is to protect against an electrical startle reaction that may lead to injuries when engaging in inherently dangerous activities, such as climbing a ladder, handling hot liquids, or handling power cutting tools. The experimental basis for the UL startle criteria is described in Sect. 7.11. The values listed under "let-go" are intended to protect against grip tetanus with continuous 50/60 Hz currents, as described in Sects. 8.5 and 8.6. The UL limit of 5 mA applies to an especially sensitive small child at approximately the 0.5% percentile rank of a distribution of let-go thresholds applying to children.

In some cases, such as with an electric fencer, the energizing voltage may be pulsed. In such cases, grip tetanus may not be a credible hazard. Although it is possible to eliminate a let-go hazard with a pulsed current, one might still retain a cardiac fibrillation hazard if the current during the on-time were great enough. Consequently, UL and IEC have separate ventricular fibrillation criteria as shown in Table 11.12. The listed values apply to exposure durations of one or more seconds for a small child. The weight of a child that is just ambulatory was estimated by UL to be 8.4 kg (18.5 lb) based on anthropomorphic data (Skuggevig, 1992). Applying that weight to Dalziel's regression formula (Eq. 6.5), one obtains a median fibrillation current of 59.4 mA for an 8.4-kg child. And from the data of Table 6.12, the fibrillating current at the 0.5 percentile rank is 25.5 mA. Thus, we see that at the 20 mA UL limit, the probability of delivering a fibrillating current to a very small child would be less than 0.5%.

The UL limit has been developed using a body weight scaling relationship. On the other hand, the European standards do not recognize a body weight law, and instead use data on dogs to provide a conservative model for human safety applications (see Sect. 6.7). The IEC threshold of Table 11.12 is intended as a conservatively low fibrillation value for dogs and does not include an additional safety factor that may be applied to a safety standard.

Hazard Criteria Vesus Current Duration and Frequency

The relationship between a hazardous current and the duration of exposure is important in many safety applications, such as determining the speed of operation of a current interrupting device. Figure 11.10 (adapted from Skuggevig, 1992) illustrates UL and IEC ventricular fibrillation criteria as a function of the duration of current. The figure shows duration from 0.1 ms to 10 s, although the criteria may be extended by another decade to the right, and the UL curve may be extended along the same slope two additional decades to the left. The rationale for these curves is given in Sect. 6.5, in connection with Fig. 6.12. The vertical axis represents the rms value of pulsed 50/60 Hz current during its on-time. For durations beyond 18 to 20 ms, waveforms at the power frequency would include one or more individual cycles of oscillation. For durations less than a single cycle, the pulsed waveform could represent a single monophasic or biphasic event. For a monophasic pulse (the worst case for brief stimuli), the rms value of the pulse is the same as its peak value.

Figure 11.11 illustrates hazard criteria of IEC and UL as a function of frequency for ventricular fibrillation, let-go, and startle reaction. The IEC curves are intended as minimum reaction thresholds, and the UL curves are recommended as maximum limits. These curves have a minimum plateau at

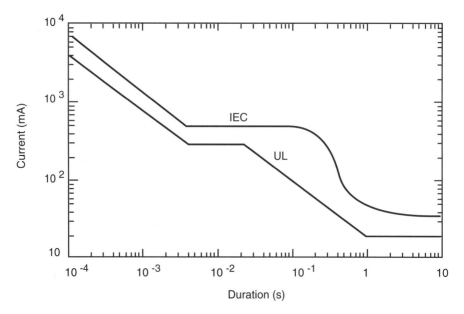

FIGURE 11.10. Ventricular Fibrillation criteria of UL and IEC. For durations above approx. 20 ms, criteria apply to 50/60 Hz AC waveforms. Below 20 ms, criteria apply to biphasic or monophasic nonrepetitive currents.

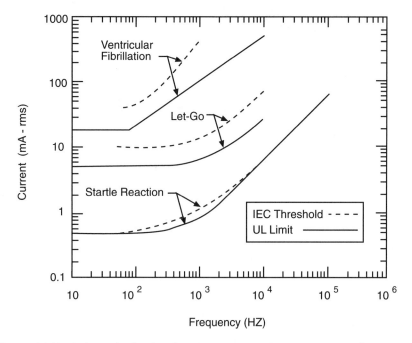

FIGURE 11.11. Safety criteria of IEC and UL as a function of frequency. (After IEC, 1984; UL, 1988.)

low frequencies, which include 50 and 60 Hz power frequencies. These curve shapes are based on neurosensory, neuromuscular, and cardiac muscle reactions to alternating current as described in previous chapters (cf., Figs. 6.4, 6.8, 7.13, and 8.10)

If hazard criteria were based solely on nerve and muscle excitation, then the hazard curves of Fig. 11.11 would rise without limit in proportion to frequency above 1 kHz or so. However, we have seen in Sect. 7.5 that tissue heating will limit the maximum current that can be tolerated. The temperature rise that will cause tissue damage depends on the duration of the thermal event. Figure 11.12 provides a strength-duration curve for cutaneous tissue damage. For instance, a temperature of 45 °C must be maintained for approximately 2 hours in order to sustain tissue damage to the porcine skin, which is considered a good thermal model for human skin. It is no coincidence that 45 °C is the approximate temperature at which a thermal stimulus to the human skin will be judged as painful. Although it would take prolonged exposure at that temperature to actually damage skin, a painful reaction would serve as a healthy warning of an potentially noxious event.

The heating potential of an electrical stimulus is largely determined by the rms current, almost independent of frequency. For a small finger touch

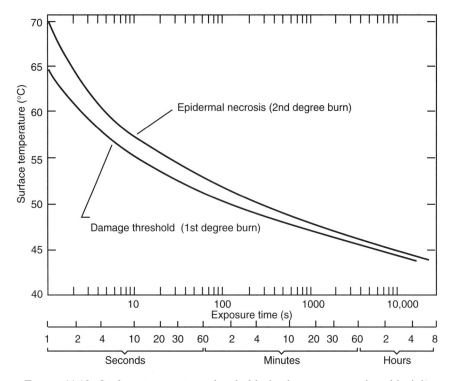

FIGURE 11.12. Surface temperature thresholds for human or porcine skin injury. (From Moritz & Henriques, 1947.)

contact of 25 mm², a perceptible thermal reaction is reached at 37 mA-rms, and a painful thermal reaction is reached at about 45 mA-rms over the frequency range 100 Hz to 3 MHz (see Chatterjee et al., 1986 and Fig. 7.12). Greater thresholds would apply to larger areas of contact. As noted in Chapter 10, the temperature rise is approximately proportional to the product of the duration of exposure, current density-squared, and tissue resistivity. Although the duration of current was not measured in Chatterjee's tests, one can surmise a duration of about 0.5 to 1s from the experimental protocol.

In order to protect against thermal burns, UL recommended a limit of 25 mA rms (UL, 1981); more recently, this recommended limit has been increased to 70 mA rms—a value consistent with that recommended by the IEC for application to electrical equipment for measurement, control, and laboratory use (IEC, 1990a). A more conservative limit of 70 mA peak has been adopted by the IEC for application to switching of radio frequency energy (IEC, 1986).

Limits for Capacitor Discharges

Voltage limits on capacitor discharges have been developed by UL as shown in Fig. 11.13 (UL, 1987). These limits have been derived from the UL ventricular fibrillation (VF) limits shown on Fig. 11.10, which can be interpreted in terms of the voltage limits for capacitor discharges by noting that the charge, Q, for a monophasic stimulus is $Q = It$, where I is the peak current (Amperes), t is the duration of a square wave pulse (seconds) or the time constant of a exponential decay, and Q is in Coulombs. For instance, at $t = 10^{-4}$s, the UL limit curve in Fig. 11.10 is equivalent to a charge of $400\,\mu C$. The discharge time constant for a capacitor discharge to a pure resistance, R, is given by $t = RC$. In deriving capacitor discharge limits, UL assumes $R = 500\,\Omega$, except for low-voltage discharges, where the resistance is assumed to be higher. The stored charge in Fig. 11.10 is given by $Q = CV$. For instance, at $C = 0.1\,\mu F$, the UL limit curve is equivalent to a charge of $290\,\mu C$, and a time constant of $50\,\mu s$. We contrast the VF limits to perception and pain thresholds for capacitor discharges as noted in Sects. 7.3 and 7.4. For instance, at low capacitance values, one can perceive a capacitor discharge by touching a charged electrode at about $0.25\,\mu C$ (Fig. 7.4), and the pain threshold would be at about $0.9\,\mu C$ (Table 7.2).

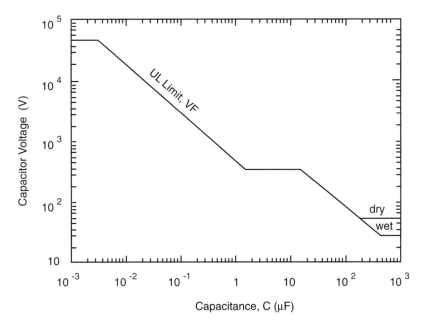

FIGURE 11.13. Capacitor discharge criteria. UL limits based on ventricular fibrillation.

Ground Fault Current Interrupter[2]

The ground-fault circuit interrupter (GFCI) is probably one of the most significant devices to be developed in the field of electric safety. It was introduced in the United States in the early 1960s and has since been required by the National Electrical Code to be installed in certain branch circuits and as part of certain products. In Europe, the equivalent device is called a residual-current device (RCD). The RCD operates on the same principle as the GFCI but was developed independently and with a few differences that will be explained presently.

The two-conductor GFCI (and RCD) continuously compare the magnitude of current in the energized conductor to that in the grounded conductor. If the current in the two conductors differs by more than a specified amount, the device rapidly trips and opens the circuit. Typically, a GFCI trips in less than 25 ms. The presumption is that the current flowing through an unintended path to ground might be through a person's body as illustrated in Fig. 11.14. If the person's body were to touch parts that are connected across the line rather than between a live part and ground, the resulting body current would flow through the same pathway as normal electric load current, and no protection against electric shock would be provided. However, when a person simultaneously touches a live part and ground, the device would sense differential current, and can be effective against most electric shock scenarios.

A GFCI acts independently from other safety mechanisms that might also be used. When it is used to protect a product that is equipped with an equipment grounding conductor or that is double insulated, each system contributes to protect the user against electric shock—one system does not interfere with or depend on the other. Being current sensitive, it protects the user regardless of the body impedance. Most GFCIs in the United States are rated at 5 mA differential current and are intended to prevent inability to let go as well as ventricular fibrillation. In Europe, an RCD with a rating as high as 30 mA is considered suitable for protection of people against electric shock. At this rating, the protection provided by the RCD would be against ventricular fibrillation, but not against inability to let go.

An advantage of the 30-mA-rated RCD in Europe is that it can trip satisfactorily using only the power available in the fault current itself; it does not need auxiliary power to open the contacts. As such, it is inherently invulnerable to being inoperative if the grounded line conductor is inadvertently open. The GFCIs in the United States take power from the line at the input to the device to open the contacts when the device is tripped. The power available from 5 mA at 120 V is not sufficient by itself to cause the rapid tripping required in this application. If the GFCI is not a type that is

---

[2] Adapted from Skuggevig (1992).

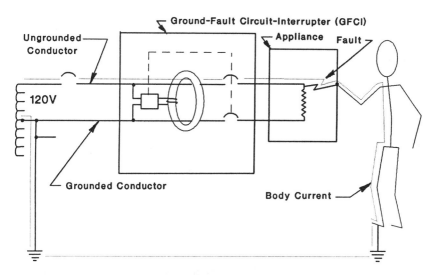

FIGURE 11.14. Ground-fault circuit interrupter (or residual-current device) sensing body current by comparing the magnitudes of the currents in the conductors serving the appliance. If the difference exceeds approximately 5 mA, the ground-fault circuit interrupter opens the circuit. (From Skuggevig, 1992.)

likely to be located where the continuity of the grounded conductor is sufficiently reliable, the device is required by UL to protect regardless of the continuity of the grounded conductor. Many listed GFCIs satisfy this requirement by including circuitry that will not energize the output unless there is voltage across the input terminals.

If a GFCI is used to protect against electric shock from a product that itself generates high voltage, such as with an internal autotransformer, it might not trip fast enough to prevent ventricular fibrillation because of the higher body current. For example, GFCIs typically contain mechanical contacts that interrupt the circuit when the device trips. These contacts require a certain minimum time to separate–on the order of 15 to 20 ms. If the body impedance is as low as 500 Ω, and the high-voltage source is capable of applying 1,000 volts across the body, then the body current would be 2 A. According to the current limit as a function of duration in Fig. 11.10, the duration of body current of this magnitude should be limited to 264 μs, which is beyond the tripping-speed capability of the typical GFCI.

References

Abdeen, M.A., and M.A. Stuchly (1994). Modeling of magnetic field stimulation of bent neurons. *IEEE Trans. Biomed. Eng.* 41(11): 1092–1099.

Accornero, N., G. Bini, G.L. Lenzi, and M. Manfredi (1977). Selective activation of peripheral nerve fibre groups of different diameter by triangular shaped stimulus pulses. *J. Physiol.* 273: 539–560.

ACGIH (1996). 1996 TLVs and BEIs. American Conference Governmental and Industrial Hygienists, Cincinnati, OH.

Ackmann, J.J., and M.A. Seiti (1984). Methods of complex impedance measurements in biologic tissue. *CRC Crit. Rev. Bioeng.* 11(4): 281–311.

Adair, R.K. (1991). Constraints on biological effects of weak extremely-low-frequency electromagnetic fields. *Physical Rev. A* 43(2): 1039–1048.

Adrian, D. (1977). Auditory and visual sensations stimulated by low-frequency currents. *Radio Sci.* 12(65 S): 243–250.

Agnew, W.F., and D.B. McCreery, eds. (1990). *Neural Prostheses: Fundamental Studies.* Prentice-Hall, Englewood Cliffs, NJ.

Agnew, W.F., D.B. McCreery, T.G.H. Yuen, and L.A. Bullara (1989). Histologic and physiologic evaluation of electrically stimulated peripheral nerve: Considerations for the selection of parameters. *Annals Biomed. Eng.* 17: 39–60.

Ahrenholz, D.H., W. Schubert, and L.D. Solem (1988). Creatine kinase as a prognostic indicator in electrical injury. *Surgery* 104(4): 741–747.

Al-Rabiah, S.M., D.B. Archer, R. Millar, A.D. Collins, and W.F. Shepherd (1987). Electrical injury of the eye. *Int. Ophthamol.* 11(1): 31–40.

Albuquerque, E.X., and S. Thesleff (1968). A comparative study of membrane properties of innervated and chronically denervated fast and slow skeletal muscles of the rat. *Acta. Physiol. Scand.* 73: 471–480.

Alekseev, S.I., and M.C. Ziskin (1994). Millimeter microwave effect on ion transport across lipid bilayer membranes. *Bioelectromagnetics* 16: 124–131.

Alekseev, S.I., M.C. Ziskin, N.V. Kochetkova, and M.A. Bolshakov (1997). Millimeter waves thermally alter the firing rate of the Lymnaea pacemaker neurons. *Bioelectromagnetics* 18: 89–98.

Allessie, M.A., F.I.M. Bonke, and F.J.G. Schopman (1973). Circus movement in rabbit atrial muscle as a mechanism of tachycardia. *Circ. Res.* 33: 54–62.

Allessie, M.A., F.I.M. Bonke, and F.J.G. Schopman (1976). Circus movement in rabbit atrial muscle as a mechanism of tachycardia. II. The role of nonuniform recovery of excitability in the occurrence of unidirectional block, as studied with multiple microelectrodes. *Circ. Res.* 39: 168–177.

Allessie, M.A., F.I.M. Bonke, and F.J.G. Schopman (1977). Circus movement in rabbit atrial muscle as a mechanism of tachycardia. III. The "leading circle" concept: A new model of circus movement in cardiac tissue without the involvement of an anatomical obstacle. *Circ. Res.* 41: 9–18.

Alon, G., J. Allin, and G.F. Inbar (1983). Optimization of pulse duration and pulse charge during transcutaneous electrical nerve stimulation. *Austral. J. Physiotherapy* 29(6): 195–201.

Altman, K.W., and R. Plonsey (1989). Excitation in a model for time-dependent electrical nerve bundle stimulation. *Proc. 11th IEEE-EMBS Conf.:* 975–976.

Amassian, V.E., R.Q. Cracco, and P.J. Maccabee (1988). Basic mechanisms of magnetic coil excitation of nervous system in humans and monkeys and their applications. *Proc. 10th Ann. Conf. IEEE-EMBS*: 10–17.

Amassian, V.E., J.B. Cracco, R.Q. Cracco, L. Eberle, P.J. Maccabee, and A. Rudell (1989a). Suppression of visual perception by magnetic coil stimulation of human occipital cortex. *EEG Clin. Neurophysiol.* 74: 458–462.

Amassian, V.E., R.Q. Cracco, and P.J. Maccabee (1989b). A sense of movement elicited in paralyzed distal arm by focal magnetic coil stimulation of human motor cortex. *Brain Res.* 479: 355–360.

Anderson, A.B., and W.A. Munson (1951). Electrical stimulation of nerves in the skin at audio frequencies. *J. Acoust. Soc. Am.* 23(2): 155–159.

Anderson, N.H. (1970). Functional measurement and psychophysical judgment. *Psychol. Rev.* 77: 153–170.

ANSI (1982). *Safety Levels with Respect to Human Exposure to Radio Frequency Electromagnetic Fields, 300 kHz to 100 GHz.* Document ANSI C95.1-1982, published by The Institute of Electrical and Electronics Engineers, New York.

ANSI (1986). Leakage current for appliances. ANSI C101.1-1986, American National Standards Institute.

ANSI (1987). Procedures for the development and coordination of American national standards. American National Standards Institute.

Antoni, H. (1979). What is measured by the so-called threshold for fibrillation? *Prog. Pharmacol.* (Stuttgart) 2(4): 5–12.

Antoni, H. (1985). Pathophysiological basis of ventricular fibrillation. In J.F. Bridges, G.L. Ford, I.A. Sherman, and M. Vainberg (eds.), *Electrical Shock Safety Criteria*, Pergamon, New York: 33–43.

Antoni, H. (1996). Electrophysiology of the heart at the single cell level and cardiac rhythmogenesis. In R. Greger and U. Windhorst (eds.), *Comprehensive Human Physiology*. Springer-Verlag, Berlin: 1825–1842.

Antoni, H., J. Toppler, and R. Krause (1970). Polarization effects of sinusoidal 50-cycle alternating current on membrane potential of mammalian cardiac fibers. *Pflugers Arch.* 314: 274–291.

Antzelevitch, C., and G.K. Moe (1981). Electrotonically mediated delayed conduction and reentry in relation to "slow responses" in mammalian ventricular conducting tissue. *Circ. Res.* 49: 1129–1139.

Artz, C.P. (1974). Changing concepts of electrical injury. *Am. J. Surg.* 128(5): 600–602.

Artz, C.P. (1979). Electrical injury. In C.P. Artz, J.A. Moncrief, and B.A. Pruitt, (eds.), *Burns: A Team Approach*, W.B. Saunders, Philadelphia.

ASAE (1993). Dimensions of livestock and poultry, ASAE Standards D321.2, American Soc. Ag. Eng., St. Joseph, MI: 511–517.

Ashton, W.D. (1972). *The Logit Transformation.* Hafner, New York.

Astumian, R.D. et al. (1995). Rectification and signal averaging of weak electric fields by biological cells. *Proc. Natl. Acad. Sci.* 92: 3740–3743.

Athey (1992). Current FDA guidance for MR patient exposure and consideration for the future. In R.L. Magin, R.L. Liburdy, and B. Persson (eds.), *Biological Effects and Safety Aspects of Nuclear Magnetic Resonance Imaging*, New York Academy of Sciences, New York: 242–257.

Atkinson, W.H. (1982). A general equation for sensory magnitude. *Perception & Psychophys.* 31: 26–40.

Attneave, F. (1962). Perception and related areas. In S. Koch (ed.), *Psychology: A Study of a Science*, McGraw-Hill, New York: 619–659.

Babkoff, R. (1976). Magnitude estimation of short electrocutaneous pulses. *Psychol. Res.* 39: 39–49.

Baeten, C.G.M.I., B.P. Geerdes, E.M.M. Adang, et al. (1995). Anal dynamic graciloplasty in the treatment of intractable fecal incontinence. *N. Engl. J. Med.* 332: 1600–1605.

Bajzek, T.J., and R.J. Jaeger (1987). Characterization and control of muscle response to electrical stimulation. *Annals Biomed. Eng.* 15: 485–501.

Baker, M.D., and C. Chiaviello (1989). Household electrical injuries in children; Epidemiology and identification of avoidable hazards. *Am. J. Dis. Child.* 143(1): 59–62.

Banks, R.S., and T. Vinh (1984). An assessment of the 5-mA 60-Hz contact current safety level. *IEEE Trans. Pwr. Sys.* PAS-103(12): 3608–3614.

Barker, A.T. (1994). Magnetic nerve stimulation principles, advantages, and disadvantages. In S. Ueno (ed.), *Biomagnetic Stimulation*, Plenum, New York: 9–28.

Barker, A.T., I.L. Freeston, R. Jalinous, P.A. Merton, and H.B. Morton (1985). Magnetic stimulation of the human brain. *J. Physiol.* 369: 3P.

Barker, A.T., I.L. Freeston, B. Jalinous, and J.A. Jarratt (1987). Magnetic stimulation of the human brain and peripheral nervous system: An introduction and the results of an initial clinical evaluation. *Neurosurgery*, 20: 100–109.

Barker, A.T., C.W. Garnham, and I.L. Freeston (1991). Magnetic nerve stimulation: The effect of waveform on efficiency, determination of neural membrane time constant, and the measurement of stimulator output. In W.J. Levy, R.Q. Cracco, A.T. Barker, and J. Rothwell (eds.), *Magnetic Motor Stimulation: Basic Principles and Clinical Experience (EEG Supplement 43)*: 227–237.

Barlow, H.B., H.I. Kohn, and E.G. Walsh (1947a). Visual sensations aroused by magnetic fields. *Am. J. Physiology* 148: 372–375.

Barlow, H.B., H.I. Kohn, and E.G. Walsh (1947b). The effect of dark adaptation and of light upon the electric threshold of the human eye. *Am. J. Physiol.* 148: 376–381.

Barnard, M.D., and J.A. Boswick, Jr. (1976). Electrical injuries of the upper extremity. *Rocky Mt. Med. J.* 73: 20–24.

Barnes, F.S. (1996). Interaction of DC and ELF electric fields with biological materials and systems. In C. Polk and E. Postow (eds.), *Handbook of Biological Effects of Biological Effects of Electromagnetic Fields, Second Edition*, CRC Press, Boca Raton, FL: 107–147.

Barta, E., D. Adam, E. Salant, and S. Siderman (1987). 3-D ventricular myocardial electrical excitation: A minimal orthogonal pathways model. *Annals Biomed. Eng.* 15: 443–456.

Basser, P.J., and B.J. Roth (1991). Stimulation of a myelinated nerve axon by electromagnetic induction. *Med. Biol. Eng. Comp.* 29(3): 261–268.

Baumgardt, E. (1951). Sur le seuil phosphène électrique. Quantité liminaire et pseudo-chronaxie. *Comptes Rendeu Soc. Biol.* (Nov. 24): 1654–1657.

Bauwens, P. (1971). Introduction to electrodiagnostic procedures. In S.H. Licht (ed.), *Electrodiagnosis and Electromyography* (3rd ed.), E. Licht, New Haven, CT, chap. 7.

Bawin, S.M., A.R. Sheppard, M.D. Mahoney, and W.R. Adey (1984). Influences of sinusoidal electric fields on excitability in the rat hippocampal slice. *Brain Research*, 323: 227–237.

Bawin, S.M., A.R. Sheppard, M.D. Mahoney, M. Abu-Assal, and W.R. Adey (1986). Comparison between the effects of extracellular direct and sinusoidal currents on excitability in hippocampal slices. *Brain Res.* 362: 350–354.

Baxter, C.R. (1970). Present concepts in the management of major electrical injury. *Surg. Clin. North Am.* 50(6): 1401–1418.

Beck, C.S., W.H. Pritchard, and S.H. Feil (1947). Ventricular fibrillation of long duration abolished by electric shock. *J. Am. Med. Assoc.* 135: 985–995.

Beck, C., and B.S. Rosner (1968). Magnitude scales and evoked potentials to percutaneous electrical stimulation. *Physiol. Behav.* 3: 947–953.

Beeler, G.W., and H. Reuter (1977). Reconstruction of the action potential of ventricular myocardial fibers. *J. Physiol.* 268: 177–210.

Benz, R. and U. Zimmerman (1980). Relaxation studies on cell membranes and lipid bilayers. *Bioelectrochem. Bioenerg.* 7: 723–739.

Bergeron, J.A., M.R. Hart, J.A. Mallick, and L.H. String (1995). Strength-duration curve for human electro- and magnetophosphenes. *Proc. Bioelectromagnetics Soc. Annual Meeting*, Boston.

Bernhardt, J.H. (1985). Evaluation of human exposures to low frequency fields. In *The Impact of Proposed Frequency Radiation Standards on Military Operations*, Lecture Series 138, Advisory Group for Aerospace Research and Development (NATO), Surseine, France.

Bernhardt, J.H. (1988). The establishment of frequency dependent limits for electric and magnetic fields and evaluation of indirect effects. Radiat. Env. Biophys. 27: 1–27.

Biegelmeier, G. (1978). Report on the electrical impedance of the human body. Presented to the International Electrotechnical Commission, Technical Committee No. 23, Austria.

Biegelmeier, G. (1982). Report on the electrical impedance of the human body. IEC Technical Committee No. 64, *Electrical Installations of Buildings*, Austin, Texas.

Biegelmeier, G. (1985a). New knowledge of the impedance of the human body. In J.E. Bridges, G.L. Ford, I.A. Sherman, and M. Vainberg (eds.), *Electrical Shock Safety Criteria*, Pergamon, New York: 115–132.

Biegelmeier, G. (1985b). New experiments with regard to basic safety measures for electrical equipment and installations. In J.E. Bridges, G.L. Ford, I.A. Sherman, and M. Vainberg (eds.), *Electrical Shock Safety Criteria*, Pergamon, New York: 161–172.

Biegelmeier, G. (1985c). The impedance of the human body. *Rev. Gen. L'Electricite* 11: 817–832.

Biegelmeier, G. (1986). *Wirkungen des Elektrischen Stroms auf Menschen und Nutztiere.* Vde-Verlag, Berlin.

Biegelmeier, G. (1987). Effects of current passing through the human body and the electrical impedance of the human body: A guide to IEC-Report 469. ETZ Report 20, VDE-Verlag, Berlin.

Biegelmeier, G., and E. Homberger (1982). Über die wirkungen von unipolaren impulsatromen auf den menschlicken körper. *Bulletin ASE/USE* 73, Sept. 18.

Biegelmeier, G., and W.R. Lee (1980). New considerations on the threshold of ventricular fibrillation for a.c. shocks at 50–60 Hz. *IEE Proc.* 127(2), Pt. A: 103–110.

Biegelmeier, G., and J. Miksch (1980). Effect of the skin on the body impedance of humans (translated from German). *Electrotechnik and Maschinenbau, mit Industrieller Electronik und Nachrichtentchnik* 97(9): 369–378.

Biegelmeier, G., and K. Rotter (1971). Electrical resistances and currents in the human body. *Electrotechnic and Maschinenbau*: 104–114.

Bigland-Ritchie, B., R. Johansson, O.C. J. Lippold, S. Smith, and J.J. Woods (1983). Changes in motoneuron firing rate during sustained maximal voluntary contractions. *J. Physiol.* 340: 335–346.

Bigland-Ritchie, B., F. Bellemare, and J.J. Woods (1986). Excitation frequencies and sites of fatigue. In N.L. Jones, N. McCartney, and A.J. McComas (eds.), *Human Muscle Power*, Human Kinetics Publishers, Champaign, IL: 197–211.

Bini, G., G. Cruccu, K.E. Hagbarth, W. Schady, and E. Torebjörk (1984). Analgesic effect of vibration and cooling on pain induced by intraneural electrical stimulation. *Pain* 18: 239–248.

Birks, R., R.E. Huxley, and B. Katz (1960). The fine structure of the neuromuscular junction of the frog. *J. Physiol.* 150: 134–144.

Bishop, G.H. (1943). Responses to electrical stimulation of single sensory units of skin. *J. Neurophysiol.* 6: 361–382.

Bishop, G.H. (1946). Neural mechanisms of cutaneous sense. *Psychol. Rev.* 26: 77–102.

Blackman, C.F., et al. (1988). Influence of Electromagnetic fields on the efflux of calcium ions from brain tissue in vitro. *Bioelectromagnetics* 9: 215–227.

Blackman, C.F., J.P. Blanchard, S.G. Benane, and D.E. House (1994). Empirical tests of an ion parametric resonance model for magnetic field interactions with PC-12 cells. *Bioelectromagnetics* 15: 239–260.

Blair, E.A., and J. Erlanger (1933). A comparison of the characteristics of axons through their individual electrical responses. *Am. J. Physiol.* 106: 524–564.

Blair, H.A. (1932a). On the intensity–time relations for stimulation by electric currents. I. *J. Gen. Physiol.* 15: 709–729.

Blair, H.A. (1932b). On the intensity–time relations for stimulation by electric currents. II. *J. Gen. Physiol.* 15: 731–755.

Blanchard, J.P. and C.F. Blackman (1994). Clarification and application of an ion parametric resonance model for magnetic field interactions with biological systems. *Bioelectromagnetics* 14: 217–238.

Blank, M. (ed.) (1993). *Electricity and Magnetism in Biology and Medicine.* San Francisco Press, San Francisco.

Bo, W.J., N.T. Wolfman, W.A. Krueger, and I. Meschan (1990). *Basic Atlas of Sectional Anatomy with Correlated Imaging.* 2nd edition, W.B. Saunders, Philadelphia.

Bostock, H. (1983). The strength–duration relationship for excitation of myelinated nerve: Computed dependence on membrane parameters. *J. Physiol.* 341: 59–74.

Bostock, H., T.A. Sears, and R.M. Sherratt (1983). The spatial distribution of excitability and membrane current in normal and demyelinated mammalian nerve fibers. *J. Physiol.* 341: 41–58.

Bouman, M.A., and R.A. van der Velden (1947). The two-quanta explanation of the dependence of the threshold values and visual acuity on the visual angle and the time of observation, *J. Opt. Soc. Am.* 37: 908–919.

Bourland, J.D., W.A. Tacker, and L.A. Geddes (1978). Strength–duration curves for trapezoidal waveforms of various tilts for transchest defibrillation in animals. *Med. Instrum.* 12: 38–41.

Bourland, J.D., J.A. Nyenhuis, G.A. Mouchawar, L.A. Geddes, and D.J. Schaefer (1990). Human peripheral nerve stimulation from z-gradients. *Society for Magnetic Resonance in Med., Proc. 9th Annual Meeting,* Aug. 18–24, New York: 1157.

Bourland, J.D., J.A. Nyenhuis, G.A. Mouchawar, L.A. Geddes, D.J. Schaefer and M.E. Riehl (1991a). Z-gradient coil and eddy-current stimulation of skeletal and cardiac muscle in the dog. *Society for Magnetic Resonance in Med., Proc. 10th Annual Meeting,* Aug. 10–16, San Francisco.

Bourland, J.D., J.A. Nyenhuis, G.A. Mouchawar, T.Z. Elabbady, L.A. Geddes, D.J. Schaefer, and M.E. Riehl (1991b). Physiologic indicators of high MRI gradient-induced fields. *Society for Magnetic Resonance in Med., Proc. 10th Annual Meeting,* San Francisco: 1276.

Bourland, J.D., J.A. Nuyenhuis, G.A. Mouchawar, L.A. Geddes, D.J. Schaefer, and M.E. Riehl (1991c). Effect of body position on z-gradient coil cardiac stimulation. Cardiac arrhythmias induced in the dog. *Society for Magnetic Resonance in Med., Proc. 10th Annual Meeting,* Berkeley, San Francisco.

Bourland, J.D., J.A. Nyenhuis, D.J. Schaefer, K.S. Foster, W.E. Schoelein, T.Z. Elabbady, L.A. Geddes, and M.E. Rielh (1992). Gated, gradient-induced cardiac stimulation in the dog: Absence of ventricular fibrillation. *Society for Magnetic Resonance in Med., Proc. 11th Annual Meeting,* Berlin: 4804.

Bourland, J.D., J.A. Nyenhuis, W.A. Noe, D.J. Schaefer, K.S. Foster, and L.A. Geddes (1996). Motor and sensory strength–duration curves for MRI gradient fields. Proc. Soc. Mag. Res. Med., 4th Annual. Meeting, NY, Apr. 27–May 3: 1724.

Bourland, J.D., J.A. Nyenhuis, K.S. Foster, L.A. Geddes (1997). Threshold and pain strength–duration curves for MRI gradient fields. *Proc. Soc. Mag. Res. Med., 5th Ann. Meeting,* Vancouver, Apr. 12–18: 1974.

Bowman, B.R., and R.C. Erickson (1985). Acute and chronic implantation of coiled wire interaneural electrodes during cyclical electrical stimulation. *Annals. Biomed. Eng.* 13: 75–93.

Bowman, B.R., and D.R. McNeal (1986). Response of single alpha motoneurons to high-frequency pulse trains. *Appl. Neurophysiol.* 49: 121–138.

Boxtel, A. (1977). Skin resistance during square-wave electrical pulses of 1 to 10 mA. *Med. Biol. Eng. Comp.* 15: 679–687.

Boyd, I.A., and M.R. Davey (1968). *Composition of Peripheral Nerves,* E & S Livingstone Ltd, Edinburgh.

Bracken, T.D. (1976). Field measurements and calculations of electrostatic effects of overhead transmission lines. *IEEE Trans. Pwr. Apparat. Sys.* PAS-95: 494–502.

Brazier, M.A. (1977). *Electrical Activity of the Nervous System.* Williams & Wilkins, Baltimore.

Brengelmann, G., and A.C. Brown (1965). Temperature regulation. In T.C. Ruch and H.D. Patton (eds.), *Physiology and Biophysics*, 19th ed., W.B. Saunders, Philadelphia: 1050–1068.

Bridges, J.E. (1985). Potential distributions in the vicinity of the hearts of primates arising from 60 Hz limb-to-limb body currents. In J.E. Bridges, G.L. Ford, I.A. Sherman, and M. Vainberg (eds.), *Electrical Shock Safety Criteria*, Pergamon, New York: 61–70.

Bridges, J.E., G.L. Ford, I.A. Sherman, and M. Vainberg (eds.) (1985). *Electrical Shock Safety Criteria*, Pergamon, New York.

Bridges, J.E., M. Vainberg, and M.C. Wills (1987). Impact of recent developments in biological electrical shock safety criteria. *IEEE Trans. Pwr. Del.* PWRD-2(1): 238–248.

Brindley, G.S. (1955). The site of exectrical excitation of the human eye. *J. Physiol.* 127: 189–200.

Brindley, G.S., and W.S. Lewin (1968). The sensations produced by electrical stimulation of the visual cortex. *J. Physiol.* 196: 479–493.

Brindley, G.S., and D.N. Rushton (1977). Observations of the representation of the visual field of the human occipital cortex. In F.T. Hambrecht and J.B. Reswick, Marcel Dekker, New York: 261–276.

Brooke, M.H., and K.K. Kaiser (1970). Muscle fiber types: How many and what kind? *Arch. Neurol.* 23: 369–379.

Brooks, C.M., B.F. Hoffman, E.E. Suckling, and O.O. Orias (1955). *Excitability of the Heart.* Grune & Stratton, New York.

Budinger, T.F. (1992). Emerging nuclear magnetic resonance technologies: Health and safety. In R.L. Magin, R.L. Liburdy, and B. Persson (eds.), *Biological Effects and Safety Aspects of Nuclear Magnetic Resonance Imaging*, New York Academy of Sciences, New York (1950).

Budinger, T.F., C. Cullander, and R. Bordow (1984). Switched magnetic field thresholds for the induction of magnetophosphenes. *Proc. Annual Meet. Soc. Mag. Res. Med.*, New York: 118.

Budinger, T.F., R. Fischer, D. Hentschel, R.E. Reinfelder, and F. Schmitt (1990). Neural Stimulation dB/dt thresholds for frequency and number of oscillations using sinusoidal magnetic gradient fields. *Society for Magnetic Resonance in Med., Proc. 9th Annual Meeting*, Aug, 18–24, NY: 276.

Budinger, T.F., H. Fischer, D. Hentschel, H. Reinfelder, and F. Schmitt (1991). Physiological effects of fast oscillating magnetic field gradients. *J. Computer Assisted Tomography* 15(6): 909–904.

Buechler, D.N., Durney, C.H., and Christensen, D.A. (1997). Calculation of electric fields induced near metal implants by magnetic resonance imaging switched-gradient magnetic field. *Magnetic Resonance Imaging.*

Burke, R.E. (1968). Firing patterns of gastrocnemius motor units in the decerebrate cat. *J. Physiol.* 196: 631–654.

Burke, R.E. (1981). Motor units: Anatomy, physiology and functional organization. In V.B. Brooks (ed.), *Handbook of Physiology Section 1: The Nervous System. Vol.* III. *Motor Systems*, American Physiology Society, Bethesda, MD: 345–422.

Burke, R.E. (1986). The control of muscle force: Motor unit recruitment and firing patterns. In N.L. Jones, N. McCartney, and A.J. McComas (eds.), *Human Muscle Power*, Human Kinetics Publishers, Champaign, IL: 97–106.

Burke, R.E., D.N. Levine, P. Tsairis, and F.E. Zajac (1973). Physiological types of histochemical profiles in motor units of the cat gastrocnemius. *J. Physiol.* 234: 723–748.

Burton, C.E., R.M. David, W.M. Portnoy, and L.A. Akers (1974). The application of Bode analysis to skin impedance. *Psychophysiology* 11(4): 517–525.

Bütikoffer, R., and P.D. Lawrence (1978). Electrocutaneous nerve stimulation—I: Model and experiment. *IEEE Trans. Biomed Eng.* BME-25(6): 526–531.

Bütikoffer, R., and P.D. Lawrence (1979) Electrocutaneous nerve stimulation—II: Stimulus waveform selection. *IEEE Trans Biomed Eng.* BME-26(2): 69–75.

Butler, E.D., and T.D. Gant (1978). Electrical injuries, with special reference to the upper extremities—A review of 182 cases. *Am. J. Surg.* 134: 95–101.

Cabanes, J., and C. Gary (1981). La perception directe du champ electrique. Paper 233-08. *CIGRE Symp. 22–81*, Stockholm: 1–6.

Caldwell, C.W., and J.B. Reswick (1975). A percutaneous wire electrode for chronic research use. *IEEE Trans, Biomed. Eng.* 22: 429–432.

Campbell, J.N., and R.A. Meyer (1983). Sensitization of unmyelinated nociceptive afferents in monkey varies with skin type. *J. Neurophysiol.* 49(1): 98–110.

Campbell, J.N., R.A. Meyer, and R.H. LaMotte (1979). Sensitization of myelinated afferents that innervate monkey hand. *J. Neurophysiol.* 42(6): 1669–1679.

Carmeliet, E. (1977). Repolarization and frequency in cardiac cells. *J. Physiol.* (Paris). 73: 903–923.

Carmeliet, E. (1992). Potassium channels in cardiac cells. *Cardiovasc. Drugs Ther.* 6: 305–312.

Carpentier, A. and J.C. Chachques (1985). Myocardial substitution with a stimulated skeletal muscle: First successful clinical case. *Lancet* 8440: 1267.

Carpentier, A., J.C. Chachques, and P.A. Grandjean (eds.) (1991). *Cardiomyoplasty*. Futura Publishing, Mount Kisco, NY.

Carstensen, E.L. (1985). Sensitivity of the human eye to power frequency electric fields. *IEEE Trans. Biomed. Eng.* BME-32(8): 561–565.

Carstensen, E.L. (1987). *Biological Effects of Transmission Line Fields*. Elsevier, New York.

Carter, A.O., and R. Morley (1969a). Electrical current flow through human skin at power frequency voltages. *Br. J. Ind. Med.* 26: 217–223.

Carter, A.O., and R. Morley (1969b). Effects of power frequency voltages on amputated human limb. *Br. J. Ind. Med.* 26: 224–230.

Carter, B.L., J. Morehead, S.M. Wolpert, S.B. Hammerschlag, H.J. Griffiths, and P.C. Kahn (1977). *Cross-Sectional Anatomy: Computed Tomography and Ultrasound Correlation*. Appleton-Century-Crofts, New York.

Caruso, P.M., J.A. Pearce, and D.P. DeWitt (1979). Temperature and current density distributions at electrosurgical dispersive electrode sites. *Proc. 7th N. Engl. Bioeng. Conf.*, Troy, NY, March 22–23, 1979: 373–376.

Cattell, M.K., and R.W. Gerard (1935). The "inhibitory" effect of high frequency stimulation and excitation state of nerve. *J. Physiol.* 83: 407–415.

CENELEC (1995). Human exposure to electromagnetic fields: Low-frequency (0 to 10kHz). European Prestandard ENV 50166-1: 1995 E, European Committee for Electrotechnical Standardization, Brussels.

Chachques, J.C., M. Radermercker, J. Tolan, E.I.C. Fischer, P.A. Grandjean, and A.F. Carpentier (1996). Aortomyoplasty counterpulsation: Experimental results and early clinical experience. *Ann. Thorac. Surg.* 61: 420–425.

Chakravarti, K., and G.J. Pontrelli (1976). The measurement of carpet static. *Textile Res. J.* 46(2): 129–134.

Chatterjee, I., D. Wu, and O.P. Gandhi (1986). Human body impedance and threshold currents for perception and pain for contact hazard analysis in the VLF-MF band. *IEEE Trans. Biomed. Eng.* BME-33(5): 486–494.

Chen, J.Y., and O.P. Gandhi (1988). Thermal implications of high SAR's in the body extremities at the ANSI-recommended MF-VHF safety levels. *IEEE Trans. Biomed. Eng.* 35(6): 435–441.

Chen, P., G.H. Myers, V. Parsonnet, K. Chatterjee, and P. Katz (1975). Relationship between pacemaker fibrillation thresholds and electrode area. *Med. Instrum.* 9(4): 165–170.

Chilbert, M.A., A. Sances, J.B. Mykelbust, T. Swiontek, and T. Prieto (1983). Postmortem resistivity studies at 60 Hz. *J. Clin. Eng.* 8(3): 219–224.

Chilbert, M., D. Maiman, A. Sances Jr., J. Myklebust, T.E. Prieto, T. Swiontek, M. Heckman, and K. Pintar (1985a). Measure of tissue resistivity in experimental electrical burns. *J. Trauma* 25(3): 209–215.

Chilbert, M., T. Swiontek, T. Prieto, A. Sances, J. Myklebust, J. Ackmann, C. Brown, and J. Szablya (1985b). Resistivity changes of tissue during the application of injurious 60 Hz currents. In J.E. Bridges, G.L. Ford, I.A. Sherman, and M. Vainberg (eds.), *Electrical Shock Safety Criteria*, Pergamon, New York: 193–201.

Chilbert, M., D.J. Moretti, T. Swiontek, J.B. Myklebust, T. Prieto, A. Sances, and C. Leffingwell (1988). Instrumentation design for high-voltage electrical injury studies. *IEEE Trans. Biomed. Eng.* 35(7): 565–568.

Chilbert, M., T. Swiontek, C. Leffingwell, T. Prieto, J. Myklebust, A. Sances, and T. Schneider (1989). Fibrillation induced at high current levels. *IEEE Trans. Biomed. Eng.* 36(8): 864–868.

Chiu, M.C. (1982). Fuel ignition by high voltage capacitive discharges. The Johns Hopkins University Applied Physics Laboratory, JHU PPSE T-18.

Chiu, S.Y., and J.M. Ritchie (1981). Evidence for the presence of potassium channels in the paranodal region of acutely demyelinated mammalian single nerve fibers. *J. Physiol.* 313: 415–437.

Chiu, S.Y., J.M. Ritchie, R.B. Rogart, and D. Stagg (1979). A quantitative description of membrane currents in rabbit myelinated nerve. *J. Physiol.* 292: 149–166.

Chizeck, H.J., R. Kobetic, E.B. Marsolais, J.J. Abbas, I.R. Donner, and E. Simon (1988). Control of functional neuromuscular stimulation systems for standing and locomotion in paraplegics. *Proc. IEEE* 76(9): 1155–1165.

Chokroverty, S., and J. DiLullo (1989). Percutaneous magnetic stimulation of the human lumbosacral spinal column: Physiological mechanism and clinical application. *Neurology* 39: 376.

Christensen, J.A., R.T. Sherman, G.A. Balis, and J.D. Wuamett (1980). Delayed neurologic injury secondary to high-voltage current, with recovery. *J. Trauma* 20(2): 166–168.

Clairmont, B.A., G.B. Johnson, L.E. Zafanella, and S. Zelingher (1989). The effect of HVAC-HVDC line separation in a hybrid corridor. *IEEE Trans. Pwr. Del.* 4(2): 1338–1350.

Clar, E.J., C.P. Her, and C.G. Sturelle (1975). Skin impedance and moisturization. *J. Soc. Cosmet. Chem.* 26: 337–353.

Clark, W.C., and S.B. Clark (1980). Pain responses in Nepalese porters. *Science* 209: 410–412.

Clerc, L. (1976). Direction differences of impulse spread in trabecular muscle from mammalian heart. *J. Physiol.* (London). 255: 335–346.

Coburn, B. (1989). Neural modeling in electrical stimulation. *Critical Reviews in Biomed. Eng.* 17(2): 133–178.

Coers, C. (1955). Les variations structurelles normales et pathologiques de la jonction neuromusculaire. *Acta Neurol. Psychiatr. Belg.* 55: 741.

Cohen, M.S., R.M. Weisshoff, R.R. Rzedzian, and H.C. Kantor (1990). Sensory stimulation by time-varying magnetic fields. *Magnetic Resonance in Medicine* 14: 409–414.

Cook, M.R., C. Graham, H.D. Cohen, and M. Gerkovich (1992). A replication study of human exposure to 60 Hz fields. *Bioelectromagnetics* 13: 261–285.

Cooley, J.W., and F.A. Dodge (1966). Digital computer solutions for excitation and propagation of the nerve impulse. *Biophys. J.* 6: 583–599.

Cowburn, J., and R.H. Fox (1974). A technique for studying thermal perception. *J. Physiol.* (London). 239: 77–78.

Cracco, R.Q. (1987). Evaluation of conduction in central motor pathways: Techniques, pathophysiology, and clinical interpretation. *Neurosurgery* 20: 199–203.

Craggs, J.D. (1978). High-frequency breakdown of gases. In J.M. Meek and J.D. Craggs (eds.), *Electrical Breakdown of Gases*. John Wiley & Sons, Chichester, UK: 689–715.

Crago, P.E., P.R. Peckham, J.T. Mortimer, and J.P. Van Der Meulen (1974). The choice of pulse duration for chronic electrical stimulation via surface, nerve, and intramuscular electrodes. *Annals Biomed. Eng.* 2: 252–264.

Crago, P.E., P.R. Peckham, and G.B. Thorpe (1980). Modulation of muscle force by recruitment during intramuscular stimulation. *IEEE Trans. Biomed. Eng.* 27: 679–684.

Crago, P.E., N. Lan, P.H. Veltink, J.J. Abbas, and C. Kantor (1996). New control strategies for neuroprosthetic systems. *Rehab. Res. Develop.* 33(2): 158–172.

Craig, K.D., and S.M. Weiss (1971). Vicarious influences on pain-threshold determinations. *J. Personality and Social Psychol.* 19: 53–59.

Cranefield, P.F., B.F. Hoffman, and A.A. Siebens (1957). Anodal excitation of cardiac muscle. *Am. J. Physiol.* 190(2): 383–390.

Cranefield, P.F., R.O. Klein, and B.F. Hoffman (1971). Conduction of the cardiac impulse: I. Delay, block, and one-way block in depressed Purkinje fibers. *Circ. Res.* 28: 199–219.

Creasey, G., J. Elifteriades, A. DiMarco, P. Talonen, M. Bijak, W. Girsch, and C. Kantor (1996). Electrical stimulation to restore respiration. *J. Rehab. Res. Develop.* 33(2): 123–132.

Critchley, M. (1934). Neurological effects of lightning and of electricity. *Lancet.* 1: 68–72.

Crochetiere, W.J., L. Vodovnik, and J.B. Reswick (1967). Electrical stimulation of skeletal muscle—A study of muscle as an actuator. *Med. Biol. Eng.* 5: 111–125.

Currence, H.D., B.J. Stevens, D.F. Winter, W.K. Dick, and G.F. Krause (1990). Dairy cow and human sensitivity to short duration 60 Hz currents. *App. Eng. in Agriculture*, 6(3): 349–353.

Dalziel, C.F. (1938). Danger of electric shock. *Electrical West* 80(4): 30–31.

Dalziel, C.F. (1943). Effect of wave form on let-go currents, *AIEE Trans.* 62: 739–744.

Dalziel, C.F. (1953). A study of the hazards of impulse current. *Trans. AIEE, Pt. III*, 72: 1032–1043.

Dalziel, C.F. (1954). The threshold of perception current. *Trans. AIEE, Pt. III* B: 990–996.

Dalziel, C.F. (1959). The effects of electric shock on man. *IRE Trans. Med. Elect.* PGME-5: 44–62.

Dalziel, C.F. (1960). Threshold 60-cycle fibrillating currents. *Trans. AIEE, Pt. III*, 79: 667–673.

Dalziel, C.F. (1968). Reevaluation of lethal electric currents. *IEEE Trans. Ind. Appl.* IGA-4(5): 467–476.

Dalziel, C.F. (1972). Electric shock hazard. *IEEE Spectrum* (9): 41–50.

Dalziel, C.F. (1978). Recent developments in ground fault circuit interrupters and ground fault receptacles. *Profession. Safety*, November: 31–40.

Dalziel, C.F., and W.R. Lee (1968). Reevaluation of lethal electric currents. *IEEE Trans. Ind. Appl.* IGA-4(5): 467–476.

Dalziel, C.F., and W.R. Lee (1969). Lethal electric currents. *IEEE Spectrum,* Feb.: 44–50.

Dalziel, C.F., and T.H. Mansfield (1950a). Perception of electric currents. *Electrical Eng.* 69: 794–800.

Dalziel, C.F., and T.H. Mansfield (1950b). Effects of frequency on perception currents. *AIEE Trans.* 69: 1162–1168.

Dalziel, C.F., and F.P. Massoglia (1956). Let-go currents and voltages. *AIEE Trans. P. II, Appl. Ind.* 75: 49–56.

Dalziel, C.F., J.B. Lagen, and J.L. Thurston (1941). Electric shock. *AIEE Trans.* 60: 1073–1079.

Dalziel, C.F., E. Ogden, and C.E. Abbott (1943). Effect of frequency on let-go currents. *AIEE Trans.* 62: 745–750.

Darien-Smith, I. (1982). Touch in primates. *Ann. Rev. Psychol.* 33: 155–194.

Darien-Smith, I., K.O. Johnson, C. LaMotte, P. Kenins, Y. Shigenaga, and V.C. Ming (1979). Coding of incremental changes in skin temperature by single warm fibers in the monkey. *J. Neurophysiol.* 42: 1316–1331.

Davey, K., K.C. Kalaitzakis, and C. Epstein (1988). Transcranial magnetic stimulation of the cerebral cortex. *Proc. 10th Ann. Conf. IEEE-EMBS:* 922–923.

Davis, H. (1923). The relationship of the "chronaxie" of muscle to the size of the stimulating electrode. *Proc. Physiol. Soc.,* July 7: lxxxi–lxxxii.

de Mello, W.C. (1972). The healing-over process in cardiac and other muscle fibres. In *Electrical Phenomena of the Heart*, Academic Press, New York: 323–325.

Dean, D., and P.D. Lawrence (1983). Application of phase analysis of the Frankenhaeuser-Huxley equations to determine threshold stimulus amplitudes. *IEEE Trans. Biomed. Eng.* BME-30(12): 810–818.

Dean, D., and P.D. Lawrence (1985). Optimization of neural stimuli based upon a variable threshold potential. *IEEE Trans. Biomed. Eng.* 32(1): 8–14.

Déléze, J. (1970). The recovery of resting potential and input resistance in sheep heart injured by knife or laser. *J. Physiol.* (London) 208: 547–562.

Deno, D.W. (1975a). Calculating electrostatic effects of overhead transmission lines. *IEEE Trans. Pwr. Apparat. Sys.* PAS-93(5): 1458–1471.

Deno, D.W. (1975b). Electrostatic effect induction formulae. *IEEE Trans. Pwr. Apparat. Sys.* PAS-94(5): 1524–1536.

Deno, D.W. (1977). Currents induced in the human body by high voltage transmission line electric field—measurement of calculation of distribution and dose. *IEEE Trans. Pwr. Apparat. Sys.* PAS-96(5): 1517–1527.

Deno, D.W. (1978). Electrostatic and electromagnetic effects of ultrahigh-voltage transmission lines. Report EL-802. Electric Power Research Institute, Palo Alto, CA.

Deno, D.W., and L.E. Zaffanella (1982). Field effects of overhead transmission lines and stations. In *Transmission Line Reference Book*, Electric Research Institute, Palo Alto, CA: 329–419.

DiFrancesco, D., and D. Noble (1985). A model of cardiac electrical activity incorporating ionic pumps and concentration changes. *Phil. Trans. R. Soc. London,* B: 307: 353–398.

DiMarco, A.F., M.D. Altose, A. Cropp, and D. Durand (1987). Activation of the inspiratory intercostal muscles by electrical stimulation of the spinal cord. *Am. Rev. Respir. Dis.* 136: 1385–1390.

DiMarco, A.F., G.S. Supinski, J.A. Petro, and Y. Takaoka (1994). Evaluation of intercostal pacing to provide artificial respiration in quadriplegics. *Amer. J. Respir. Crit. Care Med.* 150: 934–940.

Dixon, G.F. (1983). The evaluation and management of electrical injuries. *Crit. Care Med.* 11: 384–387.

Dominguez, G., and H.A. Fozzard (1970). Influence of extracellular K+ concentration on cable properties and excitability of sheep cardiac Purkinje fibers. *Circ. Res.* 26: 565–574.

Donly, K.J., and A.J. Nowak (1988). Oral electrical burns: Etiology, manifestations, and treatment. *Gen. Dent.* 36(2): 103–107.

Dorfman, Y.G. (1971). Physical phenomena occurring in live objects under the effect of constant magnetic fields. In Y.A. Kholodov (ed.), *Influence of Magnetic Field on Biological Objects*, National Technical Information Service, Springfield, VA, Report JPRS63038: 11–19.

Dowling, J.E., and B.B. Boycott (1966). Organization of the primate retina: Electron miscroscopy. *Proc. Royal Soc. London*, Series B, 166: 80–111.

Dowse, C.M., and C.E. Iredell (1920). The effective resistance of the human body to high frequency currents. *Arch. Radiol. Electrother.* 25(2): 34–46.

Drouhard, J.P., and F.A. Roberge (1982a). A simulation study of the ventricular myocardial action potential. *IEEE Trans. Biomed. Eng.* BME-29(7): 494–502.

Drouhard, J.P., and F.A. Roberge (1982b). The simulation of repolarization events of the cardiac Purkinje fiber action potential. *IEEE Trans. Biomed. Eng.* BME-29(7): 481–493.

DuBois, E.F., and D. DuBois (1916). Formula to estimate approximate surface area if height and weight be known. *Arch. Intern. Med.* 17: 863.

Dudel, J. (1989). Transmission of excitation from cell to cell, Chap. 3 (pp 41–60) in R.F. Schmidt and G. Therw (eds.), *Human Physiology*, Springer-Verlag, Berlin.

Durney, C.F., C.C. Johnson, and H. Massoudi (1975). Long wavelength analysis of plane wave irradiation of a prolate spheroid model of a man. *IEEE Trans. Microwave Theory*, MTT-23(2): 246–253.

Durney, C.F., H. Massoudi, and M.F. Isadander (1985). *Radiofrequency Radiation Handbook*, Report USAFSAM-TR-85-73, USAF School of Medicine, Brooks Air Force Base, Texas.

Eaton, H.A.C. (1992). The electric field induced in a spherical volume conductor from arbitrary coils: Application to magnetic stimulation and MEG. *Med. Biol. Eng. Comp.* 30: 433–440.

Ebihara, L., and E.A. Johnson (1980). Fast sodium current in cardiac muscle. *Biophys. J.* 32: 779–790.

Edelberg, R. (1971). Electrical properties of the skin. In R.R. Elden (ed.), *Biophysical Properties of the Skin*, Wiley-Interscience, New York: 513–550.

Edwards, R.H.T., D.K. Hill, D.A. Jones, and P.A. Merton (1977). Fatigue of long duration in human skeletal muscle after exercise. *J. Physiol.* 272: 769–778.

Edwards, R.R.T. (1981). Human muscle function and fatigue. In R. Porter and J. Whelan (eds.), *Human Muscle Fatigue: Physiological Mechanisms*, Pitman Medical, London: 1–18.

Eifler, W.J., and R. Plonsey (1975). A cellular model for the simulation of activation in the ventricular myocardium. *J. Electrocardiol.* 8(2): 117–128.

El-Sherif, N., B.J. Scherlag, and R. Lazzara (1975). Electrode catheter recordings during malignant ventricular arrhythmias following experimental acute myocardial ischemia. *Circulation,* 51: 1003–1014.

Elden, R.R. (ed.) (1971). *Biophysical Properties of the Skin*, Wiley-Interscience, New York.

Elharrar, V., P.R. Forster, T.L. Jirak, W.E. Gaum, and D.P. Zipes (1977). Alterations in canine myocardial excitability during ischemia. *Circ. Res.* 40: 98–105.

Engelmann, T.W. (1875). Leitung und Erregung im Herzmuskel, *Pflugers Arch.* 11: 465–480.

Epstein, B.R., and K.R. Foster (1983). Anisotropy in the dielectric properties of skeletal muscle. *Med. Biol. Eng. Comp.* 21: 51–55.

Evans, B.A., W.J. Litchy, and J.R. Daube (1988). The utility of magnetic stimulation for routine peripheral nerve conduction studies. *Muscle & Nerve*, 11: 1074–1078.

Fang, Z.P., and J.T. Mortimer (1987). A method for attaining natural recruitment order in artificially activated muscles. *Proc. 9th IEEE-EMBS Conf.*: 657–658.

FDA (1992). FDA safety parameter action levels. In R.L. Magin, R.L. Liburdy, and B. Persson (eds.), *Biological Effects and Safety Aspects of Nuclear Magnetic Resonance Imaging*, New York Academy of Sciences, New York: 399–400.

FDA (1995). MRI guidance update. Draft statement, R.A. Phillips (Chief), Computed Imaging Devices Branch, ODE/CDRH, Rockville, MD (Nov. 11, 1995).

Ferris, L.P., B.G. King, P.W. Spence, and H.B. Williams (1936). Effect of electric shock on the heart. *AIEE Trans.* 55: 498–515.

Fisun, O.I. (1993). 2D plasmon excitation and nonthermal effects of microwaves on biological membranes. *Bioelectromagnetics* 14: 57–66.

Fitts, R.H. (1994). Cellular mechanisms of muscle fatigue. *Physiol. Rev.* 74(1): 49–94.

Fitzhugh, R. (1962). Computation of impulse initiation and saltatory conduction in a myelinated nerve fiber. *Biophys. J.* 2: 11–21.

Fitzhugh, R. (1966). Theoretical effects of temperature on threshold in the Hodgkin–Huxley nerve model. *J. Gen. Physiol.* 49: 989–1005.

Flottropp, O. (1953). Effect of different types of electrodes in electrophonic hearing. *J. Acoust. Soc. Amer.*, 25: 236–245.

Forbes, T.W., and A.L. Bernstein (1935). The standardization of sixty-cycle electric shock for practical use in psychological experimentation. *J. Gen. Psychol.* 12/13: 436–441.

Foster, K.R., and R.P. Schwan (1996). Dielectric properties of tissues. In C. Polk and E. Postow (eds.), *CRC Handbook of Biological Effects of Electromagnetic Fields*, CRC Press, Boca Raton, FL: 25–102.

Fozzard, H.A. (1979). Conduction of the action potential. In R.M. Berne, N. Sperelakis, and S. Geiger (eds.), *The Cardiovascular System. Handbook of Physiology*, vol. I, sec. 2, American Physiological Society, Bethesda, MD: 335–356.

Fozzard, R.A., and M.F. Arnsdorf (1986). Cardiac electrophysiology. In R.A. Fozzard, E. Haber, R.B. Jenings, A.M. Katz, and R.E. Morgan (eds.), *The Heart and Cardiovascular System*, Raven, New York: 1–30.

Fozzard, R.A., and M. Schoenberg (1972). Strength–duration curves in cardiac Purkinje fibers: Effects of liminal length and charge distribution. *J. Physiol.* 226: 593–618.

Frankel, R.B., and R.P. Liburdy (1996). Biological effects of static magnetic fields. In C. Polk and E. Postow (eds.), *Biological Effects of Electromagnetic Fields*, CRC Press, Boca Raton, FL.

Frankenhaeuser, B., and A.F. Huxley (1952). A quantitative description of membrane currents and its application to conduction and excitation in nerve. *J. Physiol.* 117: 500–544.

Frankenhaeuser, B., and A.F. Huxley (1964). The action potential in the myelinated nerve fiber of Xenopus laevis as computed on the basis of voltage clamp data. *J. Physiol.* 171: 302–315.

Freiberger, R. (1934). *Der elektrische Widerstand des menschlichen Körpers gegen technischen Gleich-und Wechselstrom*, Springer-Verlag, Berlin. Translation TR 79-45. The electrical resistance of the human body to commercial direct and alternating currents. Translated from German by Allen Translation Service, Maplewood, NJ, Item no. 9005. Published by Bell Laboratories, Maplewood, NJ.

Frey, A.H. (ed.) (1994). *On the Nature of Electromagnetic Field Interactions*. R.G. Landes Co., Austin, TX.

Friedli, W.G., and M. Meyer (1984). Strength–duration curve: A measure for assessing sensory deficit in peripheral neuropathy. *J. Neurol. Psychiat.* 47: 184–189.

Furman, S., B. Parker, and D.J.W. Escher (1967). Endocardial electrical threshold of human cardiac response as a function of electrode surface area. *Digest 7th Int. Conf. Med. and Biol. Eng.*, Stockholm, Sweden Aug. 14–19, p. 71.

Furman, S., J. Garvey, and P. Hurzeler (1975). Pulse duration variation and electrode size as factors in pacemaker longevity. *J. Thorac. Cardiovas. Surg.* 69(3): 382–389.

Furnary, A.P., J.C. Chachques, L.F. Moreira, G.L. Grunkemeier, J.S. Swanson, N. Stolf, S. Haydar, C. Acar, A. Starr, A.D. Jatene, and A.F. Carpentier (1996). Long-term outcome, survival analysis, and risk stratification of dynamic cardiomyoplasty. *J. Thorac. Cardiovasc. Surg.* 112(6): 1640–1650.

Gaffey, C.T., and T.S. Tenforde (1981). Alterations in the rat electrocardiogram induced by a stationary magnetic field. *Bioelectromagnetics* 2: 357–370.

Galvani, L. (1791). *Commentary on the Effect of Electricity on Muscular Motion*, translated by R.M. Green. Waverly, Baltimore, MD: 1953.

Gandhi, O.P. (1988). Advances in dosimetry of radiofrequency radiation and their past and projected impact on safety standards. *Proc. Instrumentation and Technology Conf.*, San Diego, California, April 20–22.

Gandhi, O.P. (ed.) (1990). *Biological Effects and Medical Applications of Electromagnetic Energy*. Prentice-Hall, Englewood Cliffs, NJ.

Gandhi, O.P., and J. Chen (1992). Numerical dosimetry at power line frequencies using anatomically based models. *Bioelectromagnetics*, Supplement 1: 43–60.

Gandhi, O.P., J.F. DeFord, and R. Kanai (1984). Impedance method for calculation of power deposition patterns in magnetically induced hyperthermia. *IEEE Trans. Biomed. Eng.* BME-31(1O): 644–651.

Garrey, W.E. (1914). The nature of fibrillary contractions. Its relation to tissue mass and form. *Am. J. Physiol.* 33: 397–402.

Gauger, J.R. (1985). Household appliance magnetic field survey. *IEEE Trans. Pwr. Apparat. Sys.* PAS-104(9): 2436–2444.

Gaylor, D.C. (1989). Physical mechanism of cellular injury in electrical injury. In R.C. Lee (ed.), *Electric Trauma: Biophysical Mechanisms of Tissue Injury and Clinical Concepts*, Chicago, IL, July 13–14.

Gaylor, D.C., K. Prakah-Asante, and R.C. Lee (1988). Significance of cell size and tissue structure in electrical trauma. *J. Theor. Biol.* 133: 223–237.

Gaylor, D.C., D.L. Bhatt, and R.C. Lee (1992). Skeletal muscle cell membrane electrical breakdown in electrical trauma. In R.C. Lee, E.G. Gravalho, and J.F. Burke (eds.) (1992) *Electrical Trauma*. Cambridge University Press, Cambridge/ New York: 401–425.

Geddes, L.A. (1972). *Electrodes and the Measurement of Bioelectric Events*. Wiley-Interscience, New York.

Geddes, L.A. (1985). The conditions necessary for the electrical induction of ventricular fibrillation. In J.E. Bridges, G.L. Ford, I.A. Sherman, and M. Vainberg (eds.), *Electrical Shock Safety Criteria*, Pergamon, New York: 45–59.

Geddes, L.A. (1987). Optimal stimulus duration for extracranial cortical stimulation. *Neurosurgery* 20: 94–99.

Geddes, L.A., and H. Antoni (panel chairmen) (1985). Panel meeting on physiology of electrical shocks. In J.E. Bridges, G.L. Ford, I.A. Sherman, and M. Vainberg (eds.), *Electrical Shock Safety Criteria*, Pergamon, New York: 89–112.

Geddes, L.A., and J.D. Bourland (1985). Tissue stimulation: Theoretical considerations and practical applications. *Med. Biol. Eng. Comp.* 23: 131–137.

Geddes, L.A., and L.F. Baker (1967). The specific resistance of biological material—A compendium for the biomedical engineer and physiologist. *Med. Biol. Eng.* 5: 271–293.

Geddes, L.A., and L.E. Baker (1971). Response to passage of electric current through the body. *J. Assoc. Adv. Med. Instrum.* 5(1): 13–18.

Geddes, L.A., and L.E. Baker (1975). *Principles of Applied Biomedical Instrumentation*, 2nd ed. John Wiley & Sons, New York.

Geddes, L.A., L.E. Baker, A.G. Moore, and T.W. Coulter (1969). Hazards in the use of low frequencies for the measurement of physiological events by impedance. *Med. Biol. Eng.* 7: 289–296.

Geddes, L.A., L.E. Baker, P. Cabler, and D. Brittain (1971). Response to passage of sinusoidal current through the body. In R.N. Wolfson and A. Sances (eds.), *The Nervous System and Electric Currents*, Vol. II, Plenum, New York.

Geddes, L.A., N. Morehouse, and A. Surawiez (1972). Effect of premature depolarization on the duration of action potentials in Purkinje and ventricular fibers of moderator band of the pig heart. *Circ. Res.* 30: 55–66.

Geddes, L.A., P. Cabler, A.G. Moore, J. Rosborough, and W.A. Tacker (1973). Threshold 60-Hz current required for ventricular fibrillation in subjects of various body weights. *IEEE Trans. Biomed. Eng.* BME-20: 465–468.

Geddes, L.A., W.A. Tacker, and P. Cabler (1975). A new hazard associated with the electrocautery. *Med. Instr.* 9: 112–113.

Geddes, L.A., M.J. Niebauer, C.F. Babbs, and J.D. Bourland (1985a). Fundamental criteria underlying the efficacy and safety of defibrillating current waveforms. *Med. Biol. Eng. Comp.* 23: 122–130.

Geddes, L.A., W.D. Voorhees, C.F. Babbs, and J.A. DeFord (1985b). Electro-ventilation. *Amer. J. Emerg. Med.* 3: 337–339.

Geddes, L.A., J.D. Bourland, and G. Ford (1986). The mechanism underlying sudden death from electric shock. *Med. Instrum.* 20(6): 303–315.

Geddes, L.A., W.D. Voorhees, R. Lagler, C. Riscili, K. Foster, and J.D. Bourland (1988). Electrically produced artificial ventilation. *Med. Instrum.* 22(5): 263–271.

Geddes, L.A., W.D. Voorhees, J.D. Bourland, and C.E. Riscili (1990). Optimum stimulus frequency for contracting the inspiratory muscles with chest-surface electrodes to produce artificial respiration. *Annals Biomed. Eng.* 18: 103–108.

Geddes, L.A., G. Mouchawar, J.D. Bourland, and J. Nyenhuis (1991). Inspiration produced by bilateral electromagnetic cervical phrenic nerve stimulation in man. *IEEE Trans. Biomed. Engin.* 38(9): 1047–1048.

Geldard, F.A. (1972). *The Human Senses.* John Wiley & Sons, New York.

George, M.S., E.M. Wassermann, W.A. Williams, A. Callahan, T.A. Ketter, P. Basser, M. Hallett, and R.M. Post (1995). Daily repetitive transcranial magnetic stimulation (rTMS) improves mood in depression. *Neuroreport* (Rapid Science Publishers), 6(14): 1853–1856.

Gerst, P.H., W.H. Fleming, and J.R. Malm (1996). Increased susceptibility of the heart to ventricular fibrillation during metabolic acidosis. *Circulation Res.* 19: 63–70.

Gescheider, G.A., and J.H. Wright (1968). Effects of sensor adaptation on the form of the psychophysical magnitude function for cutaneous vibration. *J. Exp. Psychol.* 77: 308–313.

Gibson, R.R. (1968). Electrical stimulation of pain and touch. In D.R. Kenshalo (ed.), *The Skin Senses*, Charles C. Thomas, Springfield, IL: 223–261.

Gilbert, T.C. (1939). What is a safe voltage? *The Electrical Review* 119(3062): 145–146.

Girvin, J.P., L.E. Marks, J.L. Antunes, D.O. Quest, M.D. O'Keefe, P. Ning, and W.H. Dobelle (1982). Electrocutaneous stimulation—I. The effects of stimulus parameters on absolute threshold. *Perception & Psychophys.* 32(6): 524–528.

Glenn, W.W.L., and M.L. Phelps (1985). Diaphragm pacing by electrical stimulation of the phrenic nerve. *Neurosurgery* 17: 974–984.

Glenn, W.W.L., et al. (1986). Twenty years of experience in phrenic nerve stimulation to pace the diaphragm. *PACE* 9: 780–784.

Glenn, W.W.L., et al. (1988). Fundamental considerations in pacing of the diaphragm for chronic respiratory insufficiency: A multi-center study. *PACE* 11: 2121–2127.

Goellner, D., B. Zackheim, and M. Bockleman (1993). Safety of high speed guided ground transportation systems: Review of existing EMF guidelines, standards, and regulations. U.S. Dept. of Transportation, Washington, DC, Report DOT/ FRA/ORD-93/27.

Goff, G.D., B.S. Rosner, T. Detre, and D. Kennard (1965). Vibration perception in normal man and medical patients. *J. Neurol. Neurosurg. Psychiatr.* 18: 503–509.

Gorman, P.H., and J.T. Mortimer (1983). The effect of stimulus parameters on the recruitment characteristics of direct nerve stimulation. *IEEE Trans. Biomed. Eng.* BME-30(7): 407–414.

Gracely, R.H., R. Dubner, and P.A. McGrath (1979). Narcotic analgesia: Fentanyl reduces the intensity but not the unpleasantness of painful tooth pulp sensations. *Science* 203(3): 1261–1263.

Graham, C., M.R. Cook, H.D. Cohen, and M.M. Gerkovich (1994). Dose response study of human exposure to 60 Hz electric and magnetic fields. *Bioelectromagnetics* 15: 447–463.

Grandjean, P.A., and J.T. Mortimer (1986). Recruitment properties of monopolar and bipolar epimysial electrodes. *Annals Biomed. Eng.* 14: 53–66.

Grandjean, P.A., R. Leinders, and I. Bourgeois (1991). Implantable stimulation systems for systolic and diastolic biomechanical cardiac assistance. *Sem. Thorac. Cardiovasc. Surg.* 3(2): 119–123.

Grandjean, P., M. Acker, R. Madoff, N.S. Williams, J. Woloszko, and C. Kantor (1996). Dynamic myoplasty: Surgical transfer and stimulation of skeletal muscle for functional substitution or enhancement. *J. Rehab. Res. Develop.* 33(2): 133–144.

Grandori, F., and P. Ravazzani (1991). Magnetic stimulation of the motor cortex—theoretical considerations. *IEEE Trans. Biomed. Eng.* 38(2): 180–191.

Graupe, D. (1989). EMG pattern analysis for patient-responsive control of FES in paraplegics for walker-supported walking. *IEEE Trans. Biomed. Eng.* 36(7): 711–719.

Gray, H. (1985). *Gray's Anatomy*, 30th ed. C.D. Clemente (ed.), Lea & Febiger, Philadelphia.

Gray, R. (1989). *Gray's Anatomy*, 37th ed., P.L. Williams, R. Warwick, M. Dyson, and L.R. Bannister (eds.), Churchill Livingstone, New York.

Grayson, A.C. (1931). Treat low voltages with respect. *National Safety News* 23: 32, 34, 64.

Green, B.G. (1977). The effect of skin temperature on vibrotactile sensitivity. *Perception & Psychophys.* 21: 243–248.

Green, D.M. (1960). Psychoacoustics and detection theory. *J. Acoust. Soc. Am.* 32: 1189–1203.

Green, D.M., and J.A. Swets (1966). *Signal Detection Theory and Psychophysics.* John Wiley & Sons, New York.

Green, R.L., E.B. Rafferty, and I.C. Gregory (1972). Ventricular fibrillation threshold of healthy dogs to 50 Hz current in relationship to earth leakage currents of electromedical equipment. *Biomed. Eng.* 7: 408–414.

Green, R.L., J. Ross, and P. Kurn (1985). Danger levels of short (1 msec. to 15 msec.) electrical shocks from 50-Hz supply. In J.F. Bridges, G.L. Ford, I.A. Sherman, and M. Vainberg (eds.), *Electrical Shock Safety Criteria*, Pergamon, New York: 259–272.

Greenberg, A.W. (1940). Experimental radiological observations on the action of electrical current upon the respiratory and circulatory organs. I. Respiratory organs. *J. Ind. Hyg. Toxicol.* 22: 104–110.

Grimnes, S. (1983a). Dielectric breakdown of human skin *in vivo. Med. Biol. Eng. Comp.* 21: 379–381.

Grimnes, S. (1983b). Electrovibration, cutaneous sensation of microampere currents. *Acta Physiol. Scand.* 118(1): 19–25.

Grissom, C.B. (1995). Magnetic field effects in biology: A survey of possible mechanisms with emphasis on radical-pair mechanisms. *Chem. Rev.* 95: 3–24.

Guinard, J.P., R. Chiolero, E. Buchser, A. Delaloye-Bischof, M. Payot, A. Grbic, S. Krupp, and J. Freeman (1987). Myocardial injury after electrical burns: Short and long term study. *Scand. J. Plast. Reconstr. Surg. Hand Surg.* 21(3): 301–302.

Gumbel, E.J. (1958). *Statistics of Extremes.* Columbia University Press, New York.

Gustafson, R.J., T.M. Brennan, and R.D. Appleman (1985). Behavioral studies of dairy cow sensitivity to AC and DC electric currents. *Trans. ASAE,* 28(5): 1680–1685.

Gustafson, R.J., Z. Sun, and T.D. Brennan (1988). Dairy cow sensitivity to short duration electrical currents. ASAE Paper No. 88-3522, Amer. Soc. Ag. Eng, St. Joseph, MI.

Gybels, J., R.O. Handwerker, and J.N. Hees (1979). A comparison between the discharges of human nociceptive nerve fibers and the subject's ratings of his sensations. *J. Physiol.* 292: 193–206.

Haberal, M. (1986). Electrical burns: A five-year experience—1985 Evans lecture. *J. Trauma* 26(2): 103–109.

Hahn, J.F. (1958). Cutaneous vibratory thresholds for square-wave electrical pulses. *Science* 127: 879–880.

Hallgren, R. (1973). Inductive Neural Stimulator. *IEEE Trans. Biomed. Eng.* BME-20(6): 470–472.

Hammond, E., and T.D. Robson (1955). Comparison of electrical properties of various cements and concretes. *The Engineer (London),* 199(5156): 78–80 and 199(5166): 114–115.

Hammond, J.S., and C.G. Ward (1988). High-voltage electrical injuries: Management and outcome of 60 cases. *South. Med. J.* 81(11): 1351–1352.

Han, J. (1969). Ventricular vulnerability during acute coronary occlusion. *Am. J. Cardiol.* 24: 857–864.

Han, J., and G.K. Moe (1964). Nonuniform recovery of excitability in ventricular muscle. *Circulation Res.* 14: 44–60.

Han, J., G. Dejalon, and G.K. Moe (1966). Fibrillation threshold of premature ventricular responses. *Circulation Res.* 18: 18–25.

Handa, Y., N. Hoshimiya, Y. Iguchi, and T. Oda (1989). Development of percutaneous intramuscular electrode for multichannel FES system. *IEEE Trans. Biomed. Eng.* 36(7): 705–710.

Hardy, J.D., R.G. Wolff, and R. Goodbell (1952). *Pain Sensations and Reactions.* Rafner, New York.

Hare, P.R. (1968). Detection threshold for electric shock in psychopaths. *J. Abnormal Psychol.* 73(3): 268–272.

Harkins, S.W., and C.R. Chapman (1976). Detection and decision factors in pain perception in young and elderly men. *Pain* 2: 253–264.

Harkins, S.W., and C.R. Chapman (1977). The perception of induced dental pain in young and elderly women. *J. Gerontol.* 32: 428–435.

Harkness, R.D. (1971). Mechanical properties of skin in relation to its biological function and its chemical components. In R.R. Elden (ed.), *Biophysical Properties of the Skin*, Wiley-Interscience, New York: 393–436.

Harris, R. (1971). Chronaxy. In S. Licht (ed.), *Electrodiagnosis and Electromyography*, 3rd ed., E. Licht, New Haven, CT, chap. 9.

Hart, F.X. (1992). Numerical and analytical methods to determine the current density distributions produced in human and rat models by electric and magnetic fields. *Bioelectromagnetics*, Supplement 1: 27–42.

Hart, W.F. (1985). A five-part resistor-capacitor network for measurement of voltage and current levels related to electric shock and burns. In J.E. Bridges, G.L. Ford, I.A. Sherman, and M. Vainberg (eds.), *Electrical Shock Safety Criteria*, Pergamon, New York: 183–192.

Hauf, G., K. Haap, M. Lay, and R. Antoni (1977). Beziehungen zwischen der Richtung des Stromdurchgangs und der Flimmerschwelle bei perfundierten Tierherzen. In R. Rauf (ed.), *Beiträge Zur Ersten Hilfe*, Forschungsstelle fur Elektropathologie, Freiburg i. Br.: 146–163.

Hauf, R. (1986). *Beiträge zur Ersten Hilfe und Behandlung von Unfällen durch elektrischien Strom.* Proceedings of Conference on Electropathology, Sept. 11–13, 1986, Forschüngsstelle für Elektropathologie, Freiburg, Germany.

Havel, W.J., J.A. Nyenhuis, J.D. Bourland, K.S. Foster, L.A. Geddes, G.P. Graber, M.S. Waniger, and D.J. Schaefer (1997). Comparison of rectangular and damped sinusoidal dB/dt waveforms in magnetic stimulation. *IEEE Trans. Magnetics*, 33(5): 4269–4271.

Hawkes, G.R., and J.S. Warm (1960). The sensory range of electrical stimulation of the skin. *Am. J. Psychol.* 73: 485–487.

Hawkes, G.R. (1962). Effect of skin temperature on absolute threshold for electrical current. *J. Appl. Physiol.* 17: 110–112.

Heckmann, J.R. (1972). Excitability curve: A new technique for assessing human peripheral nerve excitability *in vivo. Neurology* 22: 224–230.

Heft, M. (1982). Conjoint measurement analysis of verbal category judgments for electrocutaneous stimulation. Ph.D. dissertation, American University, Washington, DC.

Heinz, M., and F. Lippay (1928). Uber die beziehungen zwischen der unterscheidsempfindlichkeit und der zahl der erregten sinneselemente: I. *Pflugers Arch. ges. Physiol. Menschien Tiere* 218: 437–447.

Henneman, E., and L.M. Mendell (1981). Functional organization of motoneuron pool and its inputs. In V.B. Brooks (ed.), *Handbook of Physiology, Section 1: The Nervous System. Vol. II. Motor Control, Part II*, American Physiological Society, Bethesda, MD: 423–507.

Henneman, E., G. Somjen, and D.O. Carpenter (1965a). Functional significance of cell size in spinal motoneurons. *J. Neurophys.* 28: 581–598.

Henneman, E., G. Somjen, and D.O. Carpenter (1965b). Excitability and inhibitability of motoneurons of different sizes. *J. Neurophys.* 28: 599–620.

Henriquez, C.S. (1993). Simulating the electrical behavior of cardiac tissue using the bidomain model. *Crit. Rev. Biomed. Eng.* 21(1): 1–77.

Henriquez, C.S., and R. Plonsey (1987). Effect of resistive discontinuities on waveshape and velocity in single cardiac fiber. *Med. Biol. Eng. Comput.* 25: 428–438.

Henriques, F.C. (1947). Studies of thermal injury v. the predictability and the significance of thermally induced rate processes leading to irreversible epidermal injury. *Arch. Pathol.* 43: 489–502.

Henriques, F.C., and A.R. Moritz (1947). Studies of thermal injury I. The conduction of heat to and through skin and the temperatures attained therein. A theoretical and experimental investigation. *Am. J. Pathol.* 23: 531–549.

Hensel, R. (1973). Cutaneous thermoreceptors. In A. Iggo (ed.), *Handbook of Sensory Physiology: Somatosensory System,* vol. 2, Springer-Verlag, New York: 79–110.

Henshaw, D., A.N. Ross, A.P. Fews, and A.W. Preece (1996). Enhanced deposition of radon daughter nuclei in the vicinity of power frequency electromagnetic fields. *Int. J. Radiat. Biol.* 69(1): 25–38.

Héroux, P. (1992). Thermal damage: Mechanisms, patterns, and detection in electrical burns. In R.C. Lee, E.G. Gravalho, and J.F. Burke (eds.) (1992). *Electrical Trauma,* Cambridge University Press, Cambridge/New York: 189–215.

Hess, C.W., K.R. Mills, N.M.F. Murray, and T.N. Schriefer (1987). Magnetic brain stimulation: Central motor conduction studies in multiple sclerosis. *Annals Neurol.* 22: 744–752.

Higashiyama, A., and G.B. Rollman (1991). Perceived locus and intensity of electrocutaneous stimulation. *IEEE Trans. Biomed. Eng.* 38(7): 679–686.

Higgins, J.D., B. Tursky, and G.E. Schwartz (1971). Shock-elicited pain and its reduction by concurrent tactile stimulation. *Science* 172: 866–867.

Hill, A.V., R.S. Fullerton, B. Katz, and D.Y. Solandt (1937). Nerve excitation by alternating current. *Proc. R. Soc. London, Ser. B* 121: 74–132.

Hille, B. (1984). *Ionic Channels in Excitable Membranes,* Sinauer Associates, Sunderland, M.A. Hodgkin, A.L., and A.F. Huxley (1952). A quantitative description of membrane current and its application to conduction and excitation in nerve. *J. Physiol.* 117: 500–544.

Hodgkin, A.L., and B. Katz (1949). The effect of temperature on the electrical activity of the giant axon of the squid. *J. Physiol* (London). 109: 240–249.

Hoffer, J.A., R.B. Stein, M. Haugland, T. Sinkjaer, W.K. Durfee, A.B. Schwartz, G.E. Loeb, and C. Kantor (1996). Neural signals for command control and feedback in functional neuromuscular stimulation. *Rehab. Res. Develop.* 33(2): 145–157.

Hoffman, B.F., and P.F. Cranefield (1960). *Electrophysiology of the Heart.* McGraw-Hill, New York.

Hoffmeister, B., W. Jänig, and S.J. Lisney (1991). A proposed relationship between circumference and conduction velocity of unmyelinated axons from normal and regenerated cat hindlimb cutaneous nerves. *Neuroscience* 42(2): 603–601.

Hohnloser, S., S. Weirich, and R. Antoni (1982). Influence of direct current on the electrical activity of the heart and on its susceptibility to ventricular fibrillation. *Basic Res. Cardiol.* 77: 237–249.

Holle, J., M. Frey, R. Gruber, R. Kern, R. Stohr, and R. Thoma (1984). Functional electrostimulation of paraplegics: Experimental investigations and first clinical experience with an implantable stimulation device. *Orthop.* 7: 1145–1155.

Hooker, D.R., W.B. Kouwenhoven, and O.R. Langworthy (1932). The effect of alternating currents on the heart. *Am. J. Physiol.* 103: 444–454.

Hooshmand, H., F. Radfar, and E. Beckner (1989). The neurophysiological aspects of electrical injuries. *Clin. Electroencephalogr.* 20(2): 111–120.

Hoque, M., and O.P. Gandhi (1988). Temperature distributions in the human leg for VLF-VHF exposures at the ANSI-recommended safety levels. *IEEE Trans. Biomed. Eng.* 35(6): 442–449.

Horowitz, L.N., J.F. Spear, M.E. Josephson, J.A. Kastor, and E.N. Moore (1979). The effects of coronary artery disease on the ventricular fibrillation threshold in man. *Circulation Res.* 60: 792–797.

Hoshimiya, N., A. Naito, M. Yajima, and Y. Handa (1989). A multichannel FES system for the restoration of motor functions in high spinal cord injury patients: A respiration-controlled system for multijoint upper extremity. *IEEE Trans. Biomed. Eng.* 36(7): 754–760.

Hosono, A., T. Andoh, T. Goto, T. Kawakami, F. Okumura, K. Takayama, T. Takenaka, S. Ueno, M. Yamaguchi, and I. Yamamoto (1992). Effective combination of stimulating coils for magnetic heart stimulation. *Jpn. J. Appl. Phys.* 3(Pt. 1, no. 11): 3759–3762.

Howatson, A.M. (1965). *An Introduction to Gas Discharges.* Pergamon, Oxford.

HSRI (1977). Anthropometry of infants, children, and youths to age 18 for product safety design. Final Report, May 31, 1977, prepared for the U.S. Consumer Product Safety Commission by the Highway Safety Research Institute, University of Michigan Contract CPSC-C-75-0068.

Hufeland, C.W. (1783). Usum uis electriciae in asphyxia experimentis illustratum. In: *Dissertatio Inauguralis Medica.* Gottingen, Germany.

Hunt, J.L. (1979). The use of technetiun-99m stannous pyrophosphate scintigraphy to identify muscle damage in acute electric burns. *J. Trauma.* 19(6): 409–413.

Hunt, J.L., R.M. Sato, and C.R. Baxter (1980). Acute electric burns. Current diagnostic and therapeutic approaches to management. *Arch. Surg.* 115: 434–438.

Huxley, R.E., and J. Hanson (1954). Changes in the cross-atriation of muscle during contraction and stretch and their structural interpretation. *Nature* 173: 973–976.

Hylten-Cavallius, N. (1975). Certain ecological effects of high voltage power lines. Institut de recherche de l'Hydro-Quebec (IREQ), IREQ-1160.

IAEI News Bulletin (Anon.) (1940). Oregon's first death from an electric fence. Vol. 12, p. 70.

ICNIRP (1994). Guidelines on limits of exposure to static magnetic fields. *Health Physics* 66(1): 100–106.

ICNIRP (1998). Guidelines limiting exposure to time-varying electric, magnetic, and electromagnetic fields (up to 300 Ghz). *Health Physics* 74(3): 494–522.

Ichioka, S., M. Iwasaka, M. Shibata, K. Harii, A. Kamiya, and S. Veno (1998). Biological effects of static magnetic fields on the micoirculatory blood flow in vivo: a preliminary report. *Med. Biol. Eng. Comput.* 36: 91–95.

IEC (1982). Report of Technical Committee 64: Electrical installation of buildings. International Electrotechnical Commission, Geneva, Switzerland.

IEC (1984). Effects of current passing through the human body, Part 1: General Aspects. Publication 479-1, International Electrotechnical Commission, Geneva, Switzerland.

IEC (1985). Meeting Minutes TC74/WGS, Everett, Washington, April 2, 3, 4, 1985. International Electrotechnical Commission, Geneva, Switzerland, July 16, 1985.

IEC (1986). Safety of information technology equipment including electrical business equipment. Publication 950 (including Amendment No. 1, Nov., 1988).

IEC (1987). Effects of current passing through the human body, Part 2: Special Aspects. Publication 479-2, International Electrotechnical Commission, Geneva, Switzerland.

IEC (1990a). Safety requirements for electrical equipment for measurement, control, and laboratory use. Publication 1010-1, International Electrotechnical Commission, Geneva, Switzerland.

IEC (1990b). Methods of measurement of touch current and protective conductor current. Report of TC74/WGS, Publication 990, International Electrotechnical Commission, Geneva, Switzerland.

IEC (1993). *Diagnostic Imaging Equipment.* IEC/TC 62B(Secretariat)145, International Electrotechnical Commission, Geneva, Switzerland.

IEC (1995). Medical Electrical Equipment—Part 2: Particular Requirements for the Safety of Magnetic Resonance Equipment for Medical Diagnosis. International Electrotechnical Commission Publication 601-2-33, Geneva, Switzerland.

IEEE (1982). *IEEE Recommended Practice for Grounding of Industrial and Commercial Power Systems.* Std. 142-1982. Institute of Electrical and Electronics Engineers, New York.

IEEE (1986). Guide for safety in ac substation grounding. ANSI/IEEE Std 80-1986, Institute of Electrical and Electronics Engineers, New York.

IEEE (1990). National Electrical Safety Code. ANSI C2-1990, Institute of Electrical and Electronics Engineers, New York.

IEEE (1992). *IEEE Standard for Safety Levels with Respect to Human Exposure to Radio Frequency Electromagnetic Fields, 3 kHz to 300 GHz.* Document IEEE C95.1-1991, published by Institute of Electrical and Electronics Engineers, New York.

Inancsi, W., and T.L. Guidotti (1987). Occupation-related burns: Five-year experience of an urban burn center. *J. Occup. Med.* 29(9): 730–733.

Irnich, W. (1973). Physikalische Überlegungen zur Elektrostimulation (Physical consideration on electrostimulation). *Biomed. Technik.* 18: 97–104.

Irnich, W. (1980). The chronaxie time and its practical importance. *PACE*, 3: 292–301.

IRPA (1988). Guidelines on limits of exposure to radiofrequency electromagnetic fields in the frequency range from 100 kHz to 300 Ghz. Health Physics, 54(1): 115–123.

IRPA (1990). Interim guidelines an limits of exposure to 50/60 Hz electric and magnetic fields. *Health Physics* 58(1): 113–122.

IRPA (1991). Protection of the patient undergoing a magnetic resonance examination. *Health Physics* 61(6): 923–928.

Irwin, D., S. Rush, R. Everling, E. Lepeschkin, D.B. Montgomery, and R.J. Weggel (1970). Stimulation of cardiac muscle by a time-varying magnetic field. *IEEE Trans. Magnet.* MAG-6(2): 321–322.

ITT (1979). *Reference Data for Radio Engineers.* Howard W. Sams, New York.

Jack, J.B., D. Noble, and R.W. Tsien (1983). *Electric Current Flow in Excitable Cells.* Oxford University Press (Clarendon). London/New York.

Jackson, J.D. (1962). *Classical Electrodynamics*, John Wiley & Sons, New York.

Jackson, T.A., and B.F. Riess (1934). Electric shock with different size electrodes. *J. Gen. Psychol.* 45: 262–266.

Jacobsen, J., S. Buntenkutter, and R.J. Reinhard (1975). Experimentelle untersuchungen an schweinen zur frage der mortalität durch sinusformige, phasengeschnittene sowie gleichgerichtete elektrische ströme. *Biomed. Technik* 20: 99–107.

Jaeger, R.J., G.M. Yarkony, and R.M. Smith (1989). Standing the spinal cord injured patient by electrical stimulation: Refinement of a protocol for clinical use. *IEEE Trans. Biomed. Eng.* 36(7): 720–728.

Jalife, J., and G.K. Moe (1976). Effect of electrotonic potentials on pacemaker activity of canine Purkinje fibers in relation to parasystole. *Circ. Res.* 39: 801–808.

Jalife, J., and G.K. Moe (1981). Excitation, conduction, and reflection of impulses in isolated bovine and canine cardiac Purkinje fibers. *Circ. Res.* 49: 233–247.

Janse, M.J., F.J.L. Van Capelle, R. Morsink, A.G. Kleber, F. Wilms-Schopman, R. Cardinal, C. Naumann D'Alnoncourt, and D. Durrer (1980). Flow of "injury" current and patterns of excitation during early ventricular arrhythmias in acute regional myocardial ischemia in isolated porcine and canine hearts. *Circ. Res.* 47: 151–165.

Janzekovic, Z. (1975). The burn wound from the surgical point of view. *J. Trauma* 15: 42.

Jeheson, P., D. Duboc, T. Lavergne, L. Guize, F. Guerin, M. Degeorges, and A. Syrota (1988). Change in human cardiac rhythm induced by a 2-T static magnetic field. *Radiology* 166: 227–230.

Jex-Blake, A.J. (1913). *Br. Med. J.* 1: 425.

Johna, R. (1989). Elektrophysiologische Eigenschaften gekoppelter Herzmuskelzellen in Zellkultur. Dissertation, Universität Freiburg i. Br., Germany.

Johnston, F.E., and R.M. Malina (1966). Age changes in the composition of the upper arm in Philadelphia children. *Human Biol.* 38: 1–21.

Jones, M., and L.A. Geddes (1977). Strength–duration curves for cardiac pacemaking and ventricular fibrillation. *Cardiovasc. Res. Bull.* 15(4): 101–112.

Jones, R.O., J.C. Wright, W.T. Jones, and A. Berger (1983). A case study of high voltage electrical injury. *J. Am. Podiatry Assoc.* 73: 638–642.

Joyner, R.W., F.R. Ramon, and W. Moore (1975). Simulation of action potential propagation in an inhomogeneous sheet of electrically coupled excitable cells. *Circ. Res.* 36: 654–661.

Kaczmarek, K.A., J.G. Webster, and R.G. Radwin (1992). Maximal dynamic range electrotactile stimulation waveforms. *IEEE Trans. Biomed. Eng.* 39(7): 701–715.

Kalmijn, A.J. (1990). Transduction of nanovolt signals: Limits of electric-field detection. *Bioelectromagnetics Soc. Newsl.* Jan/Feb. issue, no. 92: 1–7.

Kandel, E.R., and J.B. Schwartz (1981). *Principles of Neural Science.* Elsevier/North-Rolland, New York.

Kandel, R., J.H. Schwartz, and T.M. Jessel (1991). *Principles of Neural Science*, 3rd edition. Elsevier, New York.

Kanitkar, S., and A.R. Roberts (1988). Paraplegia in an electrical burn: A case report. *Burns Incl. Therm. Ini.* 14(1): 49–50.

Kantrowitz, A. (1990). Autologous muscle to assist the failing heart: First experiments. *J. Heart Transp.* 9: 146–150.

Kaplan, E.B. (1984). *Kaplan's Functional and Surgical Anatomy of the Hand*, 3rd ed. Morton Spinner, Philadelphia.

Katims, J.J., E.R. Naviasky, M.S. Randell, K.Y. Lorenz, and M.L. Bleecker (1987). Constant current sinewave transcutaneous nerve stimulation for the evaluation of peripheral neuropathy. *Archives Phys. Med. Rehab.* 68: 210–213.

Katims, J.J., D.N. Taylor, and S.A. Wesely (1991). Sensory perception in uremic patients. *ASAIO Transactions* 37(3): M370–M372.

Kato, M., S. Ohta, T. Kobayashi, and G. Matsumoto (1986). Response of sensory receptors of the cat's hindlimb to a transient, step function DC electric field. *Bioelectromagnetics* 7: 395–404.

Kato, M., S. Ohta, K. Shimizu, Y. Ysuchida, and G. Matsumoto (1989). Detection-threshold of 50-Hz electric fields by human subjects. *Bioelectromagnetics* 10: 319–327.

Katz, B. (1939). Nerve excitation by high-frequency alternating current. *J. Physiol.* 96: 202–224.

Katz, B. (1966). *Nerve, Muscle, and Synapse.* McGraw-Hill, New York.

Katz, B., and R. Miledi (1967). The study of synaptic transmission in the absence of nerve impulses. *J. Physiol. (London)* 192: 407–436.

Kaune, W.T. (1981). Power frequency electric fields averaged over the body surfaces of grounded humans and animals. *Bioelectromagnetics* 2: 403–406.

Kaune, W.T., and W.C. Forsythe (1985). Current densities measured in human models exposed to 60-Hz electric fields. *Bioelectromagnetics* 6(1): 13–32.

Kaune, W.T., and M.F. Gills (1981). General properties of the interaction between animals and ELF electric fields. *Bioelectromagnetics* 2: 1–11.

Kaune, W.T., and R.D. Phillips (1980). Comparison of grounded humans, swine and rats to vertical, 60-Hz electric fields. *Bioelectromagnetics* 1: 117–129.

Kaune, W.T., R.G. Stevens, N.J. Callahan, R.K. Steverson, and D.B. Thomas (1987). Residential magnetic and electric fields. *Bioelectromagnetics* 8: 315–335.

Keith, M.W., K.L. Kilgore, P.H. Peckham, K.S. Wuolle, G. Creasey, and M. Lemay (1996). Tendon transfers and functional electrical stimulation for restoration of hand function in spinal cord injury. *J. Hand Surg. Am.* 21(1): 89–99.

Keltner, J.R., M.S. Roos, P.R. Brakeman, and T.F. Budinger (1990). Magnetohydrodynamics of blood flow. *Magnetic Resonance in Medicine* 16: 139–149.

Kenshalo, D.R. (1979). Aging effects on cutaneous and kinesthetic sensibilities. In S.S. Han and D.R. Coon (eds.), *Special Senses in Aging: A Current Biological Assessment,* Institute of Gerontology, University of Michigan, Ann Arbor, MI: 189–217.

Kiang, N.Y.S. (1965). *Discharge patterns of single fibers in the cat's auditory nerve.* M.I.T. Press, Cambridge, MA.

Kieback, D. (1988). International comparison of electrical accident statistics. *J. Occup. Accidents* 10: 95–106.

Kieffer, S.A., and E.R. Heitzman (eds.) (1979). An Atlas of Cross-Sectional Anatomy: *Computed Tomography, Ultrasound, Radiography, Gross Anatomy.* Harper & Row, New York.

Kinnunen, E., M. Ojala, H. Taskinen, and E. Matikainen (1988). Peripheral nerve injury and Raynaud's syndrome following electric shock. *Scand. J. Work Environ. Health* 14(5): 332–333.

Kirk, R.E. (1982). *Experimental design: Procedures for the behavioral sciences,* 2nd ed. Brooks/Cole, Belmont, CA.

Kirschvink, J.L. (1989). Magnetic biomineralization and geomagnetic sensitivity in higher animals. *Bioelectromagnetics* 10: 239–259.

Kirschvink, J.L. (1996). Microwave absorption by magnetite: A possible mechanism for coupling nonthermal levels of radiation to biological systems. *Bioelectromagnetics* 17: 187–194.

Kiselev, A.P. (1963). Threshold values of safe current at commercial frequency (in Russian). *Vopr. Elektoborud, Elekt-snabzh, i Elekt. Izmerenii Sob. MITT* 17: 47–58. CEGB Information Services, Translation no. 1167.

Klauenberg, F.J., M. Gradolfo, and D.N. Erwin (eds.) (1995). *Radiofrequency Radiation Standards,* Plenum, New York, 1995.

Kloss, D.A., and E.L. Carstensen (1982). Effects of ELF electric fields on isolated frog heart. *IEEE Trans. Biomed. Eng.* BME-30(6): 347–348.

Klump, D., and M. Zimmerman (1980). Irreversible differential block of A- and C-fibers following local nerve heating in the cat. *J. Physiol.* (London) 298: 471–482.

Knickerbocker, G.G. (1973). Fibrillating parameters of direct and alternating (20-Hz) currents separately and in combination—an experimental study. *IEEE Trans. Comm.* COM-21(9): 1015–1027.

Knighton, R.W. (1975a). An electrically evoked slow potential of the frog's retina: I. Properties of response. *J. Neurophysiol.* 38: 185–197.

Knighton, R.W. (1975b). An electrically evoked slow potential of the frog's retina: II Identification with PII component of Electroretinogram. *J. Neurophysiol.* 38: 198–209.

Knisley, S.B., W.M. Smith, and R.E. Idecker (1992). Effect of intrastimulus polarity reversal on electric field stimulation thresholds in frog and rabbit myocardium. *J. Cardiovascular Electrophysiology* 3(3): 239–254.

Kolin, A. (1945). An alternating field induction flow meter of high sensitivity. *Rev. Sci. Inst.* 16: 109–116.

Kolin, A. (1952). Improved apparatus and techniques for electromagnetic determination of blood flow. *Rev. Sci. Inst.* 23: 235–242.

Koniarek, J.P. (1989). Mechanical and electrical effects of high-frequency and high-intensity stimulation of muscle. *Bioelectromagnetics* 10: 335–345.

Koning, G., R. Schneider, and A.J. Hoelen (1975). Amplitude–duration relation for direct ventricular defibrillation with rectangular current pulses. *Med. Biol. Eng.* May: 388–395.

Kopeliowitch, J. (1946). The physiological effects of an alternating current and the danger of shock in ac electrical installations. *Assoc. Eng. Architects Palestine J.* (Tel Aviv, Palestine) 7(6): 2–8.

Kouwenhoven, W.B. (1949). Effects of electricity on the human body. *Elec. Eng.* 68: 199–203.

Kouwenhoven, W.B. (1956). Effect of capacitor discharges on the heart. *AIEE Trans. Pwr Apparat. Sys.* 75, part 3, no. 23.

Kouwenhoven, W.B. (1964). The effects of electricity on the human body. *Bulletin Johns Hopkins Hosp.* 114: 425.

Kouwenhoven, W.B., and O.R. Langworthy (1931). Effects of electric shock—II. *Trans. AIEE* 50: 1165–1171.

Kouwenhoven, W.B., and W.R. Milnor (1958). The effects of high-voltage, low-capacitance electrical discharges in the dog. *IRE Trans. Biomed. Eng.* PGME 11: 41–45.

Kouwenhoven, W.B., D.R. Hooker, and O.R. Langworthy (1932). Heart injury from electric shock. *Trans. AIEE* 51: 242–244.

Kouwenhoven, W.B., D.R. Hooker, and E.L. Lotz (1936). Electric shock effects of frequency. *AIEE Trans.* 55: 384–386.

Kouwenhoven, W.B., G.G. Knickerbocker, R.W. Chestnut, W.R. Milnor, and D.J. Sass (1959). AC shocks of varying parameters affecting the heart. *Trans. AIEE* 73, part III: 163–169.

Kralj, A., T. Bajd, R. Turk, J. Krajnik, and R. Benko (1983). Gait restoration in paraplegic patients: A feasibility demonstration using multichannel surface electrode FES. *J. Rehabil. R & D* 20: 3–20.

Kralj, A., T. Bajd, R. Turk, and R. Benko (1986). Posture switching for prolonging functional electrical stimulation standing in paraplegic patients. *Paraplegia* 24: 221–230.

Kraus, K.H., W.J. Levy, L.D. Gugino, R. Ghaly, V. Amassian, and J. Cadwell (1994). Clinical application of transcranial magnetic stimulation for intraoperative mapping of the motor cortex. In S. Ueno (ed.), *Biomagnetic Stimulation*, Plenum, New York: 59–73.

Krugilikov, L.L., and H. Dertinger (1994). Stochastic resonance as a possible mechanism of amplification of weak electric signals in living cells. *Bioelectromagnetics* 14: 539–547.

Kugelberg, J. (1976). Electrical induction of ventricular fibrillation in the human heart. *Scand. J. Cardiovasc. Surg.* 10: 237–240.

LaCourse, J.R., M.C. Vogt, W.T. Miller, and S.M. Selikowitz (1985). Effect of high-frequency current on nerve and muscle tissue. *IEEE Trans. Biomed. Eng.* 32: 82–86.

LaCourse, J.R., M.C. Vogt, W.T. Miller, and S.M. Selikowitz (1988). Spectral analysis interpretation of electrosurgical generator nerve and muscle stimulation. *IEEE Trans. Biomed. Eng.* 35(7): 505–509.

Lamb, J.F., C.G. Ingram, I.A. Johnston, and R.M. Pitman (1984). *Essentials of Physiology.* Blackwell, Oxford.

LaMotte, R.H., R.E. Torebjörk, C.J. Robinson, and J.G. Thalhammer (1984). Time-intensity profiles of cutaneous pain in normal and hyperalgesic skin: A comparison with C-fiber nociceptor activities in monkey and human. *J. Neurophysiol.* 51(6): 1434–1450.

LaMotte, R.R., and J.N. Campbell (1978). Comparison of warm and nociceptive C-fiber afferents in monkey with human judgments of thermal pain. *J. Neurophysiol.* 41(6): 509–528.

Lane, J.F., and T.J. Zebo (1967). Volume potential fields developed in cats' limbs during the passage of constant current pulses. *Digest 7th Int. Conf. Med. and Biol. Eng.*, Stockholm, Sweden, Aug. 14–19: 207.

Lapicque, L. (1907). Recherches quantitatives sur l'excitation électrique des nerfs traitée comme une polarization. *J. Physiol. Paris* 9: 620–635.

Larkin, W.D., and J.P. Reilly (1984). Strength/duration relationships for electrocutaneous sensitivity: Stimulation by capacitive discharges. *Perception Phychophys.* 36(1): 68–78.

Larkin, W.D., and J.P. Reilly (1986). Electrocutaneous sensitivity: Effect of skin temperature. *Somatosensory Res.* 3(3): 261–271.

Larkin, W.D., J.P. Reilly, and L.B. Kittler (1986). Individual differences in sensitivity to transient electrocutaneous stimulation. *IEEE Trans. Biomed. Eng.* BME-33(5): 494–504.

Lassek, A.M. (1942). The human pyramidal tract. *Jl Comparative Neurology* 76: 217–225.

Lawrence, A.F., and W.R. Adey (1982). Non-linear wave mechanisms in interactions between excitable tissue and electromagnetic fields. *J. Neurol.* 4: 115.

Lazzara, R., N. El-Sherif, and B.J. Scherlag (1975). Disorders of the cellular electrophysiology produced by ischemia of the canine His bundle. *Circ. Res.* 36: 444–454.

Leake, P.A., D.K. Kessler, and M.M. Merzenich (1990). Application and safety of cochlear protheses. In W.F. Agnew and D.B. McCreery (eds.), *Neural Protheses*, Prentice-Hall, Englewood Cliffs, NJ: 253–296.

Lednev, V.V. (1991). Possible mechanisms for the influence of weak magnetic fields on biological systems. *Bioelectromagnetics* 12: 71–75.

Lee, R.C. (1990). Biophysical injury mechanisms in electrical shock victims. *IEEE Int. Conf. Eng. Med. Biol.* 12(4): 1502–1504.

Lee, R.C. (1992). The pathophysiology and clinical management of electrical injury. In R.C. Lee, E.G. Gravalho, and J.F. Burke (eds.) (1992). *Electrical Trauma*, Cambridge University Press, Cambridge/New York: 33–77.

Lee, R.C., and M.S. Kolodney (1987a). Electrical injury mechanisms: Electrical breakdown of cell membranes. *Plast. Reconstr. Surg.* 80(5): 672–679.

Lee, R.C., and M.S. Kolodney (1987b). Electrical injury mechanisms: Dynamics of the thermal response. *Plast. Reconstr. Surg.* 80(5): 663–671.

Lee, R.C., D.C. Gaylor, D. Bhatt, and D.A. Israel (1988). Role of cell membrane rupture in the pathogenesis of electrical trauma. *J. Surg. Res.* 44(6): 709–719.

Lee, R.C., E.G. Gravalho, and J.F. Burke (eds.) (1992). *Electrical Trauma*, Cambridge University Press, Cambridge/New York.

Lee, R.C., M. Capelli-Schellpfeffer, and K.M. Kelley (eds.) (1994a). *Electrical Injury: A Multidisciplinary Approach to Therapy, Prevention, and Rehabilitation.* Annals New York Academy of Sciences, vol. 720, New York.

Lee, R.C., A. Myerov, and C.P. Maloney (1994b). Promising therapy for cell membrane damage. In R.C. Lee, M. Capelli-Schellpfeffer, and K.M. Kelley (eds.) (1994), *Electrical Injury: A Multidisciplinary Approach to Therapy, Prevention, and Rehabilitation.* Annals New York Academy of Sciences, vol. 720, New York: 239–245.

Lee, W.R. (1961). A clinical study of electrical accidents. *B. J. Ind. Med.* 18: 260–269.

Lee, W.R. (1964). Electrophysiology. In *Proc. Int. Symp. on Electrical Accidents*, International Labour Office, Geneva, Switzerland, Chap. 2.

Lee, W.R. (1966). Death from electric shock. *Proc. IEEE* 111(1): 144–148.

Lee, W.R., and S. Zoledziowski (1964). Effects of electric shock on respiration in the rabbit. *B. J. Ind. Med.* 21: 135–144.

Lefcourt, A. (1982). Behavioral responses of dairy cows subjected to controlled voltages, *J. Dairy Sci.* 65(4): 672–674.

Lefcourt, A.M. (ed.) (1991). Effects of electrical voltage/current on farm animals: How to detect and remedy problems. U.S. Dept. of Agriculture, Agriculture Handbook no. 696.

Lefcourt, A.M., and R.M. Akers (1982). Endocrine responses of cows subjected to controlled voltages during milking. *J. Dairy Sci.* 65: 2125–2130.

Levitt, H. (1971). Transformed up–down methods in psychoacoustics. *J. Acoust. Soc. Am.* 49: 467–477.

Levy, W.J. (1987). Clinical experience with motor and cerebellar evoked potential monitoring. *Neurosurgery* 20: 169–182.

Lewis, T.H., R.S. Feil, and W.D. Stroud (1920). Observations upon flutter and fibrillation. *Heart* 7: 191–233.

Leyden, J.G. (1990). Death in the hot seat: A century of electrocutions. *The Washington Post*, Aug. 5, p. D5.

Li, C.L., and A. Bak (1976). Excitability characteristics of the A- and C-fibers in a peripheral nerve. *Exp. Neurol.* 50: 67–79.

Liberson, W.T. (1971). Progressive and alternating currents. In S.R. Licht (ed.), *Electrodiognosis and Electromyography.* E. Licht, New Haven, CT: 272–285.

Libet, B., W.W. Alberts, E.W. Jr. Wright, L. DeLattre, G. Levin, and V. Feinstein (1964). Production of threshold levels of conscious sensation by electrical stimulation of human somatosensory cortex. *J. Neurophysiol.* 27: 546–578.

Liboff, A.R. (1985). Geomagnetic cyclotron resonance in living cells. *J. Biol. Phys.* 13: 99–102.

Liboff, A.R., and W.C. Parkinson (1991). Search for ion–cyclotron resonance in an Na^+-transport system. *Bioelectromagnetics* 12: 77–83.

Liboff, A.R. et al. (1993). Calcium oscillations and the ELF magnetic field interactions. *Abstracts: BEMS 15th annual meeting*, Los Angeles, June 14.

Licht, S.H. (1971). History of electrodiagnosis. In S.H. Licht (ed.), *Electrodiagnosis and Electromyography*. E. Licht, New Haven, CT: 1–23.

Lin, J.C. (1978). *Microwave Auditory Effects and Applications*. Charles C. Thomas, Springfield, IL.

Lin, J.C. (ed.) (1989). *Electromagnetic Interaction with Biological Systems*. Plenum, New York.

Lin, J.C. (1990). Auditory perception of pulsed microwave radiation. In O.P. Gandhi (ed.), *Biological Effects and Medical Application of Electromagnetic Energy*, Prentice-Hall, 1990.

Lindermans, F.W., and J.J. Danier van der Gon (1978). Current thresholds and liminal size in excitation of heart muscle. *Cardiovasc. Res.* 12: 477–485.

Litovitz, T.A., C.J. Montrose, P. Doiniv, K.M. Brown, and M. Barber (1994a). Superimposing spatially coherent electromagnetic noise inhibits field-induced abnormalities in developing chick embryos, *Bioelectromagnetics* 15: 105–113.

Litovitz, T.A., D. Krause, C.J. Montrose, and J.M. Mullins (1994b). "Temporally incoherent magnetic fields mitigate the response of biological systems to temporally coherent magnetic fields", *Bioelectromagnetics* 15: 399–409.

Lochner, J.P.A., and J.F. Burger (1961). Form of the loudness function in the presence of masking noise. *J. Acoust. Soc. Am.* 33: 1705–1707.

Lord, F.M., and M.R. Novick (1968). *Statistical Theories of Mental Test Scores*. Addison-Wesley, Reading, MA.

Lövsund, P., P.A. Öberg, S.A. Nilson, and T. Reuter (1980a). Magnetophosphenes: A quantitative analysis of thresholds. *Med. Biol. Eng. Comput.* 18: 326–334.

Lövsund, P., P.A. Öberg, and S.E. Nilson (1980b). Magneto- and electrosphosphenes: A comparative study. *Med. Biol. Eng. Comput.* 18: 758–764.

Lövsund, P., P.A. Öberg, and S.E. Nilson (1982). ELF magnetic fields in electrosteel and welding industries. *Radio Sci.* 17(5S): 35S–38S.

Luce, E.A., and S.E. Gottlieb (1984). True high-tension electrical injuries. *Ann. Plast. Surg.* 12: 321–326.

Luce, E.A., W.L. Dowden, C.T. Su, and J.E. Hoopes (1978). High tension electrical injury of the upper extremity. *Surg. Gynecol. Obstet.* 147: 38.

Lundborg, G.L. (1988). *Nerve Injury and Repair*. Churchill Livingstone, New York.

Lykken, D.T. (1971). Square-wave analysis of skin impedance. *Psychophysiology* 7(2): 262–275.

Maccabee, P.J., V.E. Amassian, R.Q. Cracco, and J.A. Cadwell (1988a). Analysis of peripheral motor stimulation in humans using the magnetic coil. *EEG Clin. Neurophysiol.* 70: 524–533.

Maccabee, P.J., V.E. Amassian, R.Q. Cracco, J.B. Cracco, and B.J. Anziska (1988b). Intracranial stimulation of facial nerve in humans with the magnetic coil. *EEG Clin. Neurophysiol.* 70: 350–354.

Magovern, G.J., S.B. Park, R.L. Kao, I.Y. Christlieb, and G.J. Magovern, Jr. (1990). Dynamic cardiomyoplasty in patients. *J. Heart Transp.* 9: 258–263.

Malina, R.M. (1975). *Growth* and *Development: The First Twenty Years in Man.* Burgess, Minneapolis, MN.

Mansfield, P., and P.R. Harvey (1993). Limits to neural stimulation in echo-planar imaging. *Magnetic Resonance in Medicine* 29: 746–758.

Marg, E. (1991). Magnetostimulation of vision: Direct noninvasive stimulation of the retina and the brain. *Optometry and Vision Science* 69(6): 427–440.

Marks, L.E. (1974). *Sensory Processes: The New Psychophysics.* Academic Press, New York.

Marsolais, E.B., and R. Kobetic (1987). Functional electrical stimulation for walking in paraplegia. *J. Bone Joint Surg.* 69A: 728–733.

Martin, J.H. (1991). Coding and processing of sensory information. In Kandel, R., J.H. Schwartz, and T.M. Jessell (eds.), *Principles of Neural Science,* 3rd edition. Elsevier, New York: 329–340.

Martin, J.H., and T.M. Jessell (1991). Modality coding in the somatic sensory systems. In R. Kandel, J.H. Schwartz, and T.M. Jessell (eds.), *Principles of Neural Science,* 3rd edition. Elsevier, New York: 341–352.

Martin, J.R. (1985). Receptor physiology and submodality coding in the somatic sensory system. In E.R. Kandel and J.R. Schwartz (eds.), *Principles of Neural Science,* 2nd edition, Elsevier, New York.

Mason, J.L., and N.A.M. Mackay (1976). Pain sensations associated with electrocutaneous stimulation. *IEEE Trans. Biomed. Eng.* BME-23(5): 405–409.

Massello, W., 3d (1988). Lightning deaths. *Med. Leg. Bulletin* 37(1): 1–7.

McAllister, R.E., D. Noble, and R.W. Tsien (1975). Reconstruction of the electrical activity of cardiac Purkinje fibres. *J. Physiol.* 251: 1–59.

McCammon, R.W. (1970). *Human Growth and Development.* Chartes C. Thomas, Springfield, IL.

McCarroll, G.D., and B.A. Rowley (1979). An investigation of the existence of electrically located acupuncture points. *IEEE Trans. Biomed. Eng.* BME-26(3): 177–181.

McConville, J.T., T.D. Churchill, L. Kaleps, C.E. Clauser, and J. Cuzzi (1980). *Anthropometric Relationships of Body and Body Segment Moments of Inertia.* Air Force Aerospace Medical Research Lab., Wright–Patterson Air Force Base, Ohio.

McCreery, D.B., and W.F. Agnew (1990). Mechanisms of stimulation-induced neural damage and their relation to guidelines for safe stimulation. In W.F. Agnew and D.B. McCreery (eds.), *Neural Prostheses: Fundamental Studies,* Prentice-Hall, Englewood Cliffs, NJ: 297–317.

McKinlay, A.F., S.G. Allen, P.J. Dimbylow, C.R. Muirhead, and R.D. Saunders (1993). Restrictions on human exposure to static and time varying electromagnetic fields and radiation. *Documents NRPB* 4(5): 7–64.

McNeal, D.R. (1976). Analysis of a model for excitation of myelinated nerve. *IEEE Trans. Biomed. Eng.* BME-23: 329–337.

McNeal, D.R. (1977). 2000 years of electrical stimulation. In T.F. Hambrecht and J.B. Reswick (eds.), *Functional Electrical Stimulation,* Marcel Dekker, New York: 3–35.

McNeal, D.R., and L.L. Baker (1988). Effects of joint angle, electrodes and wave-form in electrical stimulation of the quadriceps and hamstrings. *Annals Biomed. Eng.* 16: 299–310.

McNeal, D.R., and B.R. Bowman (1985a). Peripheral neuromuscular stimulation. In J.B. Mykleburst, J.F. Cusick, A. Sances, and S.J. Larsons (eds.), *Neural Stimulation*, vol. II. CRC Press, Boca Raton, FL: 95–118.

McNeal, D.R., and B.R. Bowman (1985b). Selective activation of muscles using peripheral nerve electrodes. *Med. Biol. Eng. Comput.* 23: 249–253.

McNeal, D.R., and D.A. Teicher (1977). Effect of electrode placement on threshold and initial site of excitation of a myelinated nerve fiber. In T.F. Hambrecht and J.B. Reswick (eds.), *Functional Electrical Stimulation*. Marcel Dekker, New York: 405–412.

McNeal, D.R., B.R. Bowman, and W.L. Momsen (1973). Peripheral block on motor activity. In M. Gavrilovic and A.B. Wilson (eds.), *Advances in External Control of Human Extremities*, Yugoslav Committee for Electronics and Automation, Belgrade: 575–583.

McNeal, D.R., R. Waters, and J. Reswick (1977). Experience with implanted electrodes. *Neurosurgery* I: 228–229.

McRobbie, D., and M.A. Foster (1984). Thresholds for biological effects of time-varying magnetic fields. *Clin. Phys. Physiol. Measurements* 5(2): 67–78.

McRobbie, D., and M.A. Foster (1985). Cardiac response to pulsed magnetic fields with regard to safety in NMR imaging. *Phys. Med. Biol.* 30(7): 695–702.

Melzak, R., and P.D. Wall (1965). Pain mechanisms: *A* new theory. *Science* 150(3699): 971–979.

Memberg, W.D., P.H. Peckham, and M. Keith (1994). Surgically-implanted intramuscular electrode for an implantable neuromuscular stimulation system. *IEEE Trans. Rehab. Engin.* 2(2): 80–91.

Merton, P.A., and H.B. Morton (1986). A magnetic stimulation for the human motor cortex. *J. Physiol.* 381: 10P.

Meyer, R.A., and J.N. Campbell (1981). Myelinated nociceptive afferents accounts for the hyperalgesla that follows a burn to the hand. *Science* 213: 1527–1529.

Mild, K.H., and M. Sandström (1994). Health aspects of electric and magnetic fields from VDTs. In J.C. Lin (ed.), *Advances in Electromagnetic Fields in Living Systems*, vol. 1, Plenum, New York: 155–183.

Mills, K.R., and N.M.F. Murray (1985). Corticospinal tract conduction time in multiple sclerosis. *Annals Neurol.* 18: 601–610.

Mines, G.R. (1914). On circulating excitations in the heart muscle and their possible relation to tachycardia and fibrillation. *Trans. R. Soc. Can.* 8: 43.

Mogul, D.I., N.V. Thakor, J.R. McCullogh, G.A. Meyers, R.E. Teneick, and D.R. Siniger (1984). Modified Beeler–Reuter model yields improved simulation of myocardial action potentials. *Proc. IEEE Conf, Computers in Cardiology*, Salt Lake City, Utah, Sept. 18–21: 159–162.

Moncrief, J.A., and B.A. Pruitt, Jr. (1971). Hidden damage from electrical injury. *Geriatrics* 26(4): 84–85.

Monster, A.W., and R. Chan (1977). Isometric force produced by motor units of extensor digitorum communis muscle in man. *J. Neurophysiol.* no. 40: 1432–1443.

Montagu, M.F.A. (1960). *A Handbook of Anthropometry*. Charles C. Thomas, Springfield, IL.

Moreira, L.F., E.A. Bocchi, N.S. Stolf, G. Bellotti, and A.D. Jatene (1996). Dynamic cardiomyoplasty in the treatment of dilated cardiomyopathy: Current results and perspectives. *J. Card. Surg.* 11(3): 207–216.

Moriarty, B.J., and J.N. Char (1987). Electrical injury and cataracts—an unusual case. *West Indian Med. J.* 36(2): 114–116.

Moritz, A.R., and F.C. Henrigves (1947). Studies of thermal injury II. The relative importance of time and surface temperature in the causation of cutaneous burns. *Am. J. Pathol.* 23: 695–720.

Mortimer, J.T. (1981). Motor prostheses. In J.M. Brookhart, V.B. Mountcastle, V.B. Brooks, and S.R. Geiger (eds.), *Handbook of Physiology, Section 1: The Nervous System. Vol. II. Motor Control, Part I*, Am. Physiol. Soc., Bethesda, MD.

Mortimer, J.T., W.F. Agnew, K. Horch, P. Citron, G. Creasy, and C. Kantor (1995). Perspectives on new electrode technology for stimulating peripheral nerves with implantable motor prostheses. *IEEE Trans. Rehab. Engin.* 3(2): 145–154.

Morton, D.J. (1944). *Manual of Human Cross Section Anatomy*. Williams & Wilkins, Baltimore.

Morton, R., and K.A. Provins (1960). Finger numbness after acute local exposure to cold. *J. Appl. Physiol.* 15: 149–154.

Moskowitz, R.R., B. Scharf, and J.C. Stevens (1974). *Sensation and Measurement*. D. Reidel, Boston.

Motz, H., and F. Rattay (1986). A study of the application of the Hodgkin–Huxley and the Frankenhaeuser–Huxley model for electrostimulation of the acoustic nerve. *Neuroscience* 18: 699–712.

Mouchawar, G.A., L.A. Geddes, J.D. Bourland, and J.A. Pearce (1989). Ability of the Lapicque and Blair strength–duration curves to fit experimentally obtained data from the dog heart. *IEEE Trans. Biomed. Eng.* 36(9): 971–974.

Mouchawar, G.A., J.D. Bourland, L.A. Geddes, and J.A. Nyenhuis (1991). Magnetic electrophrenic nerve stimulation to produce inspiration. *Annals Biomed. Eng.* 19: 219–221.

Mouchawar, G.A., J.D. Bourland, J.D. Nyenhuis, J.A. Geddes, L.A. Foster, K.S. Jones, and G.P. Graber (1992). Closed chest cardiac stimulation with a pulsed magnetic field. *Med. Biol. Eng. Comput.* 30: 162–168.

Mueller, E.F., R. Loeffel, and S. Mead (1953). Skin impedance in relation to pain threshold testing by electrical means. *J. Appl. Physiol.* 5: 746–752.

Murphy, K.P., Y. Zhao, and M. Kawai (1996). Molecular forces involved in force generation during skeletal muscle contraction. *J. Exp. Biol.* 199(12): 2565–2571.

MWN (1995). Swedish VDT emissions standard goes international. *Microwave News*, March/April: 8–9.

Myklebust, I., A. Sances, M. Chilbert, T. Prieto, and T. Swiontek (1985). Capacitive discharge studies. In J.E. Bridges, G.L. Ford, I.A. Sherman, and M. Vainberg (eds.), *Electrical Shock Safety Criteria*, Pergamon, New York: 71–76.

Nagarajan, S.S., D.M. Durand, and E.N. Warman (1993). Effects of induced electric fields on finite neuronal structures: A simulation study. *IEEE Trans. Biomed. Eng.* 40(11): 1175–1188.

Nahin, P.J. (1987). *Oliver Heaviside, Sage in Solitude*. IEEE Press, New York.

Nannini, N., and K. Horch (1991). Muscle recruitment with intrafascicular electrodes. *IEEE Trans. Biomed. Engin.* 38: 769–776.

Naples, G.G., J.T. Mortimer, A. Scheiner, and J.D. Sweeney (1988). A spiral nerve cuff electrode for peripheral nerve stimulation. *IEEE Trans. Biomed. Eng.* BME-35: 905–916.

Naples, G.G., J.T. Mortimer, and T.G.R. Yuen (1990). Overview of peripheral nerve electrode design and implantation. In W.F. Agnew and D.B. McCreery (eds.), *Neural Prostheses: Fundamental Studies*, Prentice-Hall, Englewood Cliffs, NJ: 107–145.

National Radiological Protection Board (1983). Revised guidance on acceptable limits of exposure during nuclear magnetic resonance clinical imaging. *Brit. J. Radiol.* 56: 974–977.

NAVSEA (1982). Electromagnetic radiation hazards. NAVSEA OP 3565/ NAVAIR 16-1-529/NACELEX 0967-LP-624-6010, vol. I, 5th revision. Published by Naval Sea Systems Command, Washington, DC.

NCRP (1986). *Biological Effects and Exposure Criteria for Radio-frequency Electromagnetic Fields*, Pub. No. 86, National Council on Radiation Protection and Measurements, Washington, D.C.

Netherlands (1997). Radio frequency electromagnetic fields (300 Hz–300 GHz). Health Council of the Netherlands, Radiation Committee, Rijswijk, publication No. 1997/01.

Nethken, R.T., and M.A. Bulot (1967). Threshold of electrical signals on the upper arm. *Proc. IEEE Region III Convention*, Jackson, MI: 81–83.

Neumann, E., A.E. Sowers, and C.A. Jordan (1989). *Electroporation and Electrofusion in Cell Biology*. Plenum, New York.

Newman, A.L. (1984). Self-injurious behavior inhibiting system. *Johns Hopkins APL Tech. Digest* 5(3): 290–295.

NFPA (1990). *National Electrical Code.* ANSI/NFPA 70-1990, National Fire Protection Association.

Nicholson, P.W. (1965). Specific impedance of cerebral white matter. *Exp. Biol.* 13: 386–401.

Niinami, H., K. Greer, H. Koyanagi, and L. Stephenson (1996). Skeletal muscle ventricles: Another alternative for heart failure. *J. Card. Surg.* 11(4): 280–287.

Noble, D. (1962). A modification of the Hodgkin–Huxley equations applicable to Purkinje fibre action and pace-maker potentials. *J. Physiol* 160: 317–353.

Noble, D. (1984). The surprising heart: A review of recent progress in cardiac electrophysiology. *J. Physiol.* 353: 1–50.

Noble, D., and R.B. Stein (1966). The threshold conditions for initiation of action potentials by excitable cells. *J. Physiol.* 187: 129–162.

Noble, D., and S.J. Noble (1984). A model of sino-atrial node electrical activity based on a modification of the DiFrancesco–Noble (1984) equations. *Proc. R. Soc. London* B222: 295–304.

Nochomovitz, M. (1983). Electrical activation of respiration. *EMBS Magazine* 2(3): 27–31.

Nordén, J., and C. Ramel (1992). *Interaction Mechanisms of Low-Level Electromagnetic Fields in Living Systems*, Oxford University Press, Oxford, UK.

Norell, R.J., R.J. Gustafson, R.D. Appleman, and J.B. Overmeire (1983). Behavioral studies of dairy cattle; sensitivity to electrical currents. *Trans. ASAE*, 26: 1506–1511.

Notermans, S.L.H., and M.M.W.A. Tophofif (1975). Sex differences in pain tolerance and pain apperception. In M. Weisenberg (ed.), *Pain: Clinical and Experimental Perspectives*, Mosby, St. Louis, MO: 111–116.

Notermans, S.L.R. (1966). Measurement of the pain threshold determined by electrical stimulation and its clinical application, Part I: Method and factors possibly influencing the pain threshold. *Neurology* 16: 1071–1086.

Notermans, S.L.R. (1967). Measurement of the pain threshold determined by electrical stimulation and its clinical application, Part II: Clinical applications in neurological and neurosurgical patients. *Neurology* 17: 58–73.

NRPB (1991). Board Statement: Principles for the protection of patients and volunteers during clinical magnetic resonance diagnostic procedures. *Documents of the NRPB* 2(1): 1–5.

NRPB (1993). Documents of the NRPB, 4(5), National Radiological Protection Board, Chilton, UK.

Nute, R. (1985). Dynamic aspects of body impedance. In J.E. Bridges, G.L. Ford, I.A. Sherman, and M. Vainberg (eds.), *Electrical Shock Safety Criteria*, Pergamon, New York: 173–181.

Nyenhuis, J.A., J.D. Bourland, G.A. Mouchawar, T.Z. Elabbady, L.A. Geddes, D.J. Schaefer, and M.E. Riehl (1990). Comparison of stimulation effects of longitudinal and transverse MRI gradient coils. *Society for Magnetic Resonance in Med., Proc. 10th Annual Meeting*, San Francisco: 1275.

Nyenhuis, J.A., J.D. Bourland, D.J. Schaefer, K.S. Foster, W.E. Schoelein, G.A. Mouchawar, T.Z. Elabbady, L.A. Geddes, and M.E. Riehl (1992). Measurement of cardiac stimulation thresholds for pulsed z-gradient fields in a 1.5-T magnet. *Society for Magnetic Resonance in Med., Proc. 11th Annual Meeting*, Berlin: 586.

Nyenhuis, J.A., J.D. Bourland, G. Mouchawar, L. Geddes, K. Foster, J. Jones, W. Schoenlein, G. Garber, and T. Elabbady (1994). Magnetic stimulation of the heart and safety issues in magnetic resonance imaging. In S. Ueno (ed.), *Biomagnetic Stimulation*, Plenum, New York: 75–98.

Nyenhuis, J.A., J.D. Bourland, and D.J. Schaefer (1997). Analysis from a stimulation perspective of the field patterns of magnetic resonance imaging gradient coils. *J. Appl. Phys.* 81(8): 4314–4316.

Oester, Y.T., and S.H. Licht (1971). Routine electrodiagnosis. In S.H. Licht (ed.), *Electrodiagnosis and Electromyography*. E. Licht, New Haven, CT: 201–217.

Oh, J.H., V. Badhwar, and R.C. Chiu (1996). Mechanisms of dynamic cardiomyoplasty: Current Concepts. *J. Card. Surg.* 11(3): 194–199.

Olsen, R.W., L.J. Hayes, E.R. Wissler, R. Nikaidoh, and R.C. Eberhart (1985). Influence of hypothermia and circulatory arrest on cerebral temperature distributions. *ASME J. Biomech. Eng.* 107: 354–360.

Omura, Y. (1977). Critical evaluation of the methods of measurement of "tingling threshold" and "pain tolerance" by electrical stimulation. *Acupuncture & Electrotherapy Res. Int. J.* 2: 161–236.

Osepchuk, J. (1989). Panel discussion on standards. In J.C. Lin (ed.), *Electromagnetic Interaction with Biological Systems*, Plenum, New York: 281–289.

Osepchuk, J. (1994). Impact of public concerns about low-level electromagnetic fields (EMF) on interpretation of EMF/radiofrequency (RFR) data base. In B.J. Klauenberg, M. Grandolfo, and D.N. Erwin (eds.), *Radiofrequency Radiation Standards*, Plenum, New York: 415–426.

Osypka, P. (1963). Quantitative investigation of current strength, duration, and routing in ac electrocution accidents involving human beings and animals. *Elektromedizen 8*, Sonderdruck Fachverlag Schiele and Schön, Berlin, Translation by SLA Translations Center, TT 66-1588 & TT-11470.

Paintal, A.S. (1967). A comparison of the nerve impulses of mammalian nonmedullated nerve fibers with those of the smallest diameter medullated fibers. *J. Physiol.* 193: 523–533.

Paintal, A.S. (1973). Conduction in mammalian nerve fibers. In J.E. Desmedt (ed.), *New Developments in Electromyography and Clinical Neurophysiology*, vol. 2, Karger, Basel, Switzerland: 19–41.

Palin, W.E., Jr., A.M. Sadove, J.E. Jones, W.F. Judson, and R.D. Stambaugh (1987). Oral electrical burns in a pediatric population. *J. Oral Med.* 42(1): 17–21, 34.

Panescu, D., K.P. Cohen, and J.G. Webster (1993). The mosaic characteristics of the skin. *IEEE Trans. Biomed. Eng.* 40(5): 434–439.

Panescu, D., J.G. Webster, and R.A. Stratbucker (1994a). A nonlinear electrical–thermal model of the skin. *IEEE Trans. Biomed Eng.* 41(7): 671–680.

Panescu, D., J.U.G. Webster, and R.A. Stratbucker (1994b). A nonlinear finite element model of the electrode–electrolyte–skin system. *IEEE Trans. Biomed. Eng.* 41(7): 681–688.

Papoulis, A. (1965). *Probability, Random Variables, and Stochastic Processes.* McGraw-Hill, New York.

Parry, C.B.W. (1971). Strength–duration curves. In S.H. Licht (ed.), *Electrodiagnosis and Electromyography*, E. Licht, New Haven, CT: 241–271.

Parshley, P.F., J. Kilgore, J.F. Pulito, P.W. Smiley, and S.R. Miller (1985). Aggressive approach to the extremity damaged by electric current. *Am. J. Surg.* 150(1): 78–82.

Pearce, J.A., J.D. Bourland, W. Neilsen, L.A. Geddes, and M. Voelz (1982). Myocardial stimulation with ultrashort duration current pulses. *PACE* 5: 52–58.

Peckham, P.R. (1983). Restoration of upper extremity function. *EMBS Magazine* 2(3): 30–32.

Peckham, P.R. (1987). Functional electrical stimulation: Current status and future prospects of applications to the neuromuscular system in spinal cord injury. *Paraplegia* 25: 279–288.

Peckham, P.H., and D.B. Gray (1996). Single topic issue: Functional neuromuscular stimulation (FNS). *J. Rehab. Res. Develop.* 33(2): ix–xi.

Peleska, B. (1963). Cardiac arrhythmias following condenser discharges and their dependence upon the strength of current and phase of cardiac cycle. *Circ. Res.* 13: 21–32.

Peleska, B. (1965). Cardiac arrhythmias following condenser discharges led through an inductance. *Circ. Res.* 16: 11–18.

Pelzer, D., and W. Trautwein (1987). Currents through ionic channels in multicellular cardiac tissues and single heart cells. *Experientia* 43: 1153–1162.

Pennes, R.R. (1948). Analysis of issue and arterial blood temperatures in the resting human forearm. *J. Appl. Physiol.* 1: 93–122.

Persson, B.R., and F. Stahlberg (1989). *Health and Safety of Clinical NMR Examinations*, CRC Press, Boca Raton, FL.

Peter, J.B., R.J. Barnard, V.R. Edgerton, C.A. Gillespie, and K.E. Stempel (1972). Metabolic profiles of three fiber types of skeletal muscle in guinea pigs and rabbits. *Biochemistry* 11: 2627–2633.

Petersen, R.C. (1991). Radiofrequency/microwave protection guides. *Health Physics* 61(1): 59–67.

Petersen, R.C. (1995). Safety Standards Setting. Course book from Rutgers University short course: *Management of Electromagnetic Energy Hazards*, Las Vegas, Oct. 16–19.

Peterson, D.K., M. Nochomovitz, A.F. DiMarco, and J.T. Mortimer (1986). Intramuscular electrical activation of the phrenic nerve. *IEEE Trans. Biomed. Engin.* 33(3): 342–351.

Peterson, D.K., M.L. Nochomovitz, T.A. Stellato, and J.T. Mortimer (1994a). Long-term intramuscular electrical activation of the phrenic nerve: Safety and reliability. *IEEE Trans. Biomed. Engin.* 41(12): 1115–1126.

Peterson, D.K., M.L. Nochomovitz, T.A. Stellato, and J.T. Mortimer (1994b). Long-term intramuscular electrical activation of the phrenic nerve: Efficacy as a ventilatory prosthesis. *IEEE Trans. Biomed. Engin.* 41(12): 1127–1135.

Pethig, R. (1979). *Dielectric and Electronic Properties of Biological Materials.* John Wiley & Sons Chichester, UK.

Petrofsky, J.S. (1978). Control of the recruitment and firing frequencies of motor units in electrically stimulated muscles in the cat. *Med. Biol. Eng. Comput.* 16: 302–308.

Petrofsky, J.S., and C.A. Phillips (1983). Computer controlled walking in the paralyzed individual. *J. Neurol. Orthoped. Surg.* 4: 153–164.

Pette, D., and G. Vrobova (1985). Invited review: Neural control of phenotypic expression in mammalian muscle fibers. *Muscle and Nerve* 8: 676–689.

Pfeiffer, E.A. (1968). Electrical stimulation of sensory nerves with skin electrodes for research, diagnosis, communication, and behavioral conditioning: A survey. *Med. Biol. Eng.* 6: 637–651.

Plonsey, R. (1969). *Bioelectric Phenomena.* McGraw-Hill, New York.

Plonsey, R., and R.C. Barr (1986). A critique of impedance measurements in cardiac tissue. *Annals Biomed. Eng.* 14: 307–322.

Plonsey, R., and R.C. Barr (1988). *Bioelectricity.* Plenum, New York.

Plonsey, R., and R.C. Barr (1995). Electric field stimulation of excitable tissue. *IEEE Trans. Biomed. Eng.* 42(4): 329–336.

Polk, C. (1986). Introduction. In C. Polk and E. Postow (eds.), *CRC Handbook of Biological Effects of Electromagnetic Fields.* CRC Press, Boca Raton, FL: 1–24.

Polk, C. (1993). Physical mechanisms for biological effects of ELF low-intensity electric and magnetic fields: Thermal noise limit and counterion polarization. In M. Blank (ed.), *Electricity and Magnetism in Biology and Medicine,* San Francisco Press, San Francisco: 543–546.

Polk, C., and E. Postow (eds.) (1996). *Handbook of Biological Effects of Electro-magnetic Fields,* second edition, CRC Press, Boca Raton, FL.

Polson, M.J.R., A.T. Barker, and I.L. Freeston (1982). Stimulation of nerve trunks with time-varying magnetic fields. *Med. Biol. Eng. Comput.* 20: 243–244.

Polson, M.J.R., A.T. Barker, and S. Gardiner (1982). The effect of rapid rise-time magnetic fields on the ECG of the rat. *Clin. Physiol. Measurements* 3: 231–234.

Ponten, B., U. Erikson, S.R. Johansson, and L. Olding (1970). New observations on tissue changes along the pathway of the current in an electrical injury. Case report. *Scand. J. Plast. Reconstr. Surg.* 4(1): 75–82.

Postow, E., and M.L. Swicord (1996). Modulated fields and "window" effects. In C. Polk and E. Postow (eds.), *Handbook of Biological Effects of Electromagnetic Fields,* second edition, CRC Press, Boca Raton, FL: 535–580.

Prieto, T., A. Sances, J. Myklebust, and M. Chilbert (1985). Analysis of cross-body impedance at household voltage levels. In J.E. Bridges, G.L. Ford, I.A. Sherman, and M. Vainberg (eds.), *Electric Shock Safety Criteria,* Pergamon, New York: 151–160.

Procacci, B.G. (1968). A study on the cutaneous pricking pain threshold in normal man. In A. Soulairac, J. Cahn, and J. Charpentier (ed.), *Pain,* Academic Press, London.

Procacci, P., M. Zoppi, M. Maresca, and S. Romano (1974). Studies on the pain threshold in man. In J.J. Bonica (ed.), *Advances in Neurology*, vol. 4, *International Symposium on Pain*, Raven, New York.

Project UHV (1982). *Transmission Line Reference Book, 345 kV and Above*, 2nd ed. The Electric Power Research Institute, Palo Alto, CA.

Provins, K.A., and R. Morton (1960). Tactile discrimination and skin temperature. *J. App. Physiol.* 15: 155–160.

Pruna, S., C. Ionescu-Tirogoviste, E. Popa, and I. Mincu (1989). Measurement of perception threshold to an electrical stimulus using a phase-sensitive technique in normal and diabetic subjects. *Med. Biol. Eng. Compt.* 27: 111–116.

Quiring, D.P. (1944). Surface area determination. In O. Glasser (ed.), *Medical Physics*, vol. 1, Year Book Publishers, Chicago: 1490–1494.

Rafferty, E.B., H.L. Green, and I.C. Gregory (1975a). Disturbances of heart rhythms produced by 50 Hz leakage currents in dogs. *Cardiovasc. Res.* 9: 256–262.

Rafferty, E.B., R.L. Green, and M.R. Yacoub (1975b). Disturbances of heart rhythm by 50 Hz leakage currents in human subjects. *Cardiovasc. Res.* 9: 263–265.

Rail, W. (1977). Core conductor theory and cable properties of neurons. In *Handbook of Physiology: A Critical, Comprehensive Presentation of Physiological Knowledge and Concepts*, vol. 1, American Physiological Society, Bethesda, MD: 39–97.

Ranck, J.B. (1963). Specific impedance of rabbit cerebral cortex. *Exp. Neurol.* 7: 144–152.

Ranck, J.B. (1975). Which elements are excited in electrical stimulation of mammalian central nervous system: A review. *Brain Res.* 98: 417–440.

Rasch, P.J. (1989). *Kinesiology and Applied Anatomy*, 7th ed., Lea & Febiger, Philadelphia.

Rattay, F. (1986). Analysis of models for external stimulation of axons. *IEEE Trans. Biomed. Eng.* 33: 974–977.

Rattay, F. (1988). Modeling the excitation of fibers under surface electrodes. *IEEE Trans. Biomed. Eng.* 35(3): 199–202.

Rattay, F. (1989). Analysis of models for extracellular fiber stimulation. *IEEE Trans. Biomed. Eng.* 36(3): 676–682.

Reilly, J.P. (1978a). Electric field induction on sailboats and vertical poles. *IEEE Trans. Pwr. Apparat. Sys.* PAS-97(4): 1373–1381.

Reilly, J.P. (1978b). Electric and magnetic coupling from high voltage AC power transmission lines—classification of short-term effects on people. *IEEE Trans. Pwr. Apparat. Sys.* PAS-97(6): 2243–2252.

Reilly, J.P. (1979a). Electric field induction on long objects—A methodology for transmission line impact studies. *IEEE Trans. Pwr. Apparat. Sys.* PAS-98(6): 1841–1849.

Reilly, J.P. (1979b). An approach to the realistic-case analysis of electric field induction from AC transmission lines. Third Int. Symp. High Voltage Engineering, Milan, Italy.

Reilly, J.P. (1980). Spark discharge characteristics of vehicles energized by AC electric field. JHU PPSE T-16, The Johns Hopkins University Applied Physics Laboratory, Laurel, MD.

Reilly, J.P. (1982). Characteristics of spark discharges from vehicles energized by AC electric fields. *IEEE Trans. Pwr. Apparat. Sys.* PAS-101(9): 3178–3186.

Reilly, J.P. (1988). Electrical models for neural excitation studies. *Johns Hopkins APL Tech. Digest* 9(1): 44–58.

Reilly, J.P. (1989). Peripheral nerve stimulation by induced electric currents: Exposure to time-varying magnetic fields. *Med. Biol. Eng. Comput.* 27: 101–110.

Reilly, J.P. (1991). Magnetic field excitation of peripheral nerves and the heart: A comparison of thresholds. *Med. Biol. Eng. Comput.* 29(6): 571–579.

Reilly, J.P. (1992). *Electrical Stimulation and Electropathology.* Cambridge University Press, Cambridge, 1992.

Reilly, J.P. (1993). Safety considerations concerning the minimum threshold for magnetic excitation of the heart. *Med. Biol. Eng. Comput.* 31: 651–654.

Reilly, J.P. (1994). Transient current effects in stray voltage exposure: Biophysical principles and mechanisms. Paper No. 943594, Amer. Soc. Ag. Eng. International Meeting, Atlanta.

Reilly, J.P. (1995). Nerve stimulation of cows and other farm animals by time-varying magnetic fields. *Trans. Am. Soc. Ag. Eng.* 38(5): 1487–1494.

Reilly, J.P. (1998). Maximum Pulsed Electromagnetic Field Limits Based on Peripheral Nerve Stimulation: Application to IEEE/ANSI C95.1 Electromagnetic Field Standards. *IEEE Trans. Biomed. Eng.* 45(1): 137–141.

Reilly, J.P., and R.H. Bauer (1987). Application of a neuroelectric model to electrocutaneous sensory sensitivity: Parameter variation study. *IEEE Trans. Biomed. Eng.* BME-34(9): 752–754.

Reilly, J.P., and M. Cwiklewski (1978). A realistic-case analysis of electric field induction on vehicles near AC transmission lines. IEEE Can. Conf. Communications and Power, Montreal.

Reilly, J.P., and M. Cwiklewski (1981). Rain gutters near high-voltage power lines: A study of electric field induction. *IEEE Trans. Pwr. Apparat. Sys.* PAS-100(4): 2068–2076.

Reilly, J.P., and A.M. Diamant (1997). Theoretical evaluation of peripheral nerve stimulation during MRI with an implanted spinal fusion stimulator, *Mag. Res. Imaging* 15(10): 1145–1156.

Reilly, J.P., and W.D. Larkin (1983). Electrocutaneous stimulation with high voltage capacitive discharges. *IEEE Trans. Biomed. Eng.* BME-30: 631–641.

Reilly, J.P., and W.D. Larkin (1984). Understanding electric shock. *Johns Hopkins APL Tech. Digest* 5(3): 296–304.

Reilly, J.P., and W.D. Larkin (1985a). Human reactions to transient electric currents—summary report. PPSE T-34(NTIS No. PB 86-117280/AS), The Johns Hopkins University Applied Physics Laboratory, Laurel, MD.

Reilly, J.P., and W.D. Larkin (1985b). Mechanisms for human sensitivity to transient electric currents. In J.E. Bridges, G.L. Ford, I.A. Sherman, and M. Vainberg (eds.), *Electrical Shock Safety Criteria*, Pergamon, New York: 241–249.

Reilly, J.P., and W.D. Larkin (1987). Human sensitivity to electric shock induced by power frequency electric fields. *IEEE Trans. Electromagnetic Compatibility* EMC-29(3): 221–232.

Reilly, J.P., W. Larkin, R.J. Taylor, and V.T. Freeman (1982). Human reactions to transient electric currents, annual report, July 1981–July 1982. CPE-8203(NTIS No. PB83 204628), The Johns Hopkins University Applied Physics Laboratory, Laurel, MD.

Reilly, J.P., W. Larkin, R.J. Taylor, V.T. Freeman, and L.B. Kittler (1983). Human reactions to transient electric currents, annual report, July 1982–June 1983.

Report CPE-8305(NTIS No. PB84-112895), The Johns Hopkins University Applied Physics Laboratory, Laurel, MD.

Reilly, J.P., W.D. Larkin, L.B. Kittler, and V.T. Freeman (1984). Human reactions to transient electric currents—annual report, July 1983–June 1984. Report CPE-8313(NTIS No. PB84–231463), The Johns Hopkins University Applied Physics Laboratory, Laurel, MD.

Reilly, J.P., V.T. Freeman, and W.D. Larkin (1985). Sensory effects of transient electrical stimulation—evaluation with a neuroelectric model. *IEEE Trans. Biomed. Eng.* BME-32(12): 1001–1011.

Reilly, J.P., H.A. Eaton, and D. Gluck (1993). A novel method for improving focality of magnetic stimulation of the brain. *Abstracts: Bioelectromagnetics Society Annual Meeting*, Los Angeles CA, June 14–17.

Reinemann, D.J., L.E. Stetsen, and N. Laughlin (1994). Effects of frequency and duration on the sensitivity of dairy cows to transient voltages. Paper No. 943597, Amer. Soc. Ag. Eng. International Meeting, Atlanta.

Reinemann, D.J., L.E. Stetson, J.P. Reilly, N.K. Laughlin, S. McGuirk, and S.D. LeMire (1996). Dairy cow sensitivity and aversion to short duration transient currents. Paper 963087, American Society Agricultural Engineers International Meeting, Phoenix, AZ.

Reinemann, D.J., L.W. Stetson, J.P. Reilly, and N.K. Laughlin (1998). Sensitivity of dairy cows to short duration currents. *Trans. Amer. Soc. Ag. Eng.*

Riscili, C.E., K.S. Foster, W.D. Voorhees, J.D. Bourland, and L.A. Geddes (1988). Electroventilation in the baboon. *Amer. J. Emerg. Med.* 6: 561–565.

Robblee, L.S., and T.L. Rose (1990). Electrochemical guidelines for selection of protocols and electrode materials for neural stimulation. In W.F. Agnew and D.B. McCreery (eds.), *Neural Prostheses: Fundamental Studies*, Prentice-Hall, Englewood Cliffs, NJ: 25–66.

Rodgers, S.J. (1981). Radiofrequency burn hazards in the MF/HF band. In J.C. Mitchell (ed.), *Proceeding: Workshop on the protection of personnel against radiofrequency electromagnetic radiation*, Review 3-81, USAF School of Medicine, Brooks AFB, Texas: 76–89.

Rollman, G.B. (1969). Electrocutaneous stimulation: Psychometric functions and temporal integration. *Perception Psychophys.* 5(5): 289–293.

Rollman, G.B. (1974). Electrocutaneous stimulation. In F.A. Geldard (ed.), *Conference on Cutaneous Communication Systems and Devices*, Monterey, CA, 1973, The Psychonomic Society, Austin, TX: 38–51.

Rollman, G.B. (1975). Behavioral assessment of peripheral nerve function. *Neurology* 26: 339–342.

Rollman, G.B., and G. Harris (1987). The detectability and perceived magnitude of painful electrical shock. *Perception Phychophys.* 42(3): 257–268.

Ronner, S.F. (1990). Electrical excitation of CNS neurons. In W.F. Angew and D.B. McCreery (eds.), *Neural Protheses*, Prentice-Hall: 169–196.

Rosenberg, D.B., and M. Nelson (1988). Rehabilitation concerns in electrical burn patients: A review of the literature. *J. Trauma* 28(6): 808–812.

Rosenblueth, A., and J. Garcia Ramos (1947). Studies on flutter and fibrillation. II. The influence of artificial obstacles on experimental, auricular flutter. *Am. Heart J.* 33: 677.

Rösler, K.M., C.W. Hess, R. Reckmann, and H.P. Ludin (1989). Significance of shape and size of the stimulating coil in magnetic stimulation of the human motor cortex. *Neurosci. Lett.* 100: 347–352.

Rosner, B.S. (1961). Neural factors limiting cutaneous spatio-temporal discrimination. In W.A. Rosenblith (ed.), *Sensory Communication*, MIT Press, Cambridge, MA: 725–737.

Rosner, B.S., and W.R. Goff (1967). Electrical response of the nervous system and subjective scales of intensity. In W.D. Neff (ed.), *Contributions to Sensory Physiology*, vol. 2, Academic Press, New York.

Roth, B. (1989). Interpretation of skeletal muscle four-electrode impedance measurements using spatial and temporal frequency-dependent conductivities. *Med. Biol. Eng. Comput.* 27: 491–495.

Roth, B.J. (1995). A mathematical model of make and break electrical stimulation of cardiac tissue by a unipolar anode or cathode. *IEEE Trans. Biomed. Eng.* 42(12): 1174–1184.

Roth, B.J., J.M. Saypol, M. Hallett, and L.G. Cohen (1991). A theoretical calculation of the electric field induced in the cortex during magnetic stimulation. *Electroencephalog. Clin. Neurophysiol.* 81: 47–56.

Rothberger, C.J., and R. Winterberg (1941). Über Vorhofflimmern und Vorhofflattern. *Pflügers Arch.* 160: 42–90.

Rouse, R.G., and A.R. Dimick (1978). The treatment of electrical injury compared to burn injury: A review of pathophysiology and comparison of patient management protocols. *J. Trauma* 18: 43–47.

Roy, O.Z. (1980). Summary of cardiac fibrillation thresholds for 60-Hz currents and voltages applied directly to the heart. *Med. Biol. Eng. Comput.* 18: 657–659.

Roy, O.Z., J.R. Scott, and G.C. Park (1976). 60-Hz ventricular fibrillation and pump failure thresholds versus electrode area. *IEEE Trans. Biomed. Eng.* BME-23(1): 45–48.

Roy, O.Z., G.C. Park, and J.R. Scott (1977). Intracardiac catheter fibrillation thresholds as a function of the duration of 60-Hz current and electrode area. *IEEE Trans. Biomed. Eng.* BME-24(5): 430–435.

Roy, O.Z., A.J. Mortimer, B.J. Trollope, and E.J. Villeneuve (1985). Electrical stimulation of the isolated rabbit heart by short duration transients. In J.E. Bridges, G.L. Ford, I.A. Sherman, and M. Vainberg (eds.), *Electrical Shock Safety Criteria*, Pergamon, New York: 77–86.

Roy, O.Z., J.R. Scott, and B.J. Trollope (1986). 60 Hz ventricular fibrillation thresholds for large-surface-area electrodes. *Med. Biol. Eng. Comput.* 24: 471–474.

Roy, O.Z., B.J. Trollope, and J.R. Scott (1987). Measurement of regional cardiac fibrillation thresholds. *Med. Biol. Eng. Comput.* 25: 165–166.

Rozman, J., B. Sovinec, M. Trlep, and B. Zorko (1993). Multielectrode spiral cuff for ordered and reversed activation of nerve fibres. *J. Biomed. Engin.* 15: 113–120.

Rubenstein, J.T. (1993). Axon termination conditions for electrical stimulation. *IEEE Trans. Biomed. Eng.* 40(7): 654–663.

Ruch, S., J.A. Abildskov, and R. McFee (1963). Resistivity of body tissues at low frequencies. *Circ. Res.* 12: 40–50.

Ruch, T.C. (1979). Somatic sensation: Receptors and their axons. In T.C. Ruch and H.D. Patton (eds.), *Physiology and Biophysics*, W.B. Saunders, Philadelphia: 157–200.

Ruch, T.C., and H.D. Patton (1979). *Physiology and Biophysics*. W.B. Saunders, Philadelphia.

Ruch, T.C., R.D. Patton, J.W. Woodbury, and A.L. Towe (1968). *Neurophysiology.* W.B. Saunders, Philadelphia.

Ruiz, E.V., J.A. Russo, G.V. Savino, and M.E. Valentinuzzi (1985). Ventricular fibrillation threshold in the dog determined with defibillating paddles. *Med. Biol. Eng. Comp.* 23: 281–284.

Rush, S., and D.A. Driscoll (1968). Current distribution in the brain from surface electrodes. *Current Researches* 47(6): 717–723.

Sachs, R.M., J.D. Miller, and K.W. Grant (1980). Perceived magnitude of multiple electrocutaneous pulses. *Perception Psychophys.* 28: 255–262.

Sackeim, H.A., P. Decina, S. Portnoy, P. Neeley, and S. Maliz (1987). Studies of dosage, seizure threshold, and seizure duration in ECT. *Biol. Psychiatry* 22: 249–268.

Sagan, L.A. (1996). *Electric and Magnetic Fields: Invisible Risks?* Gordon & Breach, Australia.

Sagan, P.M., M.E. Stell, G.K. Bryan, and W.R. Adey (1987). Detection of 60-Hertz vertical electric fields by rats. *Bioelectromagnetics* 8: 303–313.

Sances, A., S.J. Larson, J. Myklebust, and J.F. Cusick (1979). Electrical injuries. *Surg. Gynecol. Obstet.* 149(1): 97–108.

Sances, A., J.B. Myklebust, S.J. Larson, J.C. Darin, T. Swiontek, T. Prieto, M. Chilbert, and J.F. Cusick (1981a). Experimental electrical injury studies. *J. Trauma* 21(8): 589–597.

Sances, A., J.B. Myklebust, J.F. Szablya, T.J. Swiontek, S.J. Larson, M. Chilbert, T. Prieto, and J.F. Cusick (1981b). Effects of contacts in high voltage injuries. *IEEE Trans. Pwr. Apparat. Sys.* PAS-100(6): 2987–2992.

Sances, A., J.F. Szyblya, J.D. Morgan, J.B. Myklebust, and S.J. Larson (1981c). High voltage powerline injury studies. *IEEE Trans. Pwr. Apparat. Sys.* PAS-100(2): 552–558.

Sances, A., J.B. Myklebust, J.F. Szablya, T.J. Swiontek, S.J. Larson, and M. Chilbert, et al. (1983). Current pathways in high-voltage injuries. *IEEE Trans. Biomed. Eng.* BME-30(2): 118–124.

Sandove, A.M., J.E. Jones, T.R. Lynch, and P.W. Sheets (1988). Appliance therapy for perioral electrical burns: A conservative approach. *J. Burn Care Rehabil.* 9(4): 391–395.

Sanguinetti, M.C., and N.K. Jurkiewicz (1991). Delayed rectifier outward K+ current is composed of two currents in guinea pig atrial cells. *Amer. J. Physiol., Heart and Circulatory Physiol.* 260: H393–H399.

Sasyniuk, G.I., and C. Mentez (1971). A mechanism for reentry in canine ventricular tissue. *Circ. Res.* 28: 3–15.

Sato, M., and J. Ushiyama (1950). On the relation of strength–frequency curve in excitation by low frequency AC to the minimal gradient of the nerve fiber. *Jpn. J. Physiol.* 1: 141–146.

Saunders, F.A. (1974). Electrocutaneous displays. In F.A. Geldard (ed.), *Conference on Cutaneous Communication Systems and Devices*, Monterey, CA, 1973, Psychonomic Society, Austin, TX: 20–26.

Saunders, R.D. (1991). Limits on patient and volunteer exposure during clinical magnetic resonance diagnostic procedures: Recommendations for the practical implementation of the Board's statement. *Documents of the NRPB* 2(1) 5–29.

Schaefer, D.J. (1992). Dosimetry and effects of MR exposure to RF and switched magnetic fields. In R.L. Magin, R.P. Liburdy, and B. Persson (eds.), *Biological*

Effects and Safety Aspects of Nuclear Resonance Imaging and Spectroscopy, New York Academy of Sciences, New York.

Schaefer, D.J., J.D. Bourland, J.A. Nyenhuis, K.S. Foster, W.F. Wirth, L.A. Geddes, and M.E. Riehl (1994). Determination of gradient-induced human peripheral nerve stimulation thresholds for trapezoidal pulse trains. *Soc. Mag. Res., Proc. 2nd Annual Meeting,* San Francisco: 101.

Schaefer, D.J., J.D. Bourland, J.A. Nuyenhuis, K.S. Foster, P.E. Licato, and L.A. Geddes (1995). Effects of simultaneous gradient combinations on human peripheral nerve stimulation thresholds. *Society for Magnetic Resonance in Med., Proc. 12th Annual Meeting,* Nice, France, Poster No. 1220.

Schenk, J.F., C.L. Dumoulin, C.L. Redington, R.W. Kressel, H.Y. Elliot, and I.L. McDougall (1992). Human exposure to 4.0 Tesla magnetic fields in a whole-body scanner. *Medical Physics* 19(4): 1089–1098.

Scherf, D. (1947). Studies on auricular tachycardia caused by aconitine administration. *Proc. Soc. Exp. Biol. Med.* 64: 233–239.

Schludermann, E., and J.P. Zubeck (1962). Effect of age on pain sensitivity. *Perceptual and Motor Skills.* 14: 295–301.

Schmid, E. (1961). Temporal aspects of cutaneous interaction with two-point electrical stimulation. *J. Exp. Psychol.* 67: 191–192.

Schmidt, R. (ed.) (1978). *Fundamentals of Sensory Physiology.* Springer-Verlag, New York.

Schmidt-Nielsen, K. (1984). *Scaling: Why Is Animal Size So Important?* Cambridge University Press, Cambridge.

Schmitt, F.P. Wielopolski, H. Fischer, and R.R. Edelman (1994). Peripheral stimulations and their relation to gradient pulse shapes. *Proc. Soc. Magnetic Resonance in Medicine, 2nd Annual Meeting,* San Francisco: 102.

Schriefer, T.N., K.R. Mills, N.M. Murray, and C.W. Hess (1988). Evaluation of proximal facial nerve conduction by transcranial magnetic stimulation. *J. Neurol. Neurosurg. Psychiatr.* 51: 60–66.

Schwan, H.P. (1954). Die elektrischen Eigenschaften von muskelgewebe bie Niederfrequenz. *Z. Naturforsche,* 96: 245–251.

Schwan, H. (1968). Electrical impedance of the human body. Technical Report TR-2199, U.S. Naval Weapons Laboratory, Dahlgren, VA, NTIS No. AD 842306.

Schwan, H.P. (1966). Alternating current electrode polarization. *Biophysik.* 3: 181–201.

Scott, J.P., W.P. Lee, and S. Zoledziowski (1973). Ventricular fibrillation thresholds for A.C. shock of long duration in dogs with normal acid-base states. *Br. J. Ind. Med.* 30: 155.

Scott, W.T. (1966). *The Physics of Electricity and Magnetism.* John Wiley & Sons, New York.

Sepulveda, N.G., J.P. Wikswo, and D.S. Echt (1990). Finite element analysis of cardiac defibrillation current distributions. *IEEE Trans. Biomed. Eng.* 37(4): 354–365.

Sharma, M., and A. Smith (1978). Paraplegia as a result of lightning injury. *Br. Med. J.* 12: 1464–1465.

Sharp, G.H., and R.W. Joyner (1980). Stimulated propagation of cardiac action potential. *Biophys. J.* 31: 403–424.

Shellock, F.G., and E. Kanal (1994). *Magnetic Resonance Bioeffects, Safety, and Patient Management.* Raven, New York.

Shields, C.B., H.L. Edmonds, M. Paloheimo, J.R. Johnson, and R.T. Holt (1988). Intraoperative use of transcranial magnetic motor-evoked potentials. *Proc. 10th Ann. Conf IEEE-EM BS*: 926–927.

Shimada, Y., K. Sato, E. Abe, H. Kagaya, K. Ebata, M. Oba, and M. Sato (1996). Clinical experience of functional electrical stimulation in complete paraplegia. *Spinal Cord* 34(10): 615–619.

Silny, J. (1986). The influence of threshold of the time-varying magnetic field in the human organism. In J.H. Bernhardt (ed.), *Biological Effects of Static and Extremely Low Frequency Magnetic Fields*, MMV Medzin Verlag, Munchen, Germany.

Silva, M., N. Hummon, D. Ruttor, and C. Hooper (1989). Power frequency magnetic fields in the home. *IEEE Trans. Pwr. Del.* 4(1): 465–478.

Silverglade, D. (1983). Splinting electrical burns utilizing a fixed splint technique: A report of 48 cases. *ASDC J. Dent. Child.* 50(6): 455–458.

Silversides, J. (1964). The neurological sequelae of electrical injury. *Calif. Med. Assoc. J.* 91: 195–204.

Skoog, T. (1970). Electrical injuries. *J. Trauma.* 10: 816–830.

Skuggevig, W. (1992). Standards and protective measures. Chap. 11 in J.P. Reilly, *Electrical Stimulation and Electropathology*, Cambridge University Press, Cambridge/New York.

Slager, C.J., J.C. Schuurbiers, J.A. Oomen, and N. Bom (1993). Electrical nerve and muscle stimulation by radio frequency surgery: Role of direct current loops around the active electrode. *IEEE Trans. Biomed. Eng.* 40(2): 182–187.

Smith, B.T., M.J. Mulcahey, and R.R. Betz (1996). Development of an upper extremity FES system for individuals with C4 tetraplegia. *IEEE Trans. Rehab. Engin.* 4(4): 264–270.

Smith, L. (1990). Electrocutions involving consumer products. Memorandum of March 30, 1990, U.S. Consumer Product Safety Commission, Washington, DC.

Smoot, A.W. (1985). The seventh meeting of IEC TC74/WG5. Memorandum of April 11, 1985, Underwriters Laboratories.

Smoot, A.W., and J. Stevenson (1968a). Investigation of reaction current. Technical Report of April 1968, Underwriters Laboratories.

Smoot, A.W., and J. Stevenson (1968b). Report on investigation of reaction currents. Technical Report of Dec. 3, 1968, Underwriters Laboratories.

Smyth, P.D., P.P. Tarjan, E. Chernof, and N. Baker (1976). The significance of electrode surface area and stimulating thresholds in permanent cardiac pacing. *J. Thorac. Card. Surg.* 71(4): 559–565.

Snyder, R.G., M.L. Spencer, C.L. Owings, and L.W. Schneider (1975). Anthropometry of US infants and children. SAE Automotive Engineering Congress and Exposition, Paper No. 750423, as Referenced by A.F. Roche and R.M. Malina (1983), *Manual of Physical Status and Performance in Childhood*, Vol. lB: Physical Status, Plenum, New York.

Solem, L., R.P. Fischer, and R.G. Strate (1977). The natural history of electrical injury. *J. Trauma.* 17(7): 487–492.

Solomonow, M., E. Eldred, J. Lyman, and J. Foster (1983). Control of muscle contractile force through indirect high-frequency stimulation. *Am. J. Phys. Med.* 62: 71–82.

Song, W.J., S. Weinbaum, and L.M. Jiji (1988). A combined macro and microvascular model for whole limb heat transfer. *J. Biomed. Eng.* 110(4): 259–268.

Spach, M.S., W.T. Miller, D.B. Geselowitz, R.C. Barr, J.M. Kootsey, and E.A. Johnson (1981). The discontinuous nature of propagation in normal canine cardiac muscle. Evidence for recurrent discontinuities of intracellular resistance that affect the membrane currents. *Circ. Res.* 48: 39–54.

Spach, M.S., W.T. Miller, P.C. Dolber, J.M. Kootsey, J.R. Sommer, and C.E. Mosher (1982). The functional role of structural complexities in the propagation of depolarization in the atrium of the dog. *Circ. Res.* 50: 175–191.

Spiegel, R.J. (1976). Magnetic coupling to a prolate spheroid model of man. *IEEE Trans. Pwr. Apparat. Sys.* PAS-96(1): 208–212.

Starmer, C.F., and Whalen, R.E. (1973). Current density and electrically induced ventricular fibrillation. *Med. Instrum.* 7(1): 3–6.

Stein, R.B. (1980). *Nerve and Muscle.* Plenum, New York.

Steiner, U.E., and T. Ulrich (1989). Magnetic field effects in chemical kinetics and related phenomena. *Chem. Rev.* 89: 51–147.

Stembach, R.A., and B. Tursky (1964). On the psychophysical power function in electric shock. *Psychol. Sci.* 1: 217–218.

Stembach, R.A., and B. Tursky (1965). Ethnic differences among housewives in psychophysical and skin potential responses to electric shock. *Psychophysiology* 1: 241–246.

Sten-Knudsen, O. (1960). Is muscle contraction initiated by internal current flow? *J. Physiol.* 151: 363–384.

Stern, S., and V.G. Laties (1985). 60 Hz electric fields: Detection by female rats. *Bioelectromagnetics* 6: 99–103.

Stern, S., V.G. Laties, C.V. Stancompiano, C. Cox, and J.O. deLorge (1983). Behavioral detection of 60-Hz electric fields by rats. *Bioelectromagnetics.* 4: 215–247.

Stevens, J.C. (1980). Thermo-tactile interactions: Some influences of temperature on touch. In D.R. Kenshalo (ed.), *Sensory Functions of the Skin of Humans*, Plenum, New York: 207–222.

Stevens, J.C., J.D. Mack, and S.S. Stevens (1960). Growth of sensation on seven continua as measured by force of handgrip. *J. Exp. Psychol.* 59: 60–67.

Stevens, J.C., L.E. Marks, and D.C. Simonson (1974). Regional sensitivity and spatial summation in the warmth sense. *Physiol. Behav.* 13: 825–836.

Stevens, J.C., B.G. Green, and Krimsley (1977). Punctuate pressure sensitivity: Effects of skin temperature. *Sensory Processes* 1: 238–243.

Stevens, R.G., B.W. Wilson, and L.F. Anderson (1997). *The Melatonin Hypothesis: Breast Cancer and Use of Electric Power.* Battelle Press, Columbus.

Stevens, S.S. (1959). Cross-modality validations of subjective scales for loudness, vibration, and electric shock. *J. Exp. Psychol.* 57: 201–209.

Stevens, S.S. (1966). Matching functions between loudness and ten other continua. *Perception Psychophys.* 1: 5–8.

Stevens, S.S. (1975). *Psychophysics: Introduction to Its Perceptual Neural, and Social Prospects.* John Wiley & Sons, New York.

Stevens, W.G.S. (1963). The current-voltage relationship in human skin. *Med. Electron. Biol. Eng.* 1: 389–399.

Stevenson, J. (1969). Progress report on investigation of reaction current. Technical Report Nov. 18, 1969, Underwriters Laboratories.

Stevenson, J. (1971). Reaction to leakage currents. *UL Lab. Data* 2(1): 16–19. Underwriters Laboratories.

Stoy, R.D., K.R. Foster, and H.P. Schwan (1982). Dielectric properties of mammalian tissues from 0.1 to 100 MHz: A summary. *Phys. Med. Biol.* 27(4): 501–513.

Strasser, E.J., R.M. David, and M.J. Mehshey (1977). Lightning injuries. *J. Trauma.* 17(4): 315–319.

Struijk, J.J., J. Holsheimer, and H.B. Boom (1993). Excitation of dorsal root fibers in spinal cord stimulation: A theoretical study. *IEEE Trans. Biomed. Eng.* 40(7): 632–653.

Stuchly, M.A., and D.W. Lecuyer (1989). Exposure to electromagnetic fields in arc welding. *Health Phys.* 56(3): 297–302.

Stuchly, M.A., and S.S. Stuckly (1980). Dielectric properties of biological substances-tabulated. *J. Mic. Pwr.* 15(1): 19–26.

Stuchley, M.A., and S.S. Stuchley (1996). Experimental radio and microwave dosimetry. In C. Polk and E. Postow (eds.), *Biological Effects of Electromagnetic Fields*, CRC Press, Boca Raton, FL.

Suchi, T. (1954). Experiments on electrical resistance of the human epidermis. *Jpn. J. Physiol.* 5: 75–80.

Sugimoto, T., S.F. Schaal, and A.G. Wallace (1967). Factors determining vulnerability to ventricular fibrillation induced by 60-cps alternating current. *Circ. Res.* 21: 601–608.

Sunde, E.D. (1968). *Earth Conduction Effects in Transmission Systems*, Macmillan, New York.

Sunderland, S. (1978). *Nerves and Nerves Injuries*, Churchill Livingstone, New York.

Sweeney, J.D. (1993). A theoretical analysis of the "let-go" phenomenon. *IEEE Trans. Biomed. Eng.* 40(12): 1335–1338.

Sweeney, J.D., J.T. Mortimer, and D. Durand (1987). Modeling of mammalian myelinated nerve for functional neuromuscular stimulation. *Proc. 9th Ann. Int. Conf of the IEEE-EMBS*: 1577–1578.

Sweeney, J.D., K. Deng, E. Warman, and J.T. Mortimer (1989). Modeling of electric field effects on the excitability of myelinated motor nerve. *Proc. 11th Ann. Int. Conf of the IEEE-EMBS*: 1281–1282.

Sweeney, J.D., D. Ksienski, and J.T. Mortimer (1990). A nerve cuff technique for selective activation of peripheral nerve trunk regions. *IEEE Trans. Biomed. Eng.* BME-31: 706–715.

Sweeney, J.D., N.R. Crawford, and T.A. Brandon (1995). Neuromuscular stimulation selectivity of multiple-contact nerve cuff electrode arrays. *Med. Biol. Eng. Comput.* 33: 418–425.

Sweeney, J.D., S.F. Cogan, J.T. Mortimer, and K. Horch (1996). Electrodes, leads, and connectors. *J. Rehab. Res. Develop.* 33(2): 194–197.

Szeto, A.Y., and F.A. Saunders (1982). Electrocutaneous stimulation for sensory communication in rehabilitation engineering. *IEEE Trans. Biomed. Eng.* BME-25(4): 300–308.

Takagi, T., and T. Muto (1971). Influence upon human bodies and animals of electrostatic induction caused by 500 kV transmission lines (English transl.), Tokyo Electric Power.

Takemoto-Hambleton, R.M., W.J. Dunseath, and W.T. Joines (1988). Electromagnetic fields induced in a person due to devices radiating in the 10 Hz to 100 kHz range. *IEEE Trans. Elect. Compat.* 30(4): 529–537.

Takeuchi, A., and N. Takeuchi (1962). Electrical changes in pre- and post-synaptic axons of the giant synapse of Loligo, *J. Gen. Physiol.* 45: 1181–1193.

Tanner, J.A. (1962). Reversible blocking of nerve conduction by alternating current excitation. *Nature* 195: 712–713.

Tasaki, I. (1953). *Nervous Transmission.* Charles C. Thomas, Springfield, IL.

Tasaki, I. (1982). *Physiology and Electrochemistry of Nerve Fibers.* Academic Press, New York.

Tasaki, I., and M. Sato (1951). On the relation of the strength–frequency curve in excitation by alternating current to the strength–duration and latent addition curves of the nerve fiber. *J. Gen. Physiol.* 34: 373–388.

Tashiro, T., and A. Higashiyama (1981). The perceptual properties of electrocutaneous stimulation: Sensory quality, subjective intensity, and intensity–duration relation. *Perception Psychophys.* 30(6): 579–586.

Taussig, H. (1968). Death from lightning and the possibility of living again. *Ann. Intern. Med.* 68: 1345–1349.

Taylor, D.N., J.G. Wallace, and J.C. Masdeu (1992). Perception of different frequencies of cranial transduatneous electrical nerve stimulation in normal and HIV-positive individuals. *Perceptual and Motor Skills* 74: 259–264.

Taylor, R.J. (1985). Body impedance for transient high voltage currents. In J.E. Bridges, G.L. Ford, I.A. Sherman, and V. Vainberg (eds.), *Electric Shock Safety Criteria,* Symposium on Electric Shock Safety Criteria, Pergamon: 251–258.

Teghtsoonian, R. (1973). Range effects in psychological scaling and a revision of Stevens' law. *Am. J. Psychol.* 86: 3–27.

Teicher, D.A., and D.R. McNeal (1978). Comparison of a dynamic and steady-state model for determining nerve fiber threshold. *IEEE Trans. Biomed. Eng.* BME-25(1): 105–107.

Teissie, J., and M. Rols (1994). Manipulation of cell cytoskeleton affects the lifetime of cell membrane electropermeabilization. In R.C. Lee, M. Capelli-Schellpfeffer, and K.M. Kelley (eds.). *Electrical Injury: A Multipliciplinary Approach to Therapy, Prevention, and Rehabilitation,* Annals New York Academy of Sciences, vol. 720, New York: 98–110.

Tenforde, T.S. (1989). Electroreception and magnetoreception in simple and complex organisms. *Bioelectromagnetics* 10: 215–221.

Tenforde, T.S. (1993). Cellular and molecular pathways of extremely-low-frequency electromagnetic field interactions with living systems. In M. Blank (ed.), *Electricity and Magnetism in Biology and Medicine,* San Francisco Press, San Francisco: 1–8.

Tenforde, T.S., C.T. Gaffey, B.R. Moyer, and T.F. Budinger (1983). Cardiovascular alterations in Macaca monkeys exposed to stationary magnetic fields: Experimental observations and theoretical analysis. *Bioelectromagnetics* 4: 1–9.

Tessier-Lavigne, M. (1991). Phototransduction and information processing in the retina. In E.R. Kandell, J.H. Schwartz, and T.M. Jessell (eds.), *Principles of Neural Science,* third edition, Elsevier, New York: 400–418.

Thalen, H.J.T., J.V.P. Berg, J.N. Heide, and J. Nieveen (1975). *The Artificial Cardiac Pacemaker.* Van Gorum, Assem, The Netherlands.

Thompson, G. (1933). Shock threshold fixes appliance insulation resistance. *Electrical World* 101: 793–795.

Torebjörk, H.E., and R.G. Hallin (1973). Perceptual changes accompanying controlled preferential blocking of A and C fibre responses in intact human skin nerves. *Exp. Brain Res.* 16: 321–322.

Trautwein, W., and J. Dudel (1954). Actionspotential und Mechanogramm des Warmblüterherzmuskels als Funktion der Schlagfrequenz. *Pflügers Arch. ges. Physiol.* 260: 24–39.

Treagear, R.T. (1966). *Physical Functions of Skin.* Academic Press, London.

Triolo, R., R. Nathan, Y. Handa, M. Keith, R. Betz, S. Carroll, and C. Kantor (1996). Challenges to clinical deployment of upper limb neuroprostheses. *J. Rehab. Res. Develop.* 33(2): 111–122.

Tsuji, S., Y. Mural, and M. Yarita (1988). Somatosensory potentials evoked by magnetic stimulation of lumbar roots, cauda equina, and leg nerves. *Annals Neurol.* 24: 568–573.

Tucker, R.D., and O.H. Schmitt (1978). Tests for human perception of 60 Hz moderate strength magnetic field. *IEEE Trans. Biomed. Eng.* BME-25(6): 509–518.

Tucker, R.D., O.H. Schmitt, C.E. Sievert, and S.E. Silvis (1984). Demodulated low frequency currents from electrosurgical procedures. *Surgery, Gynecology & Obstetrics* 159: 39–43.

Tursky, B., and P.D. Watson (1964). Controlled physical and subjective intensities of electric shock. *Psychophysiology* 1(2): 151–162.

Tyler, D., and D. Durand (1993). Design and acute test of a radially penetrating interfascicular nerve electrode. *Proc. 15th Ann. Int. Conf. IEEE-EMBS* 15: 1247–1248.

Tyler, D., and D. Durand (1994). Interfascicular electrical stimulation for selectively activating axons. *EMBS Magazine* 13(4): 575–583.

Ueno, S. (ed., 1994a). *Biomagnetic Stimulation*, Plenum, New York.

Ueno, S. (1994b). Focal and vectorial magnetic stimulation of the human brain. In S. Ueno (ed.), *Biomagnetic Stimulation*, Plenum, New York: 29–47.

Ueno, S., K. Harada, C. Ji, and Y. Oomura (1984). Magnetic nerve stimulation without interlinkage between nerve and magnetic flux. *IEEE Trans. Magnet.* MAG-20(5): 1660–1662.

Ueno, S., T. Tashiro, and K. Harada (1988). Localized stimulation of neural tissues in the brain by means of a paired configuration of time-varying magnetic fields. *J. Appl. Phys.* 64: 5862–5864.

UL (1945). Measurement of electric shock hazard in radio equipment. *Bulletin of Research No. 33*, Underwriters Laboratories.

UL (1975). Method of development—revision—implementation—standards for safety. Publication F 200-46 3M575, Underwriters Laboratories.

UL (1981). Development of test equipment and methods for measuring potentially lethal and otherwise damaging current levels. Consumer Product Safety Commission, Contract Number CPSC-C-79-1034, Underwriters Laboratories.

UL (1983). Standard for Safety, double insulation systems for use in electrical equipment. UL 1097, second edition, Underwriters Laboratories.

UL (1985). Standard for safety, ground-fault circuit interrupters. UL 943, 2nd ed., Underwriters Laboratories.

UL (1987). Standard for safety, telephone equipment. UL 1459, 2nd ed., December 18, 1987, Underwriters Laboratories.

UL (1988). Electric shock-A safety seminar on theory and prevention. Underwriters Laboratories.

UL (1990). Report to the instrumentation committee of ANSI C101, Preliminary report by A.W. Smoot, J. Stevenson, W. Myrick, and W. Tuthill. Underwriters Laboratories.

Vallbo, A.B., K.A. Olsson, K.G. Westberg, and F.J. Clark (1984). Microstimulation of single tactile afferents from the human hand. *Brain* 107(3): 727–749.

van Boxtel, A. (1977). Skin resistance during square-wave electrical pulses of 1 to 10 mA. *Med. Biol. Eng. Comp.* 15: 679–687.

van den Honert, C., and J.T. Mortimer (1979a). The response of the myelinated nerve fiber to short duration biphasic stimulating currents. *Annals Biomed. Eng.* 7: 117–125.

van den Honert, C., and J.T. Mortimer (1979b). Generation of unidirectionally propagated action potentials in peripheral nerve by brief stimuli. *Science* 206: 1311–1312.

Veltink, P.H., B.K. van Veen, J.J. Struijk, J. Holsheimer, and H.B.K. Boom (1989a). A modeling study of nerve fascicle stimulation. *IEEE Trans. Biomed. Eng.* BME-36(7): 683–692.

Veltink, P.H., J.A. van Alste, and H.B.K. Boom (1988). Influences of stimulation conditions on recruitment of myelinated nerve fibers: A model study. *IEEE Trans. Biomed. Eng.* BME-35: 917–924.

Veltink, P.H., J.J. Hermens, and J.A. van Alste (1989b). Multielectrode intrafascicular and extraneural stimulation. *Med. Biol Eng. Comput.* 27: 19–24.

Veraart, C., W.M. Grill, and J.T. Mortimer (1993). Selective control of muscle activation with a multipolar nerve cuff electrode. *IEEE Trans. Biomed. Eng.* 40: 640–653.

Verrillo, R.T. (1979). Comparison of vibrotactile threshold and suprathreshold responses in men and women. *Perception Psychophys.* 26: 20–24.

Verrillo, R.T. (1982). Effects of aging on the suprathreshold responses to vibration. *Perception Psychophys.* 32: 61–68.

Verillo, R.T., and G.A. Gesheider (1979). Psychophysical measurements of enhancement, suppression, and surface gradient effects in vibrotaction. In D.R. Kenshalo (ed.), *Sensory Functions of the Skin of Humans*, Plenum, New York: 153–181.

Vodovnik, L., T. Bajd, F. Gracanin, A. Kralj, and P. Strojnik (1981). Functional electrical stimulation for control of locomotor systems. *CRC Crit. Rev. Bioeng.* 6: 63–132.

Vodovnik, L., W.J. Crochetiere, and J.B. Reswick (1967). Control of a skeletal joint by electrical stimulation of antagonists. *Med. Biol. Eng.* 5: 97–109.

Volta, A. (1800). On the electricity excited by the mere contact of conducting substances of different kinds. *Phil. Trans. R. Soc. London* 90: 403–431.

Voorhees, C.R., W.D. Voorhees, L.A. Geddes, and J.D. Bourland (1992). The chronaxie for myocardium and moter nerve in the dog with chest-surface electrodes. *IEEE Trans. Biomed. Eng.* 39(6): 624–628.

Voorhes, W.D., K.S. Foster, L.A. Geddes, and C.F. Babbs (1983). Safety factor for transchest pacing. *Proc., 36th ACEMB Conf.*, Columbus, Ohio, Sept. 12–14: 19.

Voorhees, W.D., L.A. Geddes, J.D. Bourland, and G. Mouchawar (1990). Magnetically induced contraction of the inspiratory muscles in dog. *J. Clin. Eng.* 15(5): 407–409.

Wachtel, H. (1979). Firing pattern changes and transmembrane currents produced by extremely low frequency fields in pacemaker neurons. In R.D. Phillips, M.F. Gillis, W.T. Kaune, and D.D. Mahlum (eds.), *Biological Effects of Extremely Low-Frequency Electromagnetic Fields*, Publication CONF-781016, Technical Information Center, U.S. Dept. of Energy: 132–146.

Wallace, J.G., and J.C. Masdeu (1992). Perception of different frequencies of cranial transcutaneous electrical nerve stimulation in normal and HIV-positive individuals. *Perceptual and Motor Skills* 74: 259–264.

Walleczek, J. (1994). Immune cell interactions with extremely low frequency magnetic fields: Experimental verification and free radical mechanisms. Chap. 12 in A.H. Frey (ed.), *On the Nature of Electromagnetic Field Interaction with Biological Systems*, R.G. Landes Co., Austin TX.

Walter, J.S., P. Griffith, J. Sweeney, V. Scarpine, M. Bidnar, J. McLane, and C. Robinson (1997). Multielectrode nerve cuff stimulation of the median nerve produces selective movements in a raccoon animal model. *J. Spinal Cord Med.* 20: 233–243.

Walthard, K.M., and M. Thicaloff (1971). Motor points. In S.H. Licht (ed.), Electrodiagnosis and *Electromyography*, E. Licht, New Haven, CT: 153–170.

Wang, X.W., and W.H. Zoh (1983). Vascular injuries in electrical burns—the pathologic basis for mechanism of injury. *Burns Incl. Therm. Inj.* 9(5): 335–338.

Wang, X.W., C.S. Lu, N.Z. Wang, H.C. Lin, H. Su, J.N. Wei, and W.Z. Zoh (1984), High tension electrical burns of upper arms treated by segmental excision of necrosed humerus. An introduction of a new surgical method. *Burns Incl. Therm. Inj.* 10(4): 271–281.

Wang, X.W., B.B. Roberts, R.L. Zapata-Sirvent, W.A. Robinson, J.P. Waymack, E.J. Law, B.G. MacMillan, and J.W. Davies (1985). Early vascular grafting to prevent upper extremity necrosis after electrical burns. Commentary on indications for surgery. *Burns Incl. Therm. Inj.* 11(5): 359.

Wang, X.W., E.J. Bartle, and B.B. Roberts (1987). Early vascular grafting to prevent upper extremity necrosis after electric burns: Additional commentary on indications for surgery. *J. Burn Care Rehabil.* 8(5): 391–394.

Warren, R.M. (1981). Measurement of sensory intensity. *Behav. Brain Sci.* 4: 175–223.

Waters, R.L., D.R. McNeal, and J. Perry (1975). Experimental correction of footdrop by electrical stimulation of the peroneal nerve. *J. Bone Joint Surg.* 57-A: 1047–1054.

Waters, R.L., D.R. McNeal, W. Faloon, and B. Clifford (1985). Functional electrical stimulation of the peroneal nerve for hemiplegia. *J. Bone Joint Surg.* 67-A: 792–793.

Watson, A.B., J.S. Wright, and J. Loughman (1973). Electrical thresholds for ventricular fibrillation in man. *Med. J. Austral.* 1, June 16: 1179–1182.

Weaver, J.C. (1992). Cell membrane rupture by strong electric fields: Prompt and delayed processes. In Lee, R.C., E.G. Gravalho, and J.F. Burke (eds.) (1992). *Electrical Trauma*, Cambridge University Press, Cambridge/New York: 301–326.

Weaver, J.C., and R.D. Astumian (1990). The response of living cells to very weak electric fields: The thermal noise limit. *Science* 247: 459–462.

Weaver, L., R. Williams, and S. Rush (1976). Current density in bilateral and unilateral ECT. *Biol. Physiciatry* 11(3): 303–311.

Wedensky, N. (1884). Wie Rasch ermudet der Nerv. *Z. Med. Wissen.* 22: 65–68.

Weigel, R.J., R.A. Jaffe, D.L. Lunstorm, W.C. Forsythe, and L.E. Anderson (1987). Stimulation of cutaneous mechanoreceptors by 60-Hz electric field. *Bioelectromagnetics* 8:337–350.

Wéigria, R., G.K. Moe, and C.J. Wiggers (1941). Comparison of the vulnerable periods and fibrillation threshold of normal and idioventricular beats. *Am. J. Physiol.* 132: 651–657.

Weinstein, S. (1963). The relationship of laterality and cutaneous area to breast-sensitivity in sinistrals and dextrals. *Am. J. Psychol.* 76: 475–479.

Weinstein, S. (1968). Intensive and extensive aspects of tactile sensitivity as a function of body part, sex, and laterality. In D.R. Kenshalo (ed.), *The Skin Senses*, Charles C. Thomas, Springfield, IL: 195–218.

Weinstein, S. (1978). New methods for the in-vivo assessment of skin smoothness and skin softness. *J. Soc. Cosmet. Chem.* 29: 99–115.

Weirich, J., S. Hohnloser, and H. Antoni (1983). Factors determining the susceptibility of the isolated guinea pig heart to ventricular fibrillation induced by sinusoidal alternating current at frequencies from 1 to 1,000 Hz. *Basic Res. Cardiol.* 78: 604–615.

Weirich, J., K. Haverkampf, and A. Antoni (1985). Ventricular fibrillation of the heart induced by electric current. *Rev. Gen. de l'Elect.* 11: 833–843.

Weiss, G. (1901). Sur la possibilité de rendre comparables entre eux les appareils servant a l'excitation électrique. *Arch. Ital. Biol.* 35: 413–446.

Werner, G., and V.B. Mountcastle (1968). Quantitative relations between mechanical stimuli to the skin and neural response evoked by them. In D.R. Kenshalo (ed.), *The Skin Senses*, Charles C. Thomas, Springfield, IL: 112–138.

Wessale, J.L., J.D. Bourland, W.A. Tacker, and L.A. Geddes (1980). Bipolar catheter defibrillation in dogs using trapezoidal waveforms of various tilts. *J. Electrocardiol.* 13: 359–366.

Wessale, J.L., L.A. Geddes, G.M. Ayers, and K.S. Foster (1992). Comparison of rectangular and exponential current pulses for evoking sensation. *Annals Biomed. Eng.* 20: 237–244.

Wetherill, G.B. (1963). Sequential estimation of quantal response curves. *J. R. Stat. Soc.* B25: 1–48.

Whitaker, H.B. (1939). Electric shock hazard as it pertains to the electric fence. *Underwriters Laboratories, Bulletin Res.* 14: 3–56.

Whittleson, W.G., M.M. Mullord, R. Kilgour, and L.R. Cate (1975). Electric shocks during machine milking. *New Zealand Vet. Journal* 23: 105–108.

Wiggers, C.J., and R. Wégria (1939). Ventricular fibrillation due to single, localized induction and condenser shocks applied during the vulnerable phase of ventricular systole. *Am. J. Physiol.* 128: 500–505.

Williams, D.O., B.J. Scherlag, R.R. Hope, N. El-Sherif, and R. Lazzara (1974). The pathophysiology of malignant ventricular arrhythmias during acute myocardial ischemia. *Circulation* 50: 1163–1172.

Williams, J.H., and G.A. Klug (1995). Calcium exchange hypothesis of skeletal muscle fatigue: A brief review. *Muscle & Nerve* 18(4): 421–434.

Williams, N.S., J. Patel, B.D. George, et al. (1991). Development of an electrically stimulated neoanal sphincter. *Lancet* 338: 1166–1169.

Willis, R.J., and W.M. Brooks (1984). Potential hazards of NMR imaging. No evidence of the possible effects of static and changing magnetic fields on cardiac function of the rat and guinea pig. *Magn. Resonance Imag.* 2: 89–95.

Wilson, B.W., R.G. Stevens, and L.A. Anderson (1990). *Extremely Low Frequency Electromagnetic Fields: The Question of Cancer.* Battelle Press, Columbus, OH.

Wilson, C.M., J.D. Allen, J.B. Bridges, and A.A. Adgey (1988). Death and damage caused by multiple direct current shocks: Studies in an animal model. *Eur. Heart J.* 9(11): 1257–1265.

Wit, A.L., P.E. Cranefield, and B.F. Hoffman (1972). Slow conduction and reentry in the ventricular conducting system. II. Single and sustained circus movement in networks of canine and bovine Purkinje fibers. *Circ. Res.* 30: 11–22.

Wolff, B.B., and S. Langley (1975). Cultural factors and the response to pain. In M. Weisenberg (ed.), *Pain: Clinical and Experimental Perspectives*, Mosby, St. Louis: 144–151.

Woodbury, J.W. (1968). Action potential: Properties of excitable membranes. In T.C. Ruch, H.D. Patton, J.W. Woodbury, and A.L. Towe (eds.), *Neurophysiology*, 2nd ed., W.B. Saunders, Philadelphia: 26–53.

Woodbury, J.W., A.M. Gordon, and J.T. Conard (1966). Muscle. In T.C. Ruch and H.D. Patton (eds.), *Physiology and Biophysics*, W.B. Saunders, Philadelphia: 113–152.

Woodrow, K.W., G.D. Friedman, A.P. Siegelaub, and M.F. Collen (1975). Pain tolerance: Differences according to age, sex, and race. In M. Weisenberg (ed.), *Pain: Clinical and Experimental Perspectives*, Mosby, St. Louis: 133–140. (Reprinted from *Psychosomatic Medicine*, 1972, vol. 34: 548–556.)

Woodworth, R.S., and H. Schlosberg (1954). *Experimental Psychology*. Holt, Rinehart and Winston, New York.

Worthen, E.F. (1982). Surgical treatment of electrical burns of the scalp and skull: Past and present. *Clin. Plast Surg.* 9(2): 161–165.

Wu, Y.C. (1979). Electrical injuries—a literature review. *Natl. Bur. Std.* NBSIR: 79–1710.

Wuerker, R.B., A.M. McPhedran, and E. Henneman (1965). Properties of motor units in a heterogeneous fast muscle (M. Gastrocenmius) of the cat. *J. Neurophys.* 28: 85–99.

Wynn, Parry, C.B. (1971). Strength–duration curves. In S.H. Licht (ed.), *Electrodiagnosis and Electromyography*, 3rd ed., E. Licht, New Haven, CT.

Wyss, A.M. (1963). Die Reizwirkung sinuförger Wechselströme, untersucht, bis zur oberen Grenze der Niederfrequenz (1,000 Hz). *Helv. Physiol. Pharmacol.* 21: 419–443.

Yamagata, H., S. Kuhara, Y. Seo, K. Sato, O. Hiwaki, and S. Ueno (1991). Evaluation of dB/dt thresholds for nerve stimulation elicited by trapezoidal and sinusoidal gradient fields in echo-planar imaging. *Society for Magnetic Resonance in Med., Proc. 10th Annual Meeting*, Berkeley, San Francisco: 1277.

Yamaguchi, M., T. Andoh, T. Goto, A. Hosono, T. Kawakami, F. Okumura, T. Takenaka, and I. Yamamoto (1992). Heart stimulation by time-varying magnetic fields, *Jpn. J. Appl. Phys.*, 31(7): 2310–2313.

Yamaguchi, M., T. Andoh, T. Goto, A. Hosono, T. Kawakami, F. Okumura, T. Takenaka, and I. Yamamoto (1994). Effects of strong pulsed fields on the cardiac activity of an open chest dog. *IEEE Trans. Biomed. Eng.* 41(12): 1188–1191.

Yamamoto, T., and Y. Yamamoto (1977). Analysis for the change of skin impedance. *Med. Biol. Eng. Comput.* 15: 219–227.

Yamamoto, T., Y. Yamamoto, and A. Yoshida (1986). Formative mechanisms of current concentration and breakdown phenomena dependent on direct current flow through the skin by a dry electrode. *IEEE Trans. Biomed. Eng.* BME-33(4): 396–404.

Yoshida, K., and K. Horch (1993). Selective stimulation of peripheral nerve fibers using dual intrafascicular electrodes. *IEEE Trans. Biomed. Engin.* 40: 492–494.

Young, J.S., P.E. Burns, A.M. Bowen, and R. McCutchen (1982). *Spinal Cord Injury Statistics.* Experience of the Regional Spinal Cord Injury Systems, Good Samaritan Medical Center, Phoenix, AZ.

Young, J.W., R.F. Chandler, C.C. Snow, K.M. Robinette, G.F. Zehner, and M.S. Lofberg (1983). *Anthropometric and Mass Distribution Characteristics of the Adult Female.* FAA-AM-83-16 revised ed.

Younossi, V.K., H.Z. Rüdiger, K. Haap, and H. Antoni (1973). Untersuchungen über die Flimmerschwelle dis isolierten Meerchweinchen-Herzens für Gleichstrom und sunusförmigen Wechselstrom. *Basic Res. Cardiol.* 68: 551–568.

Yuen, T.G.H., W.F. Agnew, L.A. Bullara, and D.B. McCreery (1990). Biocompatibility of electrodes and materials in the central nervous system. In W.F. Agnew and D.B. McCreery eds., *Neural Prostheses: Fundamental Studies,* Prentice-Hall, Englewood Cliffs, NJ: 197–223.

Zborowski, M. (1952). Cultural components in response to pain. *J. Social Issue* 8: 16–30.

Zelt, R.G., R.K. Daniel, P.A. Ballard, Y. Brissette, and P. Heroux (1988). High-voltage electrical injury: Chronic wound evolution. *Plast. Reconstr. Surg.* 82(6): 1027–1041.

Zipes, D.P. (1975). Electrophysiological mechanisms involved in ventricular fibrillation, *Supplement III to Circulation,* vols. 51 & 52: III-120 to III-130.

Zoll, P.M., R.H. Zoll, R.H. Flak, J.E. Clinton, D.R. Eitel, and E.M. Antman (1985). External noninvasive temporary cardiac pacing: Clinical trial. *Circulation* 71(5): 937–944.

Index